AF803709

Solid Mechanics and Its Applications

Volume 280

Series Editors

J. R. Barber, Department of Mechanical Engineering, University of Michigan, Ann Arbor, USA

Anders Klarbring, Department of Mechanical Engineering, Linköping University, Linköping, Sweden

The fundamental questions arising in mechanics are: Why?, How?, and How much? The aim of this series is to provide lucid accounts written by authoritative researchers giving vision and insight in answering these questions on the subject of mechanics as it relates to solids. The scope of the series covers the entire spectrum of solid mechanics. Thus it includes the foundation of mechanics; variational formulations; computational mechanics; statics, kinematics and dynamics of rigid and elastic bodies; vibrations of solids and structures; dynamical systems and chaos; the theories of elasticity, plasticity and viscoelasticity; composite materials; rods, beams, shells and membranes; structural control and stability; soils, rocks and geomechanics; fracture; tribology; experimental mechanics; biomechanics and machine design. The median level of presentation is the first year graduate student. Some texts are monographs defining the current state of the field; others are accessible to final year undergraduates; but essentially the emphasis is on readability and clarity.

Springer and Professors Barber and Klarbring welcome book ideas from authors. Potential authors who wish to submit a book proposal should contact Dr. Mayra Castro, Senior Editor, Springer Heidelberg, Germany, email: mayra.castro@springer.com

Indexed by SCOPUS, Ei Compendex, EBSCO Discovery Service, OCLC, ProQuest Summon, Google Scholar and SpringerLink.

Toshio Hattori · Tomohiro Naruse

Compendium for Design and Manufacturing of Bolted Joints

Mechanical Design Handbook for Failure Prevention of Products

Springer

Toshio Hattori
Gifu University
Seki, Gifu, Japan

Tomohiro Naruse
Machinery Systems Division
Hitachi Industrial Products, Ltd.
Tsuchiura, Ibaraki, Japan

ISSN 0925-0042 ISSN 2214-7764 (electronic)
Solid Mechanics and Its Applications
ISBN 978-3-031-89627-9 ISBN 978-3-031-89628-6 (eBook)
https://doi.org/10.1007/978-3-031-89628-6

© The Editor(s) (if applicable) and The Author(s), under exclusive license to Springer Nature Switzerland AG 2025

This work is subject to copyright. All rights are solely and exclusively licensed by the Publisher, whether the whole or part of the material is concerned, specifically the rights of translation, reprinting, reuse of illustrations, recitation, broadcasting, reproduction on microfilms or in any other physical way, and transmission or information storage and retrieval, electronic adaptation, computer software, or by similar or dissimilar methodology now known or hereafter developed.
The use of general descriptive names, registered names, trademarks, service marks, etc. in this publication does not imply, even in the absence of a specific statement, that such names are exempt from the relevant protective laws and regulations and therefore free for general use.
The publisher, the authors and the editors are safe to assume that the advice and information in this book are believed to be true and accurate at the date of publication. Neither the publisher nor the authors or the editors give a warranty, expressed or implied, with respect to the material contained herein or for any errors or omissions that may have been made. The publisher remains neutral with regard to jurisdictional claims in published maps and institutional affiliations.

This Springer imprint is published by the registered company Springer Nature Switzerland AG
The registered company address is: Gewerbestrasse 11, 6330 Cham, Switzerland

If disposing of this product, please recycle the paper.

Preface

In 2021, the Japanese era name will be changed to "令和; Japanese character, Reiwa; English reading". This is said to be a quote from the Japanese old poem (Japanese; 和歌 waka) in the Manyoshu:

"It's a good (令; Japanese) moon in the New Year.
The air is beautiful and the wind is gentle (和; Japanese).
The plum blossoms are white like powdered in front of a beautiful lady's mirror.
The orchids give off a fragrance like the incense that adorns the body."
I thought about the skillful composition of this waka, which Japan is proud of.

First of all, the core of the artist's expression is the joy and sorrow in her own daily life and social activities. However, if she wrote this expression directly, it would end up as a mere diary or record. So, by projecting and metaphorizing this impression on the immutable natures, such as four seasons, flowers, bird, or moon these transcends human beings, they make it possible to stimulate and resonate with the reader's imagination and give the reader many times, even dozens of times more excitement.

This wonderful method was used in the composition of this book, "Compendium for Design and Manufacturing of Bolted Joints—Mechanical Design Handbook for Failure Prevention of Products."

First, Chap. 1 summarizes the author's daily experiences, failures, successes, and impressions as an engineer. In contrast to these, natural principles, mechanics, and engineering are described item by item from Chap. 3 onward. Readers are able to master the knowledge and wisdom of fastening technologies while going back and forth between the author's or the reader's thoughts spirally and three-dimensionally.

Chapter 2 summarizes the basic technical matters of screw fastening as for bridging between Chap. 1 and Chaps. 3–11.

It seems to be opposite order to ordinal technical books, which will explain first mechanics (sciences), then design and manufacturing (engineering), and finally accident analysis and maintenance technologies. But this book's order is intended to sharpen the reader's sensibilities and to become engineers who can feel things before they think. We believe that this is the most suitable chapter order for those who are currently busy working on the front lines as manufacturing engineers. The author is confident that this book will be extremely useful for all on site engineers such as in development design, production technology, quality assurance, and maintenance of manufacturing companies, university researchers and students who need to know the actual state of corporate technology, and people in the legal profession who are responsible for making just judgments regarding product accidents.

Make Hard Engineering Revival Success with Young Engineer's Rich Sensitivity!

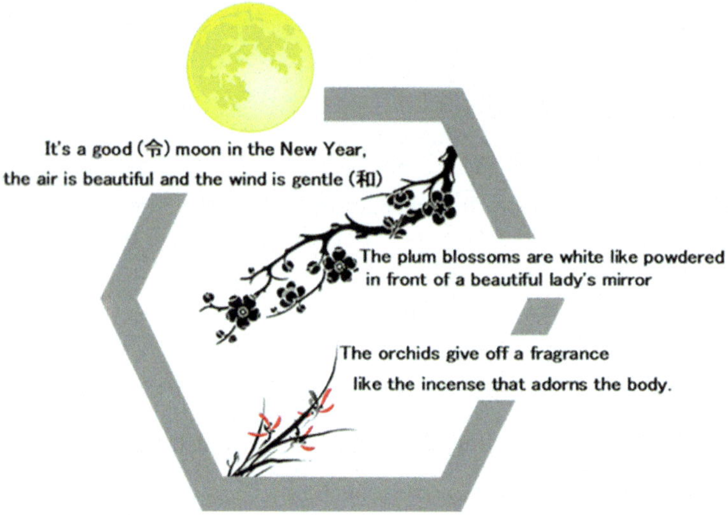

Seki, Japan
Tsuchiura, Japan

Toshio Hattori
Tomohiro Naruse

Competing Interests The authors have no competing interests to declare that are relevant to the content of this manuscript.

Introduction

Screw fastening has many advantages, including the ability to be disassembled and reassembled and the ability to obtain large fastening force with little effort. However, because they are so convenience and familiar, there is a tendency to disregard them. So, if we apply screw fastening with a naive feeling, it may lead to serious accidents such as "loosening" and "destructions".

For example, in Japan there have been many accidents that have caused social anxiety, such as the loosening and falling of motor mounting bolts on the Shinkansen Nozomi train, the breaking of roller coaster axle bolts, the breaking and falling off of trailer hub bolts, and recently, the falling off of aircraft panels and window frames.

The causes of these accidents include large variations in screw machining, variations in fastening operations, and difficulties in modeling with CAE, but the most significant reason is thought to be the lack of basic mechanical analysis techniques such as internal force coefficient, equivalent stiffness, and sliding behavior. Therefore, in this book, I will discuss the basics of mechanical analysis of bolted joints and the establishment of CAE analysis technology, design methodologies, construction, and maintenance technologies in the following order. First, in Chap. 1, past accidents involving bolted joints are discussed, and the mechanical and technical problems in the investigation of the causes are shown and the process of taking countermeasures is introduced.

In Chap. 2, mechanical analysis of bolted joints, initial tightening techniques, and the basics of design methodologies are outlined in order.

Chapter 3 details mechanical analysis and strength design under axial and bending loads.

Chapter 4 describes in detail mechanical analysis, looseness analysis, and strength design under transverse load.

Chapter 5 describes in detail mechanical analysis, looseness analysis, and strength design under axial twist loadings.

Chapter 6 details the classification of rotational loosening and non-rotating loosening and the prevention technology for loosening.

Chapter 7 introduces the application example of threaded joints on pipe fittings. And slipping analyses and loosening behaviors under various loads specific to piping thread joints are detailed.

Chapter 8 explains the mechanisms, test methods, and countermeasures for delayed fracture, which is common in high-tensile bolts.

Chapter 9 detailed mechanisms and countermeasures for fretting fatigue for each part.

Chapter 10 mechanical analyses and strength design for fastened bodies.

Chapter 11 introduces new technologies related to fastening such as plastic region tightening method, adhesion/screw hybrid fastening, effectiveness of lock washer for anti-loosening, and new strength evaluation methods for studs and pad bolted joints.

Chapter 12 overviews Japanese or international standards related to screw fastening such as international standards ISO, Japanese standards JIS, and German standards VDI2230

Chapter 13 summarizes the origin, history, and standardization status of screws as an appendix.

Contents

About the Authors

Toshio Hattori Dr. Engineering, is Professor Emeritus of Gifu University. He was born on August 13, 1948, in Aisai City, Aichi Prefecture, Japan. In March 31, 1976, he obtained Ph.D. from Tokyo Institute of Technology. His awards include prize for Japan Society of Mechanical Engineers article (April 1988) "Adhesion interface strength evaluation using the stress specific place parameter" receiving a prize object article, Hattori, Sakata, Hatsuta, Murakami, Adhesion interface; the Japan Society of Mechanical Engineers fellow (March 25, 2003); prize for Japan Society of Mechanical Engineers Strength of Materials Division achievements (August 5, 2006); and prize for Japan Society of Mechanical Engineers Mechanical Materials and Processing Division achievement (September 12, 2011). He is In Charge of research on strength and reliability technology for products and equipment at the Mechanical Engineering Research Laboratory, Hitachi, Ltd., since April 1976. He developed strength and reliability design technology for rotating electrical machinery (turbine generators and gas turbines). He moved to Gifu University in April 2004 to promote education and research mainly in the following areas: material mechanics, plastic mechanics, industrial material science, plastology, industrial English, graduation thesis, and Ph.D. thesis guidance. He developed strength of fastening, joining, and adhesive joints, promoting the following themes, focusing on joint research with academic societies and companies. He moved to Shizuoka University of Science and Technology in April 2014–2020. He is currently providing technical consultation and technical lectures to manufacturing companies and writing technical books.

Tomohiro Naruse Dr. Engineering, is Professional Engineer, Japan (Mechanical). He was born on March 18, 1970, in Mito City, Ibaraki Prefecture, Japan. In March 31, 1993, he received a Bachelor of Engineering from Osaka University. In March 31, 1995, he received a Master of Engineering from Osaka University. In January 31, 2013, he received a Ph.D. in mechanical engineering from Osaka University. He got awards from the Japan Society of Mechanical Engineers Medal for New Technology (April 2009) "Development of high head multistage pump with large capacity for large-scale water supply system", Shiro Matsui, Shinsuke Sakurai, Takahide Nagahara, Shinji Imaoka, Tomohiro Naruse, and High Pressure Institute of Japan Promotion Award for Science and Technology (May 2017) "Estimation about bolt tension tightening by using hydraulic tensioner", Yohei Tanno, Tomohiro Naruse, Taiji Hashimoto. He is In Charge of research on strength and reliability technology for products and equipment at the Mechanical Engineering Research Laboratory, Hitachi, Ltd., since April 1995. He developed strength and reliability design technology for industrial machinery (pumps, compressors, motors, generators, elevators, escalators, and automotive electronics components). He has moved to Industrial Products Group, Hitachi, Ltd. (now Machinery System Division, Hitachi Industrial products, Ltd.) in 2015. He has involved in product development of pumps, compressors, testing equipment, and logistics systems up to the present. He was participated in the research committee of the Japan Research Institute for Screw Threads and Fastener since its inception (2012). He played a significant role in the committee up to the present.

Chapter 1
Mechanical Problems of Bolted Joints; Learning from Accidents

1.1 Accidental Loosening and Drop of Shinkansen Nozomi Motor Mounting Bolts

1.1.1 Accident Situation [1]

On May 6, 1992, at 1:16 p.m., the No. 238 Hikari from Shin-Osaka to Tokyo (a new 300 series Nozomi-type train introduced in March) was traveling at 190 km/h about halfway between Nagoya and Mikawa-Anjo stations when three of the four bolts securing the motor under the floor of car No. 7 came loose, The motor vibrated, destroying parts of the drive unit connected to the motor hit the rubber hose that passes the compressed air used to power the brakes and doors and resulting in an emergency stop due to abnormal pressure.

Figures 1.1 and 1.2 show the accident situation at the motor mounting area, the starting point of the accident. The motor was fastened to the bogie with two mounting bolts each on the top and bottom. At the time of the accident, the upper two bolts had fallen off, the lower one had fallen off, and the remaining bolt's head was fallen and the thread part remained. Finally, the protruding part of the motor caught on the axle and prevented it from falling.

1.1.2 Investigation of Cause

Initially, JR Tokai was of the opinion that the bolts had been forgotten to be tightened in the accident train. It was later corrected that insufficient tightening was the cause. Finally, JR and the manufacturer agreed that "A thick coating of paint was applied to the seat surface of the motor bolt before fastening, and the bolt was tightened before this paint cured and hardened. After the paint hardened, a gap was created between

© The Author(s), under exclusive license to Springer Nature Switzerland AG 2025

T. Hattori and T. Naruse, *Compendium for Design and Manufacturing of Bolted Joints*, Solid Mechanics and Its Applications 280,
https://doi.org/10.1007/978-3-031-89628-6_1

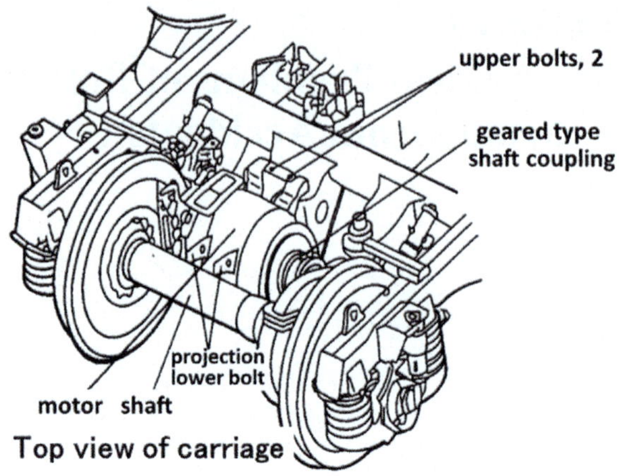

Fig. 1.1 General view of Shinkansen Nozomi motor installation

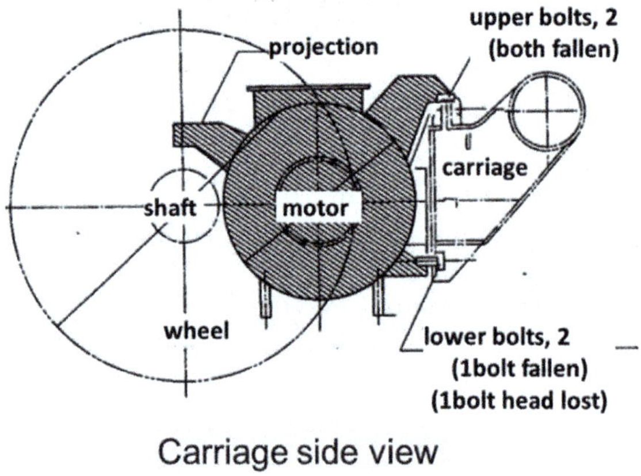

Fig. 1.2 Side view of Shinkansen Nozomi motor installation

the bolt and the seat surface, which then lead loosening, vibration and fell out" (see Fig. 1.3). However, as a practical matter, it is reasonable to assume that all wet paint at the seating interface would be pushed out if the bolts were tightened with a high torque, 25 to 30 kN–m, using a tightening device.

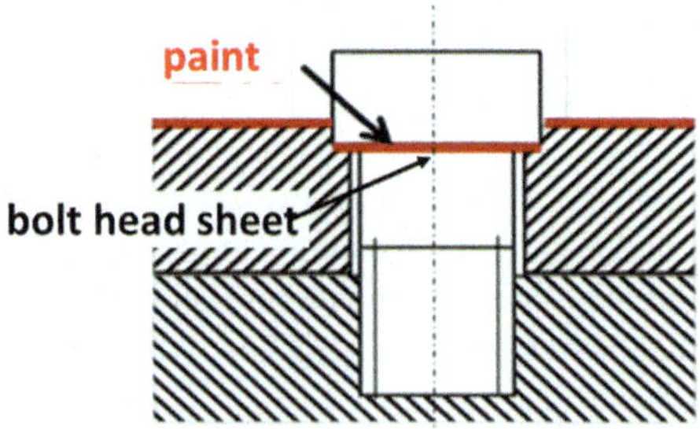

Fig. 1.3 Bolt seating surface with painting

1.1.3 Countermeasures

(A) **Need for adequate durability testing to cope with increased vibration due to weight reduction**

The new 300 series Nozomi train achieved the epoch-making speed of 270 km/h by reducing the weight by 30% through the use of aluminum car-bodies. But, this weight reduction increased vibration, so it need for sufficient durability confirmation tests.

Let's think about the accident that happened a month before of this accident, a damper mounting bolt fell off on a similar train, it is necessary to consider that this accident was not just a simple mistake made at the manufacturing site, but that there are deeper problems in the development process and design technology.

Since similar accidents like these occurred one or two months after the train was put into service in March, it should be essential to trace the cause back to the upstream side of the train.

For example, it is conceivable that weight reduction may increase vibrations of the entire car-bodies or individual car parts, resulting in an increase in dynamic loads on fastening parts.

(B) **Review of mechanical analysis due to adoption of different materials**[2]

In addition to the aforementioned weight reduction and vibration increase, special attention must be paid to mechanical analysis and review of design criteria when converting to aluminum alloys from steel. The reason is that in the case of aluminum alloy fastened body, the load on the bolt (internal force coefficient) increases compared to the case where the fastened body is steel. This mechanism can be easily predicted as shown in Fig. 1.4, if the concept of the internal force coefficient is familiar (see Subsection 2.3.1). If you are familiar with examples of weight reduction using aluminum and other materials, I urge you to calculate the internal force coefficient of the threaded fastening, evaluate its fatigue strength, and review the bolt size.

1.2 Bulkhead Failure Accident on a JAL Jumbo Passenger Jet

1.2.1 Accident Situation [3]

On August 12, 1985, Japan Airlines Flight 123, a B-747 from Haneda to Osaka, lost control of its attitude 12 min after takeoff at an altitude of 7,200 m due to the destruction of the aft pressure bulkhead and the resulting vertical tail structure. This was the world's largest single aircraft accident. Figure 1.5 shows the flight path at the time of the accident.

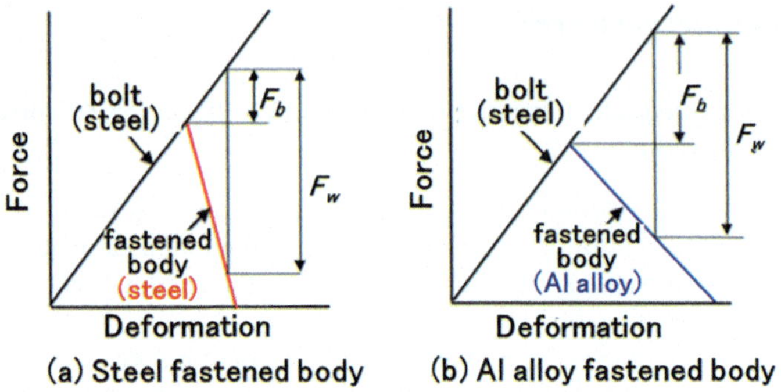

Fig. 1.4 Increase in bolt load due to the use of aluminum in the fastening body

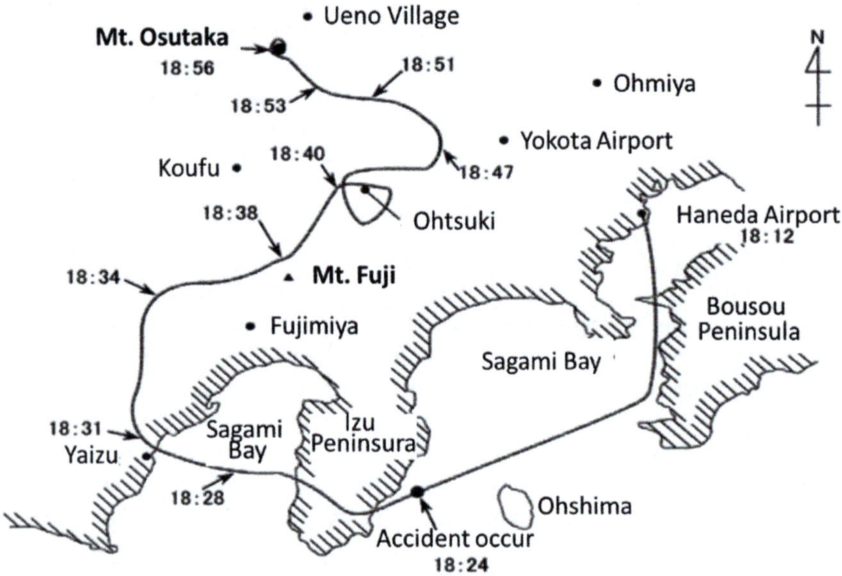

Fig. 1.5 Schematic diagram of the flight path of the accident aircraft [3]

1.2.2 Investigation of Cause

The immediate cause of the accident was the destruction of the fuselage's aft pressure bulkhead (see Fig. 1.6, ①), which resulted in the release of a large volume of high velocity air, destroying the centralized hydraulic controls (see Fig. 1.6, ②) and auxiliary engine (APU) (see Fig. 1.6, ③) located behind the pressure bulkhead. Then the vertical tail box beam (a box-shaped critical structure that supports the bending

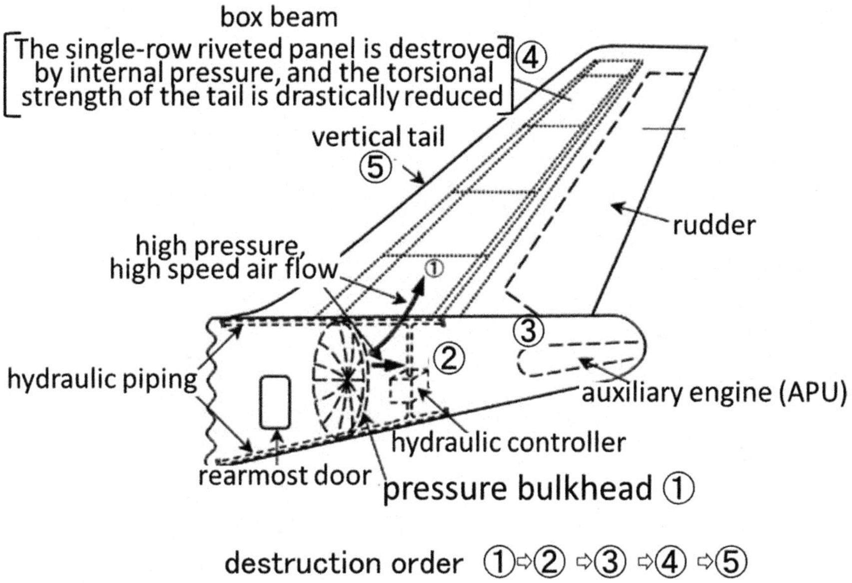

Fig. 1.6 Location of the rear pressure bulkhead at the rear of the fuselage that caused the accident and the order of destruction in the peripheral equipment [3]

and torsional loads acting on the tail) (see Fig. 1.6, ④) failed and lost torsional stiffness. And finally most of the vertical tail structure (see Fig. 1.6, ⑤) was lost, and lost control and crashed into Mt. Osutaka (see Fig. 1.5).

By the way in June 1978, seven years before the accident, the plane hit the bottom of its tail on landing at Osaka Airport and broke down in the middle of the runway (a so-called butt-striking accident) and was towed to the Haneda Airport maintenance area for repairs. The fatigue failure of the pressure bulkhead was caused by a repair error at that time. Figures 1.7 and 1.8 show the repaired bulkhead and details of the rivet joints.

The repair was performed by Boeing engineering staff, but the repair was not performed strictly according to Boeing's instructions diagrams (as shown in Fig. 1.8) and the two upper and lower plates should have been joined with two rows of fasteners through one splice plate.

The result is that only the rivets in the center play the role of load transmission, so the excessive load was applied to the center rivet row, and multiple-site cracks that originated from the rivet hole edges propagated through 12,319 cycles of applied pressure, resulting in rapid unstable failure. As shown in Fig. 1.8, the upper and lower bulk head plates should have been fastened with two rows of fasteners each via one splice plate, but because two splice plates were used and only the center fastener

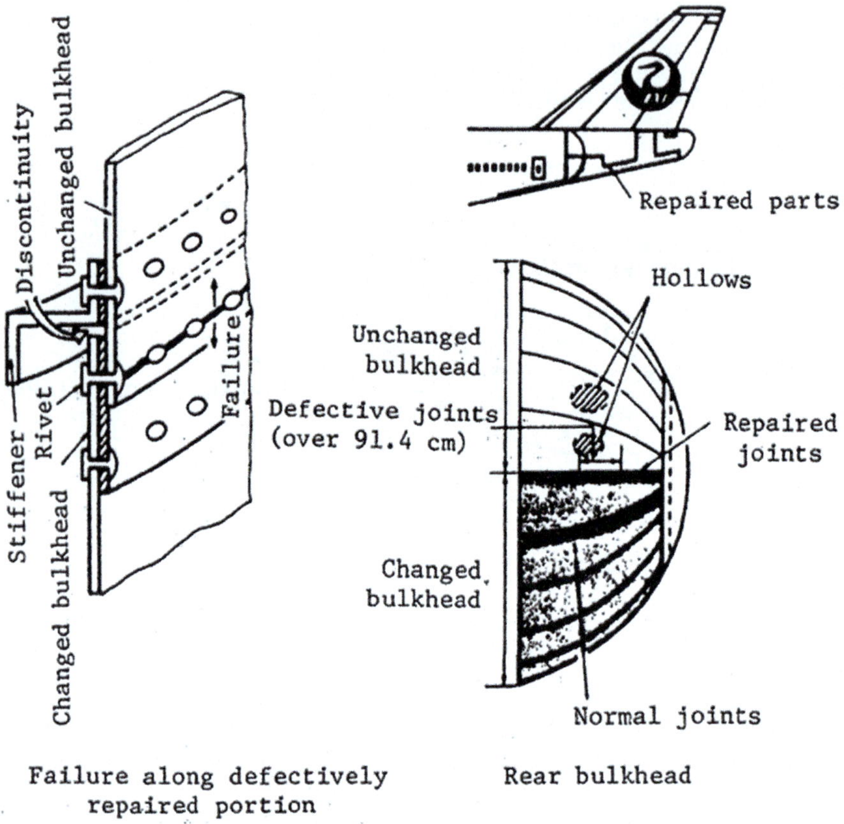

Fig. 1.7 Overview of repair of bulkhead after the butt-striking accident [4]

contributed to load transfer. It can be said that the mechanical force transmission mechanism was not understood at all by the on-site workers. In addition, the area where the repair error was made spanned two bays (approximately 1 m), which was beyond the critical crack dimension for structural instability failure, so the crack that propagated through the small cracks did not stop midway and led to rapid unstable failure. Figure 1.9 shows an SEM photograph of the fracture surface, which was the decisive factor in determining the bulkhead fatigue failure. The striation width and number of striations can explain the process of the butt-striking accident and the final fatigue failure.

It is also pointed out that, although the problem area was covered with sealant and was difficult to notice during repair, it could have been prevented if acceptance inspections or periodic inspections had been thorough.

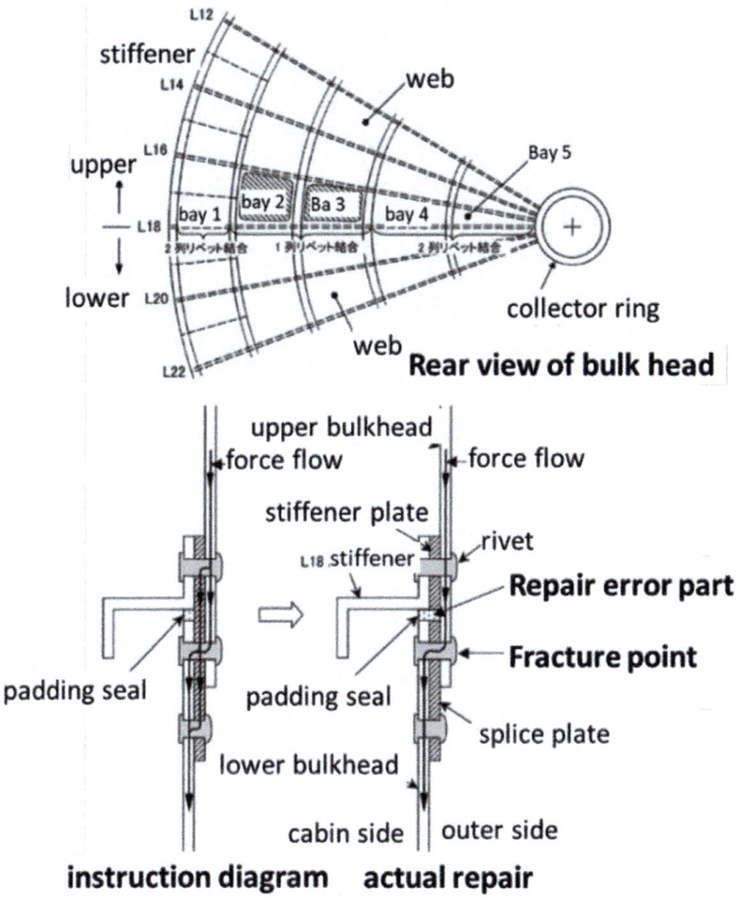

Fig. 1.8 Comparison of actual bulkhead repair situation with instruction diagram [5]

1.2.3 Countermeasures

(A) **Organizational measures**

At first glance, this may seem like a problem at the repair site, repair engineer follow with design instructions, but they are rather thought to be caused by a lack of awareness of technology throughout the organization. Generally, in manufacturing companies, a pyramid of technology sections is constructed as follows,

Development⇨Design⇨Manufacturing⇨Quality assurance⇨Maintenance

And the larger the organization, the more it tends to rely on manuals for technical communication. Regular face-to-face meetings (morning meetings, business meetings) and face-to-entity communication (drawings, on-site) are indispensable.

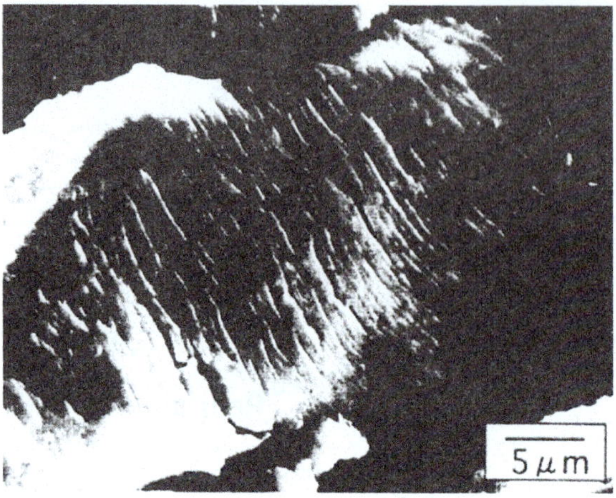

Fig. 1.9 Striation of bulkhead fatigue fracture surface [5]

(B) **Cultivate eyesight skills**

In order to analyze this type of statically indeterminate structure problem, it is most important to read the general flow of forces with one's own eye. If you are in the habit of relying on FEM without being able to grasp this general flow of force, you will not be able to acquire the eyesight skills. For example, if one has the ability to roughly understand that the load on each bolt in the bolted or riveted structure shown in Fig. 1.10 is not one-third of the total load, but that the load on the bolts at both ends is high, then one can see that even if the repair structure is complicated, the repair as shown in Fig. 1.6 right was mistake. It is essential to conduct FEM analysis (see Sects. 4.4, 4.5, and 4.6) of bolted joint structures after cultivating such eyesight skills for estimate the distribution of forces in a statically indeterminate structure.

(C) **Basic technology**

Riveting has long been used as a highly reliable permanent fastening method, but its use has been on the decline in recent years due to advances in welding technology and the widespread use of threaded friction joints. However, riveting has significant reliability advantages, such as the fact that crack propagation is expected to be prevented by the rivet holes if fatigue cracks should occur in a structure, making it a necessary technology for such aircraft.

(C-1) **Stresses in a single element of a riveted joint**

Let us take as an example the one-row overlapping joint of rivets shown in Fig. 1.7.

When this single-row lap joints are subjected to an in-plane tensile loads P, the load is transmitted by the frictional resistance between the plates while the load P is

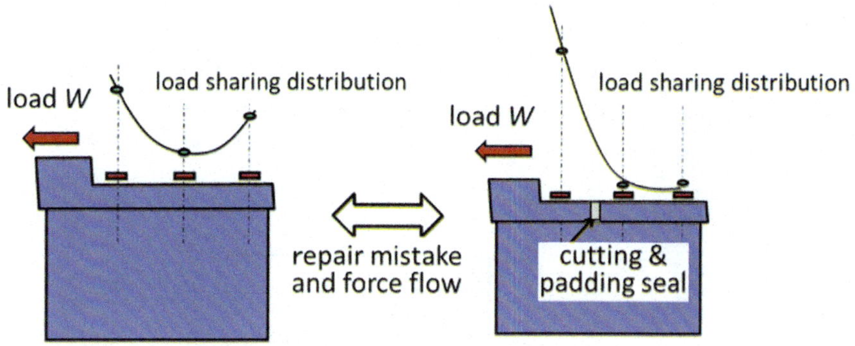

Fig. 1.10 Correspondence between force distribution and repair of bolted joint structure

small, but when the load P increases, slip occurs and the inner surface of the rivet hole contacts the outer surface of the rivet shaft to transmit the load. As a result, there is a risk of fracture in the forms shown in the Fig. 1.11a–e. The stresses corresponding to each of these fracture forms are calculated and evaluated from a material mechanics equations as follows.

(a) Shear stress of rivets

$$\tau = 4P/\pi d^2 \tag{1.1}$$

(b) Shear stress of the plate

$$\tau' = P/2e\,t \tag{1.2}$$

(c) Tensile stress in the smallest rivet hole section of the plate

$$\left(\sigma_t' = P/p - d\right)t \tag{1.3}$$

(d) Compressive stress on rivet shaft or hole surface

$$\sigma_c = \sigma' = P/d\,t \tag{1.4}$$

(e) Stress between rivet hole and plate edge

$$\sigma_b' = 2Pp/(2e - d)t^2 \tag{1.5}$$

The stresses obtained from these equations are to be compared with the strength of the plate and riveted material for strength design (see Table 1.1 or Table 1.2).

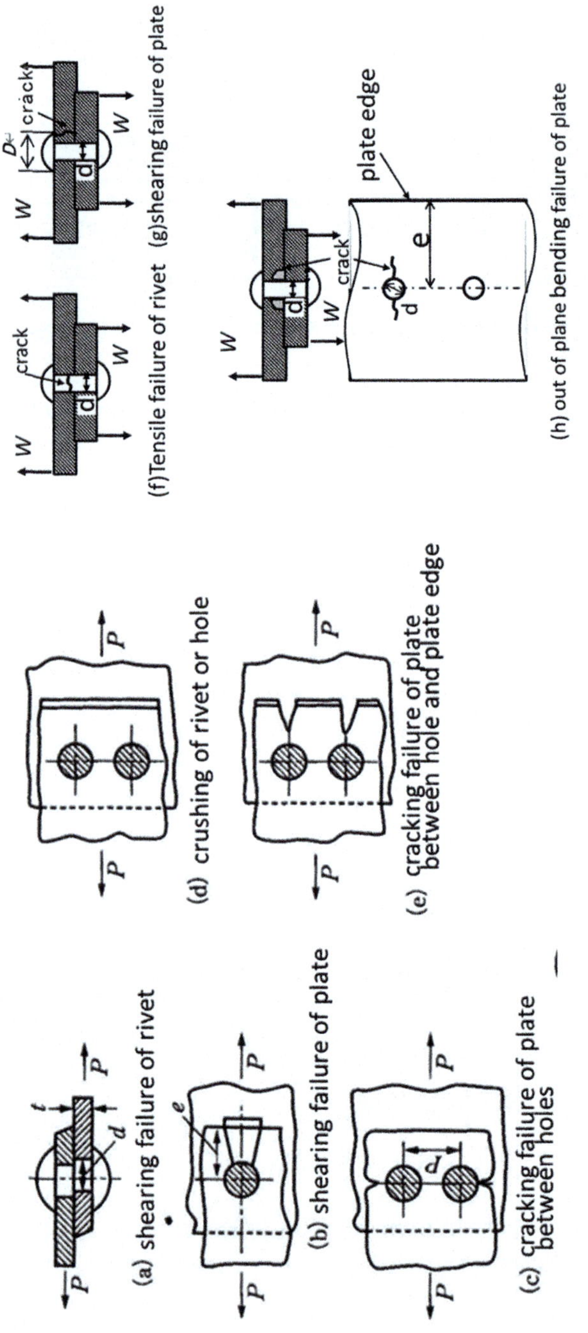

Fig. 1.11 Fracture forms of rivet joint. ((**a**)–(**e**) in-plane loadings, (**f**)–(**h**) out-of-plane loadings)

Table 1.1 Tensile strength of rivet materials [6]

Names	JIS standards of materials	Tensile strength σ_B Kgf/mm^2 (MPa)
Steel	SWCH6R~SWCH17R in JIS G3505	≧35(345)
Brass	C2600W.C2700E or C2800W in JIS H3260	≧28(275)
Cu	C1100W in JIS H3260	≧20(198)
AL, Al alloy	JIS H4210	≧8(78)

Table 1.2 Fatigue strength of riveted and bolted joints [6]

Specimens	Array	Materials	Joint surface treatments	Fatigue life strength ($\sigma_{w,N=2\times10^6}$) (Mpa); total amplitude
Rivet joint	2 rowsX2 rivets	SS400	Mill scale	196
	1 rowsX4 rivets	SS400	Mill scale	147
	1 rowsX4 rivets	Welten 50	Mill scale	127.4
Bolted joint	2 rowsX2 rivets	SS400	Mill scale	225.4
	2 rowsX2 rivets	SS400	Linsed oil	176.4
	2 rowsX2 rivets	SS400	Metallicon	225.4
	1 rowsX4 rivets	SS400	Mill scale	205.8
	1 rowsX4 rivets	SS400	Enamel	171.5
	1 rowsX4 rivets	Welten 50	Flame treatment	205.8
	Flat plate	SS400	Mill scale	225.4
Non-joint	Holed plate	SS400	Mill scale	107.8
(plate)	Rivet fastened	SS400	Mill scale	117.6
	Bolt fastened	SS400	Mill scale	156.8

Since aircraft pressure bulkheads are constructed in such a way that the rivets are not subjected to out-of-plane bending load, the strength estimation is performed selecting suitable failure mode as shown in Fig. 1.11a–e. However, out-of-plane bending failure accidents have occurred in recent years, such as panel drops in KLM aircraft in Osaka (Japan) so I'll show the out-of-plane loading failure mode in Fig. 1.11f–h and show the stress calculation formula as follows. (Details are shown in Sect. 1.6, Fig. 1.50).

(f, g) Out-of-plane tensile loading.

(f) Tensile failure of rivet

$$\sigma_b = 4W/\pi d^2 \tag{1.6}$$

(g) Shearing failure of plate

$$\tau_c = W/\pi Dt \tag{1.7}$$

(h) Out-of-plane bending stress around hole

$$\sigma_\theta = \frac{(M_\theta)_{r=\frac{d}{2}}}{\frac{t^2}{6}} \tag{1.8}$$

$$(M_\theta)_{r=\frac{d}{2}} = -\frac{w(1+\nu)}{4\pi}\left(-\frac{1-\nu}{1+\nu} + \frac{2e^2}{e^2-(d/2)^2}\ln\frac{d/2}{e}\right)$$

where

- d Hole diameter.
- D Rivet head diameter.
- t Thickness of plate.
- e Distance from rivet hole center to plate edge.
- P Distance between rivet holes.
- w Out-of-plane load.

(C-2) Load sharing in riveted joint structures

In actual riveted joints, it is rare for each individual rivet element to share the load equally. Therefore, in strength design, the load sharing acting on each rivet should be taken into consideration, and strength studies should be performed using this shearing load and the stress calculation formulas described in the previous section. Typical examples of this load sharing are shown below.

(C-2-1) Shear-type joints subjected to moment

As shown in Fig. 1.12, when a moment acts on a shear-type joint, if the center of rotation of the moment is the position of the center of gravity of the fastener, the relationship between the load P on the outermost rivet on which the maximum load

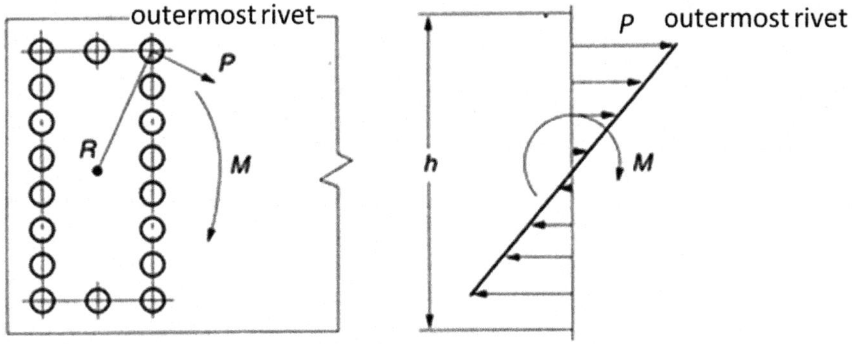

Fig. 1.12 Load sharing for moment load of shear-type rivet joint

acts and the moment "acting on the rivet joint" can be expressed using the distance R from the center of rotation to the outermost rivet as follows,

$$P = M/R \tag{1.9}$$

The stresses of rivets or plates ca be calculate by applying these P on the stand-alone evaluation formulae (1.1)–(1.5).

(C-2-2) **Tensile-type joint subjected to moment**

When a tension-type joint is subjected to a moment, the riveted body shares some of the compressive load on the compression side, so the load P acting on the rivet is determined by using 0.8 times the length of the rivet row h as the center of rotation, as shown in Fig. 1.13.

(C-3) **Strength design of rivet joints**

(C-3-1) **Static strength of rivet joints**

The riveted fastener should be designed so that the stress σ or τ calculated by the method described above or other methods does not exceed the allowable stress of the material at maximum load in joint on evaluation. For example if the main operating stress is tensile stress, the allowable stress σ_{al} can be obtained using tensile strength of the material σ_B or the yield strength σ_Y as follows is

$$\sigma_{al} = \frac{\sigma_B or \sigma_y}{S_f} \tag{1.10}$$

where

S_f safety factor (safety factor should be at least 1.4).

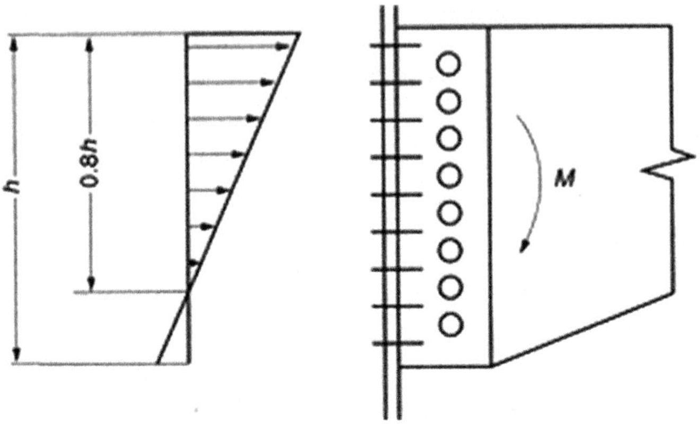

Fig. 1.13 Load sharing for moment load of tensile-type rivet joint

The tensile strength σ_B of the cold-formed rivet materials is shown in Table 1.1 [5]. Even when compressive stress is applied, Eq. (1.10) is used for the allowable compressive stress.

If the main operating stress is shearing stress τ, the allowable stress τ_{al} can be obtained using the following equation.

$$\tau_{al} = \frac{\sigma_{al}}{\sqrt{3}} \tag{1.11}$$

(C-3-2) **Fatigue strength of rivet joints**

In general riveted joints, the rivet does not fatigue fail, but the base material breaks. As with threaded fasteners, the fatigue strength of riveted joints is related to the rivet clamping force, and a low clamping force results in low fatigue strengths and large variations. Table 1.2 shows examples of fatigue strength including bolted joints [4]. As shown in Table 1.2 [5], the fatigue strength of bolted joints is higher than that of riveted joints, and fatigue strength close to that of base metal can be expected, whereas the fatigue strength of riveted joints is almost equal to that of holed plates.

In the case of fatigue, the safety factor should be considered to take 2.0 or more to account for the greater variation depending on loadings and materials. In addition, since fretting is likely to occur in riveted joints, it is necessary to consider the reduction in fatigue strength due to fretting. (See Chap. 8 Sect. 8.3 for details).

(C-4) **Related peripheral technology**

In the reliability design of transportation equipment, especially aircraft, it is important to have backup technology that will not lead to serious damage to the entire structures even if one component is damaged.

The final cause of this crash accident was the loss of attitude control due to the destruction of the vertical stabilizer. It seems essential to design a vertical stabilizer torsion box that can withstand the high-speed outflow of cabin air due to pressure bulkhead failure.

1.3 Fatigue Failure of Roller Coaster Axle Bolts (I)

1.3.1 Accident Circumstances

On August 23, 2003, while "Steel Dragon 2000" roller coaster at Nagashima Spaland in Kuwana City, Japan, was traveling at approximately 1,800 m on a 2,479 m circuit (shown in Fig. 1.14), the 12 bolts fixing of the right wheel carrier on the front axle of the lead vehicle broke (shown in Figs. 1.15 and 1.16). Figure 1.17 shows the overview photos of two of the 12 broken bolts, and Fig. 1.18 shows the fracture surface of the broken bolts. In part I of the fracture surface, a beach mark was observed, and in the SEM photograph, striations were observed as shown in Fig. 1.19, suggesting that it was a fatigue fracture surface. In the SEM photograph of the central part II, a dimple fracture surface was observed as shown in Fig. 1.20, and it was inferred that this area was the final static fracture surface area.

1.3.2 Investigation of the Cause (Refer Chap. 3)

The explanation given by the manufacturer and the verification department for this accident was that the axle bolts were originally designed with a clearance between the carrier and the axle cap so that the load on the carrier would not be applied directly to the axle bolts (see Fig. 1.21a), but during the maintenance process, a spacer was inserted as shown in Fig. 1.21b. The argument was that the spacer was inserted in the maintenance process to eliminate the backlash, as shown in Fig. 1.9b, which increased the load on the axle bolt and led to fatigue failure of the axle bolt. However, from our point of view as researchers involved in bolt fastening, we cannot accept the manufacturer's comments. Therefore, the two fastening structures shown in Fig. 1.21 will be used as teaching materials for strength design of screw fastened bodies, and mechanical and strength evaluation studies will be conducted below.

(A) **Types and sizes of loads**

The loads applied around the axle shaft and carrier, where the accident occurred, can be divided and analyzed as follows (see Fig. 1.22). It should be noted here that under the load of moment *B* and oscillatory vibration, no static load is applied to the axle joint due to the presence of the hitch ball, but only a dynamic load can be applied by the cause of mass and acceleration of the parts such as axle shaft, carrier and side wheels.

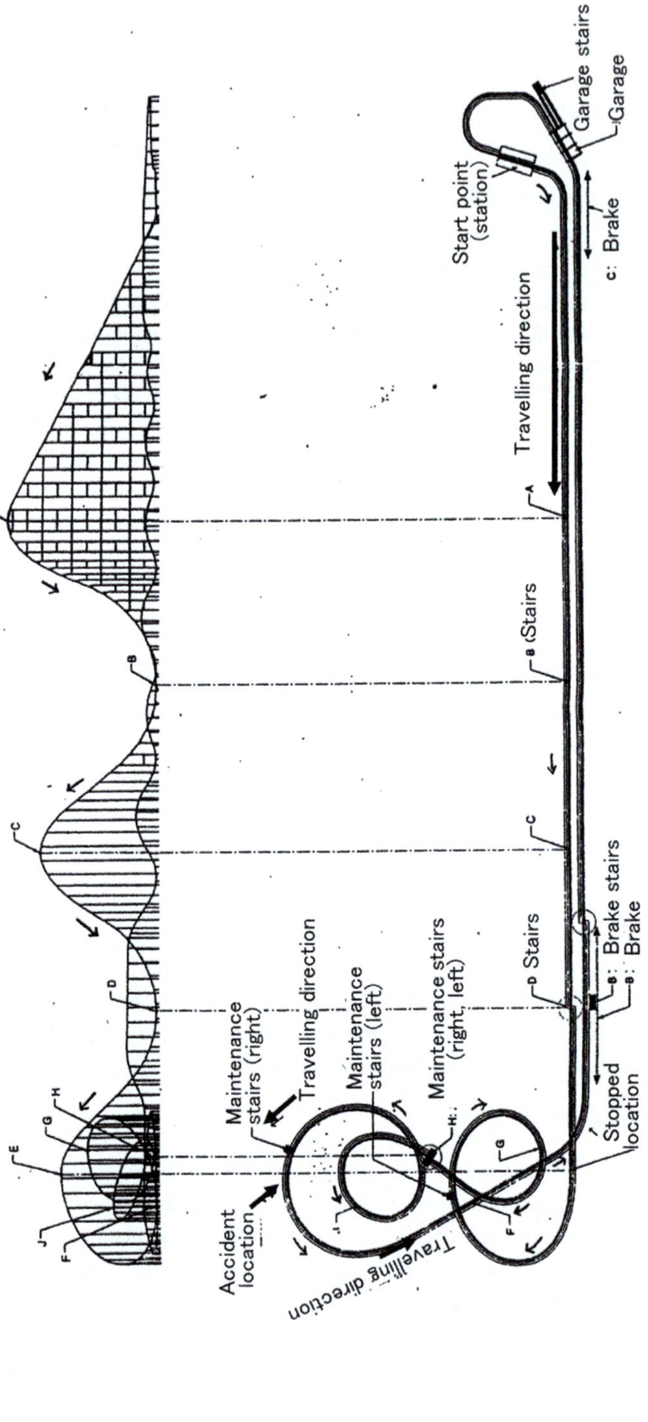

Fig. 1.14 Outline of Steel Dragon's runway [7]

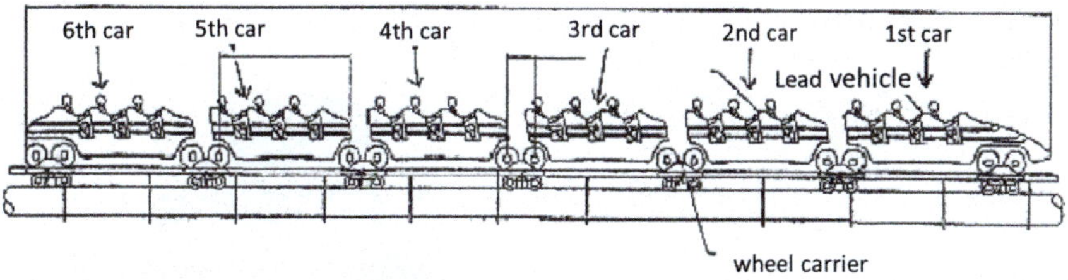

Fig. 1.15 Outline of a Steel Dragon vehicle [7]

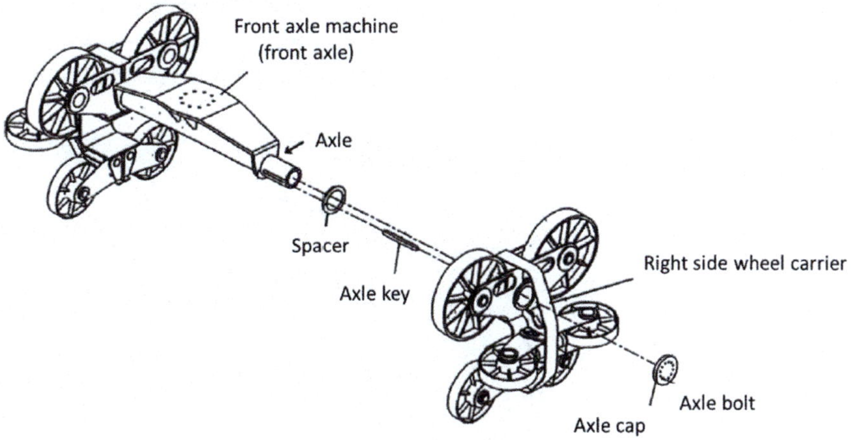

Fig. 1.16 Outline of the right side wheel carrier on the front axle in the lead car of Steel Dragon [7]

Fig. 1.17 Fractured axle bolts [7]

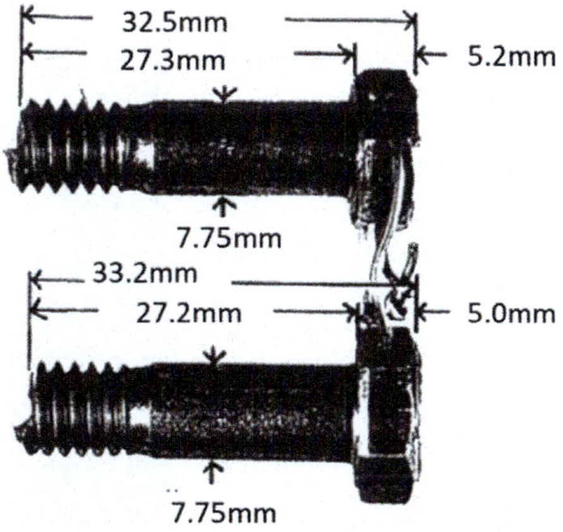

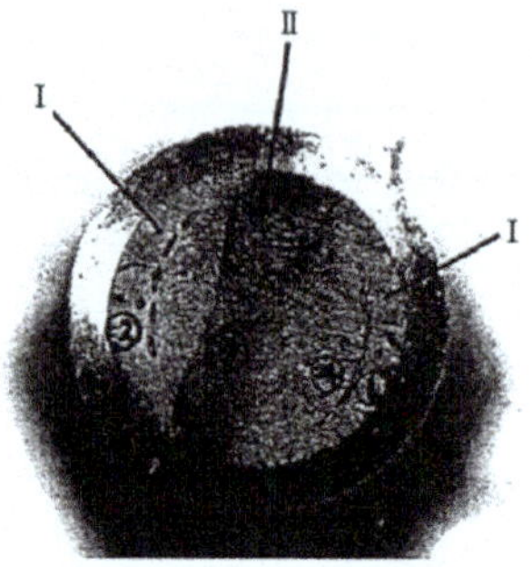

I : Fatigue fracture surface (striation)
II : Final static fracture surface (dimple)

Fig. 1.18 Photograph of fractured surface of axle bolt [7]

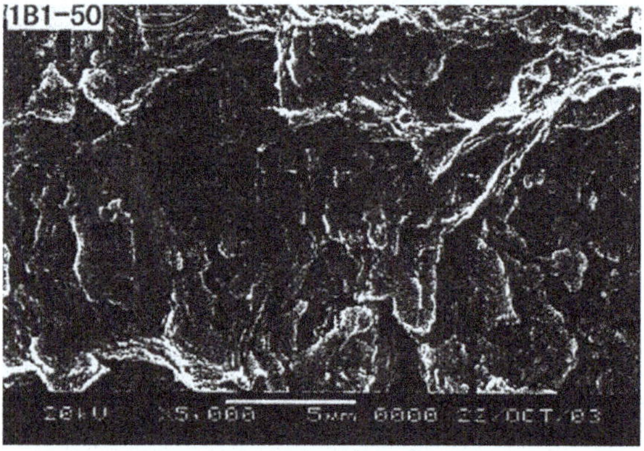

Fig. 1.19 SEM photograph of fracture surface II of axle bolt [7] (striation)

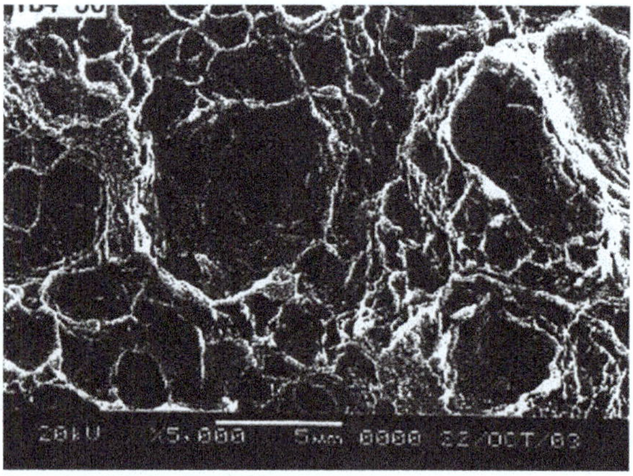

Fig. 1.20 SEM photograph of fracture surface I of axle bolt [7] (dimple)

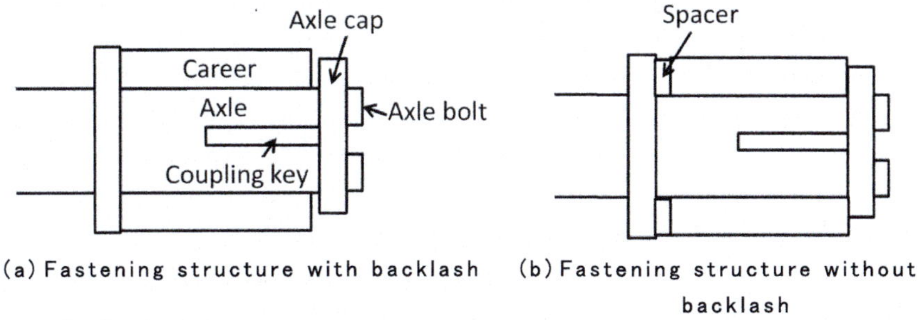

Fig. 1.21 Screw fastening structure at the end of the axle [7]

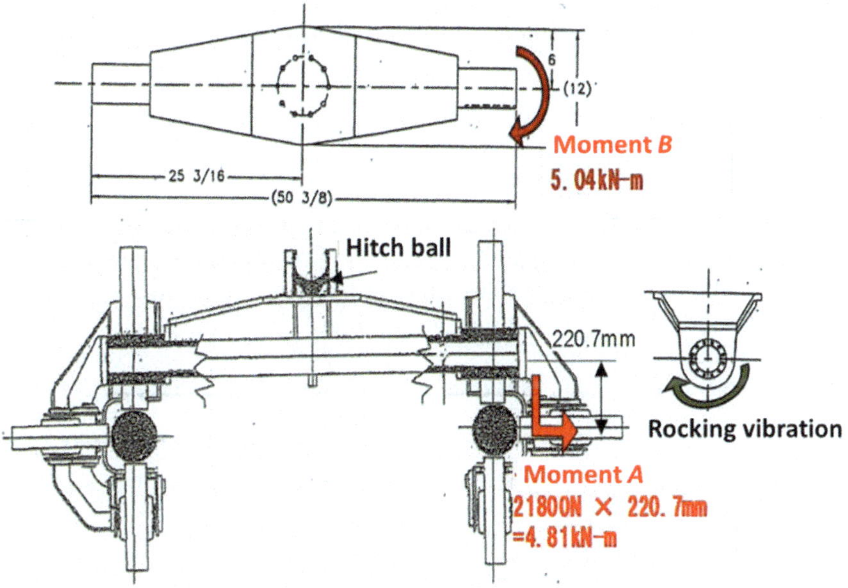

Fig. 1.22 Types and magnitudes of loads applied around the axle [7]

Moment *A*, on the other hand, is the basic load corresponding to the centrifugal force during track running shown in Fig. 1.14 and is applied as a quasi-static load in addition to the dynamic one.

(A-1) Moment $\boldsymbol{M_A}$

If a train travels on a track with a radius of curvature of 40 m at a speed of 60 km/h, each car is subjected to an acceleration of about 6.9 m/s^2 (0.7 g. If each car carries six passengers, the total mass with the car is 3146 kg and the centrifugal force on it is

$$F_C = 3146 \times 6.9 = 21800\,\text{N} \tag{1.12}$$

and the moment $\boldsymbol{M_A}$ is

$$M_A = 21800\,\text{N} \times 220.7\,\text{mm} = 4.81\,\text{kN} - \text{m} \tag{1.13}$$

(A-2) Moment $\boldsymbol{M_B}$

This is a dynamic load due to track irregularities, and is difficult to predict accurately, but actual measurements on site have confirmed that the load is 5.04 kN–m.

(A-3) Oscillating vibration (rotational torque)

This is also basically a dynamic load due to track irregularities, and is difficult to predict accurately, but actual measurements on site have confirmed also as the value to be 7.18 kN–m.

(B) **Strength evaluation**

(B-1) Moment M_A (no backlash)

In general, the design of fastening parts subjected to dynamic loads does not allow for backlash, because backlash in the structure shown in Fig. 1.21a promotes wear and tear, and thus does not maintain sufficient fastening service life. Therefore, in general bolted joints, as in the structure shown in Fig. 1.21b, spacers are used to unify the joints, and the strength design is based on the load factor Φ (ratio of load borne by bolts to total load). This is also discussed in Chap. 2, Sect. 2.2, so please refer to that section as well.

Then we perform mechanical analysis using a fastening model tightened with 12 bolts as shown in Fig. 1.23. Firstly the initial fastening force P_0 with these 12 axle bolts is set as follows,

$$P_0 = 16.9\,\text{kN} \times 12\,\text{pcs} = 202.7\,\text{kN} \tag{1.14}$$

The surface pressure of the contact surface p_0 between the carrier and the axle cap (Fig. 1.23, part can be calculated as follows,

$$p_0 = P_0/A = 202.7\,\text{kN}/7283\,\text{mm}^2 = 27.8\,\text{MPa} \tag{1.15}$$

The evaluation is necessary to evaluate whether the contact surface of this fastening unit can be kept close against the centrifugal load F_C.

The centrifugal load moment M_A can be calculated as follows,

$$M_A = 21.8\,\text{kN} \times 220.7\,\text{mm} = 4.81\,\text{kN} - \text{m} \tag{1.16}$$

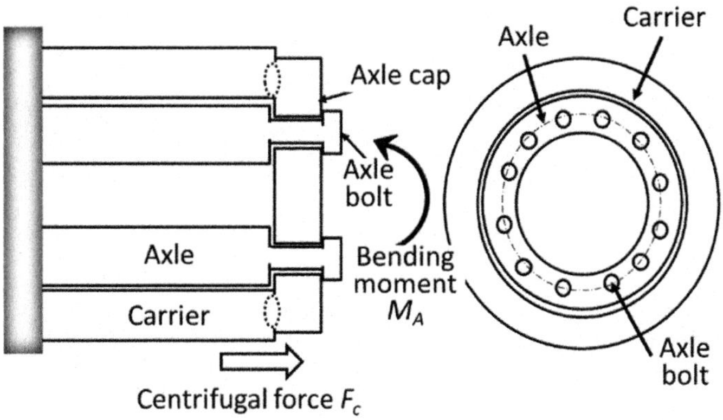

Fig. 1.23 Mechanical model of the fastening around the axle with no backlash [7]

By this bending moment following tensile stress σ_m will be applied on the contact surface. In the following equation, $Z_{carrier}$ is a section modulus of carrier as shown in Fig. 1.24.

$$\sigma_m = M_A/Z_{carrier} = 4.81\,\text{kN} - \text{m}/198289\,\text{mm}^3 = 24.2\,\text{MPa} \tag{1.17}$$

This tensile stress σ_m decrease the contact pressure p_O (Eq. (1.4)), but as thisσ_m is smaller than the above initial fastening surface pressure $p_O = 27.8$ MPa, and we can confirm the enough margin for the separation of contact surface.

If the fastener is not tightened enough and the contact surfaces separate, the bolts will receive all the bending moment M_A and axial force F_c, and the bolts stresses in that case $\sigma_{bM\,(loose)}$ and $\sigma_{bF(loose)}$ are derived as follows. In the following equation, Z_{b12} is a section modulus of bolts as shown in Fig. 1.25.

$$\sigma_{bM\,(loose)} = \frac{M_A}{Z_{b12}} = \frac{4.81\,\text{kN} - \text{m}}{12628\ \text{mm}^3} = 381\,\text{MPa} \tag{1.18}$$

Fig. 1.24 Section modulus of carrier $Z_{carrier}$[7]

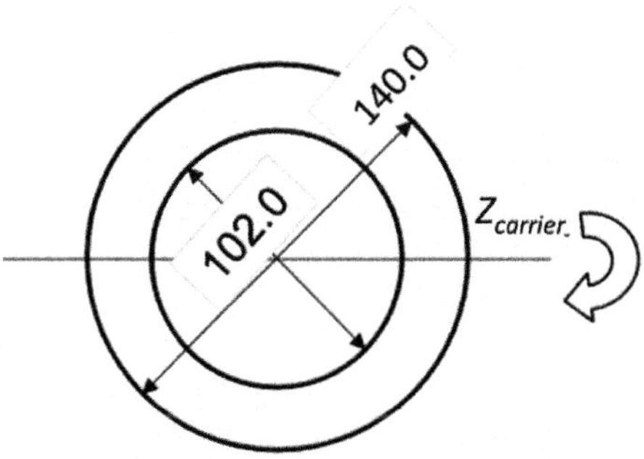

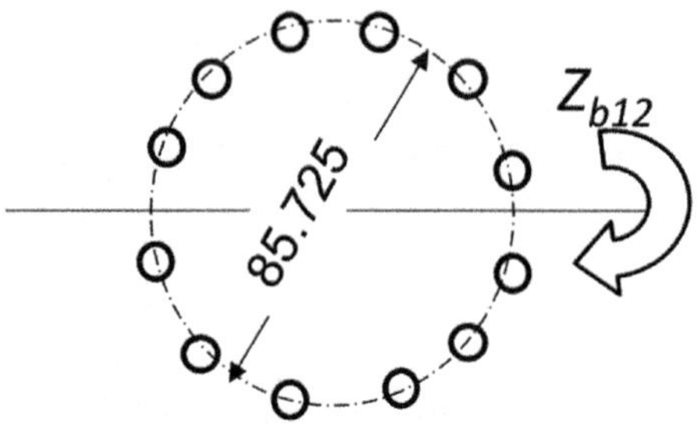

Fig. 1.25 Section modulus of bolts Z_{b12} [7]

$$\sigma_{bF(loose)} = \frac{F_c}{A_{b12}} = \frac{21.8\,\text{kN}}{603\,\text{mm}^2} = 36.2\,\text{MPa} \tag{1.19}$$

However, if the fastening force is sufficient and there is no separation on the joint surface, as in this case, the load on the axle bolt must be calculated using load factor Φ [1, 2]. And this load factor Φ can be calculate using equivalent compliance of compressed body δ_c and compliance of bolt δ_b as follows (show Fig. 1.26).

$$\delta_b = \frac{l_f}{A_b E_b} = \frac{l_f}{\frac{\pi 8^2}{4} E_b} \quad \delta_{ceq} = \frac{l_f}{A_{ceq} E_c} = \frac{l_f}{\frac{A_t + A_l}{2} E_c} = \frac{l_f}{\frac{42.6+601.6}{2} E_c}$$

$$\Phi = \frac{\delta_{ceq}}{\delta_b + \delta_{ceq}} = 0.15 \tag{1.20}$$

And the bolt load stress σ_b is calculated as follows,

$$\sigma_b = \Phi(M_A/Z_{b12} + F_c/A_{b12}) = 0.15 \times (381 + 36.2)\,\text{MPa} = 62.6\,\text{MPa} \tag{1.21}$$

The bolt size (U5/16) and bolt strength are equivalent to ISO nominal designationM8 and the ISO strength grade 10.9. So, by comparing this bolt stress σ_b with $\sigma_w = 72$ MP in Table 1.1, the safety factor S_f can be estimated as follows,

$$S_f = 72\,\text{MPa}/(62.6\,\text{MPa}/2) = 2.3 \tag{1.22}$$

and that it is safe enough.

(B-2) Moment M_A (with backlash)

Conversely, we now attempt to analyze the strength of the bolt for the case with backlash as in Fig. 1.21a, which is recommended by the accident investigation side.

In this case, the centrifugal force of 21,800 N from the side wheel causes a moment with arm length 220.7 mm + 70 mm = 290.7 mm, and this moment pushes the axle cap with arm length 2 × 70 mm = 140 mm, so the centrifugal force is multiplied by 2.1 times by the principle of leverage as follows. (See Fig. 1.27).

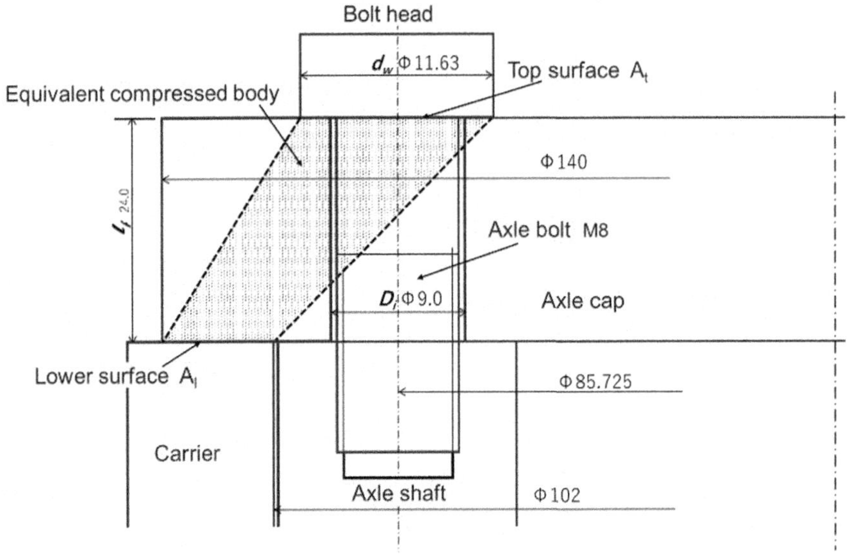

(a) Front view of equivalent fastened body

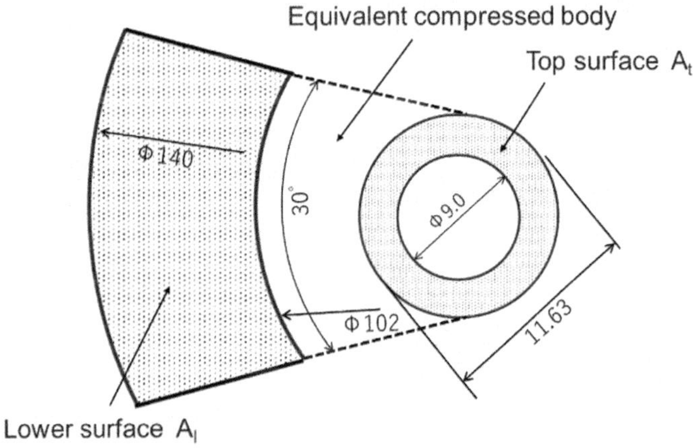

(b) Top view of equivalent fastened body

Fig. 1.26 Calculation model of fastened body around the axle with no backlash

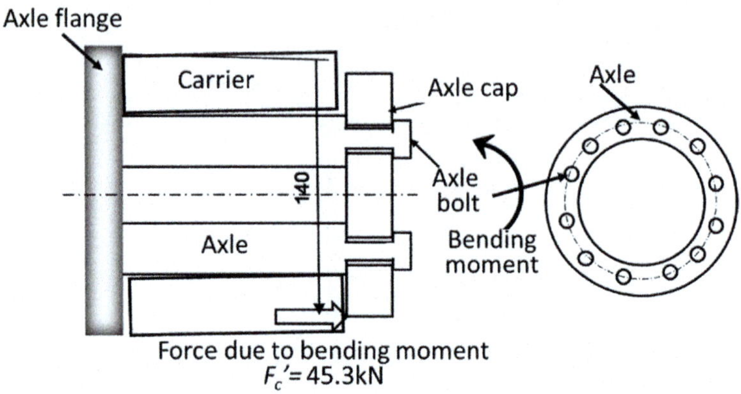

Fig. 1.27 Mechanical model of the fastening around the axle with backlash [7]

$$F_c' = 21.8\,\text{kN} \times 290.7\,\text{mm}/140\ \text{mm} = 45.3\,\text{kN} \tag{1.23}$$

So, the bolts stresses in loose condition can be calculated as follows,

$$\sigma'_{bM\,(loose)} = \frac{M_A'}{Z_{b12}} = \frac{45.3\,\text{kN} \times 70\,\text{mm}}{12628\,\text{mm}^3} = 251\,\text{MPa} \tag{1.24}$$

$$\sigma'_{bF\,(loose)} = \frac{F_c'}{A_{b12}} = \frac{45.3\,\text{kN}}{603\,\text{mm}^2} = 75.1\,\text{MPa} \tag{1.25}$$

However, if the tightening force is sufficient and there is no separation on the joint surface between axle cap and axle end, as in this no spacer case, the load on the axle bolt must be calculated using load factor Φ [1, 2]. And this load factor Φ can be calculate using equivalent compliance of compressed body $\delta_c{}'$ and compliance of bolt δ_b as follows. (Show Fig. 1.28).

$$\delta_c{}' = \frac{l_f}{A_c' E_c} = \frac{l_f}{\frac{\pi}{4}\left(\left(11.63 + \frac{24}{5}\right)^2 - 9^2\right) E_c} \tag{1.26}$$

$$\Phi' = \frac{\delta_c'}{\delta_b + \delta_c'} = 0.253 \tag{1.27}$$

And the bolt load stress $\sigma_b{}'$ is calculated as follows,

$$\sigma_b' = \Phi\left(M_A'/Z_{b12} + F_c'/A_{b12}\right) = 0.25 \times (251 + 75.1)\,\text{MPa} = 82.5\,\text{MPa} \tag{1.28}$$

And using Table 1.3, the safety factor can be estimated as follows,

$$S_f = 72\,\text{MPa}/(82.5\,\text{MPa}/2) = 1.74 \tag{1.29}$$

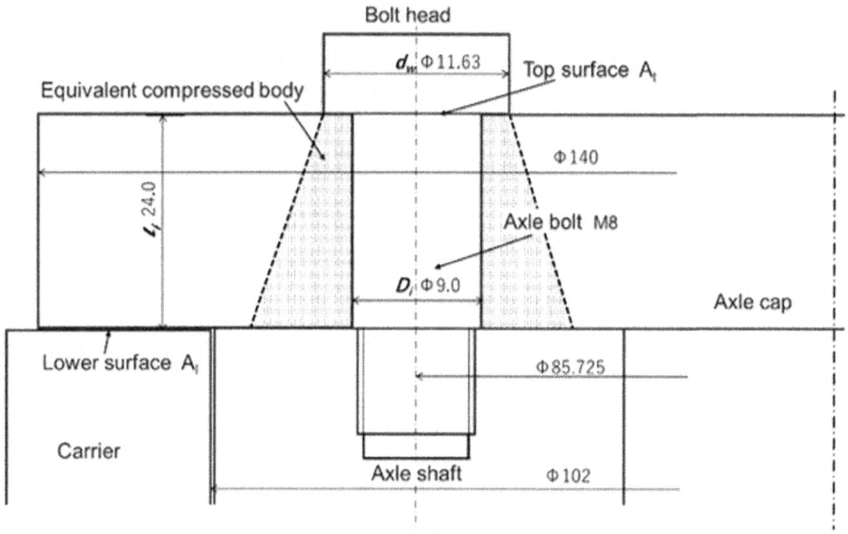

Fig. 1.28 Calculation model of fastened body around the axle with backlash

This strength estimation based on the static mechanical analyses, but in this with backlash structures the impact multiplication factor must be considered as 2–3.

Therefore, it can be seen that this apparent safety factor 1.74 cannot guarantee safety under impact loads with backlash.

(B-3) Moment B (no backlash)

As with the case of moment A, a bolt strength analysis is performed for the structure without backlash in Fig. 1.21b.

In this case, the mechanical model of the fastener is the same as Fig. 1.23 rotated by exactly 90°. Therefore, the mechanical analysis is the same as the bending moment

Table 1.3 Fatigue limit σ_w of bolt materials [8]

Nominal designation	ISO grade				
	4.8	6.8	8.8	10.9	12.9
M6	81	56	64	78	81
M8	61	49	59	72	76
M10	52	45	55	69	72
M12	48	43	53	67	71
M16	42	39	49	62	66
M20	39	37	46	61	65
M24	37	36	45	69	63
M30	24	34	43	57	61

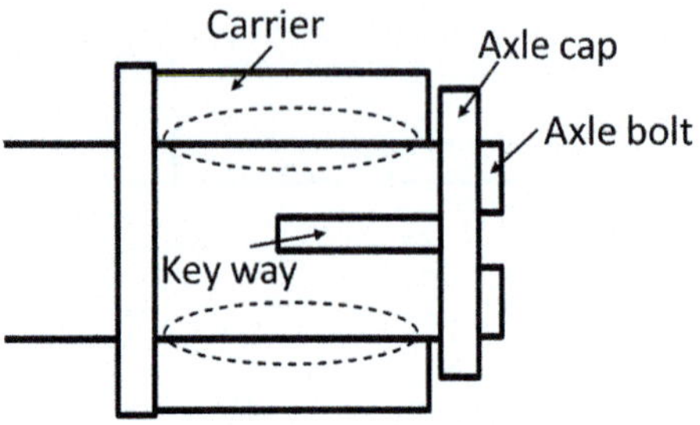

Fig. 1.29 Bending moment burdened area in a structure with backlash [7]

A described in Sect. 1.3.2 (B-1), so the details are omitted and only the results are shown,

$$\sigma_b = 16.2\ \text{MPa} \tag{1.30}$$

The safety factor S_f is

$$S_f = 72\ \text{MPa}/(16.2\ \text{MPa}/2) = 8.8 \tag{1.31}$$

and that it is safe enough.

(B-4) **Moment B (with backlash)**

Here, as in moment A, we try to analyze the strength of the bolt for the case with backlash in Fig. 1.21a.

In this case, there is no centrifugal force at all, and all of the bending moment B is borne by the part shown in Fig. 1.29, and no load is applied to the bolt.

But in the situation where the force is impacted, it is the same as if exactly the same load as moment A (with backlash) described in Sect. 1.3.2 (B-2) was applied in a 90° rotation, and the factor of safety is 6.8 when the carrier does not contact the thrust spacer at all, and about 3.2 when the carrier contacts the thrust spacer. The impact multiplier of 2 to 3 indicates that there is almost no margin for error.

(B-5) **Oscillating vibration (no backlash)**

As shown in Fig. 1.30, when the carrier and axle cap are fastened with a bolt tightening force of 202 kN, the two contact surfaces (①; between the carrier and axle cap and ②; between the carrier and spacer) and the key bear the oscillating vibration load. The rotational torque T_a that can be borne by each of these, assuming a coefficient of friction of 0.2 and an allowable surface pressure of 77 MPa for the key material AISI 1018,

Transferable torques between carrier and axle cap; ① and between carrier and spacer; ② are calculated as follows,

$$T_{a①,②} = 202\text{kN} \times 0.2 \times (102\text{mm} + 140\text{mm})/4 = 2.44\text{kN-m} \tag{1.32}$$

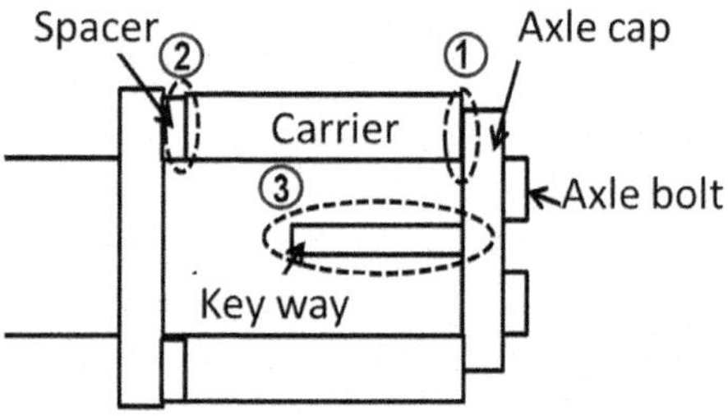

Fig. 1.30 Locations of oscillatory vibration burden in no backlash fastening structure [7]

And transferable on key;③ is calculated using key way length 150 mm, key way depth 12.7 mm as follows,

$$T_{a③} = 77\,\text{MPa} \times (12.7\,\text{mm} \times 150\,\text{mm})/2 \times 102\,\text{mm}/2 = 3.74\,\text{kN-m} \quad (1.33)$$

So the total transferable torque t_a is

$$T_a = T_{a1} + T_{a2} + T_{a3} = 2.44\,\text{kN} - \text{m} \times 2 + 3.74\,\text{kN} - \text{m} = 8.62\,\text{kN} - \text{m} \quad (1.34)$$

and the safety factor S_f is for the measured torque of 7.18 kN–m,

$$S_f = 8.62\,\text{kN} - \text{m}/7.18\,\text{kN} - \text{m} = 1.2 \quad (1.35)$$

and is therefore safe.

(B-6) **Oscillating vibration (with backlash)**

Again, we will try to analyze the strength of the case with backlash in Fig. 1.31, which is recommended by the Accident Investigation Committee.

In this case, if this centrifugal force of 21.8 kN is applied first, the oscillating vibration load is borne by the friction between the carrier and axle cap; ④, and the key; ⑤, respectively,

$$T_{a④} = 21.8\,\text{kN} \times 0.2 \times (102\,\text{mm} + 140\,\text{mm})/4 = 0.26\,\text{kN-m} \quad (1.36)$$

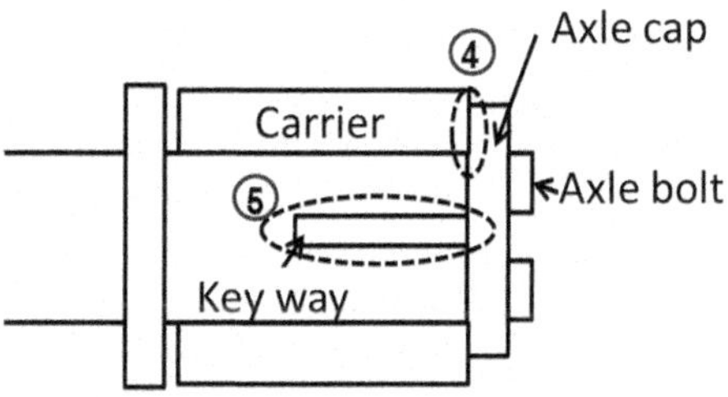

Fig. 1.31 Locations of oscillatory vibration burden in with backlash fastening structure [7]

$$T_{a⑤} = 77\,\text{MPa} \times (12.7\,\text{mm} \times 150\,\text{mm})/2 \times 102\,\text{mm}/2 = 3.74\,\text{kN-m} \tag{1.37}$$

and the transferable torque t_a is,

$$T_a = T_{a④} + T_{a⑤} = 0.26\ \text{kN} + 3.74\ \text{kN-m} = 4.0\ \text{kN-m} \tag{1.38}$$

and the safety factor S_f is for the measured torque of 7.18 kN-m,

$$S_f = 4.0\,\text{kN} - \text{m}/7.18\,\text{kN} - \text{m} = 0.56 \tag{1.39}$$

This is extremely dangerous.

Furthermore, if no centrifugal force is applied, the entire oscillating vibration load is borne by the key, and the torque that can be borne is,

$$T_a = T_{a5} = 3.74\ \text{kN} - \text{m} \tag{1.40}$$

and the safety factor S_f for the measured torque of 7.18 kN–m is,

$$S_f = 3.74\,\text{kN} - \text{m}/7.18\,\text{kN} - \text{m} = 0.52 \tag{1.41}$$

This indicates that the risk is even greater.

(C) Summary of accident causes

The report from those involved in the accident investigation that "backlash is necessary between the carrier and axle cap in the first place, and that the stress on the bolts increased due to the insertion of spacers in the maintenance process" is contradictory, and that if the fastening structure had been free of play from the beginning, a safety factor of 1 or higher would have been ensured against moment *A*, moment *B*, and oscillating vibration. However, because the structure was initially designed with backlash, the safety factor was lower than 1 for the oscillating vibration load,

First key worn and the gap occurred
↓
Then the impact loads in circumferential direction increased
↓
Finally axle bolts loosed and cause the fatigue failure

This is because the key was worn out even before the accident occurred. This is consistent with the estimation of bolt fatigue failure due to oscillating vibration, which is the most hazardous in the strength evaluation described above, because the key was worn out even before the accident, and the fatigue failure surface of the bolt was in the double-swing bending mode and fatigue occurred from the circumferential direction of the pitch circle, as shown in Fig. 1.18. This is consistent with the estimation of bolt fatigue failure due to oscillatory vibration, which is the highest risk in the above strength assessment.

(D) **Countermeasures (design technology for thread fatigue failure prevention)**

The basis of screw fastener design is to prevent opening and slippage between the contact interfaces by the sufficient tightening forces, and if these conditions are satisfied, the load borne by the bolt can reduced greatly by the internal force coefficient.

1.4 Rollercoaster Axle Bolt Fatigue (II)

1.4.1 Accident Situations

On May 5, 2007, the failure accident of the axle shaft occurred in roller coaster "FUUJINN" of the amusement park "Expo land" in Osaka Japan (see Fig. 1.32).

As shown in Fig. 1.33a, the male screw of the axle shaft failed, and the roller/shaft unit dropped out from the rail and truck flames [9]. Figure 1.33b shows a detailed view of the failed axle shaft, and Fig. 1.34 shows the failed male screw with nut.

This failed axle shaft is made of Ni–Cr steel, and the size of the screw was M32.

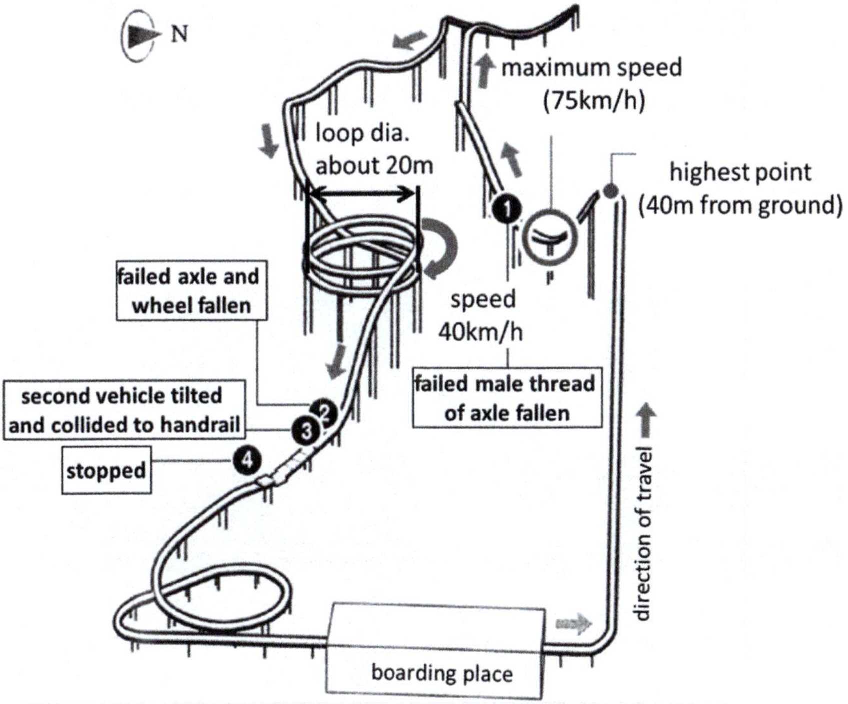

Fig. 1.32 Roller coaster track of failed "FUUJINN RAIJINN II" [9]

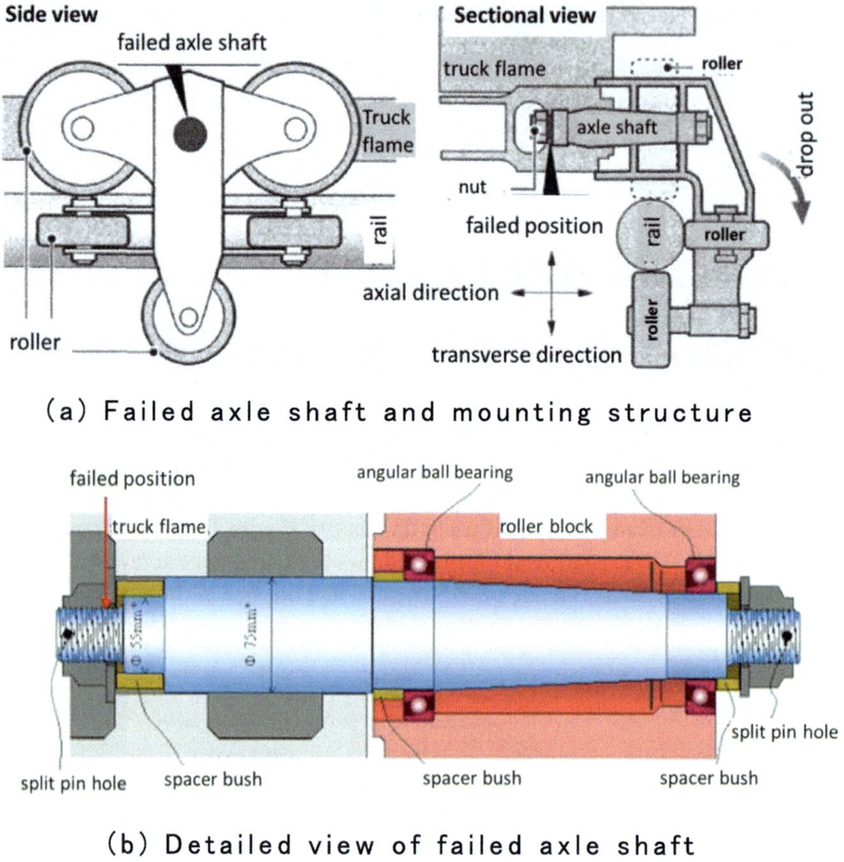

Fig. 1.33 Mounting structure and detailed view of the failed axle shaft [9]

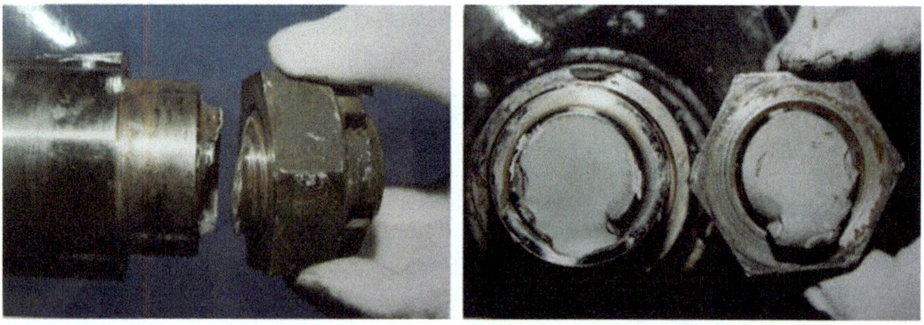

Fig. 1.34 Failed male screw of the axle shaft and nut [9]

As shown in Fig. 1.34, cracks in the broken male screw occur from near the valley bottom of the first thread of the male screw that engages with the nut, cracks propagate almost perpendicular to the axial direction, and fine stripes are observed in almost 70% of the fracture surface. This fine striped pattern can be said to be a so-called beach mark, which is a fracture surface peculiar to fatigue.

1.4.2 Investigation of the Accident Cause

Then, I will investigate the cause of the accident based on the above-mentioned accident processes and mechanics of bolt fastened structures as follows.

Dynamic analyses while running various tracks are performed, and tensile and bend loads on the screw jointed parts are derived.

(A) **Applied load analysis**

(A-1) **Load analysis while running on track**

One of the major loads on a roller coaster during operation will subject while running curved tracks. In Fig. 1.35, the schematic structure for the mechanical analysis of the centrifugal force and the load on the bolt fastening part are shown. If we estimate the radius of curvature of the track as r = 10 m, the velocity v = 50 km/h, the body mass m = 1000 kg, and the bank angle as θ, the centrifugal force and the load on the bolt fastening part can be calculated by the following formula.

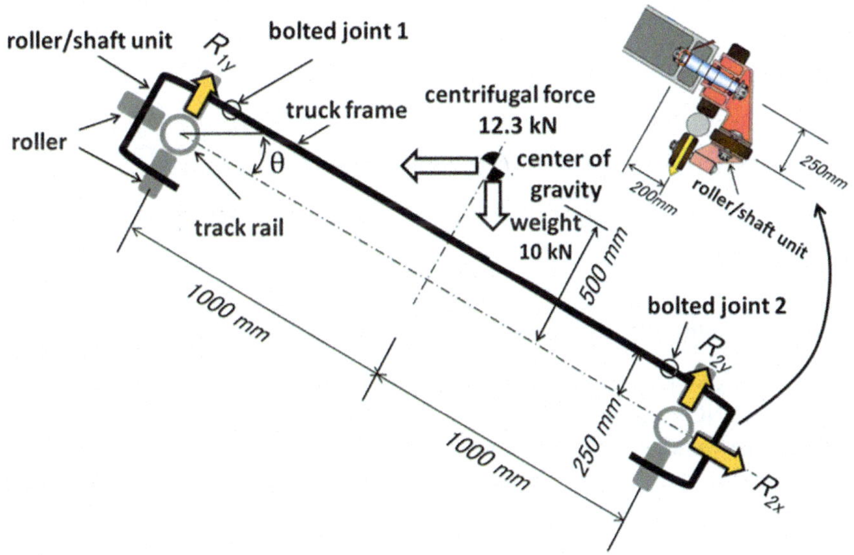

Fig. 1.35 Loading conditions of the roller coaster while running on the curved track [9]

(a) (a) Centrifugal force on the curved track

$$F = mv^2r = 12.3\,\text{kN} \tag{1.42}$$

(b) Horizontal force balance

$$R_{1y}\sin\theta + R_{2y}\sin\theta + R_{2x}\cos\theta = 12.3\text{kN} \tag{1.43}$$

(c) Vertical force balance

$$R_{1y}\cos\theta + R_{2y}\cos\theta - R_{2x}\sin\theta = 10\text{kN} \tag{1.44}$$

(d) Balance of rotational moments

$$10(500\sin\theta) - 12.3(500\cos\theta) + R_{1y}1000 - R_{2y}1000 = 0 \tag{1.45}$$

Solving Eqs. (1.43), (1.44), and (1.45), we obtain

$$R_{1y} = 3.7\sin\theta + 8.1\cos\theta. \tag{1.46}$$

$$R_{2x} = 12.3\cos\theta - 10\sin\theta \tag{1.47}$$

$$R_{2y} = 8.9\sin\theta + 1.9\cos\theta \tag{1.48}$$

When calculating each reaction force as follows by assuming a bank angle θas 45°

$$R_{1y} = 8.3\ \text{kN},\ R_{2x} = 16.3\,\text{kN},\ R_{2y} = 7.6\ \text{Kn}$$

(A-2) **Acceleration load on the swooping track**

A large vertical inertia load will be applied while running swooping tracks. If we estimate the swooping acceleration as 3G(19), the vertical load on the left and right axles is calculated as $R_{1y} = R_{2y} = 15$ kN as shown in Fig. 1.36.

(A-3) **Estimation of load based on fracture surface analysis**

By using the observed fatigue fracture surface ratio of 70% and the fracture toughness K_{IC} of the SNC steel material as $K_{IC} = 40\ \text{MPa}\sqrt{\text{m}}$ ($1265\ \text{MPa}\sqrt{\text{mm}}$)[5], we can estimate the working stress as follows.

Analysis of the crack tip stress intensity factor K at the time of final fracture is obtained by simplifying the cross section to a rectangle, as shown in Fig. 1.37. The

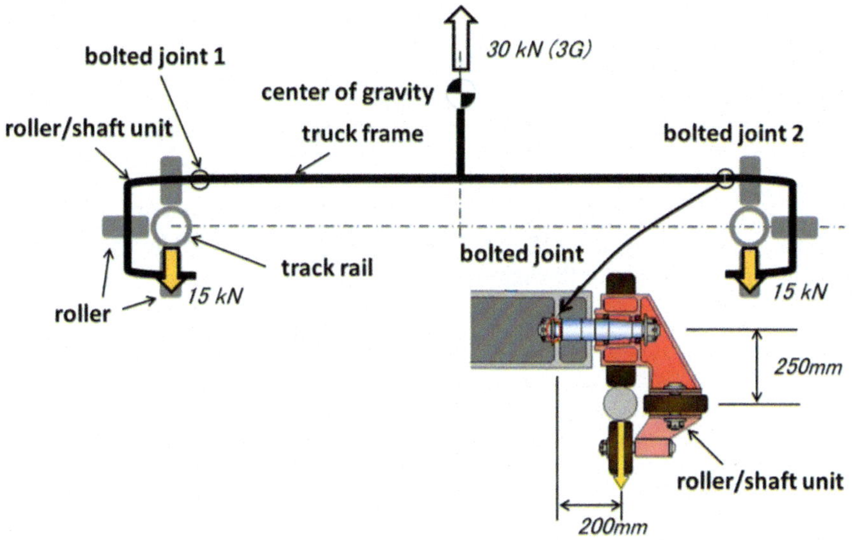

Fig. 1.36 Loading conditions of the roller coaster while running on the swooping track [9]

stress intensity factor K_I of a one-sided crack in a finite flat plate under uniform tensile stress is expressed by the following equations.

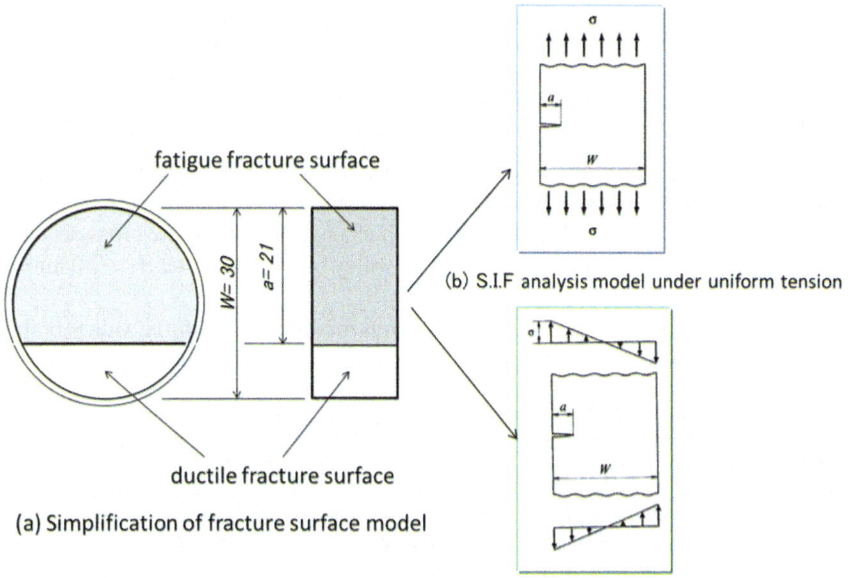

Fig. 1.37 Simplification of the fracture surface model and S.I.F calculation models

$$K_{It} = F(\alpha)\sigma\sqrt{\pi a} \tag{1.49}$$

$$\alpha = \mathrm{a/W}$$

$$\mathrm{F}(\alpha) = 1.12 - 0.23\alpha + 10.55\alpha^2 - 21.72\alpha^3 + 30.39\alpha^4 = 5.9$$

The stress intensity factor K_I of a one-sided crack in a finite-width flat plate to be bent is expressed by the following equation.

$$K_{Ib} = F(\alpha)\sigma\sqrt{\pi a} \tag{1.50}$$

$$\alpha = \mathrm{a/W}$$

$$\mathrm{F}(\alpha) = 1.122 - 1.40\alpha + 7.33\alpha^2 - 13.08\alpha^3 + 14.0\alpha^4 = 2.6$$

The working stresses at the time of final rupture are estimated as follows:
Uniform tensile stress; $\sigma_t = \frac{1265}{5.9\sqrt{\pi \times 21}} = 26.4\mathrm{MP}$
Bending stress; $\sigma_b = \frac{1265}{2.6\sqrt{\pi \times 21}} = 60.0\mathrm{MP}$

(A-4) **Summary of mechanical analysis and fracture surface analysis**

The stress value estimated from the fracture surface analysis is 60 MPa, which satisfies the safety factor $S_f = 50 \times 2/60 = 1.7$ by comparing the fatigue limit $\sigma_{wk} \approx$ 50 MPa of the SNC steel bolt (strength class of approximately 8.8) (see reference [1]). It is no wonder that it has actually endured long-term use for 15 years.

Next, the axial tensile stress σ_t and bending stress σ_b of the bolt are calculated and compared from the load obtained from the mechanical analysis of (A-3). It is difficult to calculate the load factor Φ (ratio of the load of a single bolt to the load of the entire screw connection) here because there are no detailed drawings. Therefore, here, it is roughly estimated that the load factor $\Phi = 1/3$ for the axial load and the load factor $\Phi_\beta = 1/20$ for bending. For each stress calculation, the cross-sectional area A_3 and the section modulus Z_3 with the valley bottom diameter $d_3 = 30$ mm are used.

The cross-sectional area A_3 and the section modulus Z_3 with the valley bottom diameter $d_3 = 30$ mm are used.

$$\sigma_{1b} = \frac{R_{1y} \times 200}{Z_3},\ \sigma_{2t} = \frac{R_{2x}}{A_3},\ \sigma_{2b} = \frac{R_{2x} \times 250 + R_{2y} \times 200}{Z_3} \tag{1.51}$$

The results of the calculated stress are summarized in Table 1.4. The estimated stress value due to the applied inertia load (3G) is higher than that under centrifugal load.

Table 1.4 Estimated stress in the shaft bolt

Item	R_{1y}	R_{2x}	R_{2y}	σ_{2t}	σ_{2b}	σ_2	$\sigma_{1b} = \sigma_1$
Centrifugal force (mv^2r)	8.3kN	1.63kN	7.6kN	0.77 MPa	36.4 mPa	37.2 MPa	31.3 MPa
Inertia load (3G)	−15 kN	0	−15kN	0	−56.6kN	−56.6kN	−56.6kN
Estimation by failure surface	–					-60.0kN	

(B) Investigation of fracture cause and process

All of the above analyses were performed on the premise that the screwed portion is completely tightened; in other words, they were carried out in consideration of the load factor Φ (the load bearing ratio of the bolt). In many cases of fatigue failure of bolted joints, it is presumed that the load factor Φ increased due to insufficient initial tightening or loosening. In these cases, the bolt stress becomes approximately 3 times greater for axial loads and approximately 20 times greater for bending loads, leading to a fracture. However, this explanation still raises the following questions.

- If bolt failure occurred after loosening, the ductile fracture area of the final fracture should be much greater than 30%.
- This roller coaster has been in operation for 15 years without bolt maintenance or replacement. So, mechanism in which the bolted joint slowly loosens over such a long period of time is hardly conceivable.

The answer to these questions is fretting fatigue.

The evaluation guideline of fretting fatigue failure by macroscopic stress is described as follows.

- Average surface pressure of 50 MPa or more.
- Nominal axial stress amplitude of 50 MPa or more.

The average surface pressure of the thread contact surface under a normally tightened screw connection is sufficiently 50 MPa or more.

From the estimation results in Table 1.4, the nominal stress amplitude of the right axle (subscript 2) is estimated as (37.2 + 60)/2 = ±48.6 MPa, so the risk of fretting can be predicted. From the ultralong-life fretting fatigue curve in Fig. 9.32 (in Sect. 9.9), fretting fatigue failure is predicted at $N_f > 10^8$ or more, so the validity of 15 years of non-replacement operation can be considered sufficiently.

Oku et al. [10, 11] reported on fretting fatigue at the root of the threaded joint of an oil field pipe, which is subject to repeated loads accompanied by contact surface pressure. Figure 1.38 shows the structure of the threaded joint of the oil field pipe, and Fig. 1.39 shows the fretting fatigue failure at the bottom of the thread. The root of the thread of this threaded joint is flat, and surface pressure is applied to it during fastening. When repeated axial loads were applied in this thread joint, as seen in the lower photograph of Fig. 1.39, fatigue failure occurred at corner stress concentration areas (Fig. 1.19b) in the high-load case. When the applied load is low, it is clear

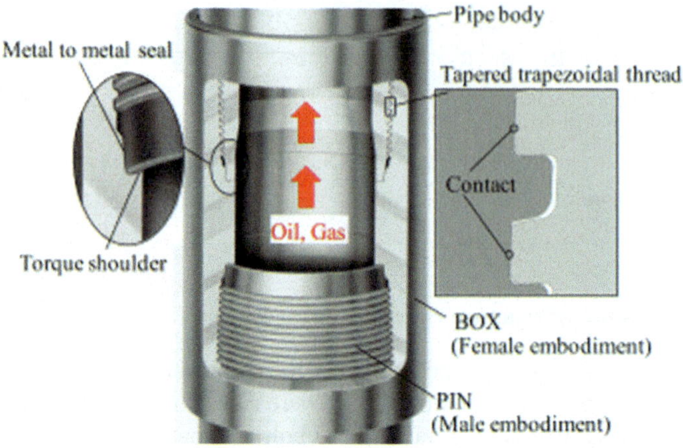

Fig. 1.38 Joint structure of oil pipe [10, 11]

that cracks occur at the contact points, which is a characteristic of fretting fatigue. Since in this oil pipe joint case the root of this screw is flat, the cracks at the stress concentration point or the crack at the surface pressure contact point could be clearly distinguished. However, in general triangular screws, it is difficult to distinguish between the valley bottom stress concentration point and the contact edge point. For this reason, it may have been presumed that in many bolted joint failure processes, the loosening of the screw → increased load on the bolt → fatigue failure at the thread valley bottom. However, due to the demand for longer product life in the future, it will be necessary to pay attention to the fretting fatigue at the thread contact surface.

1.5 Summary of Failure Analyses and Countermeasures

As shown in Table 1.4, the estimated stress value due to the inertia load (3G) is higher than that under centrifugal load, and the sign and value are close to the estimated stress value from fracture mechanics. Therefore, this inertia load (3G) causes initiations of cracks, and which will lead to progress and destruction.

In the above accident analysis, it is assumed that the fastening force of the bolted joint is completely maintained and that load factors $\varphi = 1/3$ and $\varphi_b = 1/20$ will be maintained over a long operation life. And bolt failure will be caused by fretting fatigue at the contact edge between threads.

The fatigue limit of the bolt at the stress concentration position of the thread valley bottom and the long-life fretting fatigue limit are close to approximately 50 MPa, so until now, little attention has been given to fretting in bolted joints. However, from now on, the demand for long-life use of machinery and equipment will increase, so it is thought that strength design and damage analysis must pay attention to fretting fatigue damage and failure behavior on thread joint.

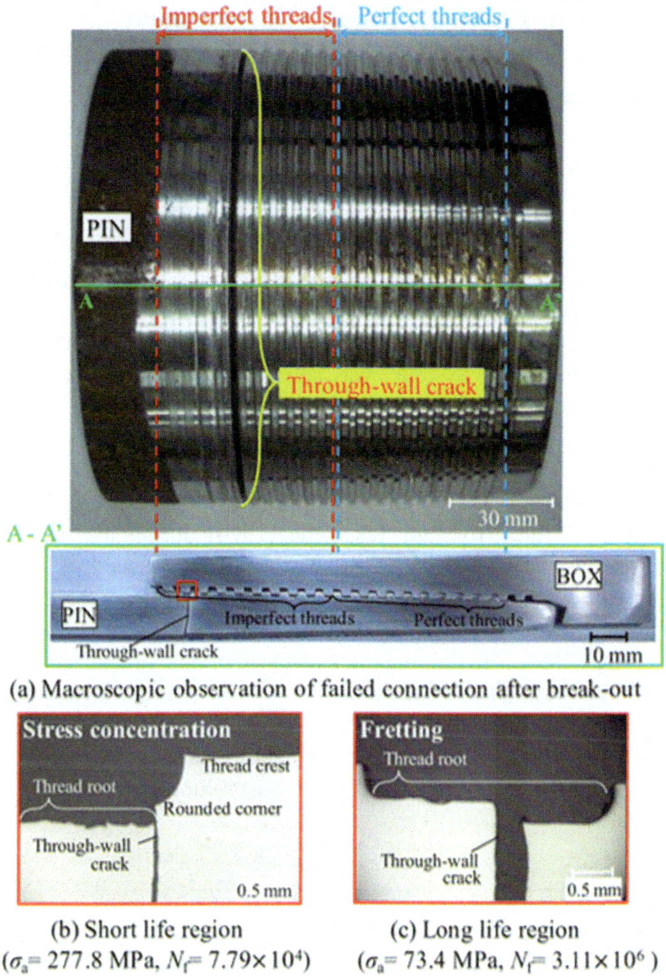

Fig. 1.39 Failure modes of oil pipe joint structures [10, 11]

The characteristic of slow crack growth on fretting fatigue is an advantage from the viewpoint of maintenance of machines or structures, and safety can be ensured by ultrasonic flaw detection and repair at periodic inspections, as introduced in the maintenance of Shinkansen axles in Chap. 9, Sect. 9.5. We hope that this comprehensive safety management method for mechanical structures, including not only strength design but also maintenance, will serve as a reference for future reliability assurance technology that is in high demand for long-life durability of machine structures.

1.6 Mechanical Problems During Development of Superconducting Linear Motor Car Ground Coil Mounting Bolts

1.6.1 Development Status [12]

The Tokaido Shinkansen has been in operation for more than 60 years since its opening in 1964, and the deterioration of the track environment has become inevitable. So, now in Japan it has become essential to develop a new high-speed railway line to replace the current Tokaido Shinkansen. Therefore, the "Superconducting Linear Motor Car System" was selected as the new high-speed railway system.

Route C shown in the diagram in Fig. 1.40 is final selected route scheduled for completion in 2027.

The development of this Superconducting Linear Motor Car System started with the construction of the Miyazaki Test Line (1974) during the former Japanese National Railways era, and in 1979, the world's highest speed of 517 km/h was achieved with the development prototype ML-500. Various improvements were made after that, and the results of these improvements have been applied to the Yamanashi Test Course since 1990. The running track distance of this Yamanashi Test Course is 18.4 km, and in November 1999, the first high-speed passing test was performed and achieved relative passing speed of 1,003 km/h, which was for the highest record in the world. Currently, the JR and equipment manufacturer company is now working to verify long-term durability and develop low-cost solutions, including maintenance (see Fig. 1.41).

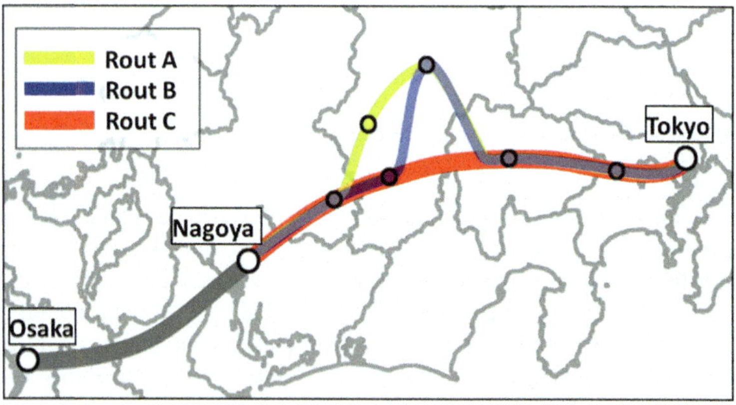

Fig. 1.40 Schematic route of the Superconducting Linear Motor Car [7]

Fig. 1.41 Superconducting Linear Motor Car on the Yamanashi Test Line and ground coil installation status [7]

A superconducting maglev train runs by levitation, propulsion, and guidance forces which were generated using the electromagnetic force between the superconducting coils on the upper side of the train and the normal electric coils on the ground side. The ground coil consists of two layers of propulsion coils and levitation coils, each of which is attached to the concrete panel with bolts as shown in Fig. 1.42.

Both the propulsion and levitation coils are made of aluminum conductor coiled and molded with resin. The propulsion coil requires electrical insulation and mechanical strength, while the levitation coil requires mainly mechanical strength, so they

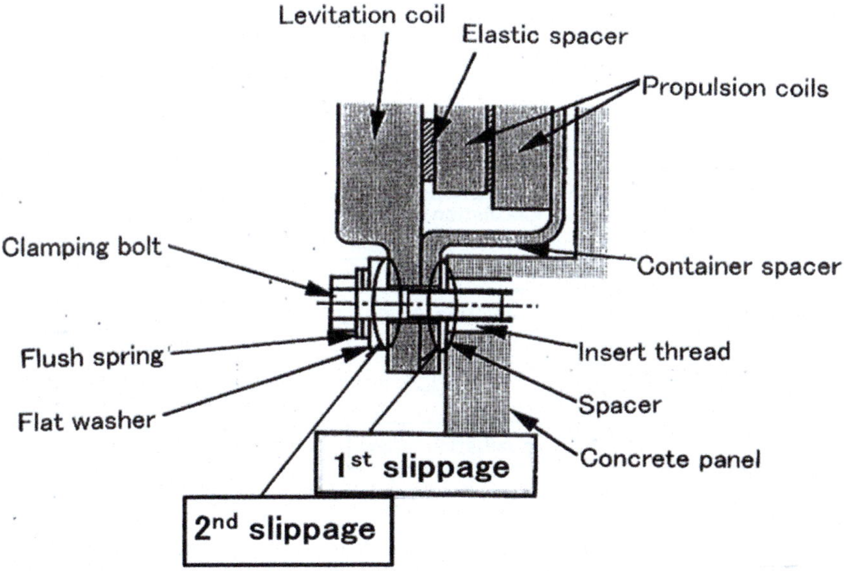

Fig. 1.42 Fastening conditions of ground coil system [7]

are made of epoxy resin and unsaturated polyester resin reinforced with glass fiber, respectively.

1.6.2 Challenges and Corresponding Technologies (See Chap. 5)

(A) **Loading and deformation conditions of ground coil systems**

The bolts shown in Fig. 1.42 subjected to a combined force of the levitation force and propulsion force in the direction perpendicular to the bolt axis and a guiding force in the axial direction of the bolt. So, the strength evaluations and loosening evaluations will be indispensable for each load direction as shown in Fig. 1.43. The axial direction load was explained in Sect. 1.1 Shinkansen motor mounting bolts. Here, we will focus on the perpendicular loading.

When designing the bolt, we initially thought that the bolt's initial fastening force F_s should be designed following Eq. (1.52), taking into consideration the friction coefficient μ of the seat surface, the combined force of levitation force and propulsive force $F_l + F_p$ and the safety factor S_f.

$$F_s \times \mu > S_f \times \left(\overleftarrow{F_l} + \overleftarrow{F_P}\right) = (1.4 \sim 2.0) \times \left(\overleftarrow{F_l} + \overleftarrow{F_P}\right) \quad (1.52)$$

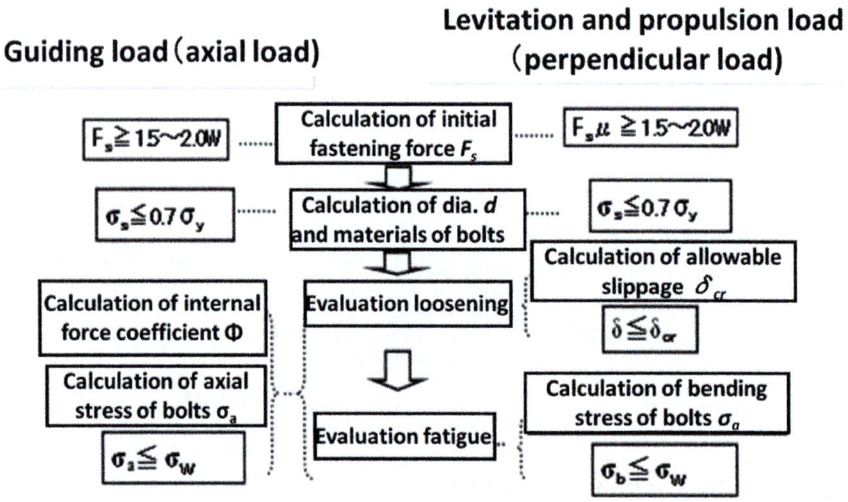

Fig. 1.43 Types of loads applied to screw fasteners, loosening evaluations and fatigue strength evaluations

Here, considering the levitation force $\overleftarrow{F_l} = 5.6\text{kN}$, propulsive force $\overleftarrow{F_P} = 4\,\text{kN}$, resultant force $\overleftarrow{F_l} + overset\_F_P = 6.9\,\text{kN}$ per bolt, friction coefficient $\mu = 0.15$, and safety factor $S_f = 2$, the required fastening force per bolt F_s is calculated to be 91.7 kN.

However, when we progressed on the environmental durability evaluation stage, it became necessary to evaluate the thermal deformation difference between coils molded with aluminum conductors and concrete panels due to differences of thermal expansion during the four seasons, day and night, and train passing and after passing. If the thermal expansion difference is calculated using a model such as that shown in Fig. 1.44 and Eq. (1.53), it will become about 0.35 mm.

$$\delta_T = \left(\alpha_c - \alpha_g\right) \times l \times \Delta T \quad (1.53)$$

The following dimensions were used for this calculation.

The distance between bolts l: 1200 mm, thermal expansion coefficient of ground coil α_c: 16.810^{-6}, thermal expansion coefficient of guide way α_g: 12.0×10^{-6}, and maximum temperature variation ⊿T: 60 °C due to four seasons, day or night, and operating or pause period.

If this thermal expansion difference is completely restrained with bolts as shown in this figure, the restraining load will be many times larger than the restraining load

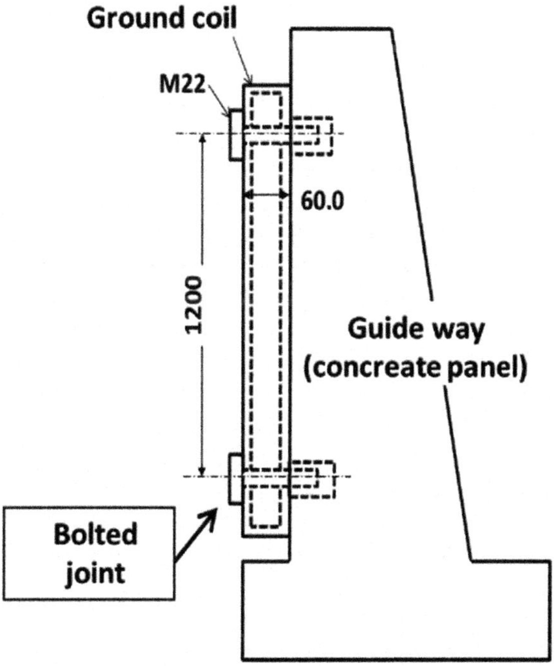

Fig. 1.44 Schematic model of ground coil fastening condition

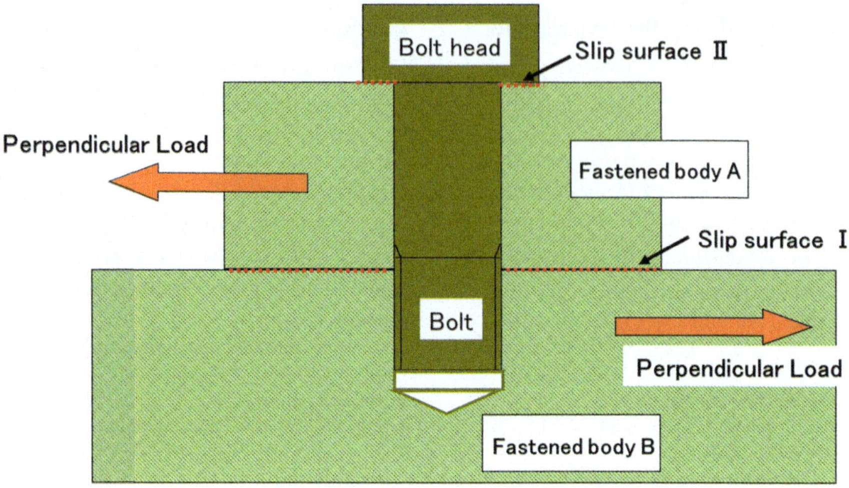

Fig. 1.45 Slip behaviors under transverse loadings

for the levitation force and propulsion force in Eq. (1.52), and it was predicted that the coil itself may be failed under this large load.

Therefore, it was necessary to design a slip allowance within a range that would not cause loosening. The outline of the design is described in detail below.

(B) **Slip behavior and critical slip distance**

When a load which is perpendicular to the axis is applied to a bolted joint, the following two contact surfaces are expected to slip, as shown in Fig. 1.45.

Slip surface I: between the fastened body A and the fastened body B.

Slip surface II: between the bolt head and the fastened body A.

Details are shown in Sect. 4.3, so I summarize the results as follows.

When the transverse load is small, slippage occurs only on slip surface I, and no loosening occurs in this stage. However, as the transverse load increases and slipping occurs on slip surface II, rotation of the bolt head begins to occur and so loosening occur. This critical slip amount at which slipping starts to occur on slip surface II is called the critical slip amount S_{cr}. And this critical amount of slip S_{cr} is shown in Eq. (1.54).

$$S_{cr} = 2\left\{ F\mu\left(\frac{l_g^3}{3EI_g} + \frac{l_p^3}{3EI_g} + \frac{l_g l_p l_n}{EI_g} + k_{wh} l_n^2\right) - M_n\left(\frac{l_g^2}{2EI_g} + \frac{l_p^2}{2EI_g} + \frac{l_g l_p}{EI_g} + k_{wh} l_n\right) + M_n k_{wn} l_n \right\} \tag{1.54}$$

S_{cr} Critical slip amount.
F_s Initial bolt fastening force.
μ Coefficient of friction at the boundary between the bolt head and the fastener.

k_{wh} Bending compliance of the contact seating surface between the bolt head and the fastened body.
k_{wn} Bending compliance of the contact seating surface between the male thread and embedded female thread.
M_n Resistance bending moment of nut.
E Modulus of elasticity of bolt.
I_g, I_p Moment of inertias of each cross-sectional area of the bolt.
α Half angle of thread.
d Nominal diameter of bolt.
d_p Valley diameter of bolt.
d_2 Effective diameter of bolt.

(C) Design example

A simulated example of a Superconducting Linear Motor Car System ground coil was shown in Fig. 1.44. Basically, the ground coils (propulsion, levitation, and guidance coil) are fastened to a concrete panel with stainless steel bolts. The evaluation of loosening due to this repeated thermal expansion difference δ_T can be made by comparing this value with the critical slip amount S_{cr} calculated by Eq. (1.53). The comparison result is shown in Fig. 1.46.

While the above evaluation was performed using a general bolt shape, a possible measure to increase this critical slip amount S_{cr} is to utilize a "wasted shank bolt" with thin diameter of the bolt's shaft section as shown in Fig. 1.47.

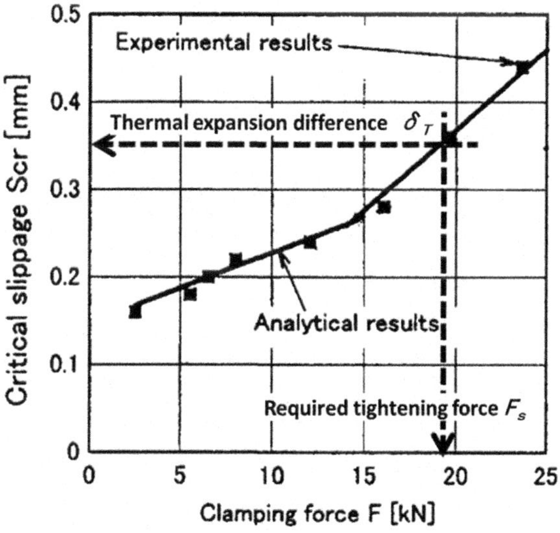

Fig. 1.46 Comparison of thermal expansion difference δ_T with the critical slip amount S_{cr} for each fastening force of bolt F_s

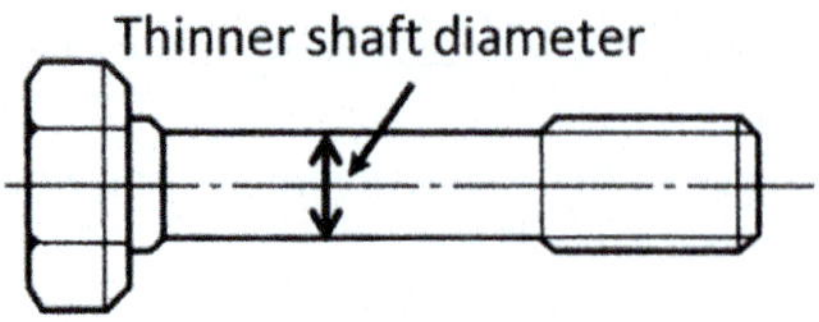

Fig. 1.47 "Wasted shank bolt" with thinner shaft diameter

1.7 KLM Royal Dutch Airlines Panel Fall Accident

1.7.1 Accident Situation

On September 23, 2017, a KLM Boeing 777 took off from Kansai International Airport and was climbing over the city of Osaka when a fuselage panel above the trailing edge of the right main wing fell off and struck a vehicle traveling in Osaka city. The bracket attached to the fallen panel was also broken. The panel was about 1.1 m long and 1.1 m wide (see Fig. 1.48).

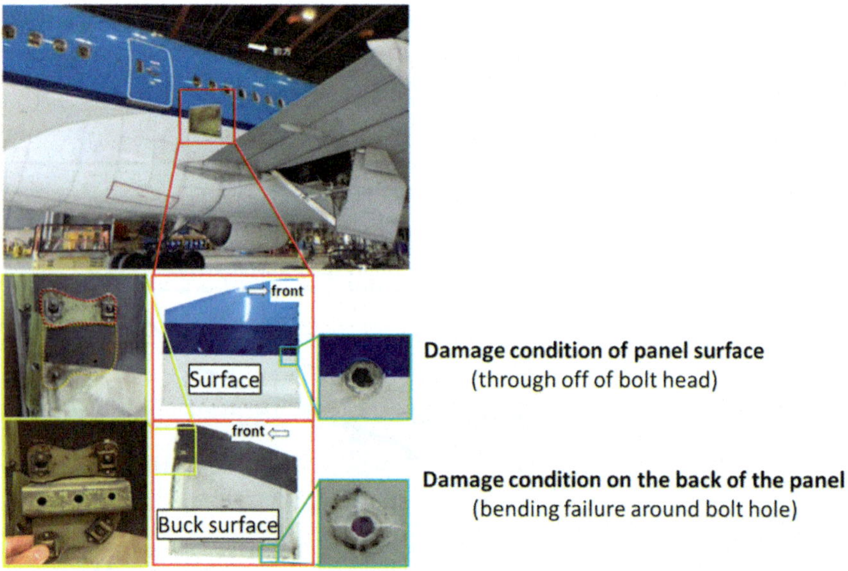

Fig. 1.48 KLM Royal Dutch Aircraft panel fallen accident(19)

1.7.2 Cause of Accident

As a result of the investigation, it was reported that slightly small-headed bolts than regular bolt (see Fig. 1.49) were used in five places to fastened the panels, which would cause the heads of the bolts to through off and the panels to fall.

However, it is not easy to imagine that the use of bolts with slightly different specifications would be enough to cause such an accident. Here we consider other design-related causes.

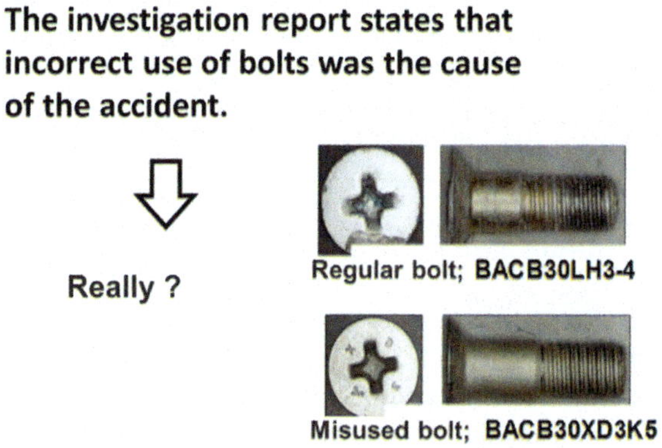

Fig. 1.49 Bolt to be used [18]

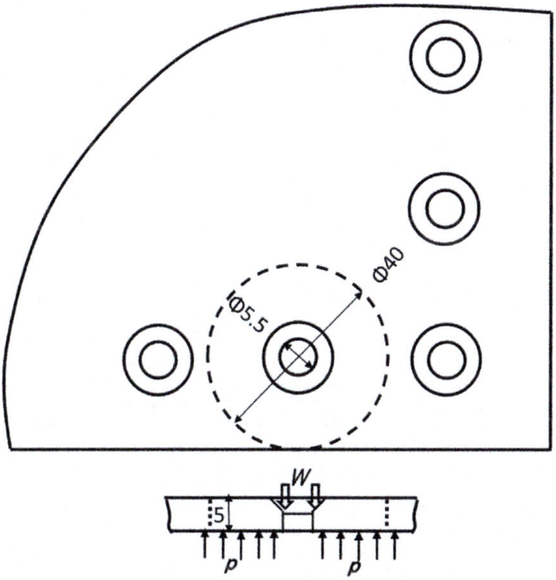

Fig. 1.50 Simplified mechanical model of a single bolt fastening [18]

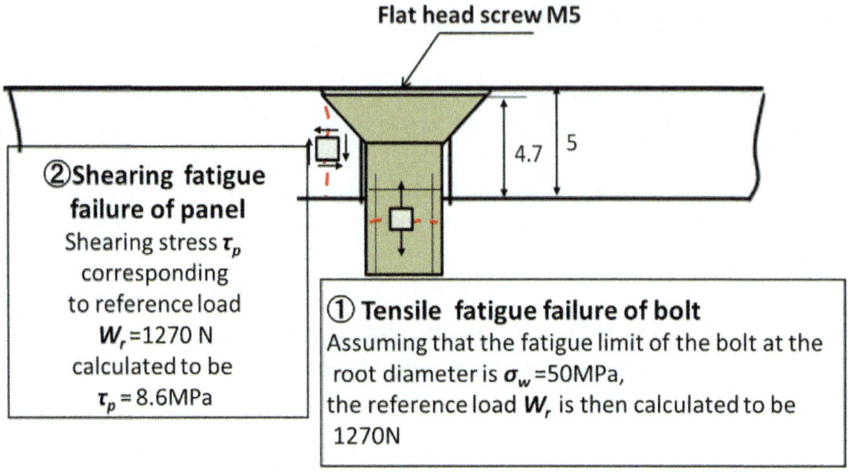

Fig. 1.51 Estimated stress and strength on each failure mode ①,② under reference external force W_r

Fastening model for each bolt is simplified and mechanically analyzed as shown in Fig. 1.50. The dimensions of each member in this model were assumed that the panel thickness is 5 mm and the mounting bolt is an M5 flat head screw, and the panel area covered by one bolt is considered to be a disk with an outer diameter of φ40 and an inner diameter of φ 5.5, as shown by the dashed line in Fig. 1.48. The cause of failure is investigated by comparing ①bolt axial stress, ② shear stress around the panel hole, and ③ panel bending stress around bolt hole when a reference external force is applied to this disk (Fig. 1.51).

In this simplified mechanical disk model, define the external force corresponding to the fatigue failure of the bolt is used as the reference external force W_r. (See Fig. 1.50).

$$W_r = A_r \times (2 \times \sigma_w) = \frac{\pi \times 4.02^2}{4} \text{mm}^2 \times (2 \times 50\text{MPa}) = 1270\text{N} \tag{1.55}$$

The shear stress τ_p of the panel under the reference external force W_r is estimated as follows,

$$\tau_p = \frac{1270N}{\pi \times 10 \times 4.7\text{mm}^2} = 8.6\text{MPa} \tag{1.56}$$

This is sufficiently low compared to the shear fatigue strength of aluminum alloy material and is unlikely to be the cause of the through off from bolt head in the accident investigation's report.

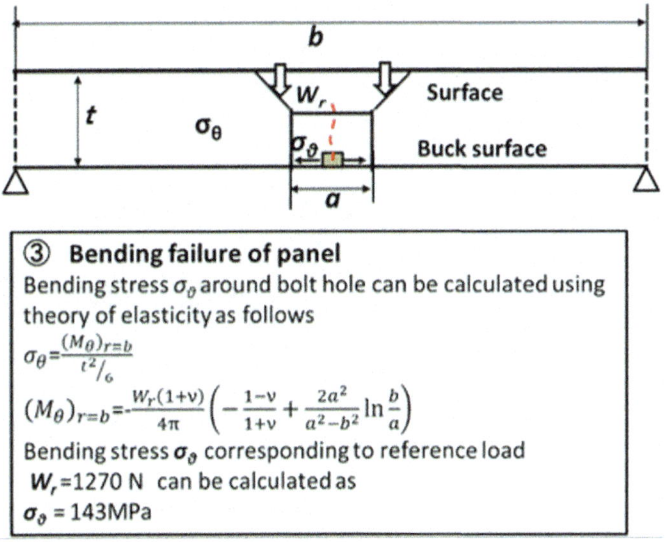

Fig. 1.52 Estimated stress on failure mode ③ under reference external force W_r

Then we will calculate the bending stress in the out-of-plane bending mode at the bottom of the inner circumference of the threaded hole in panel. These out-of-plane bending stress can be calculated using the elastic mechanical equation, as shown in Fig. 1.52 and the maximum bending stress around the rivet hole was calculated as $\sigma_\theta = 143$ MPa, which is sufficiently comparable to the fatigue limit strength of aluminum alloys. It is predicted that the failure occurred in this mode. The appearance of out of bending fatigue failure can certainly confirmed by the photo of the panel backside failure in the lower right of Fig. 1.48.

These stress analyses formula based on theory of elasticity,

$$(\sigma_\theta)_{r=b} = \frac{(M_\theta)_{r=b}}{t^2/6} \tag{1.57}$$

$$(M_\theta)_{r=b} = -\frac{W(1+\nu)}{4\pi}\left(-\frac{1-\nu}{1+\nu} + \frac{2a^2}{a^2-b^2}\ln\frac{b}{a}\right) \tag{1.58}$$

can be referred from Table 11.2 in ROARKS'S FORMULAS for STRESS and STRAIN [19], and using the equivalent disk model shown by the dashed line in Fig. 1.50, this single bolt fastening model can be replaced by the mechanical model with simple outer support and uniform inner load (W) as shown in Fig. 1.53.

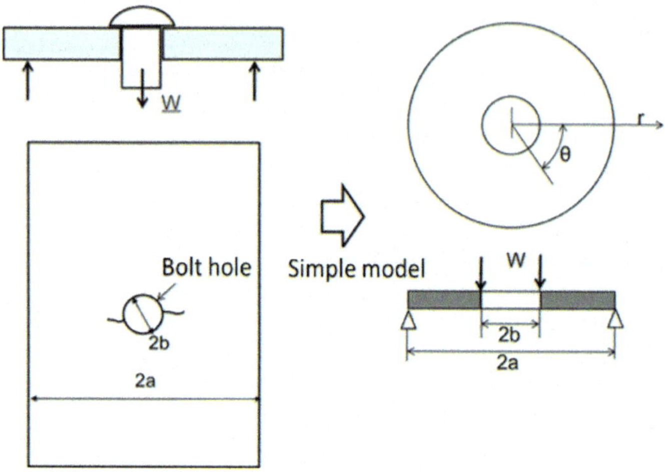

Fig. 1.53 Model for out-of-plane bending stress analysis around bolt holes

1.7.3 Similar Accident

2017/09/07,08 ANA Escape slide storage panel falling.

2017/09/27 ANA An inspection window lid (access panel) was found to have fallen onto a factory site in Inashiki, Ibaraki Prefecture. The panel was a trapezoidal white panel measuring 30 cm long, 147 cm wide, and weighing 3.14 kgf.

2017/11/30 The panel fell during a training exercise by an F-35A fighter jet deployed at the Kadena Air Base in Okinawa Prefecture. The panel was approximately 30 cm long and 60 cm wide.

2017/12/07 An object fell from a US military aircraft onto the roof of a kindergarten in Ginowan, Okinawa Prefecture. The falling object is a cylinder about 15 cm long with markings "US and other" on it.

2017/12/13 Window-like parts fell from a US military helicopter CH-53 aircraft onto the ground of an elementary school adjacent to the Futenma Air Base in Okinawa Prefecture. The falling object is 90 cm square.

2018/02/09 An air intake for the engine of an MV-22 Osprey aircraft belonging to the US military's Futenma Air Base was found on a beach on Ike Island, Uruma, Okinawa Prefecture. The discovery was made on the 9th, but it is believed to have fallen during a flight over the ocean on the 8th. The object was a semicircle, approximately 70 cm long, 100 cm wide, and 13 kgf in weight. As seen in Fig. 1.54, the damage may have been caused by a design error in the screw and rivet fasteners, similar to the panel that fell from the KLM aircraft mentioned above.

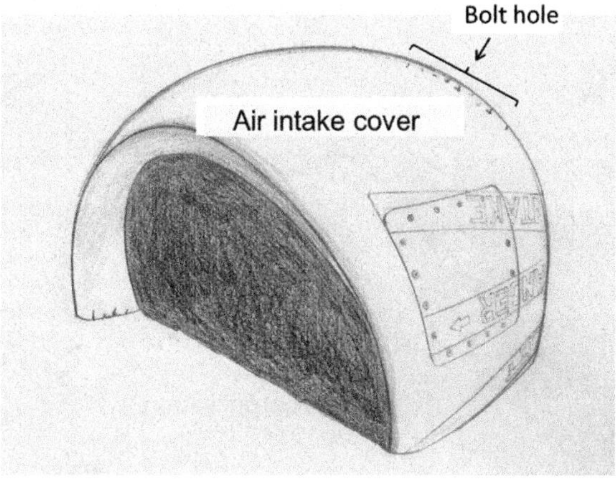

Fig. 1.54 Osprey aircraft engine air intake falling accident [18]

1.8 Fatigue Failure Accidents of Trailer and Yurikamome Hub Bolts

1.8.1 Accident Situations [20–22]

On January 10, 2002, the front left tire of a trailer fell off in Seya-ku, Yokohama, striking a pedestrian, killing one person and slightly injuring two others. After the accident, the manufacturer announced its opinion that no further wear and failure of the hub would occur as long as the vehicle was used under appropriate conditions and properly inspected and maintained. However, because there had been many accidents involving tires falling off due to broken hubs before that, the company had been voluntarily inspecting the hubs on large vehicles and replacing them free of charge, and on March 11, 2004, it announced that the front wheel hubs would be replaced under a recall.

The hub is a component that connects the axle to the wheel. It consists of an elongated cylinder and a hollow disk called a flange. The hollow part connects to the axle via bearings and both sides of the flange connect to the wheel and brake drum (see Fig. 1.55).

In the trailer where the wheel fell off, fatigue cracks started at the contact edge between flange and brake drum. This crack progressed toward the contact surface of the flange with the wheel, and finally the flange failed into slice.

Four years later, on April 14, 2006, a similar hub failure accident occurred on the Yurikamome Tokyo Waterfront New Transit Rinkai Line. Immediately after train No.1612 from Toyosu to Shimbashi, departed from Funenokagakukan Station, a wheel broke loose from the track. As in the case of the trailer, the damage was a fatigue failure at the hub bolt fastening, the details of which are shown in Fig. 1.56. Although there were no casualties among the 230 passengers involved in the accident, the train service was suspended for two days and was not fully restored until five days

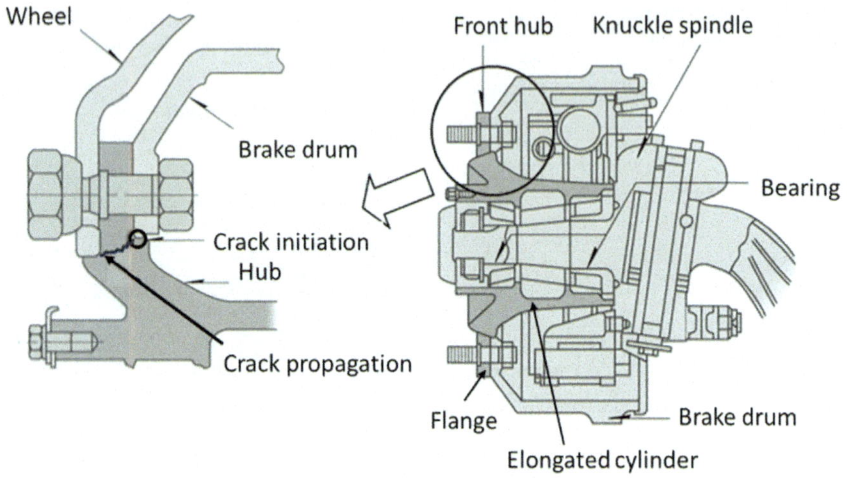

Fig. 1.55 Trailer hub and peripheral components [22]

after the accident. Ministry of Land, Infrastructure, Transport and Tourism Aviation and Traffic Accident Investigation Committee investigated the accident and released its accident investigation report on this accident at the end of February 2008 [22].

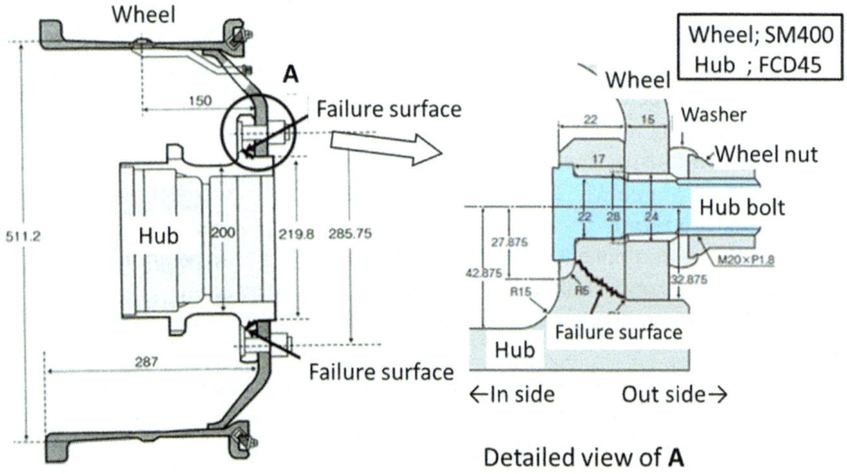

Fig. 1.56 Yurikamome hub and peripheral components [22]

1.8.2 Cause Investigation (See Chap. 9 and Sect. 10.4)

In a newspaper article on the trailer accident at the time, there was a conflict between the manufacturer's claim that they had neglected to check the hub bolt tightening force during maintenance and start-up inspections, and the user's claim that the design of the parts was defective. Defects in maintenance and start-up inspections include overloading, wear, and confirmation of bolt tightening force. Figure 1.57 compares the maximum stress in the hub stress concentration area when the hub bolt tightening force is sufficient and when the hub bolt is about 30% loose, using FEM analysis of a simple model. In the case of complete tightening force shown on the left, the bending moment load from the wheel tends to be transmitted evenly from the wheel to the hub through the eight bolts, but in the case of loosening shown on the right, the bending moment load from the wheel is concentrated on the upper and lower ends of the hub, and the maximum stress at the hub stress concentration area is 30% higher than in the case of complete tightening force. It can be seen that the maximum stress at the hub stress concentration is over 30% higher than in the case of complete tightening. Therefore, it is understandable that the possibility of damage to this hub would increase if it were used with loose bolts due to poor maintenance and inspection.

However, it is difficult to believe that such an increase in stress would lead directly to failure accident in a component to which a sufficient safety factor is applied, such as transportation equipment.

The fault on the design side is thought to be wrong choice of material for the hub, whether the wall thickness was sufficient, or whether the filet corner R at the stress concentration area was adequate.

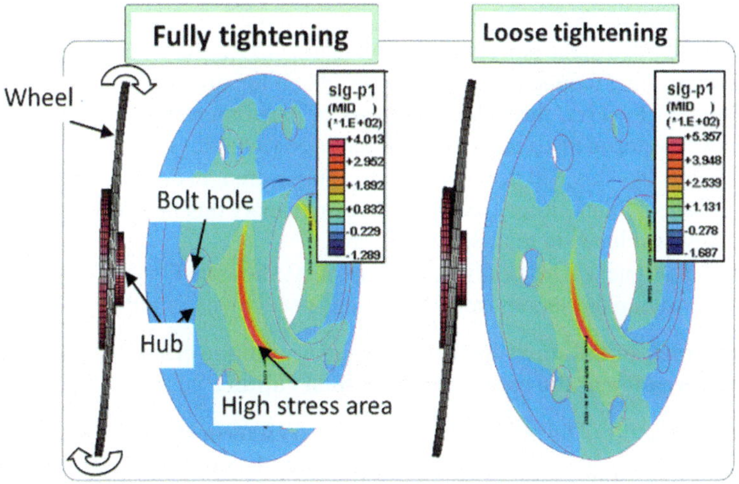

Fig. 1.57 Calculation results of maximum hub stress on each fully tightening and loose tightening condition [7]

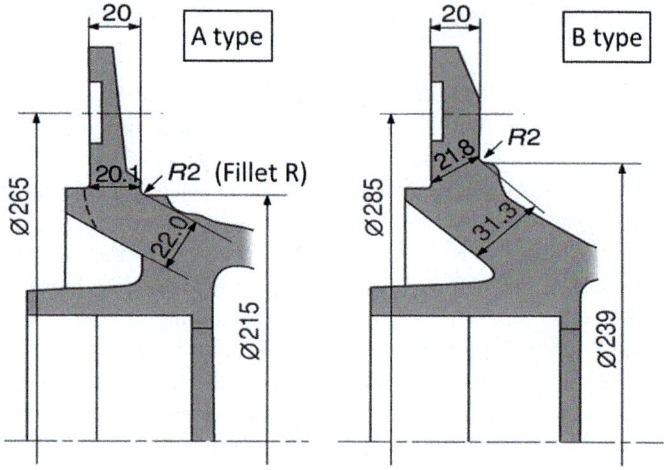

Fig. 1.58 Process of countermeasures for trailer hub structure [21]

However, it is hard to imagine that such a simple mistake could cause so many accidents in a product that has been designed based on a long history. In fact, as shown in Fig. 1.58, accidents continued to occur even after measures were taken to increase the cylinder wall thickness from A type to B type.

In the case of the Yurikamome hub, as shown in Fig. 1.59, the Accident Investigation Committee concluded that fretting wear occurred due to relative micro-slippage on the contact surface between the hub and wheel and that when the hub was re-fastened to fill the gap, bending stress (tensile stress) was generated in the hub and a crack occurred, which is more like related to maintenance problem.

However, the author concludes that it was fretting fatigue based on the fact that the reddish-brown wear powder (common name "COCOA") was occurring. In the first place, fretting fatigue has long been a problem for strength designers of railway axle fittings. However, through many years of experimentation and analysis, they have now developed an excellent countermeasure technology.

Accidents like this will occur repeatedly unless strength designers of automobile axle/hub/wheel joints closely follow, realize, and apply this fretting countermeasure technology developed in high-speed railways.

Fretting fatigue will be described in Chap. 9. As shown in Fig. 1.60, it is a complex process involving the superposition of various phenomena as follows.

Fatigue crack initiation at the contact edge → fatigue crack propagation with wear progression at the contact edge → final rupture

So the final rupture of these hubs occurs after a very long operating period. Therefore, when the contact edge of the hub/bolt and the step filet of the hub are superimposed, as in these examples of hub failure accidents, it shows the two-step S–N curve as shown in Fig. 1.61. Fatigue failure in areas with high stress amplitude occurs at stress concentration areas of step filet and exhibits normal 10^7 fatigue limits. In the

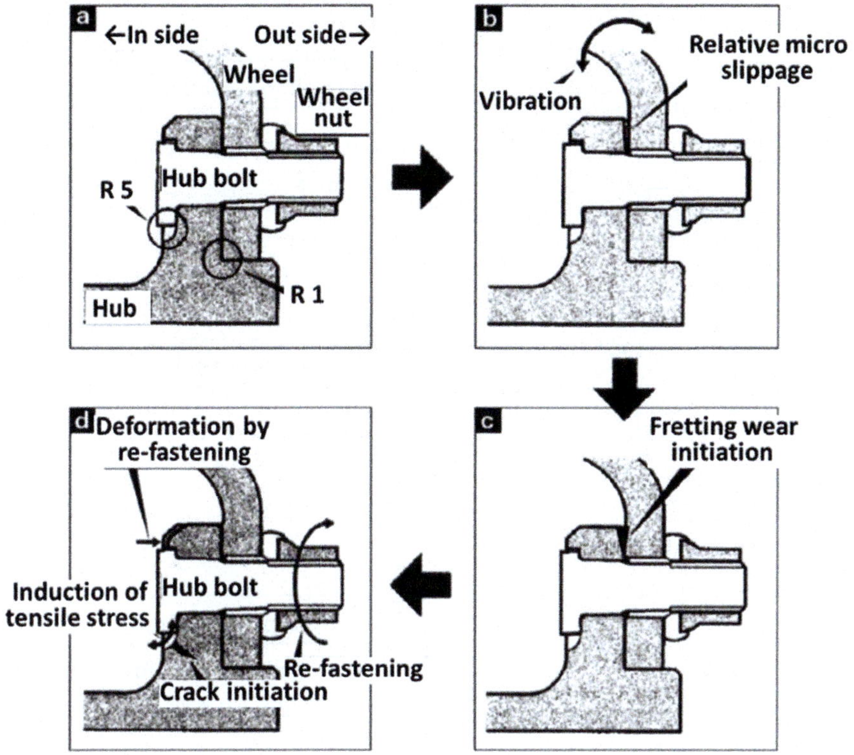

Fig. 1.59 Crack initiation mechanism of Yurikamome hub estimated in the report [22]

region of low stress amplitude, fatigue strength decreases with wear, indicating very long-life fatigue failure dominated by fretting fatigue.

This can inadvertently induce errors in the design and quality assurance engineers. If quality assurance is done with a ordinal fatigue limit verification test based on 10^7 cycles in a general factory, the customer will spend long years fretting fatigue tests in the operating field. I strongly hope you are aware of this irony.

1.9 Fretting Fatigue Failure Accidents of Roller Coaster Fastening Bolts

1.9.1 Accident Situation [24]

Date and time of occurrence: Approximately 12:25 on April 29, 2012.

Location: Fuji-Q Highland, Fujiyoshida City, Yamanashi Prefecture “Eejanaika”.

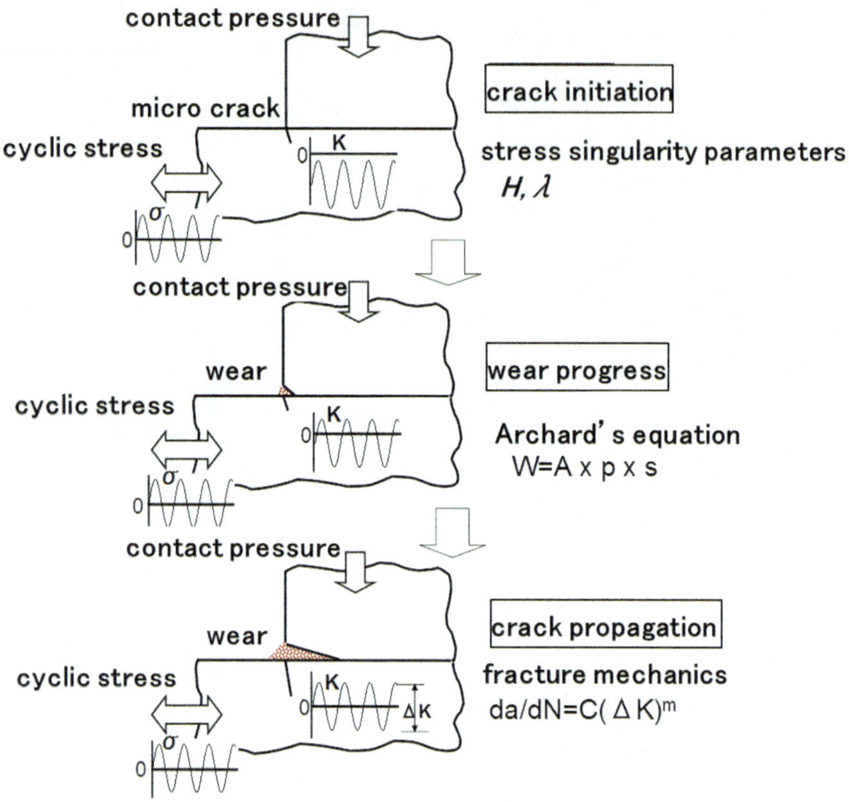

Fig. 1.60 Fretting fatigue process at wheel/hub contact edge [7]

Summary of the accident: A part of the bolt (diameter 12 mm, length 80 mm) of the car while the coaster fell out and hit a guest walking in the park. A photograph of the accident roller coaster train is shown in Fig. 1.62, and there are five cars in one train, but the bolt fell from the second car.

At the lower center of the vehicle body, a device for preventing the vehicle from running backward (hereinafter referred to as the "reverse running prevention device") is provided in the event that some kind of failure occurs during the winding-up of the vehicle. When the hook of this reverse run prevention device passes through the reverse run prevention rack, collision noise is generated due to the hitting of the rack, but to reduce the collision noise, noise control parts are provided (see Fig. 1.63), which reduce the noise by friction with the track rail. The shaft and hook of the reverse run prevention device are connected by a through-bolt, and this bolt breaks and falls out. (See Fig. 1.64).

The specifications and photographs of the broken bolt are shown in Table 1.5 and Fig. 1.65.

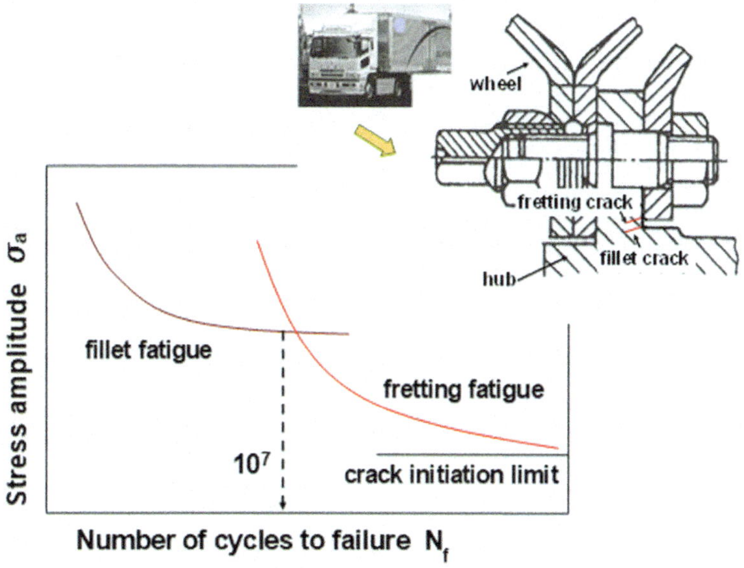

Fig. 1.61 Hub/wheel fastener structure and 2-step S–N curve [7]

Fig. 1.62 Photograph of roller coaster train [23]

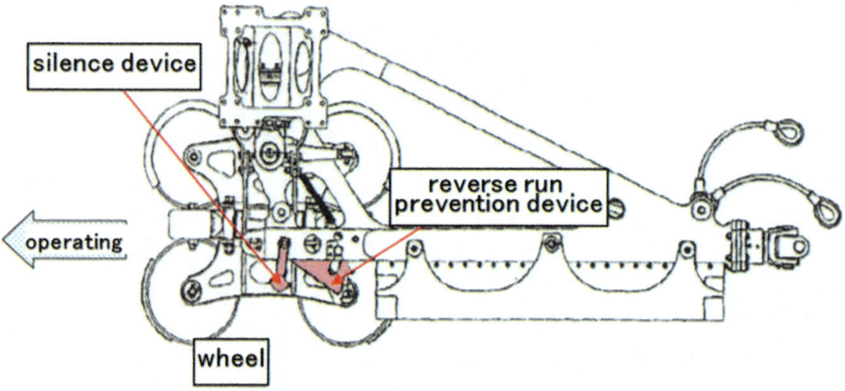

Fig. 1.63 Lower side of the center of the car body [23]

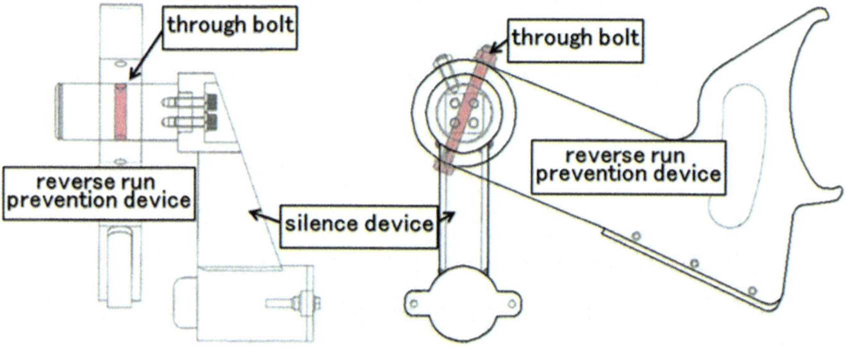

Fig. 1.64 Reverse run prevention device and silence device [23]

Table 1.5 Specifications of the broken bolt [23]

Item	Nominal diameter	Under neck length	Strength classification
Specification	M12	130 mm	12.9

Fig. 1.65 Photographs of the broken bolt [23]

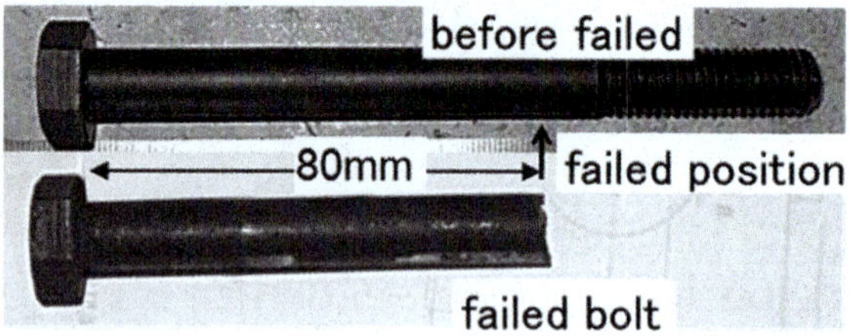

1.9.2 Investigation of Cause (See Chap. 9)

The broken position of the through bolt is 80 mm below the neck, as shown in Fig. 1.65, and this position corresponds to the outer diameter of the shaft, as shown in Fig. 1.66. If the through-bolt is properly tightened, it will be fixed perfectly in the hook hole, but the outer diameter of the shaft and the inner diameter of the hook are designed for intermediate fitting, as shown in Table 1.6, so in this case, they will easily slip each other, and when the anti-reverse device repeatedly collides with the rack of the track during winding, the shaft and hook rotate relatively by the impact force. In this case, at the above-mentioned bolt breakage position, an impact lord will be applied with each collision of the rack.

Under such a condition with high local contact pressure and a minute relative slippage, there will be a risk of fretting fatigue at the contact edge. This estimation can be confirmed from the fracture surface in Fig. 1.67. The starting point of fracture (ratchetting) is widespread along the contact edge of the bolt, which is a characteristic of fretting fatigue, as detailed in Chap. 9.

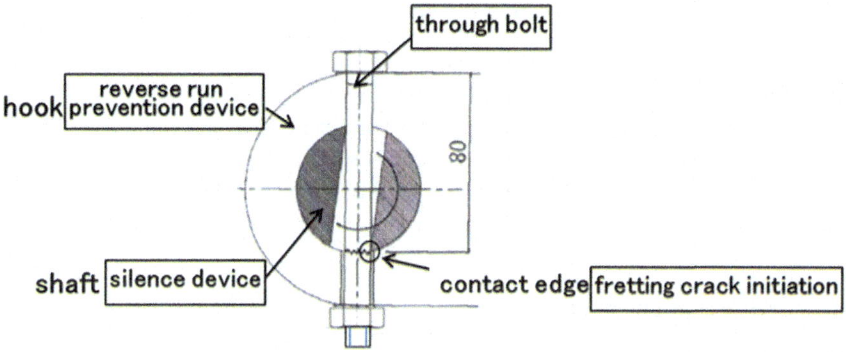

Fig. 1.66 Local contact situation between bolt and bolt hole when relative slippage is allowed [23]

Table 1.6 Gaps at each contact site [23]

Part I item	Minimum gap (mm)	Maximum gap (mm)
Axis/Hook	–0.011	0.30
Bolt/Bolt hole of shaft	0.024	1.09

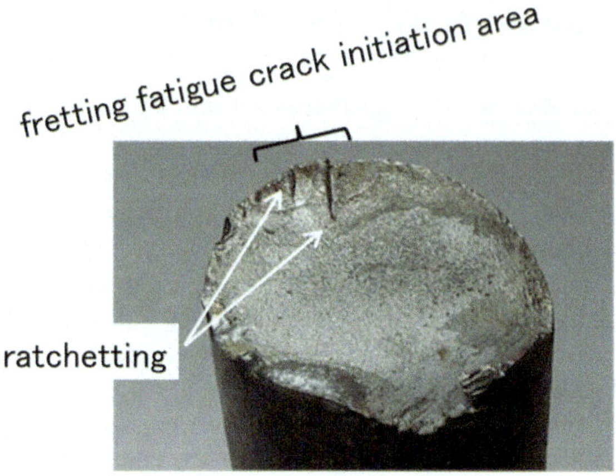

Fig. 1.67 Photograph of fracture surface observation of the through-bolt, which show the characteristics of fretting fatigue cracks [23]

1.9.3 Countermeasure Technology (See Sect. 9.4)

(A) **Details of measures taken in the accident investigation report** [23]

According to the reports from roller coaster operating company and the inspection and maintenance company, the cause of this accident was estimated as the poor torque transmission structure between through-bolt and shaft, and the rotational force is locally concentrated due to the gap between the shaft and the through bolt, as shown in Fig. 1.66. Therefore, as shown in Fig. 1.68, the shaft and the reverse run prevention device hook have been improved to a structure in which they are fitted with a D-cut joint.

(B) **Fretting fatigue measures**

Although the strength of the structure can be improved by the above measures, it takes time and effort to achieve high-precision fitting without gaps in such processing, so it is simply recommended to take measures against the problematic contact edge fretting fatigue. For general measures against fretting fatigue, refer to Chap. 9 Sect. 9.4. Among these countermeasure techniques, the countermeasures for such local contact between the bolt and the bolt hole edge are shown in Fig. 1.69. It seems appropriate to provide stress relaxation grooves as shown. Figure 1.70 shows the experimental results of improving the fretting fatigue strength when such a stress relaxation groove was provided. Due to the measures for the stress release groove, the fatigue limit was improved by approximately 1.5 times compared with that of simple fretting. It was found that more improvement can be obtained if the shape of this stress release groove (R2.5 mm in the experiment) is optimized (for example, if R is increased from 2.5 to 3 mm or 4 mm and the stress at the bottom of the groove is reduced), the fretting fatigue limit will be up to approximately 1.8 times (see Chap. 9 Sect. 9.4).

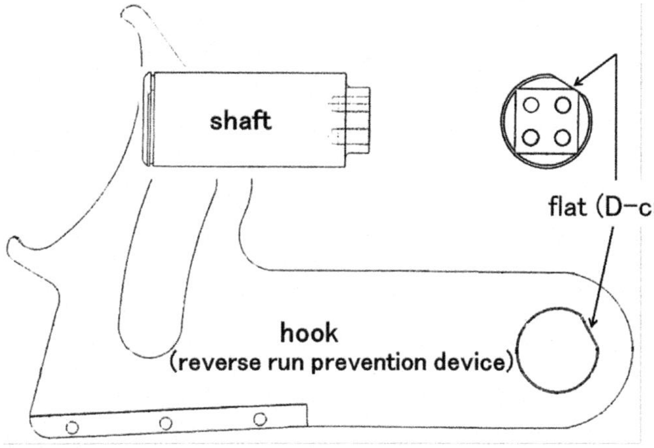

Fig. 1.68 Structure of the shaft and reverse run prevention device hook after countermeasures [23]

Fig. 1.69 Schematic diagram of stress release groove measures [18]

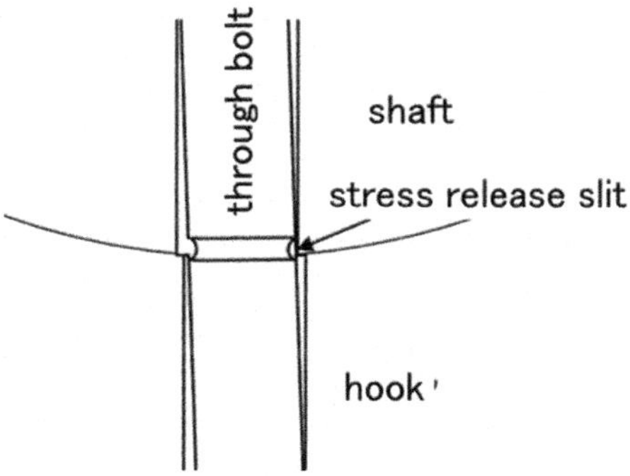

1.10 Sinking Accident of TITANIC by Brittle Fracture of Hull Fastening Rivets

1.10.1 Accident Situation

TITANIC sinking accident occurred in the North Atlantic Ocean on the night of April 14–15, 1912, the fourth day of the ship's maiden voyage from Southampton, England, to New York City, USA. The TITANIC, the largest passenger ship at the time, was carrying 2,224 passengers when it struck an iceberg at 23:40 (the local time at the site of the accident) on April 14, 1912. The ship sank at 2:20 the next day, April 15, two hours and 40 min after the accident occurred, killing more than 1,500 people. This was the largest number of fatalities in a maritime accident in 1912 (see Fig. 1.71).

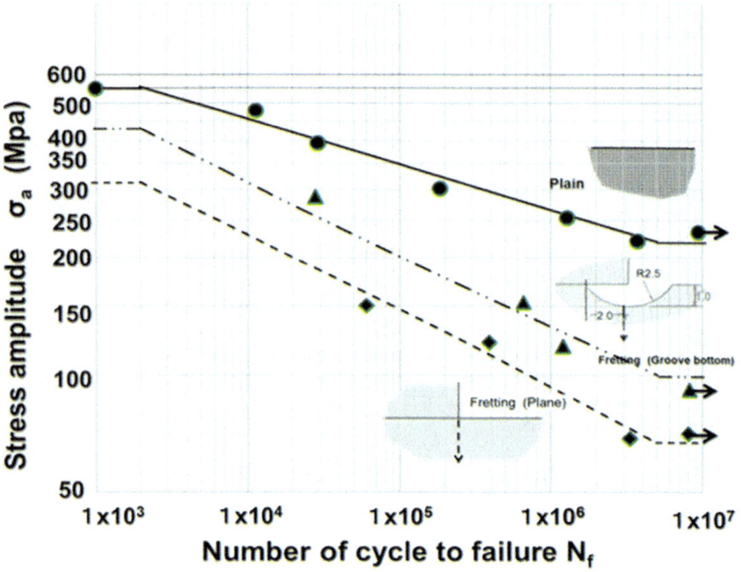

Fig. 1.70 Experimental results of fretting fatigue strength improvement by stress release groove [18]

Fig. 1.71 Situation at the time of the TITANIC's iceberg collision and sinking [24, 25]

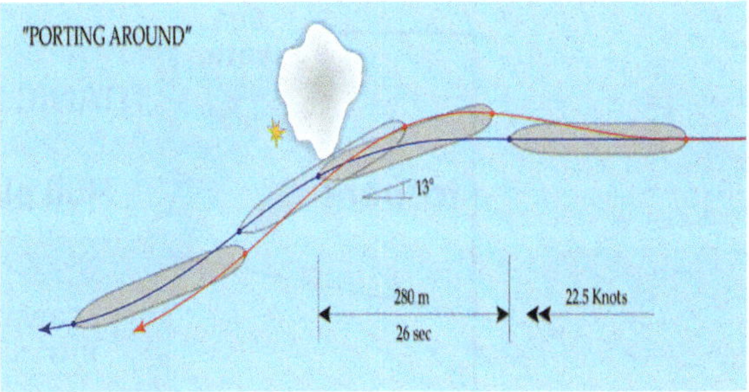

Fig. 1.72 TITANIC's course as it collides with an iceberg (blue for bow path, red for stern path) [24, 25]

Immediately before the accident, the sea surface was calm and there were no waves, so there was insufficient communication from other ships warning them "to be careful of large icebergs". At 23:39, observers spotted a large iceberg in the direction of the TTITANIC. He rang the watch bell three times and called the bridge to inform them of the situation. The captain ordered "hard-a-port around [the iceberg]" and also "full astern". As a result, a head-on collision was avoided, but a side collision occurred with the lower part of the iceberg, which was below sea level, scraped the starboard side of the ship for about seven seconds, and pieces of ice peeled off the top of the iceberg and fell to the forward deck. After a few minutes, all of TITANIC's engines stopped and the ship was forced to drift northward into the Labrador Current (see Fig. 1.72).

Initially, it was long believed that the impact of the iceberg had caused a large hole in the ship's hull. After the accident, an ultrasonic survey of the wreckage indicates that the damage consisted of six narrow holes, totaling only about 1.1 to 1.2 m^2 of the hull. The breach was up to 12 m in length and appears to have been along the hull plate. It is assumed that the steel rivets holding the plate in place were dislodged or flew open; these created a narrow rift and through which sea water entered (see Fig. 1.73).

1.10.2 Investigation of Cause

(A) **Brittle Fracture of Rivets**

The TITANIC was built with riveted joints which are popular joint technology of the time. Sixty percent of the plates in the center of the ship were joined with three rows of mild steel rivets, while the bow and stern plates were joined with two rows of wrought iron rivets. These two rows of rivets are believed to have been near their strength limits even before the collision. This is because these rivets, with their numerous slag entrapments, are brittle and can fail brittle under ordinal stress, especially in extremely cold weather. In particular, by the post-raising investigation, it found that

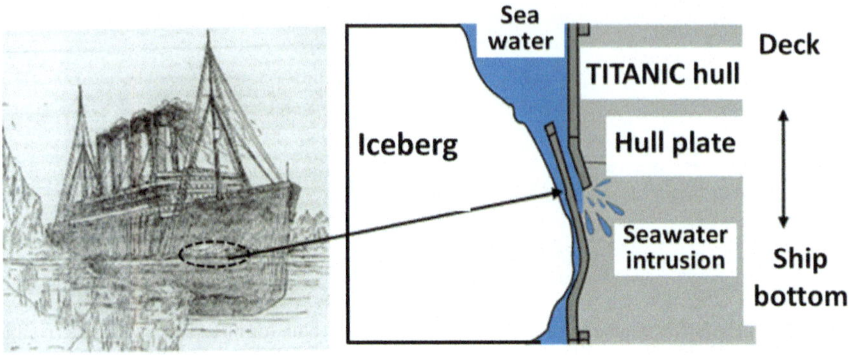

Fig. 1.73 Hull plate damaged after collision with iceberg

the rivets contained high silica-slags, as shown in Fig. 1.74, and in the area near the bottom these silica-slags were oriented perpendicular to the axis due to plastic flow during joint construction. These will lead low ductility and fracture toughness values in perpendicular to the axis, and high brittle-transition-temperature. So, these rivets were easily broken by contact with the iceberg in thus cryogenic sea area.

(B) **Hull breakage**

It was a great mystery to the engineers of the time why a crack of only 1.1 to 1.2m^2 caused by collision with an iceberg could split the hull in half and sink a huge, state-of-the-art ship of the time. The investigation into the cause of the accident will be explained along with the course of the accident as follows.

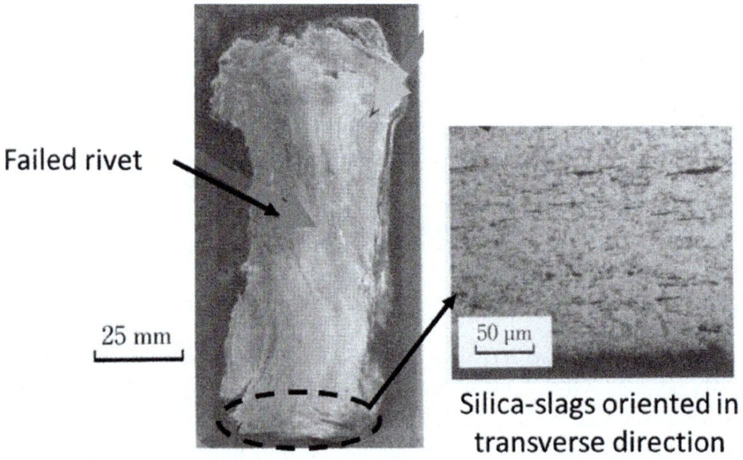

Fig. 1.74 Broken rivet and its microstructure

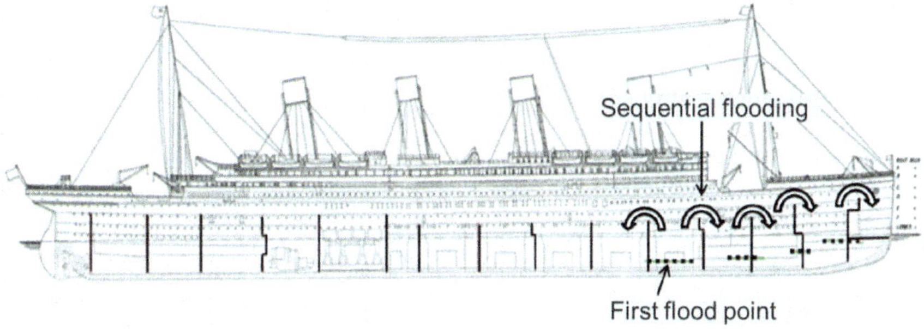

Fig. 1.75 TITANIC's hull and bulkhead arrangement

(1) The green line in the overall view of the TITANIC's hull in Fig. 1.75 is a crack caused by brittle fracture of a rivet. Although these cracks are not large, but why they lead very swift flooding of sea water and sinking of ship so rapidly.

(2) To begin with, the hull was divided into 16 compartments, each of which was separated by 15 bulkheads indicated by bold solid lines. Initially, flooding occurred from the five compartments on the bow side where there were cracks, but because the tops of these bulkheads were not closed, seawater flowed into the adjacent compartments sequentially through the gaps at the tops, causing the bow side to sink as shown in Fig. 1.76a.

(3) Finally, when the ship tilted 30–45°, the bending moment caused by the weight of the stern section, which was bounced up into the air, was concentrated around the engine room hatch, the weakest point of the ship's structure, and the ship broke in half as shown in Fig. 1.76b.

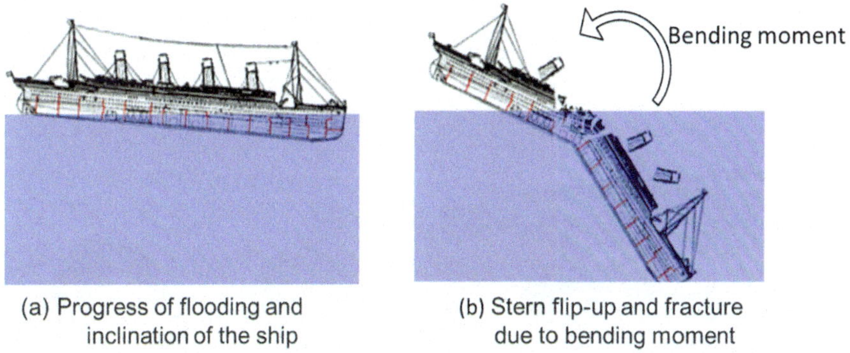

Fig. 1.76 Progress of inundation, inclination, rupture, and sinking

1.10.3 Measures

(A) **Design engineering supports**

Technically, the first step is to investigate and secure the material to prevent the reduction of fracture toughness and increase of transition temperature caused by slag defects in the rivets. In addition, bulkheads should be connected up to the top of the hull to prevent the inflow of water from neighboring compartments.

Now by these measures, the rivet failure accidents reduced drastically. (See related technology—Sect. 1.2, Fig. 1.11).

(B) **Organizational support**

The captain was not informed of the icebergs, even though he had received warnings from other ships sailing nearby. The information on organizational failures such as inadequate life-saving equipment and boats is well known and will be omitted here.

1.11 Delayed Destruction of Windmill High Tensile Bolts [26]

1.11.1 Accident Situation

On Friday, March 14, 2014, wind turbine No. 10 (manufactured by GE Windpower) at the Hibikinada wind farm (total output: 1,500 kW × 10 units) in Wakamatsu-ku, Kitakyushu City, Fukuoka Prefecture, Japan, was operating in clear weather with wind speed of 12 m/s and output of 1,500 kW when the pinion gear of the pitch gear for blade 1 fell out at 15:00 and the wind turbine automatically stopped at 15:00 (see Fig. 1.77).

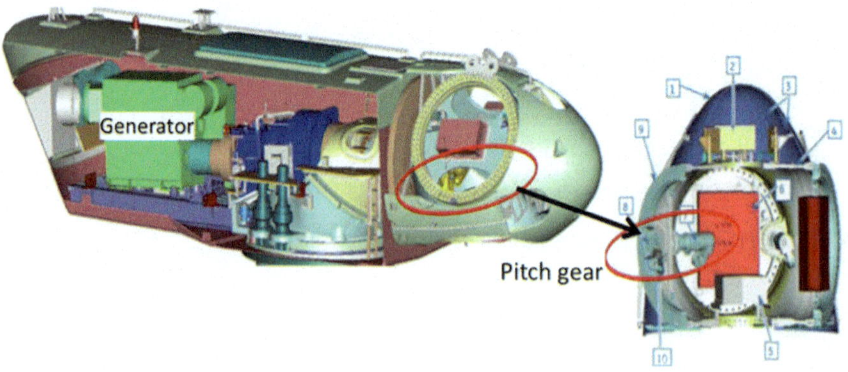

Fig. 1.77 Pitch gear for blade 1 [26]

1.11.2 Investigation of Cause

(1) The pinion gear itself that fell off was not failed, but the three bolts securing the shaft were broken (see Fig. 1.78).
(2) One of the failed three bolts (bolt 1) could be loosed by hand. (Fig. 1.78 upper left photo).
(3) The bolt that broke was M12 × 35 hexagon socket bolt with strength category 10.9. The broken parts were at the root of the bolt head on all three bolts. (Fig. 1.78, upper left photo).
(4) The chemical composition of the bolts was within the range specified for bolts of strength category 10.9 of JISB1051 as shown in Table 1.7.
(5) The approximate conversion value of tensile strength from bolt hardness measurement (average 375HV0.5) was about 1200 MPa (JIS steel HB). This strength level is considered to be an area that requires careful attention to delayed fracture, which will be discussed in detail in Chap. 8.
(6) Fig. 1.79 shows the results of macro and micro-observations of the fracture surface. The micrographs show a mixture of intergranular and intragranular

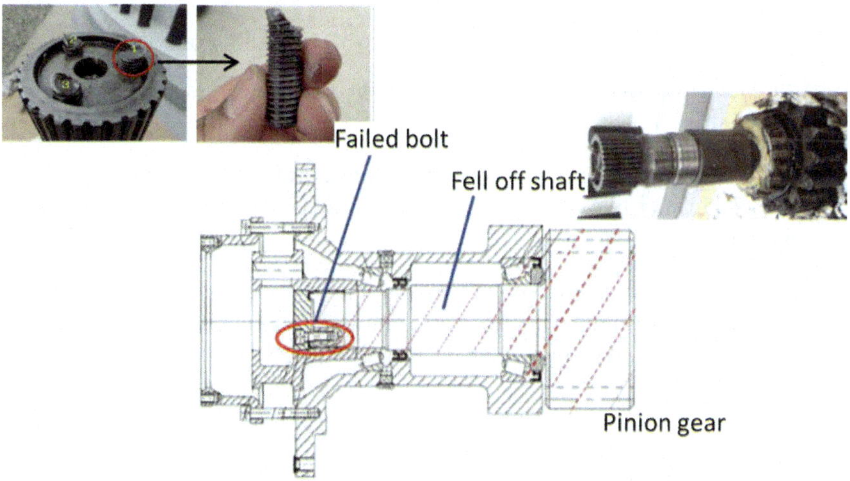

Fig. 1.78 Pitch gear dropped shaft and broken bolt [26]

Table 1.7 Chemical analysis results of bolts [26]

	C	Si	Mn	P	S	Ni	Cr	Mo	V
Analytical Results	0.24	0.11	0.91	0.01	0.005	0.02	0.31	0.01	0.003
JIS B1051	0.2–0.55			≦0.035	≦0.035		≧0.30		≦0.035
									Wt. %

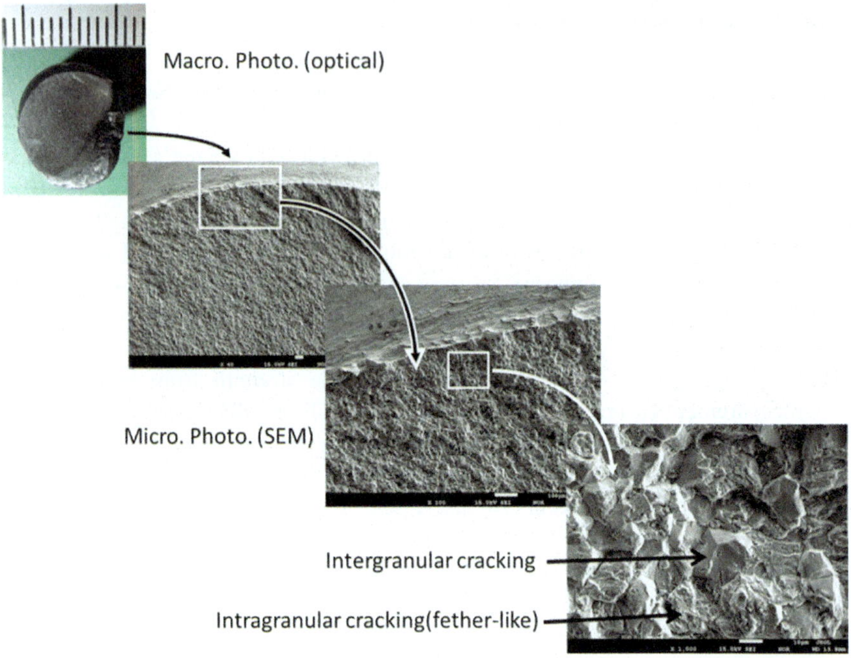

Fig. 1.79 Macro (optical) and micro (SEM) photographs of bolt fracture surface [26]

(feather-like) cracking. Intergranular cracking is a characteristic of delayed fracture, and feather-like cracking is a feature of stress corrosion cracking involving chlorides.

Based on the these natures of the material, the condition of the fracture surface and the fact that the site was on the coast, and in a bad environment, it is inferred that this accident would be caused by delayed fracture.

1.11.3 Measures

Countermeasures against delayed fracture will be discussed in detail in Chap. 8. For equipment like this that operates on the coast, environmental countermeasures are difficult, and the only way to do so is to reduce the strength of the material. The tensile strength was specified as 1000 MPa in the design standards, but it was estimated to be 1200 MPa in actual measurements. And it is necessary to take care in heat treatment to lower the tensile strength less than 1000 MPa.

1.12 Falling off Accident of Sasago Tunnel Ceiling Panel Mounting Bolts [27]

1.12.1 Accident Situation

At around 8:05 a.m. on December 2, 2012, a ceiling slab collapse accident occurred on the inbound line of the Sasago Tunnel on the Chuo Expressway, at around 1,700 m from the entrance on the Otsuki City side. Each ceiling board was made of concrete, measuring 8–9 cm thick, 5 m wide, 1.2 m long, and weighing about 1.2 tons (12 kN).

Approximately, 270 concrete slabs (including the inner wall) collapsed in a V-shape over a distance of 138 m. The tree cars were driving in that time and pinned under the concrete slabs, and two of them caught fire. The fire break out the tunnel and the smoke from the fire could not be evacuated because the smoke removal system in the tunnel did not work due to the collapse of the ceiling panel, so the tunnel was filled with hot smoke. Therefore, rescue workers were unable to reach the scene.

Five burned bodies of a man and a woman were found in an estate car that was underneath the collapse site, and the bodies of a man and a woman were found burned to death in a regular passenger car. In addition, a man was driving a truck and was caught in the collapse called his colleagues for help on his cell phone, but was already dead when he was rescued.

Rescue efforts were underway until December 3, the day after the accident, and a total of nine people were confirmed dead and two seriously injured, making it the worst accident on a highway in Japan.

1.12.2 Design and Construction Status

In February 1971, the Japan Highway Public Corporation submitted a construction plan to the Minister of Construction for the Sasago Tunnel between Otsuki JCT and Katsunuma IC on the Chuo Expressway, and the plan was approved in March 1971. The tunnel between Otsuki JCT and Katsunuma Interchange was opened to traffic on December 20, 1977.

The lateral flow ventilation system was adopted because of the tunnel length of more than 3,000 m. The ducts for both air intake and exhaust required for the lateral flow ventilation system were designed with a ceiling plate structure that utilizes the tunnel cross section. The cross section was designed based on the estimated cars speed (80 km/h), estimated traffic volume (26,000 vehicles/day), and diesel vehicle mixing rate (17%), and the specified values for soot transmission rate (50%), maximum wind speed in the roadway (8 m/s), and wind speed at the end of the duct (20 m/s) were set. Figure 1.80 shows the shape of the cross section of the collapse site, and Fig. 1.81 shows the ceiling slab and its suspension structure.

The ceiling plate was designed to be supported by 16 adhesive anchor bolts for every 6 m of CT steel. The following loads were considered to be applied to each bolt,

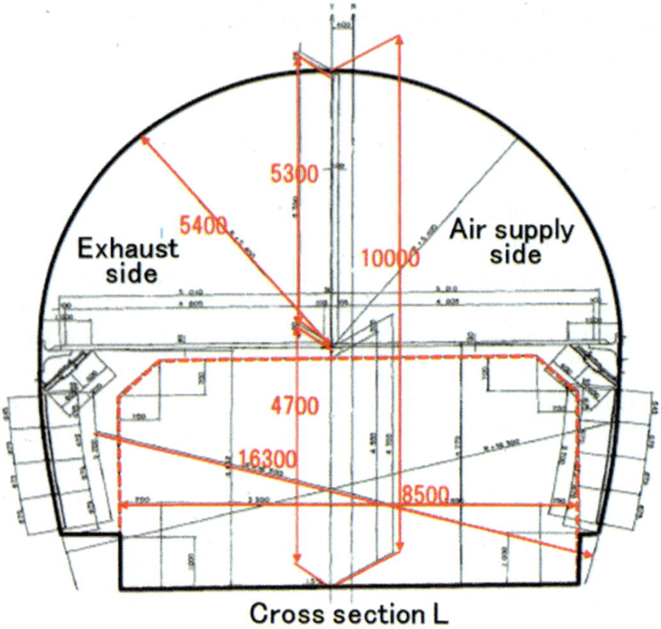

Fig. 1.80 Cross section of Sasago Tunnel on Chuo Expressway [27]

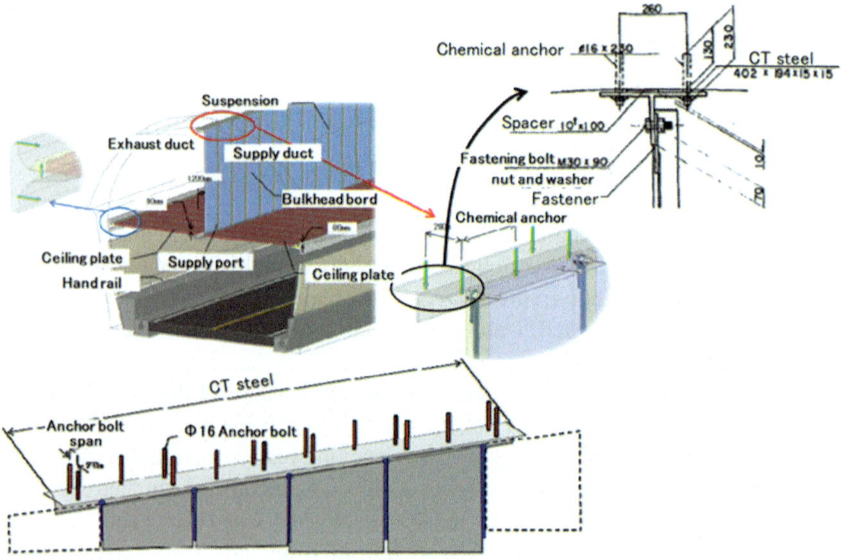

Fig. 1.81 Anchor bolts layout and ceiling slab suspension structure [27]

①;the dead weight of the ceiling plate, ②; the dead weight of the bulkhead plates, ③; the dead weight of other mortar, CT steel, etc., ④; the weight load of workers on the ceiling plate, and ⑤; wind loads applied vertically to the ceiling plate by airflow and exhaust. The design assumed that the load on the CT steel would be shared equally by the 16 adhesive anchor bolts.

The adhesive anchor bolt used here is "Chemical Anchor, Resin Capsule R-16" manufactured by Nippon Decorax. This adhesive anchor bolt (M16) is of the capsule type, and the installation method is shown in Fig. 1.82. First, a hole (drilling diameter 19 mm, drilling depth 130 mm) is drilled in the zenith mounting area, the hole is cleaned, the resin capsule is inserted, and the anchor bolt is embedded using a hammer drill or other rotating impact installation. The tightening principle is that the interfaces between hardened adhesive resin and concrete on the drilled hole surface or the bolt threads will show shear resistance (adhesion strength). Therefore, drilling a bolt hole with an appropriate diameter and casting so that the bottom of the hole and the tip of the bolt are aligned are considered prerequisites for using the product. However, the work manual shown in the product catalog at that time stated only as, "When hammering, turn off the switch as soon as the bolt reaches the bottom of the hole and do not over agitate the bolt unnecessarily", and there was no statement specifying that "The bolt tip should be aligned with the bolt hole bottom considering the drilling depth and bolt insertion length".

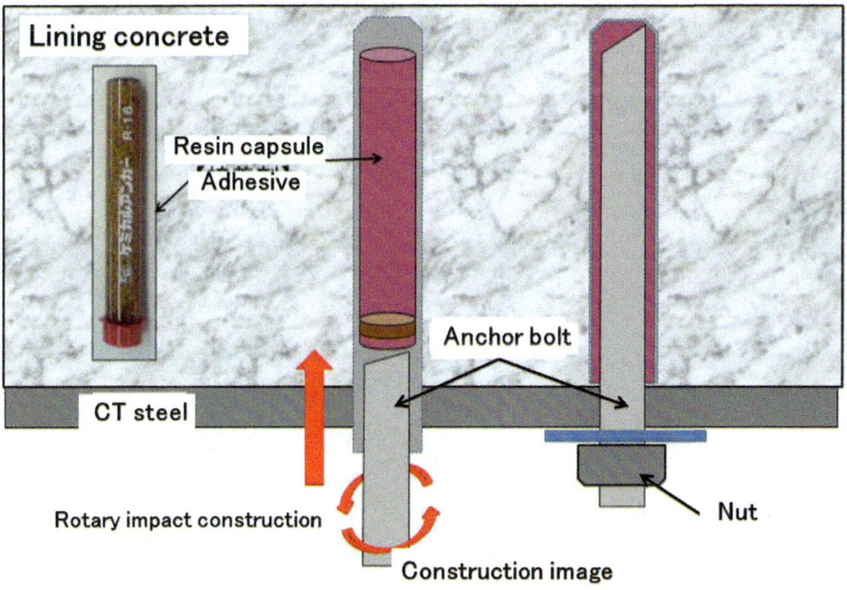

Fig. 1.82 Construction example of capsule-type adhesive anchor bolt [27, 28]

1.12.3 Status of Accident Investigation

(A) **Design-related**

From the documents at the time of design, we will reproduce the estimated forces and strength of the bonding bolts at a zenith. At the time of design, the tensile load on the adhesive bolts at the zenith was estimated by the combination of the weight of the ceiling plate and bulkhead plate ①, ②, and ③, the vertical wind load ⑤, and the worker load ④, was 12.2 kN/bolt.

The Accident Investigation Committee calculated the strength of the bulkhead against horizontal wind pressure loads, which had not been taken into account during the above design as shown in Fig. 1.83 on the right.

The study was conducted in two ways. One is to calculate the axial force on the bolt due to half of the wind pressure load applied to the bulkhead. In this calculation, 1/2 of the wind load acting on the bulkhead plate was took into account (assuming that the wind load acting on the upper half of the bulkhead plate is shared by the upper CT steel and that on the lower half by the lower CT steel), as shown in Fig. 1.83. According to this method, as shown in the bottom column of Table 1.8, the tensile force increases by 5.6 kN per bolt at the top of the zenith, resulting in a total of 17.8 kN/bolt (approximately 1.5 times the amount when only the design load is considered).

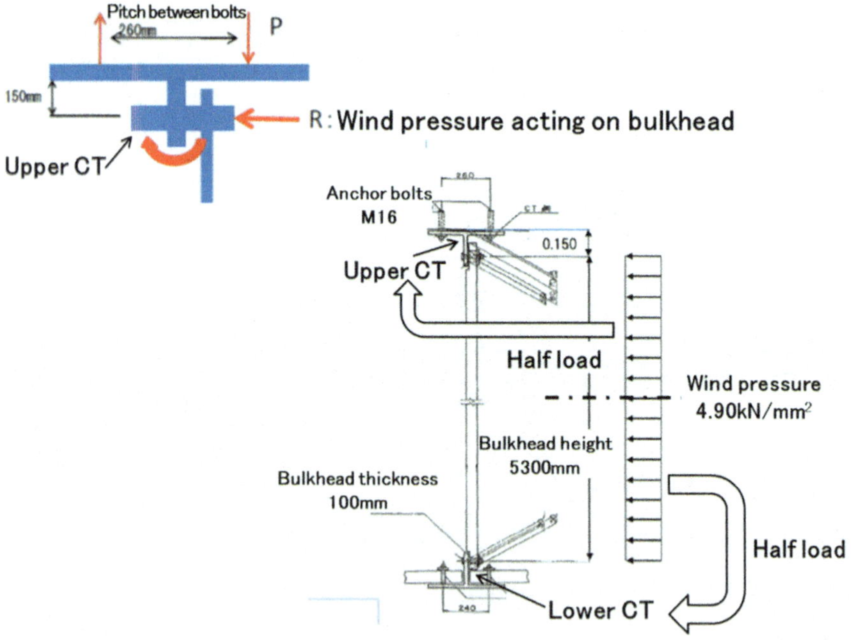

Fig. 1.83 Horizontal wind pressure load acting on the bulkhead and shared load on each CT [27]

Table 1.8 Estimated design loads on anchor bolts on each condition [27]

Construction total length		m	1235
Design bolt size		–	M16
Used bolt size		–	M16
Effective cross-sectional area		mm^2	156.7
Drilling diameter		mm	19
Drilling length		mm	130
Adhesive effective length (drilling length-1/2 bolt diameter)		mm	122
Ceiling weight ①	Tensile load P_d	kN	9.3
	Nominal shear stress τ	N/mm^2	1.28
Design load ① ② ③ ④ ⑤	Tensile load P	kN	12.2
	Nominal shear stress τ	N/mm^2	1.68
Design load + horizontal wind pressure	Tensile load P	kN	17.8 (Analysis)
			20.0 (FEM)
	Nominal shear stress τ	N/mm^2	2.44 (Analysis)
			2.74 (FEM)

① Ceiling weight, ② Bulkhead weight, ③ Other weight (Mortal, CT), ④ Worker weight, ⑤ Vertical wind pressure

The other method is a finite element calculation, as shown in Fig. 1.84, in which the ceiling plate, bulkhead plate, suspension material are modeled by shell elements, and zenith adhesive bolts are modeled by axial springs. As a calculation result of deformation, it was found that the horizontal wind load acting on the bulkhead plate is transferred to the CT steel, causing the web of the CT steel to bend, which in turn affects the tensile force on the adhesive bolts at the zenith, as shown in Fig. 1.84. The calculated results of adhesive bolts show that tensile forces of at least about 20 kN/bolt (about 1.6 times the design load only) are generated in the adhesive bolts on air supply side.

The estimated strengths of this zenith adhesive system bolts are shown in Table 1.9. By this design, the lowest failure strength of the yielding mode of the anchor bolt steel is estimated at 38.4 kN, whereas the strength of the bolt against pull-out due to shear failure along the adhesive resin and lining concrete interface or the adhesive resin and bolt interface is estimated at 52.2 kN. This estimate assumes a first design load of 12.2 kN, and estimated total load including horizontal wind pressure 17.8 kN or 20.0 kN is sufficiently safe.

(B) **Related to construction and damage conditions**

Of the total number of bolts (23 CT steel × 16 bolts = 368 bolts) in the accident Section (138 m), 39 zenith adhesive bolts remained in the lining concrete due to nuts pull-out. Other zenith adhesive bolts and CT steel that were hanging the ceiling plate along with the ceiling plate and bulkhead plate had also fallen. Examples of

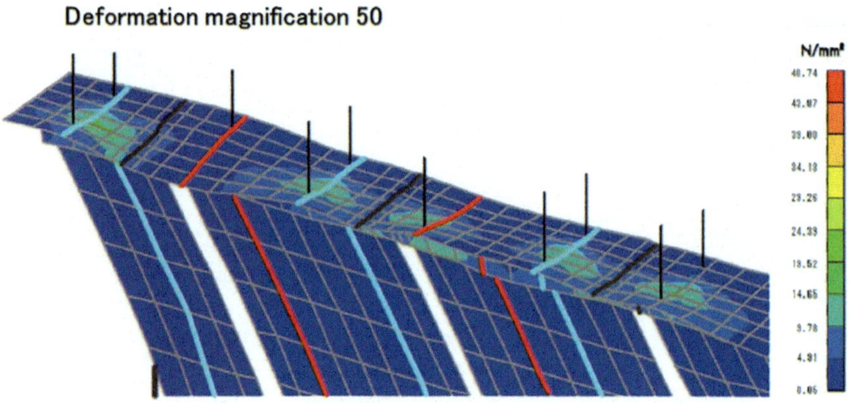

Fig. 1.84 FEM analytical model and calculated results for zenith adhesive anchor bolts [27]

Table 1.9 Estimated strength of adhesive anchor bolt for each failure mode [27]

Failure mode		Strength
Cone-shaped failure at anchor bolt fixtation site	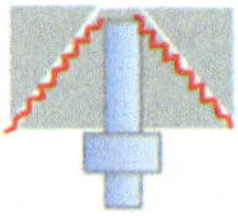	48.8 kN/bolt
Pull-out failure at adhesive site	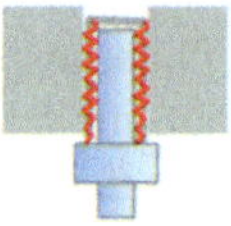	52.2 kN/bolt
Yield failure of anchor bolt	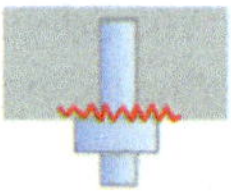	38.4 kn/bolt

photographs of the bolts that remained at the zenith of the ceiling and of the bolt holes that had pulled-off are shown in Fig. 1.85.

Observations of the deformation of these zenith adhesive bolts, the scarring in the hole of the bolt that pulled-out and the fall mark left on the pavement showed that the direction of deformation of the bolt and the direction of scarring in the hole were reversed between CT steel sections No. 1 to No. 12(Tokyo side) and No. 13 to No. 23(Nagoya side) of the accident section. In addition, trends in the depth and length of the pavement drop scar also changed. Based on these observations, it can be inferred that one or several of the CT steel Sectons 12th or 13th CT steel section originated and the fall spread from there in both the Tokyo and Nagoya directions.

Cores of the lining concrete were taken at 57 locations within the section where the ceiling plate fell, and the state of embedment of the zenith adhesive bolts were

Table 1.10 Pull-out test results for adhesive anchor bolts [27]

Pull-out strength rank		Number and ratio of bolts			Pull-out test graph results (general example)	Pull-out adhesive anchor bolt (general examples)
		First sampling test (139 bolts)	Additional test (44 bolts)	Total (183 bolts)		
A	Higher than 40 kN	59 bolts	11 bolts	70 bolts	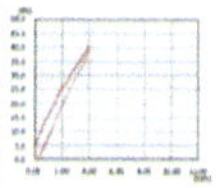	
		42%	25%	38%		
B	Higher than 12.2 kN and lower tha 40 kN	72 bolts	25 bolts	97 bolts	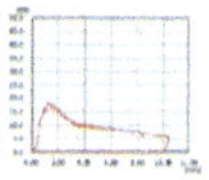	
		52%	57%	53%		
C	Lower than 12.2 kN	8 bolts	8 bolts	16 bolts	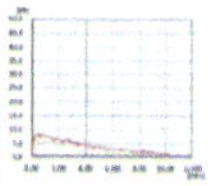	
		6%	18%	9%		

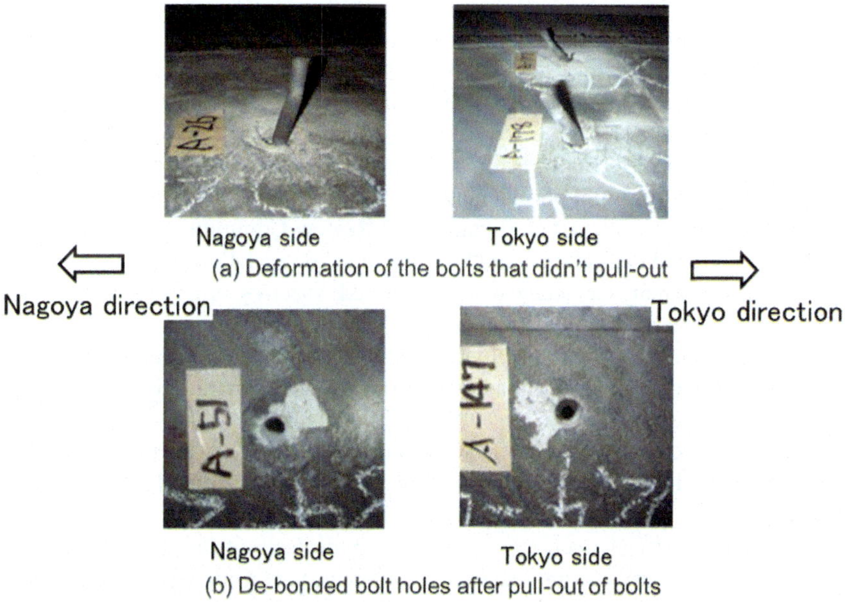

Fig. 1.85 Deformation and de-bonding observation photos [27]

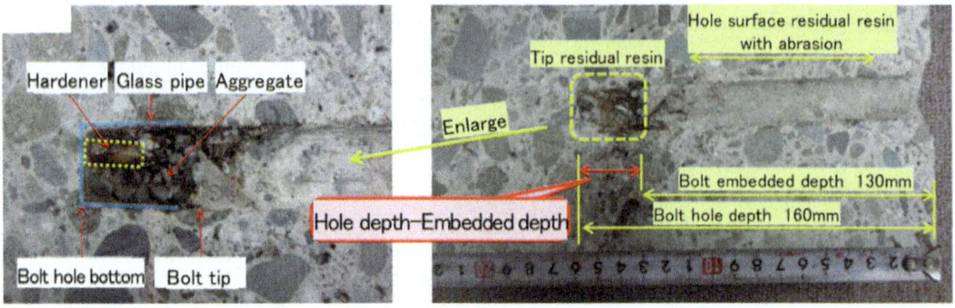

Fig. 1.86 Observation example of core sample around embedded zenith adhesive bolt [27]

observed. Figure 1.86 shows the core sample observation result. The figure shows that the bolt tip and the bottom of the bolt hole were not installed so that they were aligned, and as a result, the depth of the bolt hole was deeper than the length of the bolt embedment, and some of the glass tube, aggregate, hardener, and resin remained in the gap between the bottom of the hole and the bolt tip without being sufficiently stirred. Although normal construction requires that the drilling depth of the bolt hole match the length of the bolt embedment and that no gap be created, this standard was not met in 22 out of 57 cases. So the estimated adhesive strength shown in Table 1.9 may not have been secured.

Then, pull-out test was conducted on the zenith adhesive bolts along the whole upline, excluding the accident section. The test was loaded to a target equivalent to the yield strength of the bolts (approximately 40 kN), and the load and displacement were recorded.

Table 1.10 shows a summary of the test results. The test results showed that 70 bolts didn't pull-out with a load of 40 kN, and 113 bolts pull-out under 40 kN. In these pulled-out bolts, 16 bolts show particularly low strength below the design tensile force (<12.2 kN) on the Tokyo side.

This result indicates that the estimates of adhesive strength including degradation effect over time were highly problematic.

Therefore, the Accident Investigation and Review Committee [1] have investigated load-induced degradation analysis, chemical degradation of adhesives based on X-ray microanalyzer analysis and infrared spectroscopy, and microscopic observation of degradation damage inside adhesives.

Degradation due to loading includes dead weight loads of the ceiling plates and bulkhead plates, etc., and cyclic loading due to wind loads generated during operation of the exhaust system.

In addition, committee said there were variable loads due to the passage of large vehicles, etc., which should have been fully considered from the beginning of the design [27].

Regarding chemical material degradation, hydrolysis of unsaturated polyester resin used in adhesives in an alkaline environment is considered. Therefore, after the pull-out test, the adhesive resin is sampled and chemically analyzed by X-ray microanalyzer analysis or infrared spectroscopy.

By X-ray microanalyzer analysis, the elemental distribution in the sample core cross section of the zenith adhesive bolts remained in the lining concrete was investigated. An example of the analysis results is shown in Fig. 1.87. As a result of the analysis, calcium elements were detected in the area corresponding to the resin. From the distribution characteristics, it is considered that water containing calcium ions, hydroxide ions, and other concrete components penetrated into the adhesive resin from the interface between the adhesive resin and the lining concrete.

Infrared spectrometry is used to identify the type of adhesive resin and the degree of deterioration of the resin for the collected samples. Adhesive resin samples were collected from adhesive resin adhered to the zenith adhesive bolts and from adhesive resin adhered in the bolt holes, respectively. As a result of the analysis, as shown in Fig. 1.88, the infrared absorption spectrum had strong absorption peaks around 1730 and 1270 cm^{-1}, which are thought to be due to ester bonding. These properties confirmed that the resin of the adhesive bolts used in this tunnel was unsaturated polyester resin.

As shown in Fig. 1.89, for the adhesive resin sampled from near the lining concrete interface, there was a decrease in the absorption peak near 1730 cm^{-1}, which is thought to be due to the decomposition of ester bonds, and on the other hand, there

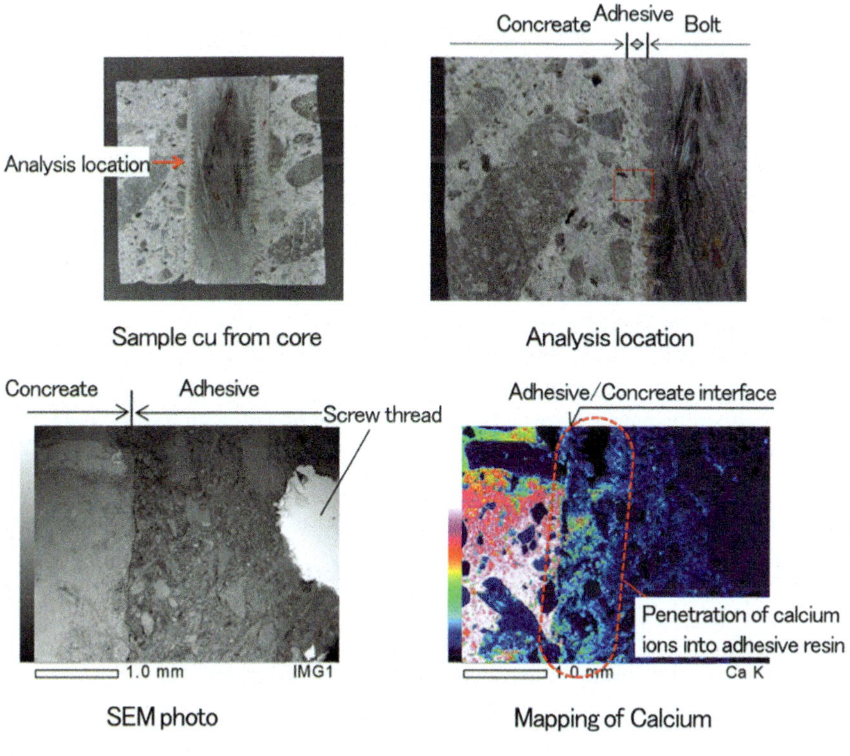

Fig. 1.87 Analysis results of X-ray microanalyzer on cross section of core [27]

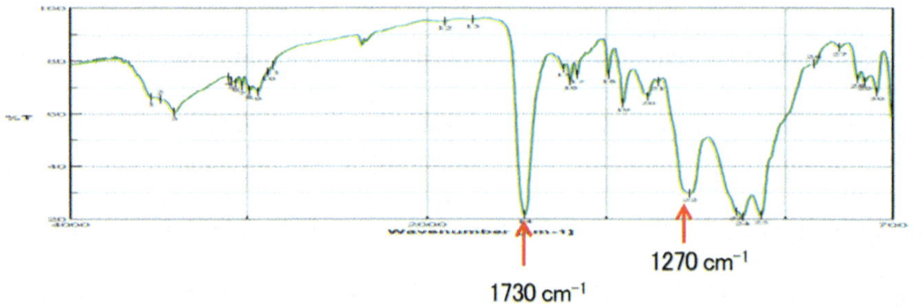

Fig. 1.88 Infrared absorption spectrum of sampled adhesive resin [27]

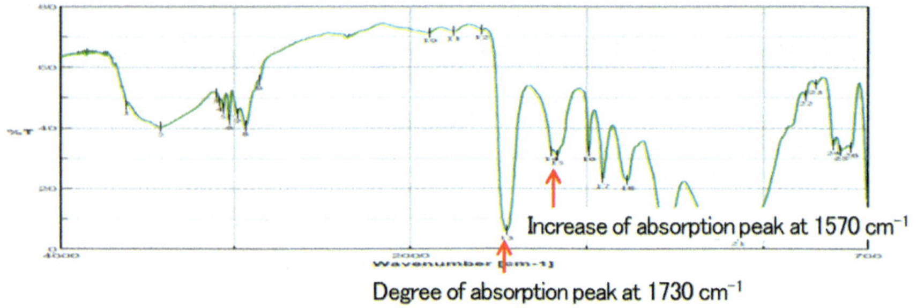

Fig. 1.89 Infrared absorption spectrum of adhesive rein near bolt [27]

were initiation and increase in the absorption peak at 1570 cm^{-1} due to the formation of carboxylate salts. This is considered to be a cause of hydrolysis. This indicates the progress of hydrolysis, especially near the lining concrete interface in the adhesive resin.

Figure 1.90 shows the results of electron microscopic observations of the adhesively joint section of a collected core sample degraded by a combination of long-term cyclic loading and chemical degradation. Numerous microscopic internal cracks and voids are observed, likely due to the combined degradation, but the cracks at the adhesive/concrete interface are particularly large and fit well with the observed areas of chemical degradation described above.

1.12.4 Investigation of the Cause (See Chap. 11, Sect. 11.2)

Based on the results of the above investigation and the author's own opinions, the cause of this tunnel ceiling collapse accident will be investigated as follows.

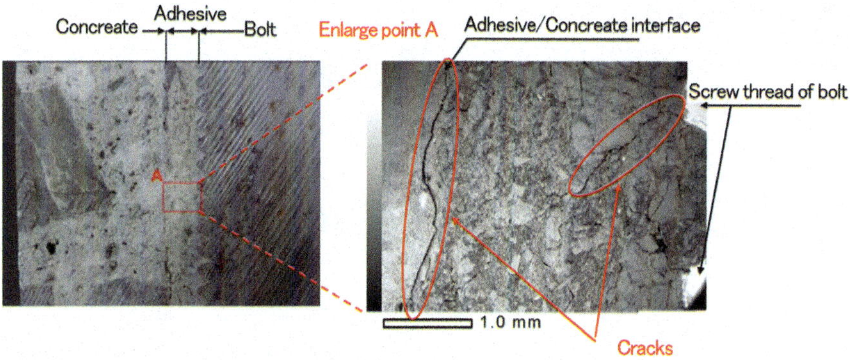

Fig. 1.90 Infrared absorption spectrum of deteriorated adhesive near the boundary [27]

(A) Summary of accident investigation and review committee report

(1) Problems in load estimation

As shown in Table 1.8, the initial estimated load per adhesive bolt was 12.2 kN (shear stress 1.68 MPa) as the static load from the dead weight of the ceiling and bulkhead plates, the load of the worker, and the vertical wind pressure, but the Accident Investigation and Review Committee estimated the increase load from the bending moment due to the horizontal wind pressure on the bulkhead. The Accident Investigation and Review Committee estimated the total load considering this load increase to be 17.8 kN (shear stress 2.44 MPa) in a simplified analysis and 20 kN (shear stress 2.74 MPa) in a detailed 3D analysis using the finite element method, and pointed out that the initial load estimation and design were incorrect.

(2) Problems in adhesive strength (deterioration) estimation

As shown in Table 1.9, the strength per bolt of this adhesive bolt was estimated to be 52.2 kN. However, the chemical degradation of the adhesive at the concrete interface and the degradation caused by over 200,000 cycles of start/stop loading during the 36-year service period as shown in Figs. 1.88 and 1.89 by X-ray microanalyzer analysis and infrared spectroscopic analysis were confirmed by the electron microscopy results shown in Fig. 1.90. And especially the occurrence of long crack damage at the adhesive/concrete interface (shown in Fig. 1.90) points out the need for estimating the degradation over a long service life.

(3) Problems on installation processes of adhesive anchor bolts

As shown in Fig. 1.86, the condition of aligning the bolt tip with the bottom of the bolt hole, which is important in the correct installation method of capsule-type adhesive anchor bolts, was not satisfied, and in 22 of the 57 bolts, the deeper depth of the bolt hole than the length of bolt embedment was observed. Aggregate, hardener, and resin remained in the gap between the bottom of the hole and the tip of the bolt

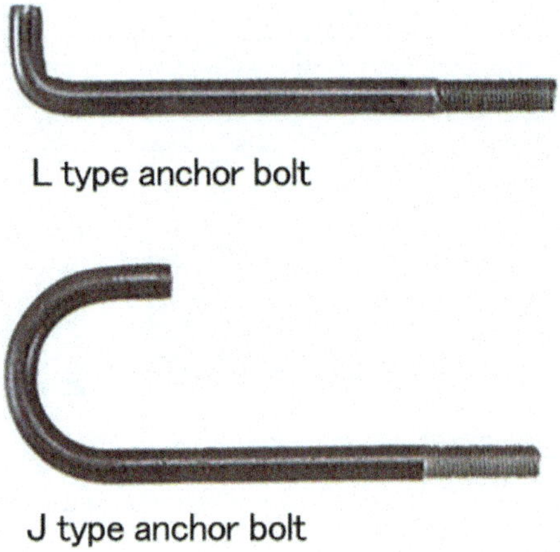

Fig. 1.91 Basic shapes of anchor bolts [29]

without being sufficiently agitated. This indicated that there was a good possibility that the adhesive strength estimates shown in Table 1.9 were not secured.

(B) (B) **Author's view (see Chap.** 11, **Sect.** 11.2)

The design and construction of anchor bolts that are subjected to axial tensile loads should originally be done by pouring L or J-shaped bolts into concrete, as shown in Fig. 1.91. The design should be based on the assumption that concrete cone failure or bolt failure will occur, as shown in Table 1.9, and it must be recognized that even if the initial adhesive strength is high, the strength of the adhesive will decrease over time due to deterioration due to organic matter.

The use of this adhesive anchor bolt in heavily loaded structures should be limited to structures subjected to non-axial loading, as shown in Fig. 1.92

As discussed in Chap. 11, Sect. 11.2.2, the greatest weaknesses of adhesive bonding are strength variation and aging/environmental degradation. Here, aging and environmental degradation are not only a decrease in average bond strength, but also an increase in variability. Therefore, it is necessary to design quantitatively using the following aging degradation equation, which takes into account the Arrhenius plot method, fracture probability, and other factors.

$$F_y = F_{fT} \times \eta_y \times D_y (Chapter\ 11, Eq.(11.2)) \tag{1.59}$$

where F_y is the adhesive strength after certain service life, F_{fT} is the initial adhesive strength, η_y is the degradation coefficient under combined environmental and stress conditions such as temperature, humidity, UV, and load, and D_y is the coefficient of variation of adhesive strength after certain service life.

In fact, according to the catalog [28] for "Chemical Anchor, Resin Capsule R-16" manufactured by Nippon Decolux, the long-term allowable tensile strength is listed as 18.9 kN, which does not satisfy the actual working load estimate of 20 kN in Table 1.9. It is the social responsibility of engineers who design and construct such

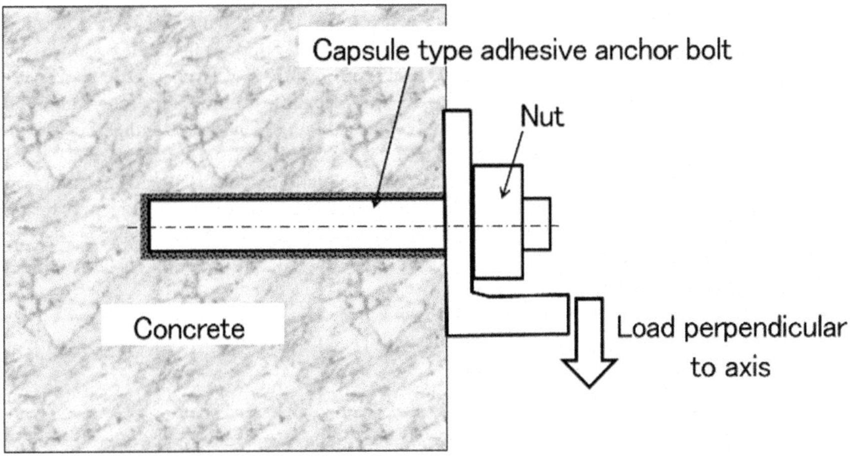

Fig. 1.92 Anchor bolt installation state under loads perpendicular to axis

critical structures to guarantee safety with taking into account environmental and load degradation over time, rather than the basic initial adhesive strength shown in Table 1.9.

References

1. The Day "Shinkansen Safety Myth" Breaks, Jun Sakurai, Kodansha, (1993).
2. Nikkei New Materials, (1992,10,5) 8.
3. Aviation Accident Investigation Committee Report 62–2, Boeing Type 747SR-100 JA8119 belonging to Japan Airlines Corporation, Yamanaka, Ueno-mura, Tano District, Gunma Prefecture, August 12, 1985, Ministry of Transport Aviation Accident Investigation Committee, June 19, 1987.
4. Kobayashi, Arai, Nakamura, Materials, 36, 1084 (1987).
5. Kobayashi, Safety Engineering, 26, 338 (1987).
6. The Materials Society of Japan, ed. Fatigue of Metals, Maruzen, (1964).
7. Toshio Hattori et al., Strength Design Handbook for Failure Prevention on Products, NTS, (2012).
8. Akira Yamamoto, Principle and Design of Screw Fastening, Yokendo, 2004, pp 1166.
9. Expoland, roller coaster accident, Failure Yearbook 2007, Failure Society.
10. D. Takasaki, Y. Kubota, R. Komoda, Y. Oku, T. Makino, M. Sugino D. Takasaki, Y. Kubota, R. Komoda, Y. Oku, T. Makino, M. Sugino, Mechanism of fretting fatigue fracture occurring inside the contact surface in oil well pipe materials, JSME 2016 Annual Meeting.
11. Y.Oku, M. . Sugino, Y. Ando, Y.Ando, T. Makino, R..Komoda, D.Takazaki, M. Kubota, Fretting fatigue on thread root of premium threaded connections, Tribology International, Vol: 108, Page: 111–120, (2017).
12. Superconducting Linear Motor Car Development Promotion Division, Railway Research Institute, Superconductivity Changes Railways - Linear Motor Car Maglev, Seibunsha, (1988).
13. S. Izumi, T. Yokoyama, A. Iwasaki, S. Sakai, Transactions of the Japan Society of Mechanical Engineers, Vol. 71, No. 702, (2005), pp. 204.

14. Satoshi Izumi, Shigetake Kimura, Shinsuke Sakai, Finite Element Analysis of Micro Loosening Behavior of Bolt and Nut Fasteners Caused by Micro Seat Slip, Transactions of the Japan Society of Mechanical Engineers (Part A) 72, 717, pp. 780–786 (2006).
15. T. Yamagishi, T. Asahina, D. Araki, H. Sano, K. Masuda and T. Hattori, Loosening and Sliding Behaviour of Bolt-Nut Fastener under Transverse Loading , Mechanical Engineering Journal, 5, 3 (2018), Journal-JSME-D-16–00622 (2018).
16. M. Nakamura, T. Hattori, S. Tsujimoto, K. Umeki, Evaluation of the rotational loosening limit of bolts subjected to a perpendicular reciprocating load, Transactions of the Japan Society of Mechanical Engineers, Vol. 67, No. 661, (2001), pp. 2976–2980.
17. M. Nakamura, T. Hattori, S. Satou, K. Umeki, Evaluation of the rotational loosening bdhavior of bolts subjected to a perpendicular reciprocating load, Transactions of the Japan Society of Mechanical Engineers, Vol. 64, No. 627, (1998), pp. 4395–4399.
18. Toshio Hattori, Material Mechanics and Strength Design Learned from Accident Examples, Nikkan kougyou Shinbun, (2018), pp. 191.
19. W. C. Young et al, ROARKS'S FORMULAS for STRESS and STRAIN (8thEdition), Table 11.2, McGraw Hill, (2012).
20. "Urgent Report: The Accident Speaks for Itself: Mitsubishi's Hub Breaking Accident," Nikkei Monozukuri, 597, 2004.6, pp. 87–93.
21. "Accident tells us Yurikamome hub rupture accident", Nikkei Monozukuri, 2008.5, 93–98.
22. "I K.K. Yurikamome, Tokyo Waterfront New Transit Rinkai Line, Train Derailment Accident at Funano Science Museum Station," RA2008–2 Railway Accident Investigation Report, Aviation and Railway Accident Investigation Committee, February 2008.
23. Yamanashi Prefecture Coaster Accident Investigation Report, Social Infrastructure Development Council, December 2014.
24. T. Foecke, Metallurgy of the RMS Titanic, NIST-IR 6118(NIST) (1968).
25. T. Shioya, Y. matsuo, T. Hattori, H. Kawada, New Ed. FRACTGRAPHY - Fracture surface analysis of various materials and its examples -, Techno System Co. Ltd., (2010) (in Japanese).
26. Ministry of Economy, Trade and Industry, 3rd Meeting, Industrial Structure Council, Safety Subcommittee, Electric Power Safety Subcommittee, Working Group on Accident Response and Structural Strength of New Energy Generation Facilities, Document 2–7, May 30, 2014. (in Japanese).
27. Report of the Investigation and Review Committee on the Accident of Falling Tunnel Ceiling Plates, Investigation and Review Committee on the Accident of Falling Tunnel Ceiling Plates, June 18, 2013.
28. http://www.marukoo.com/bb1/bb749.html, Nippon Decorax, Chemical Anchors General Catalog, 2013.
29. New Edition- Joining Technology Overview, Joining Technology Service Center, (1994).

Chapter 2
Basics of Mechanical Analysis, Initial Fastening Techniques, and Design Techniques for Bolted Joints

2.1 Introduction

Chapters 2 to 5 cover mechanical analysis and strength design techniques from basic to detailed for various loading modes, Chap. 6 covers loosening analysis and prevention technologies, Chap. 7 covers the application of thread mechanics on pipe thread joints, Chap. 8 covers delayed fracture of bolts, Chap. 9 covers fretting fatigue of bolt joints, Chap. 10 covers settling and fatigue failure of fastened bodies, and Chap. 11 describes other new fastening-related technologies. Finally, Chap. 12 describes domestic and foreign screw fastening technologies, especially the ISO, JIS, or German standard VDI 2230, which is often cited in this book.

2.2 Mechanical Analysis and Design Procedures for Thread Fasteners

For strength evaluation methods for single bolted fasteners, VDI 2230 [1–3] specified by the German Institution of Engineers has been used as the most systematic design method, with frequent revisions [4–9]. In these strength design methods, the initial clamping force is set sufficiently large for the external force, and the norm is that the seating surface and the contact surfaces between the fastened bodies do not separate, i.e., the bolt and the fastened body deform as a single body. When the bolt and fastened body are deformed as a single body, the internal force between the bolt and the fastened body, or load applied to the bolt, can be determined from the external force applied to the fastened body and the rigidities of the bolt and of the fastened body. In other words, for a single bolted fastener, the above strength evaluation method based on VDI 2230 can be applied as long as the external force applied to it and the dimensions or elastic properties of the bolt and fastened body are known.

© The Author(s), under exclusive license to Springer Nature Switzerland AG 2025

T. Hattori and T. Naruse, *Compendium for Design and Manufacturing of Bolted Joints*, Solid Mechanics and Its Applications 280,
https://doi.org/10.1007/978-3-031-89628-6_2

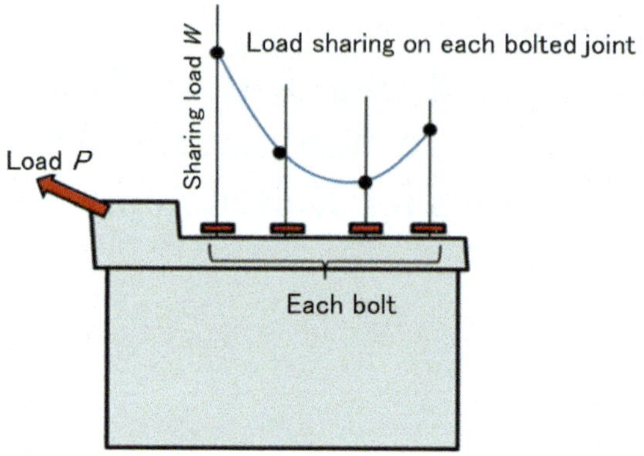

Fig. 2.1 Load sharing schematic for each bolted joint

However, in general machines and structures fastened with multiple bolts, it is first necessary to determine the external force on each single bolted fastener because each bolted fastener shares the load in a complex manner. For example, when a load P is applied to a structure fastened with four bolts as shown in Fig. 2.1, it is first important to know the shared load W applied to each bolted fastener.

This load sharing is generally analyzed by accurately determining the equivalent stiffness of each bolted joint, as shown in Fig. 2.2. Using these equivalent stiffnesses of each bolted joint, the load sharing can be calculated by solving as statically independent problem that satisfies load balance and displacement compatibility in these models.

Once the load sharing distributions are known, the strength and loosening estimations and evaluation of initial fastening force can be performed. Chapter 2 outlines these evaluation methods. Chapter 3 describes in detail the evaluation methods when axial and bending loads are applied, and Chap. 4 describes in detail the evaluation methods when loads are applied transverse to the bolt axis. Chapter 5 describes in detail the strength and loosening evaluation method for the case where torque is loaded around the bolt axis. In Chap. 6, the loosening without rotation of bolts or nuts, for example wear of seat surfaces by wear, or thermal deformation of bolt or fastened body.

In Chap. 7 we apply these methods on pipe threaded joints. Chapter 8 describes in detail the mechanisms and counter measure technologies for delayed failure.

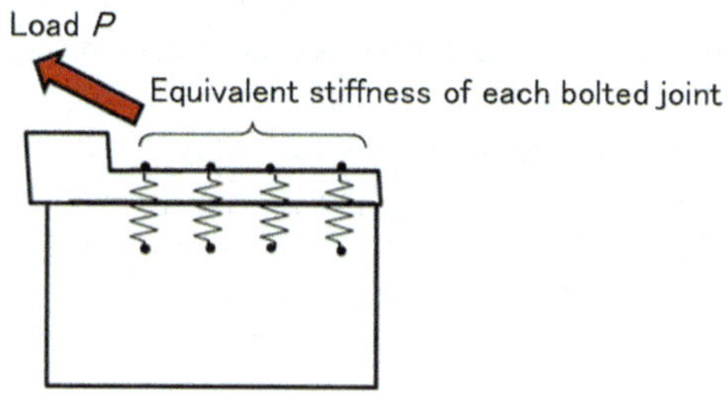

Fig. 2.2 Equivalent stiffness modeling for each bolted joint

Chapter 9 describes in detail the mechanics and countermeasures for fretting wear and fatigue problems in joint structures. Chapter 10 describes in detail the mechanical analysis and strength design for fastened body. Chapter 11 describes in detail the new technologies in fastened structures such as plastic region tightening method, screw and adhesive combined joints, and loosening prevention technology using lock washers. Finally, Chap. 12 describes in detail the international and domestic standards for screw fastening-related technologies.

2.3 Initial Provisional Settings and Final Setting of Bolt Size, Bolt Material and Initial Tightening Force

Using the mechanical model in Fig. 2.2, structural analysis can be performed to obtain the axial load, bending moment, and perpendicular load of each screw fastener. The bolt size, material, and initial tightening force are selected or set for each of the axial and perpendicular loads that dominate the strength and looseness as shown in Fig. 2.3, using the following procedure.

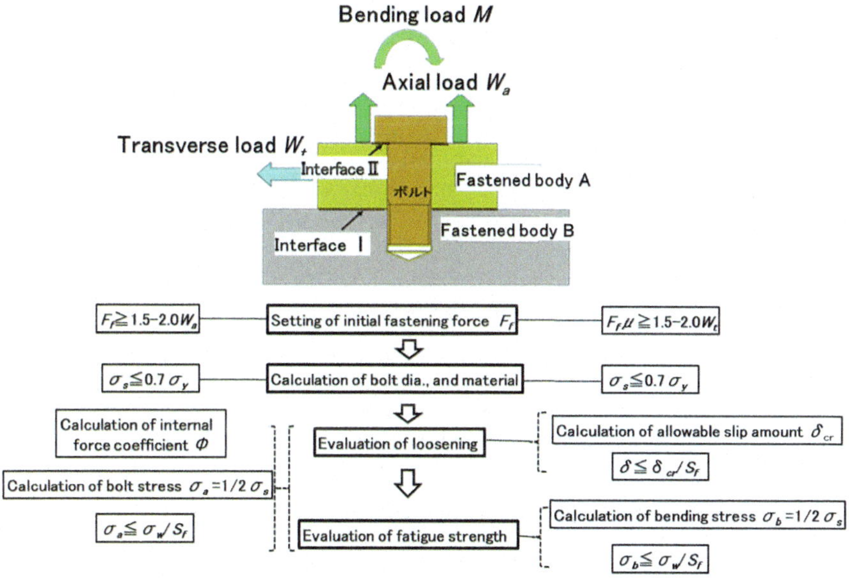

Fig. 2.3 Axial load, bending moment, and perpendicular load applied to each screw fastener and design flow

2.3.1 Design Method for Tightening with a Torque Wrench

The first step in the design of a bolted joint is the provisional setting of the initial fastening force F_f. As shown in Fig. 2.3, for axial loading of W_a,

$$F_f \geq 1.5 \sim 2.0 \text{ W}_a \tag{2.1}$$

For perpendicular load W_t to the axis direction

$$F_f \mu \geq 1.5 \sim 2.0 \text{ W}_t \tag{2.2}$$

The material and diameter of bolt is provisionally set so that the fastening bolt stress σ_s in effective diameter cross-sectional area A_s is less than 0.7 times the yield stress σ_y of the material, and strength analyses, loosening analyses, re-evaluations, and modifications are repeated to complete the final setting.

A torque wrench is often used to apply this initial fastening, but it is important to note that the torque being controlled here is not the true tightening force to be controlled.

In other words, as shown in Fig. 2.4, the torque value T_f being measured by the torque wrench is a combination of the torque T_s expended at the thread and the torque T_w expended at the bolt head seat, each of which is indicated by the following equation,

$$T_f = T_s + T_w \tag{2.3}$$

$$T_s = \frac{F_f}{2}\left(\frac{P}{\pi} + \mu_s d_2 \sec\alpha'\right) \tag{2.4}$$

$$T_w = \frac{F_f}{2}\mu_w D_w \tag{2.5}$$

It means that the thread surface friction coefficient μ_s and the seat surface friction coefficient μ_w are error-prone components. Here, the thread pitch P, the effective diameter of the bolt d_2, and the equivalent diameter of the seat surface D_w in Eqs. (2.4) and (2.5) are almost proportional to the nominal diameter d in accordance with ISO standards, so using the dimensionless torque coefficient K Eq. (2.3) can be shown as follows.

$$T_f = KF_f d \tag{2.6}$$

$$K = \frac{1}{2d}\left(\frac{P}{\pi} + \mu_s d_2 \sec\alpha' + \mu_w d_w\right) \tag{2.7}$$

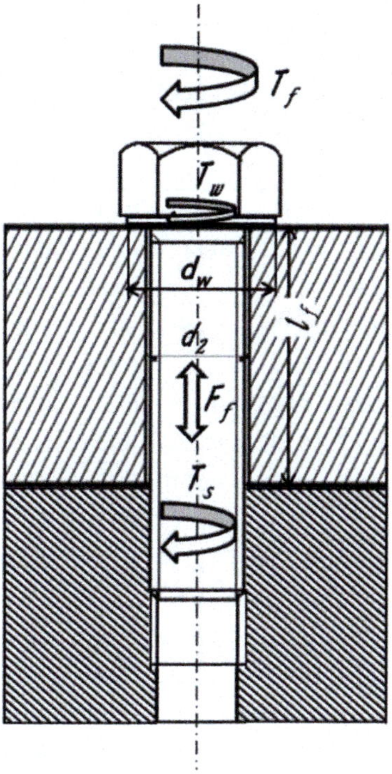

Fig. 2.4 Fastening torque and force

The relationship between the torque coefficient and the friction coefficients of the thread and head seat surfaces is shown in Table 2.1.

2.3.2 *Design Methods for Rotation Angle and Torque Gradient Method Fastening [6]*

In the initial fastening using the torque wrench described above, the fastening force varies by ±30% even when the same torque is applied due to the variation of the friction coefficient. Therefore, if the target for initial fastening force is set to the full yield stress of the material, it may result half of bolts will be threaded, so the target fastening stress must be set to 70% of the yield stress as shown in Fig. 2.3. To solve this drawback, new fastening methods such as the rotation angle method or the torque gradient method have recently been adopted. This method has come to be widely adopted in the automotive and other industries in conjunction with the spread of plastic region tightening and the emergence of automatic torque/rotation angle measurement devices, as described below.

The rotation angle method is a fastening method based on the relationship between the fastening force F_f and the fastening rotation angle θ_f as follows,

Table 2.1 Relationship between thread surface and seat surface friction coefficients μ_s and the seat surface friction coefficient μ_w and torque coefficients K [4–6]

(a) Coarse thread, hex bolt, nut

μ_w / μ_s	0.08	0.1	0.12	0.15	0.2	0.25	0.3	0.35	0.4	0.45
0.08	0.117	0.13	0.143	0.163	0.186	0.228	0.261	0.294	0.326	0.339
0.1	0.127	0.14	0.153	0.173	0.206	0.239	0.271	0.304	0.337	0.369
0.12	0.138	0.151	0.164	0.184	0.215	0.249	0.282	0.314	0.347	0.38
0.15	0.154	0.167	0.18	0.189	0.232	0.266	0.297	0.33	0.353	0.396
0.2	0.18	0.193	0.205	0.225	0.258	0.291	0.324	0.356	0.389	0.422
0.25	0.205	0.219	0.232	0.252	0.284	0.317	0.35	0.383	0.415	0.448
0.3	0.232	0.245	0.258	0.278	0.311	0.343	0.375	0.409	0.442	0.474
0.35	0.258	0.271	0.284	0.304	0.337	0.37	0.402	0.435	0.468	0.5
0.4	0.285	0.298	0.311	0.33	0.363	0.396	0.428	0.461	0.494	0.527
0.45	0.311	0.324	0.337	0.357	0.389	0.422	0.455	0.487	0.52	0.553

(b) Fine thread, hex bolt/nut

0.08	0.106	0.118	0.13	0.148	0.177	0.207	0.237	0.267	0.296	326
0.1	0.117	0.129	0.141	0.158	0.188	0.218	0.248	0.278	0.307	0.337
0.12	0.128	0.14	0.151	0.169	0.199	0.229	0.259	0.288	0.318	0.348
0.15	0.144	0.156	0.168	0.186	0.215	0.245	0.275	0.306	0.334	0.364
0.2	0.171	0.183	0.196	0.213	0.242	0.272	0.302	0.332	0.361	0.391
0.25	0.198	0.21	0.222	0.247	0.27	0.299	0.329	0.359	0.389	0.418
0.3	0.225	0.237	0.249	0.267	0.297	0.326	0.356	0.386	0.416	0.445
0.35	0.252	0.264	0.276	0.294	0.324	0.353	0.383	0.413	0.443	0.472
0.4	0.279	0.291	0.303	0.321	0.351	0.381	0.41	0.44	0.47	0.5
0.45	0.306	0.318	0.33	0.348	0.378	0.408	0.437	0.467	0.497	0.527

$$\theta_f = \frac{2\pi F_f(\delta_b + \delta_c)}{P} \tag{2.8}$$

In this equation δ_b is a compliance of bolt, δ_c is a compliance of fastened body and P is a pitch of thread.

In this case, the variation of the friction coefficient can be ignored, and it is especially effective for plastic tightening with large deformation as shown in Fig. 2.5. However, this method is inconvenient in that the rotation angle must be calculated for each bolt size (bolt diameter, fastening length and thread pitch) using Eq. (2.8).

Therefore, the torque gradient method was developed as an improvement to the rotation angle method. This torque gradient method requires a tightening tool that can automatically measure both torque T_f and rotation angle θ_f.

As shown in Fig. 2.6, by using this tool the derivative value of $\mathrm{d}T/\mathrm{d}\theta$ during tightening process can be calculated at the same time. In this figure, the region

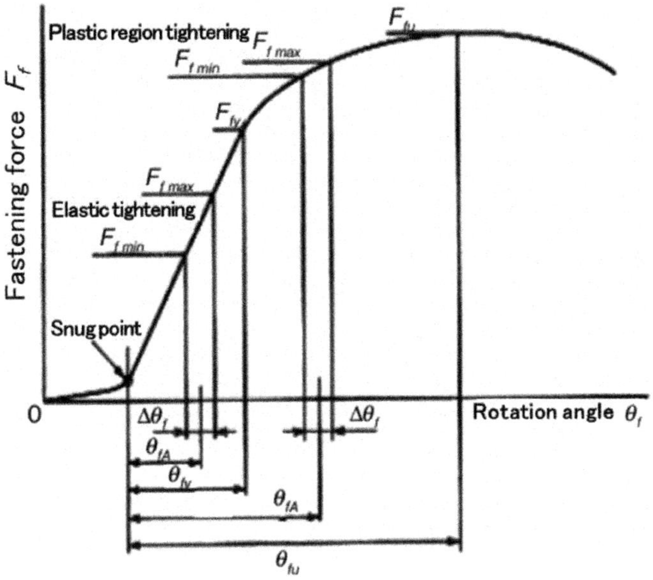

Fig. 2.5 Relationship between tightening rotation angle θ_f and tightening force F_f (rotation angle method) [6]

where the $dT/d\theta$ value is constant is the elastic region, and the region where this value decreasing is the plastic region. In the case of plastic tightening, tightening is completed when this value reaches 1/2 to 1/3 of the elastic range. This method is convenient because there is no need to change the setting depending on the bolt size, material etc.

2.4 Basis of Strength Design of Bolted Joints

In this section, the simplified mechanical compliance δ method that has been used in the past will be explained for the sake of clarity, but highly accurate analysis and design techniques that also incorporate FEM analysis have been proposed later in Chaps. 3, 4 and 5.

2.4.1 Load Factor Φ

When an axial load is applied to a bolted joints as shown in Fig. 2.7, the entire load is not borne by the bolts, and remaining load is borne by the fastened body.

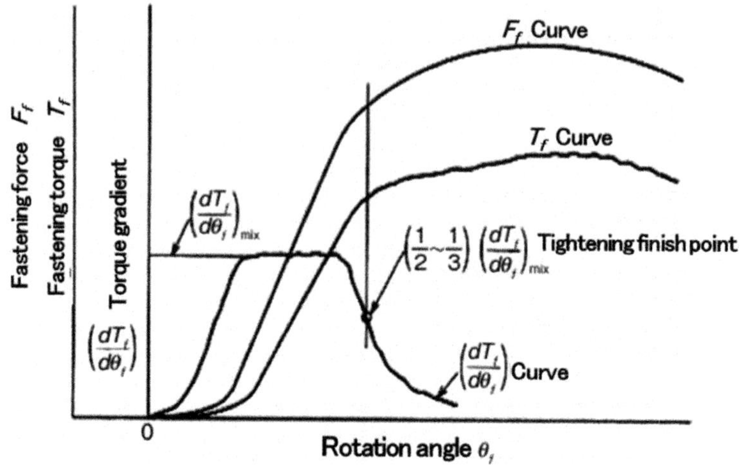

Fig. 2.6 Relationship between torque T_f and rotation angle θ_f and derivative value $dT/d\theta$ during tightening process (torque gradient method) [6]

Fig. 2.7 Bolted joint model (compliant with VDI2230 1977 [1])

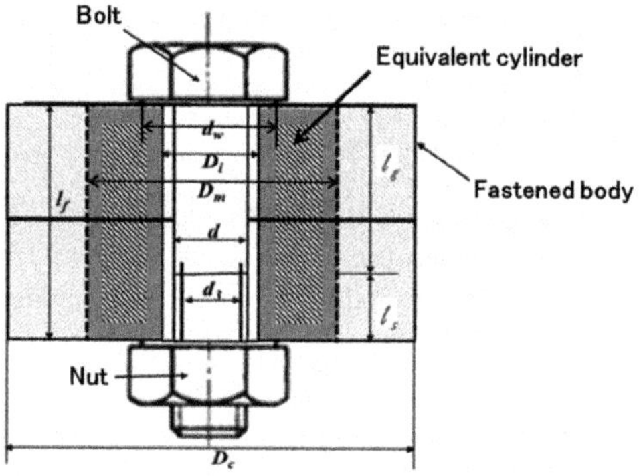

Generally, the bolt itself bears the load only 1/5 to 1/3 of the total load which is applied to the bolted joint. This mechanism works very effectively on the fatigue strength characteristics of bolts. This is the reason why bolted joints have been used as a major joint method for a long time. Therefore, the accurate prediction of fatigue strength of a bolted joint subjected to axial load W_a ultimately comes down to accurate understanding of the bolt's burden F_b (load factor $\Phi = F_b/W_a$).

The threaded fastener model for this explanation is shown in Fig. 2.7. The relationship between the force and deformation of the bolt and the fastened body in this model can be shown as Fig. 2.8, and this is called a tightening diagram (tightening triangle). Using this diagram, the load factor $\Phi = F_b/W_a$ can be obtained using the compliance of the bolt (inverse of stiffness) δ_b and the compliance of the fastened body δ_c as shown in the following equation.

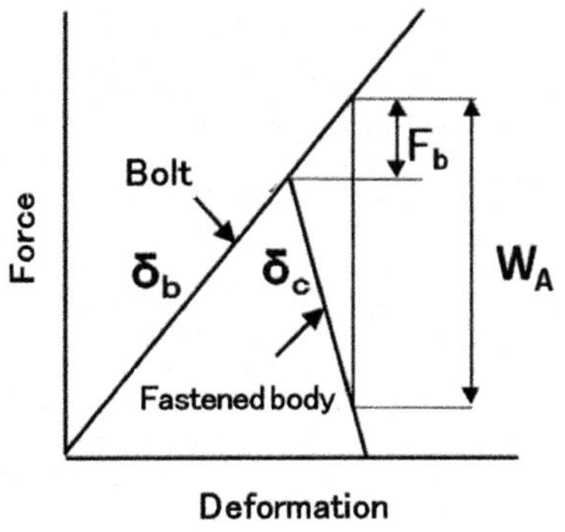

Fig. 2.8 Compliance (inverse of stiffness) of bolt δ_b and fastened body δ_c and bolt burden load F_b

$$\Phi = \frac{\delta_c}{\delta_c + \delta_b} \tag{2.9}$$

(A) Bolt Compliance δ_b

Using the bolt specifications in Fig. 2.7, the compliance of the bolt δ_b can be calculated in accordance with VDI 2230 1977 [1] as follows.

$$\delta_b = \frac{0.4d}{E_b \frac{\pi}{4} d^2} + \frac{l_a}{E_b \frac{\pi}{4} {d_g}^2} + \frac{l_s}{E_b \frac{\pi}{4} {d_3}^2} + \frac{0.4d}{E_b \frac{\pi}{4} d^2} \tag{2.10}$$

Where

E_b Young's modulus of bolted material
d Nominal diameter of bolt.
d_g Cylinder area diameter = d.
d_3 Thread valley diameter.

Also, in accordance with VDI 2230 2003 [3], using the same parameters as in Eq. (2.10), compliance of the bolt δ_b can be calculated as follows.

$$\delta_b = \frac{0.5d}{E_b \frac{\pi}{4} d^2} + \frac{l_a}{E_b \frac{\pi}{4} d_g^2} + \frac{l_s}{E_b \frac{\pi}{4} d_3^2} + \frac{0.5d}{E_b \frac{\pi}{4} d_3^2} + \frac{0.4d}{E_b \frac{\pi}{4} d^2} \tag{2.11}$$

(B) Fastened Body Compliance δ_c

The compliance δ_c of the fastened object is calculated by replacing the range subjected to compressive pressure with an equivalent cylinder, as shown by the hatching in Fig. 2.7 (VDI 2230 1977) [1]

$$\delta_c = \frac{l_f}{E_c \frac{\pi}{4} \left({D_m}^2 - {D_i}^2\right)} \tag{2.12}$$

where l_f is the fastening length and E_c is the Young's modulus of the fastened body.

The outer diameter D_m of this equivalent cylinder for each condition is calculated by the following equation:

$$\left.\begin{array}{l} \text{In the case of}: D_c \le d_w \\ D_m = D \\ \text{In the case of}: d_w \le D_c \le 3d_w \\ D_m = \sqrt{d_w^2 + \frac{l_f}{10}\left(\frac{D_c}{D_w} - 1\right)\left(d_w + \frac{l_f}{20}\right)} \\ \mathrm{D}_c > 3d_w \text{ In the case of}: \\ D_m = d_w + \frac{l_f}{10} \end{array}\right\} \tag{2.13}$$

And in accordance with VDI2230 2003 [3], the compliance is calculated using compressive equivalent area as cone, as shown by the hatching in Fig. 2.9 (VDI2230 2003) [3].

For through-bolts, the apex angle φ of the equivalent cone is shown below.

$$\tan\varphi = 0.362 + 0.032\ln\left(\frac{l_f}{2d_w}\right) + 0.153\ln\left(\frac{D_c}{d_w}\right) \tag{2.14}$$

The maximum outside diameter of the cone, $D_{A,Gr}$, is obtained as follows:

$$D_{A,Gr} = d_w + wl_f\tan\varphi \tag{2.15}$$

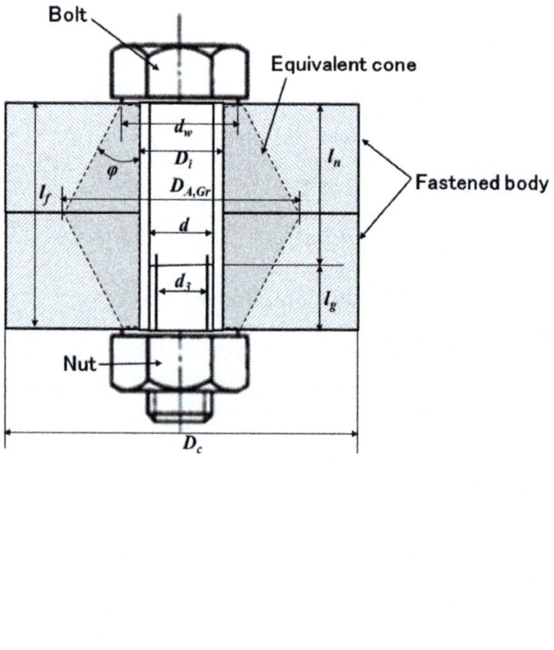

Fig. 2.9 Bolted joint model (compliant with VDI2230 2003 [3]

Here, w *is* called the fastening coefficient, and in the case of a through-bolt as shown in the figure, $w = 1$ *is* used. When the fastener is sufficiently wide $D_c > D_{A,Gr}$, the axial compliance of the fastener δ_c is obtained as follows.

$$\delta_c = \frac{2\ln\left[\frac{(d_w+D_i)(d_w+wl_f\tan\varphi-D_i)}{(d_w-D_i)(d_w+wl_f\tan\varphi+D_i)}\right]}{wE_c\pi D_i\tan\varphi} \tag{2.16}$$

Recently, a more accurate calculation formula for the compliance (2.10)–(2.16) of bolts and fastened objects has been developed, including finite element analysis, and is discussed in detail in Chap. 3.

2.4.2 Fatigue Strength Design of Bolts

Once the load factor Φ is obtained in this way, the stress amplitude of the bolt σ_a can be calculated using the valley diameter cross section of the bolt, A_3 as follows,

$$\sigma_a = \frac{1}{2}\frac{F_b}{A_3} = \frac{\Phi W_A}{A_3} \tag{2.17}$$

This stress amplitude acting on the bolt σ_a must be lower than the allowable stress amplitude which can be obtained from the fatigue limit σ_{wk} (shown in Table 2.2) of bolt and the safety factor S_f, as shown in Eq. (2.18).

$$\sigma_a \leq \frac{\sigma_{wk}}{S_f}. \tag{2.18}$$

Recently, aluminum alloys, titanium alloys, and FRP, which have low rigidity, are often used as fastening bodies for the purpose of weight reduction. In these cases, as shown in Fig. 2.10, the load on the bolt F_b will increase, making it easy to cause mistakes like the "loosening accident of Shinkansen motor mounting bolts" described in Sect. 1.1 of the previous chapter. Therefore, if readers use such materials due to the need to reduce weight of products, I would like them to carefully calculate the load factor Φ and fatigue strength.

2.4.3 Fatigue Strength Improvement Methods [7–9]

To improve the fatigue strength of bolted joints, the basic idea is to reduce the load factor Φ shown in the previous section. Specifically, the following method shown in Fig. 2.11 is taken.

i Insert cylindrical spacers to make l_f larger (Fig. 2.11a).

Table 2.2 Fatigue limit σ_{wk} of various bolts [4, 5]

Pitch series	Thread size	σ_{wk} (A_3) (Mpa)				
		Strength classification				
		4.8	6.8	8.8	10.9	12.9
Coarse thread	M16	81	56	64	78	81
	M8	61	49	59	72	76
	M10	52	45	55	69	72
	M12	48	43	53	67	71
	M16	42	39	49	62	66
	M20	39	37	46	61	65
	M24	37	36	45	59	63
	M30	34	34	43	57	61
Fine thread	M8 × 1	74	51	59	71	74
	M10 × 1.25	57	46	54	67	71
	M12 × 1.25	54	44	52	64	68
	M16 × 1.5	45	39	48	60	63
	M20 × 1.5	42	37	44	56	59
	M24 × 2	37	35	42	55	58

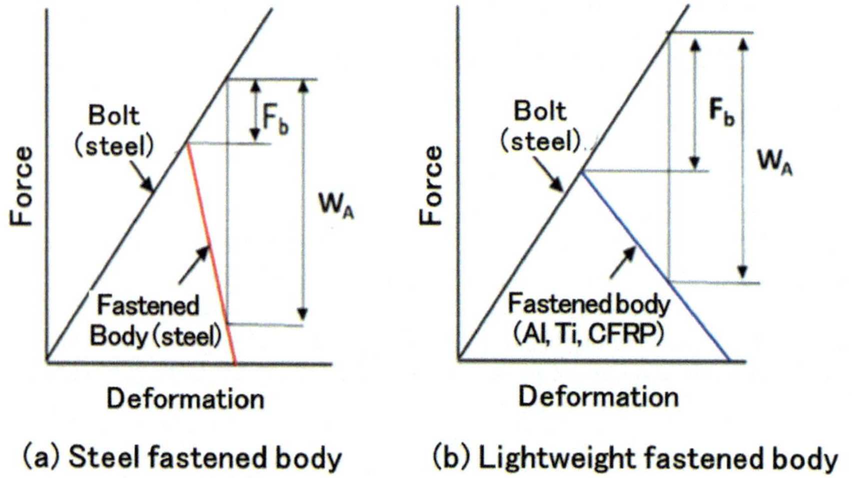

Fig. 2.10 Increase in bolt load F_b due to the use of lightweight materials for the fastening body

ii Make the bolt shaft thinner (smaller A_g) to make it an waisted-shank bolt (Fig. 2.11b).

iii For tap or stud bolts, cut out the female threads to increase l_f, (Fig. 2.11c).

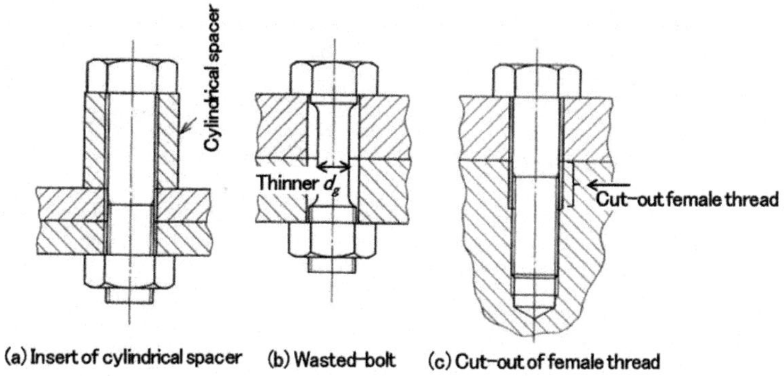

Fig. 2.11 How to improve fatigue strength by reducing the load factor Φ

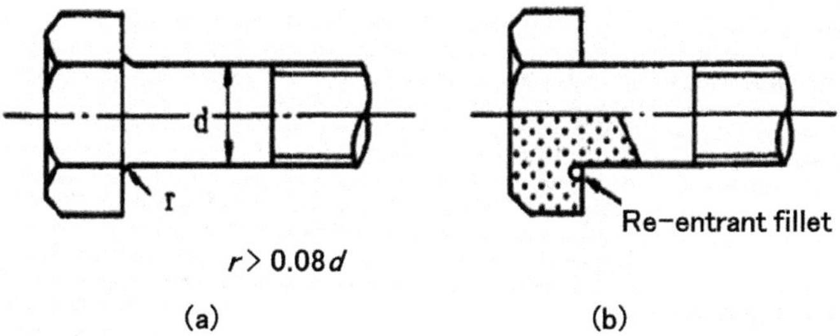

Fig. 2.12 Fatigue strength improvement method for bolt necks

In addition, the following measures have been adopted to improve the fatigue strength of the bolts themselves.

iv Make round filet more large to decrease stress concentrations (Fig. 2.12a).

This is to reduce the notch effect, and r > 0.08d is desirable. However, this increased rounding may cause problems with the rounded portion riding up on the edge of bolt mounting hole. So, in these cases the "re-entrance fillet" is suitable as shown in Fig. 2.12b.

v Measures for Incomplete Threads.

Incomplete threads are likely to cause defects at the root of the thread during machining, causing a decrease in fatigue strength, so the following countermeasures are taken.

The threading up angle of the imperfect threaded part should be as low as about 15° as shown in Fig. 2.13a. If the imperfectly threaded portion is cut out with a groove, the groove length should be $0.5d$ and the rounding radius R should be about $0.02d$, as shown in Fig. 2.13b. Furthermore, if the bolt is subjected to bending due to structural accuracy issues, or for reduced sensitivity to bolt squareness, an arc-shaped groove is suitable as shown in Fig. 2.13c.

vi Equalize the Load on the Threads.

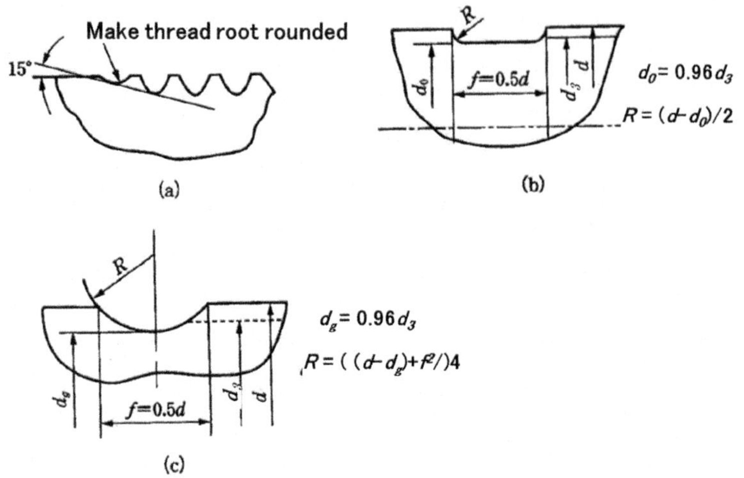

Fig. 2.13 Fatigue strength improvement method by rounding or removing incomplete threads

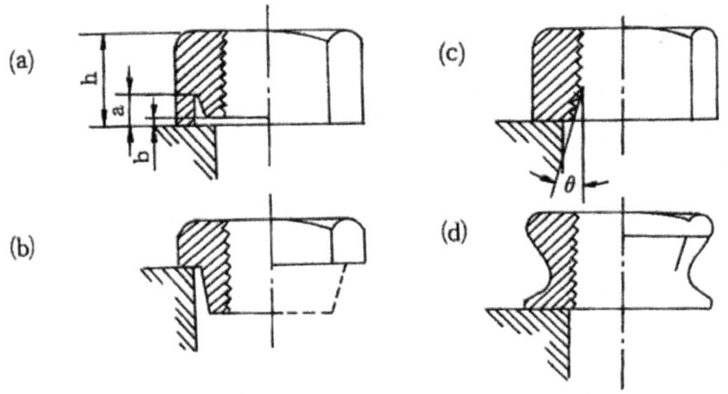

Fig. 2.14 Fatigue strength improvement method using a nut with uniform thread load sharing

Even though the threads of the nut and bolt engage evenly when there is no stress, when stress is applied, the load is not evenly applied to each thread due to the rigidity of the nut and bolt, and a large force is applied to the first thread of engagement. This is a factor that reduces the fatigue strength of the bolt, so the special nut shown in Fig. 2.14 is used to reduce this.

2.4.4 Plastic Region Tightening Method [10] (See Sect. 10.1)

When an initial tightening force is applied above the yield point of the bolt material, it is often mistakenly assumed that the bolt will continue to elongate as the load is applied to the bolted joint. If it is understood that the load acting on the bolted joint is shared not only by the bolt but also by the reduction in the compression force of the fastened body, there is no need to worry about continued bolt elongation. Figure 2.15 illustrates this. This figure shows how the eccentric external force acts on a bolted joint whose initial fastening force is in the elastic and plastic regions. The higher the initial fastening force, the lower the bolt burden load F_b for the same external force. In other words, the higher the initial fastening force, the higher the fatigue strength of bolted joint. The ultimate method to increase the initial fastening force is the plastic region fastening method. When an external force is applied to a fastener tightened in the plastic region P_3, the characteristic curve of the bolt is as shown in P_3–D. The reason why the yield point increases when an external force is applied is that the characteristic curve of the bolt O–P_3–S shows plastic behavior in initial fastening process due to the combined tensile and torsional stresses, but only tensile stress is applied for the axial external force. Then, when the external force F is removed, the characteristic curve of the bolt becomes a straight line as shown in the dashed line in the figure, and the clamping force drops to P_4 when the bolt is completely unloaded. Although the plastic strain of the bolt slightly increases and the fastening force decreases by the initial loading, the bolt shows completely elastic behavior for the second and subsequent external loadings, indicating that there is no danger of the bolt continuing to elongate.

Repeated disassembly and retightening of the bolted joints in this plastic region will, of course, gradually increase the plastic strain of the bolt, but there is sufficient

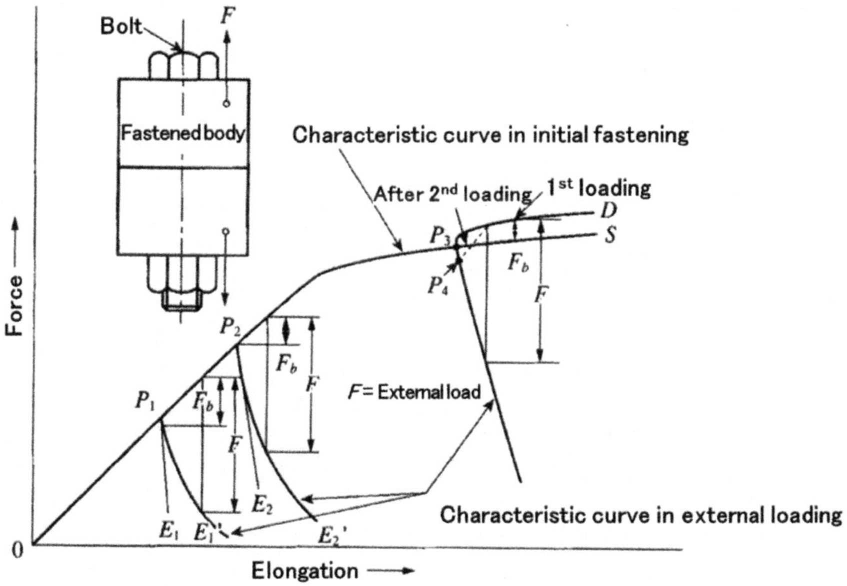

Fig. 2.15 Initial tightening force and tightening diagram for elastic and plastic region tightening [10]

margin in terms of the amount of elongation at break of the bolt, and reusability is unlikely to be a problem.

2.4.5 Basis for Strength Design Under Perpendicular Loading (See Chap. 4) [7–9]

In general, bolted joint structures are designed so that the axial direction of the bolt is aligned with the direction of the load. However, in the construction structures, friction joints are often used, in which the axial direction of the bolt is aligned perpendicular to the load. In this case, as shown in Fig. 2.3, the initial fastening force of the bolt, F_f, is designed as $F_f \mu \geq 1.5 \sim 2.0W_A$ (see Eq. (2.2)). If the bolt is designed in this way, the bolted joint is secured as a single piece under load and is safe enough against loosening and fatigue failure of this joint structure. Under these conditions, there is no need to evaluate strength or looseness. However, recently, there are many cases where electronic devices are incorporated and fastened into mechanical products, and when a thermal load is applied to this part, a load is applied to the fastened bolt in the direction perpendicular to the bolt axis due to the thermal expansion difference between the electronic device which are made of plastic, semiconductor, printed circuit board, etc., and the metal base of the machine. In such cases, the allowable stress in the electronic device often makes it difficult to design under the conditions of Eq. (2.2). Therefore, a slip-allowing design is necessary to allow a slight amount of slip (Fig. 2.3 allowable slip δ_{CR}).

These details are left to Chap. 5.

References

1. Verein Deutscher Ingenieure, VDI-Richtlinien 2230 (1977) Systematic.
2. Calculation of high duty bolted joints, 1977.
3. Verein Deutscher Ingenieure, VDI-Richtlinien 2230 (1986) Systematic calculation of high duty bolted joints, -Joints with one cylindrical bolt-, 1986.
4. Verein Deutscher Ingenieure, VDI-Richtlinien Blatt 1 2230 (2003) Systematic calculation of high duty bolted joints -Joints with one cylindrical bolt-, 2003.
5. Editorial Committee of New Edition of Junction Technology Handbook, New Edition of Junction Technology Handbook, Sangyo Gijutsu Service Center, (1992).
6. Akira Yamamoto, Principle and Design of Screw Fastening, Yokendo, (1997).
7. Isamu Yoshimoto, Key Points for Design of Screw Fasteners, Japanese Standards Association, (1992).
8. Sakai, Tomotsugu, Enlarged and supplemented outline of screw fastening, Yokendo, (2008).
9. Japan Screw Research Association Publication Committee, New Edition Screw Fastening Guidebook, (2006).
10. Bickford, J. H., Introduction to the Design and Behavior of Bolted Joints, 4th Edition Non-Gasket Joints, CRC, (2008).
11. Toshio Hattori, Toshio Nonaka, and Motoharu Taneda, Strength of Plastic Region Tightening Bolt Fasteners, Pressure Technology, 23–1, pp. 7–13 (1985).

Chapter 3
Mechanical Analysis and Strength Design Under Axial and Bending Loads

3.1 Introduction

Generally, the bolted joints in a mechanical structure are designed so that the axial direction of the bolt is fit to the direction of the external force. Therefore, mechanical analysis of bolted joints is often performed for the axial load of the bolts. However, in the case of structure with multiple bolted joints, each bolt is subject to the eccentric force or bending load, even if the load on the structure is directed toward the bolt axial direction as shown in Fig. 3.1. In this section, equivalent stiffness or compliance of bolt or clamped parts, δ, load factor Φ, and strength analysis will be discussed in detail, including the latest technology focusing on these axial and bending loads.

3.2 Equivalent Stiffness, Compliance, and Load Factor Under Axial Load

3.2.1 Equivalent Stiffness and Compliance Under Axial Load

When a bolted joint is fastened with sufficient tightening force, the stiffness of the bolted joints under axial load is the sum of the stiffness of the bolt and that of the equivalent compression body (equivalent cylinder according to the 1970 edition of VDI 2230; see Fig. 3.2a, and equivalent cone according to the 2003 edition of VDI 2230; see Fig. 3.2b, of the clamped body). These stiffnesses are the inverse of the respective compliance δ presented in Chap. 2. In multiple bolted joints, the external load is shared by each bolted joint. To calculate the proportional load of each bolted joint by using a structural analysis (e.g., finite element analysis) of a structure with multiple bolted joints, each bolted joint is modeled with a beam element of

© The Author(s), under exclusive license to Springer Nature Switzerland AG 2025

T. Hattori and T. Naruse, *Compendium for Design and Manufacturing of Bolted Joints*, Solid Mechanics and Its Applications 280,
https://doi.org/10.1007/978-3-031-89628-6_3

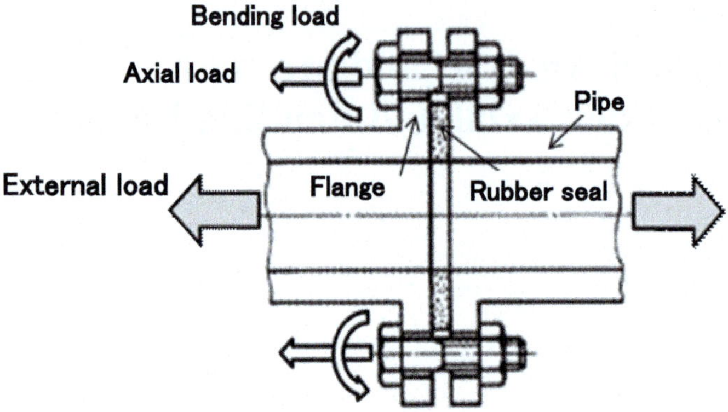

Fig. 3.1 Structure with multiple bolted joints (flange joints)

the equivalent stiffness. Using this proportional load makes it possible to perform a detailed mechanical analysis and strength design of each bolt, as shown in Fig. 3.3.

The axial compliance of the bolt, δ_b, which is the inverse of the axial stiffness of the bolt, can be obtained according to VDI 2230 (1977) [1] in Fig. 3.2a as follows:

$$\delta_b = \frac{0.4d}{E_b A_n} + \frac{l_g}{E_b A_n} + \frac{l_s}{E_b A_3} + \frac{0.4d}{E_b A_n} \tag{3.1}$$

Here, d is the nominal diameter of the bolt, E_b is the Young's modulus of the bolt, l_g is the length of the bolt cylindrical section with the nominal diameter of the bolt, l_s is the thread length in the fastening range, A_n is the cross-sectional area of the nominal diameter, and A_3 is the cross-sectional area at minor diameter d_3 (shown in Fig. 3.2a). The first and last terms are the compliance of the bolt head and nut section, which are empirically determined as an equivalent cylinder with a length of $0.4d$ and an area of A_n.

In VDI 2230 (2003) [2], the following equation is presented.

$$\delta_b = \frac{0.5d}{E_b A_n} + \frac{l_g}{E_b A_n} + \frac{l_s}{E_b A_3} + \frac{0.5d}{E_b A_3} + \frac{0.4d}{E_b A_n} \tag{3.2}$$

The stiffness of the equivalent compression body of the clamped body is modeled in VDI 2230 (1977) [1] as a hollow cylinder with an outer diameter of equivalent cylinder diameter D_m and an inner diameter of hole diameter D_i (see Fig. 3.2a, dashed line). When the diameter of the clamped body, D_c, is sufficiently large, the equivalent cylinder diameter D_m and equivalent cross-sectional area of the equivalent cylinder, A_{eq}, are obtained by Eqs. (3.3) and (3.4).

$$D_m = d_w + \frac{l_f}{10}\left(\text{when } 3d_w < D_c, l_f \leq 8d\right) \tag{3.3}$$

Fig. 3.2 Modeling of single-bolted joint

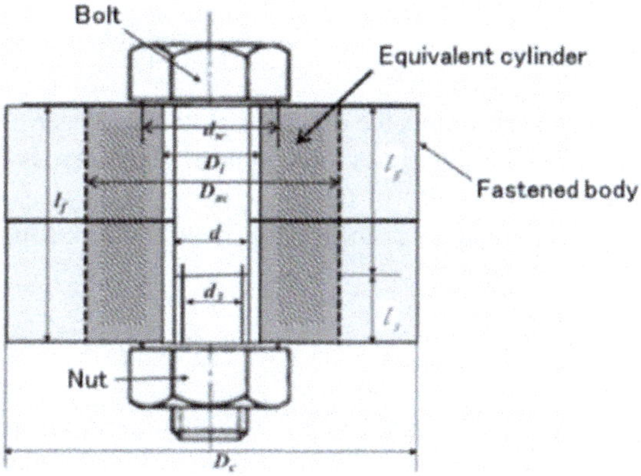

(a) VDI2230(1997)[1]

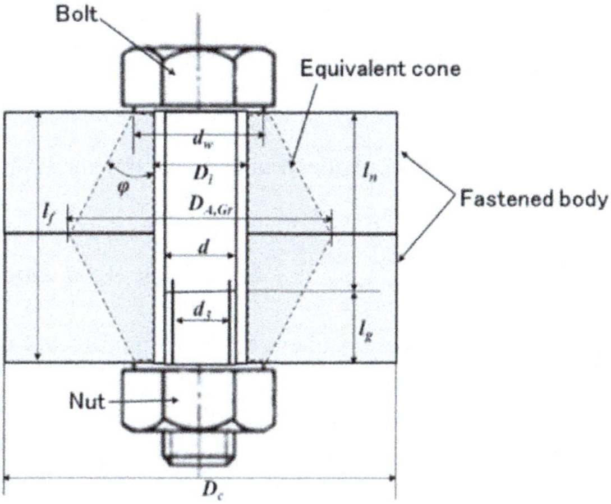

(b) VDI2230(2003)[2]

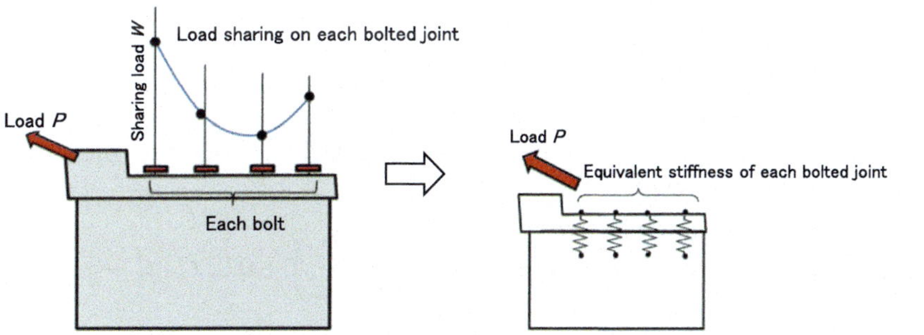

(a) Multiple bolted joints (b) Analytical model for multiple bolted joints

Fig. 3.3 Equivalent stiffness for structural analysis of multiple bolted joints

$$A_{eq} = \frac{\pi}{4}\left(D_m^2 - D_i^2\right) \tag{3.4}$$

Here, d_w represents the diameter of the head bearing surface, which for a hexagonal bolt is about 0.95 times the bolt head width across flats according to JIS B1083.

The compliance of the clamped body, δ_c, is expressed by using the equivalent cross-sectional area of this equivalent hollow cylinder, A_{eq}, as follows:

$$\delta_c = \frac{l_f}{E_c A_{eq}} \tag{3.5}$$

In VDI 2230 Blatt1 (2003) [2], the equivalent compressible body is represented by the equivalent circular cone shown by the dashed line in Fig. 3.2b.

In the case of a through-bolt, the angle of the equivalent circular cone, φ, is expressed by the following equation:

$$\tan\varphi = 0.362 + 0.032\ln\left(\frac{l_f}{2d_w}\right) + 0.153\ln\left(\frac{D_c}{d_w}\right) \tag{3.6}$$

The maximum outside diameter of the circular cone, $D_{A,Gr}$, is calculated as follows:

$$D_{A,Gr} = d_w + wl_f tan\varphi \tag{3.7}$$

Here, w is called the fastener coefficient and $w = 1$ is used for through-bolts. When the clamped plate is sufficiently wide (e.g., $D_c > D_{A,Gr}$), the axial compliance of the clamped body, δ_c, is obtained as follows:

$$\delta_c = \frac{2\ln\left[\frac{(d_w+D_i)(d_w+wl_f tan\varphi-D_i)}{(d_w-D_i)(d_w+wl_f tan\varphi+D_i)}\right]}{wE_c\pi D_i tan\varphi} \tag{3.8}$$

However, in these Eqs. (3.6)–(3.8), the clamped length l_f is only the sum of the thicknesses of the two plates, t_1 and t_2. When the two-clamped two plates have significantly different thicknesses, these VDI 2230 equations cannot precisely predict the compliance of the clamped plate. Based on the FEM analysis, Naruse [3] proposed the following modified equation for Eq. (3.6), in which the thickness ratio (t_1/t_2) is taken into consideration.

$$\tan\varphi = 0.323 + 0.032\ln\left(\frac{l_f}{2d_w}\right) + 0.153\ln\left(\frac{D_c}{d_w}\right) + 0.0717\ln\left(\frac{t_1}{t_2}\right) \tag{3.9}$$

(when $t_1 < t_2$)

Figure 3.4 shows the axial compliances δ_c calculated by Eq. (3.8) substituting this modified Eq. (3.9) and Eq. (3.6) presented in VDI 2230 Blatt1 (2003) [2] for $\tan\varphi$.

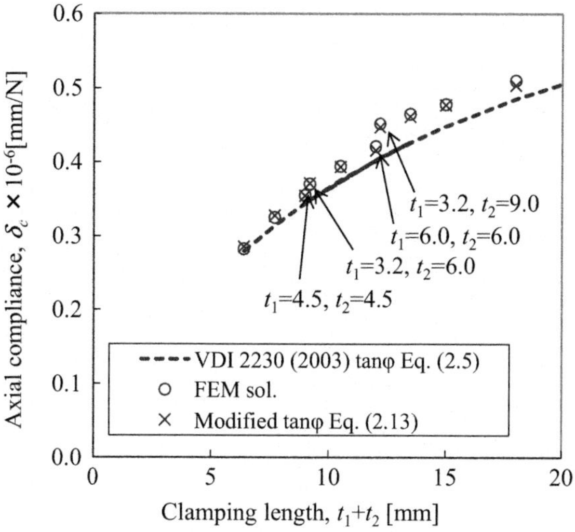

Fig. 3.4 Comparison example of calculated results of axial compliance (M10)

Here, the results by using the modified $\tan\varphi$ of Eq. (3.9), which is marked with × , are in better agreement with the FEM solution, which is marked with ○, than the results by using $\tan\varphi$ of VDI 2230 Blatt1 (2003), i.e., Eq. (3.6), which is shown by the dashed line.

The above discussion has focused on through-bolted joints; now, we shift to tapped-thread joints. We can consider a tapped-thread joint to be equivalent to half of a through-bolted joint clamping two plates of the same thickness, as shown in Fig. 3.5. First, the compliance $\delta_c(2l_f)$ of the through-bolted joints with the clamping length 2 l_f can be obtained using Eqs. (3.5) and (3.8) with reference to the right figure of Fig. 3.5. Next, we can see that half of this compliance $\delta_c(2l_f)/2$ is equivalent to the compliance of the tapped-thread joint, δ_{sc}, with reference to the left side of Fig. 3.5. Finally, we can obtain the compliance of tapped-thread joints, δ_{sc}, with clamping length l_f by using the follows equations, which are obtained in accordance with conforming to VDI 2230 (1977) [1].

$$A_{eq} = \frac{\pi}{4}\left(\left(d_w + \frac{l_f}{2}\right)^2 - D_i^2\right) \tag{3.10}$$

$$\delta_{sc} = \frac{1}{2}\frac{2l_f}{E_c A_{seq}} = \frac{l_f}{E_c A_{seq}} \tag{3.11}$$

In VDI 2230 (2003) [2], the following equation is presented by using $w = 2$ in Eq. (3.8):

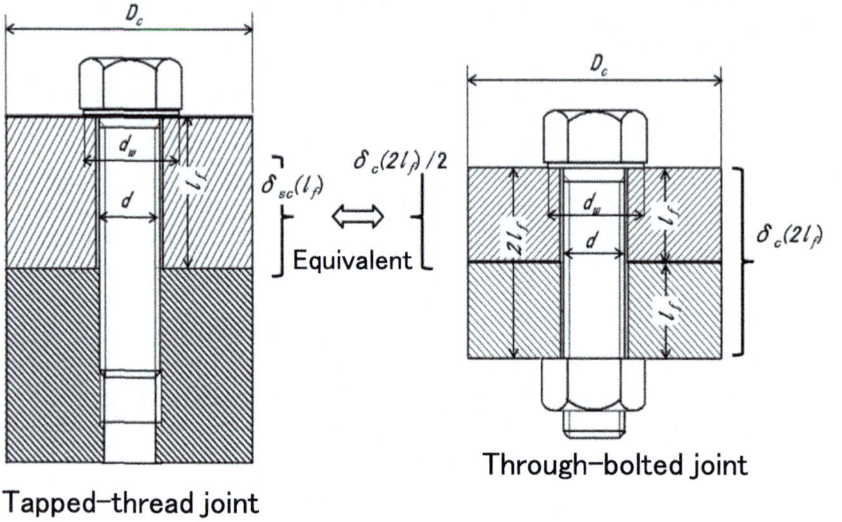

Fig. 3.5 Easy prediction method of the axial compliance of tapped-thread joints, δ_{sc}, from that of through-bolted joints, δ_c

$$\delta_c = \frac{2\ln\left[\frac{(d_w+D_i)(d_w+2l_f\tan\varphi-D_i)}{(d_w-D_i)(d_w+2l_f\tan\varphi+D_i)}\right]}{2E_c\pi D_i\tan\varphi} \tag{3.12}$$

3.2.2 *Load Factor and Load Introduction Factor Under Axial Load*

Using the stiffness of the bolt and the clamped body discussed above, we examine the state in which the bolt and the clamped body deform as a single elastic part under an external axial force. First, we assume that an external force W_a is applied on the bearing surface as an ideal case. Figure 3.6a shows a bolted joint and a tightening diagram where an external force is applied on the bearing surface. As shown in Fig. 3.6b, the bolt is elongated by ε due to the tensile internal force F_b, and the distance between the bearing surfaces of the clamped body is elongated by ε due to the reduction of the compressive internal force F_c. Therefore, from the axial compliance of the bolt, δ_b, and the axial compliance of the clamped body, δ_c, the following equation holds.

$$F_b\delta_b = F_c\delta_c = \varepsilon \tag{3.13}$$

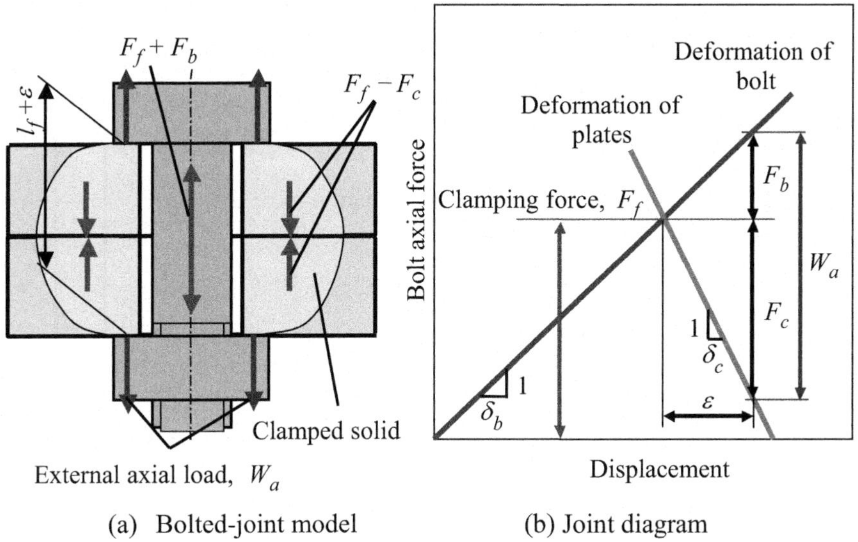

Fig. 3.6 Force balance and deformation behavior of bolted joint under external axial load (when external load is applied on bearing surfaces)

On the other hand, the elongation of the distance between the bearing surfaces, ε, is equal to the value subtracting the compressive deformation $F_b\delta_c$ caused by the reaction force of the tensile internal force of bolt, F_b, from the tensile deformation $W_a\delta_c$ caused by the external force W_a. Thus, the following equation holds.

$$W_a\delta_c - F_b\delta_c = \varepsilon \tag{3.14}$$

The following equations can be obtained from Eqs. (3.13) and (3.14).

$$W_a = F_b + F_c \tag{3.15}$$

$$\frac{F_b}{W_a} = \frac{\delta_c}{\delta_b + \delta_c} = \Phi \tag{3.16}$$

Here, Φ is called the load factor and represents the ratio of the internal tensile force of bolt, F_b, to the external force W_a.

Since the deformation state of the bolted joints changes, the load factor needs to be modified when the loading position of the external load moves from bearing surfaces [7–16]. The load factor Φ_n, which takes into consideration the external force loading position, can be obtained by multiplying the load factor Φ defined in Eq. (3.16) by the load introduction factor n, as follows.

$$\Phi_n = n\Phi \tag{3.17}$$

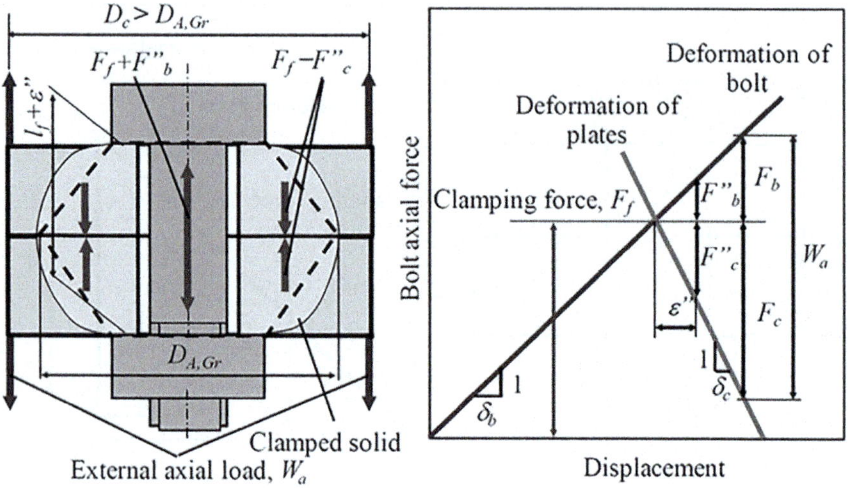

Fig. 3.7 Force balance and deformation behavior of bolted joint under external axial load (when external load is applied to the outside of the equivalent circular cone)

Among the various methods and simplified formulae for solving this load introduction factor, we here introduce the method proposed by Hagiwara [15]. Consider a case where the plate is wide enough compared to the outer diameter of the equivalent circular cone, $D_{A,\,Gr}$, and the external force W_a is also applied to the outside of the equivalent circular cone, as shown in Fig. 3.7a. In this case, as shown in Fig. 3.7b, the bolt is extended by the value of ε'' under the internal tensile force F''_b, and the clamped body between the bearing surfaces is simultaneously extended by the value of ε'' under the decrement of the internal compressive force, F''_c. Therefore, the following equation is obtained in the same way as Eq. (3.13).

$$F''_b\delta_b = F''_c\delta_c = \varepsilon'' \tag{3.18}$$

Now, the clamped body is subjected to an internal tensile force of the bolt, F''_b, on the bearing surface and an external force W_a on the outside of $D_{A,\,Gr}$ as shown in Fig. 3.8a. In this case, the joint surfaces on the outside of $D_{A,\,Gr}$ are separated even if the external force is not applied. In addition, the internal force and deformation of the clamped body are balanced in the area on the inside of $D_{A,\,Gr}$, which can be considered as a superposition of the compressive and tensile deformation of the cylinder with the outer diameter of $D_{A,\,Gr}$ and the inner diameter of D_i as shown in Fig. 3.8b, c. The compressive deformation shown in Fig. 3.8b can be considered as the deformation under the internal tensile force F''_b, which is applied on the bearing surfaces. The compliance of the clamped body for this compressive deformation is δ_c as shown in Eq. (3.8).

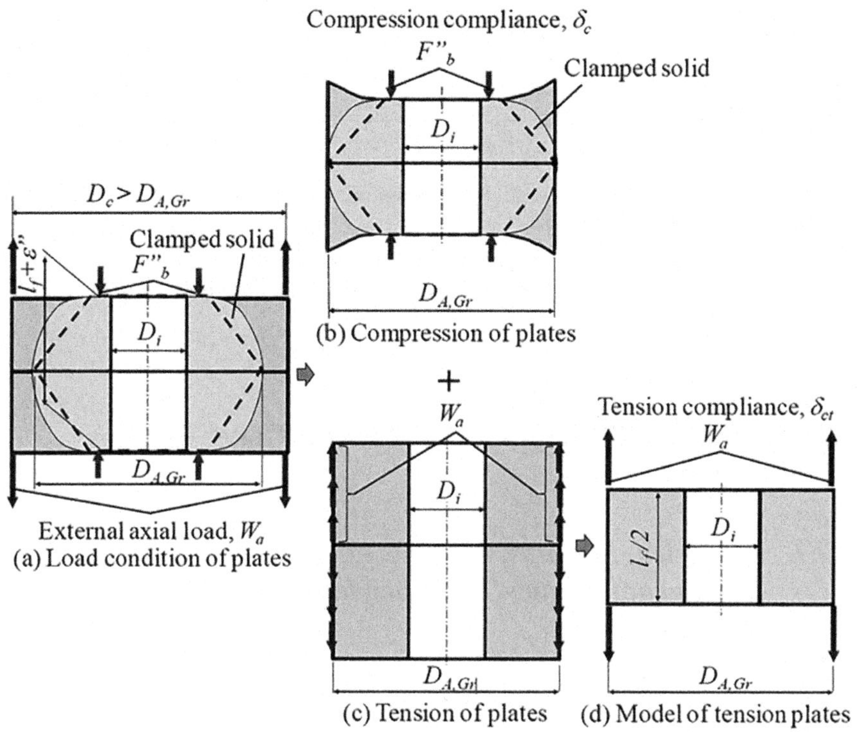

Fig. 3.8 Equilibrium of forces and deformation behavior of clamped body when an external force is applied on the outside of the outer diameter of the equivalent conical body, $D_{A,Gr}$

In contrast, when the external force W_a is applied on a sufficiently larger outer diameter than the outer diameter of the equivalent clamped body, $D_{A,Gr}$, as shown in Fig. 3.8c, the tensile deformation of the clamped body is considered to be uniformly distributed. This situation can be replaced by a model in which a concentrated force W_a is applied to the upper and lower surfaces of a cylinder with the outer diameter of $D_{A,Gr}$, the inner diameter of D_i, and the length of $l_f/2$, as shown in Fig. 3.8d. The tensile compliance δ_{ct} is obtained by the following equation.

$$\delta_{ct} = \frac{l_f/2}{E_c \frac{\pi}{4}\left(D_{A,Gr}^2 - D_i^2\right)} \tag{3.19}$$

Therefore, the elongation between the bearing surfaces, ε'', is obtained by superimposing Fig. 3.8b, c as the following equation.

$$W_a \delta_{ct} - F_b'' \delta_c = \varepsilon'' \tag{3.20}$$

The following equations can be obtained by eliminating ε" from Eqs. (3.18) and (3.20).

$$W_a\delta_{ct} = F_b^{''}\delta_c + F_b^{''}\delta_b \tag{3.21}$$

$$\Phi_n = \frac{F_b^{''}}{W_a} = \frac{\delta_{ct}}{\delta_c}\frac{\delta_c}{\delta_b + \delta_c} = n\Phi \tag{3.22}$$

Thus, when the external force W_a is applied on the outside of the equivalent clamped body as shown in Fig. 3.8, the load introduction factor n is expressed as the ratio of the tensile compliance of the clamped body, δ_{c_t}, which is shown in Eq. (3.19), and the compressive compliance of the clamped body, δ_c, which is shown in Eq. (3.12).

3.3 Equivalent Stiffness and Load Factor Under Bending Moment

3.3.1 Equivalent Stiffness, Load Factor, and Load Introduction Factor Under Bending Moment in Linear Domain

The bending stiffness of the bolted joint is also considered to be the sum of the bending stiffness of the bolt and that of the clamped body. VDI2230 (1977) [1] considers the bending compliance of the bolt, β_b, as a series connection of columnar bodies as well as the axial compliance shown in Eq. (3.1).

$$\beta_b = \frac{0.4d}{E_b I_n} + \frac{l_g}{E_b I_n} + \frac{l_s}{E_b I_3} + \frac{0.4d}{E_b I_n} \tag{3.23}$$

Here, I_n and I_3 are the cross-sectional quadratic moments of the nominal diameter and the minor diameter, respectively. The bending compliance of the bolt head and nut section was assumed to be a cylinder with the nominal diameter and a length of 0.4d.

The bending compliance of the clamped body, β_c, is also obtained by considering that the clamped body is equivalent to a hollow cylinder with an outer diameter of the equivalent diameter D_{eq} and an inner diameter of the hole diameter D_i, as shown in Eq. (3.24).

$$\beta_c = \frac{l_f}{E_c I_{eq}} = \frac{64 l_f}{\pi E_c \left(D_{eq}^4 - D_i^4\right)} \tag{3.24}$$

In VDI 2230 Blatt1 (2003) [2], the bending stiffness of the clamped body is modeled with an equivalent conical body as expressed by the dashed line in Fig. 3.2b, which has the conical angle φ of Eq. (3.6). Thus, the cross-sectional quadratic moment of the equivalent conical body, I_{Bers}, is obtained by the following equation.

$$I_{Bers} = \frac{3\pi}{64} \frac{(D_c - d_w) d_w^3 D_c^3}{D_c^3 - d_w^3} - \frac{\pi}{64} D_i^4 \tag{3.25}$$

When the clamped plates are sufficiently wide such as $D_c > D_{A,Gr}$, the bending compliance of the clamped body, β_c, is expressed by substituting $D_{A,Gr}$ for D_c as follows.

$$\beta_c = \frac{l_f}{E_c I_{Bers}} \tag{3.26}$$

As in the case of axial stiffness, these VDI equations for the bending stiffness of the clamped body do not consider the thickness difference of the clamped plates. Naruse [3] developed the modified tan φ (as shown in Eq. (3.9)), in which the thickness ratio (t_1/t_2) is taken into consideration, and the bending compliance can be obtained by substituting it into Eqs. (3.7), (3.25), and (3.26). Figure 3.9 shows a comparison of the results by this modified bending compliance of the clamped body and those of VDI 2230 Blatt1 (2003) [2] and finite element analysis results. It can be seen that the results using the modified formula (marked with ×) are in better agreement with the finite element analysis results (marked with ○) than the results calculated by VDI 2230 Blatt1 (2003) [2] (shown by the dashed lines).

The load factor for the bending moment, Φ_β, and the load introduction factor for the bending moment, n_β, can also be defined in the same way as the axial load. The

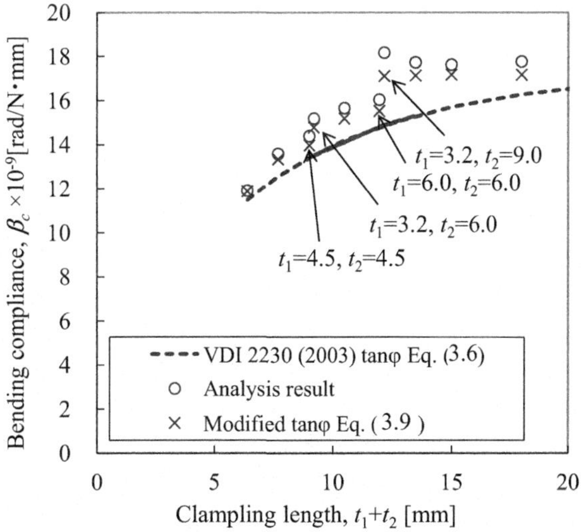

Fig. 3.9 Comparison example of calculated results of bending compliance (M10)

load factor for the bending moment, Φ_β, is obtained from the bending compliance of the bolt, β_b, shown in Eq. (3.23), and the bending compliance of the clamped body, β_c, shown in Eq. (3.26) or (3.28), as follows.

$$\Phi_\beta = \frac{\beta_c}{\beta_b + \beta_c} \tag{3.27}$$

When the bending moment M is applied on the outside of the equivalent clamped body as shown in Fig. 3.8, it is considered as a superposition of the model in which the reaction force of the bolt moment is applied on the bearing surfaces as shown in Fig. 3.8b and the model in which the moment M is applied to the outer surface of the equivalent cylinder of the clamped body as shown in Fig. 3.8c, in same way to the axial load. As an equivalent model in Fig. 3.8c, the bending compliance β_{ct} can be obtained as a cylinder with the outer diameter of $D_{A,\ Gr}$, the inner diameter of D_i, and the length of $l_f/2$, as shown in Fig. 3.8d.

$$\beta_{ct} = \frac{l_f/2}{E_c \frac{\pi}{64}\left(D_{A,Gr}^4 - D_i^4\right)} \tag{3.28}$$

The load introduction factor n_β is defined from the bending compliance of the cylinder, β_{ct}, and the bending compliance of the clamped body, β_c. In addition, the load factor $\Phi_{\beta n}$, which takes into consideration the moment loading position, can also be defined as follows.

$$n_\beta = \frac{\beta_{ct}}{\beta_c} \tag{3.29}$$

$$\Phi_{\beta n} = n_\beta \Phi_\beta \tag{3.30}$$

3.3.2 *Equivalent Stiffness, Load Factor, and Load Introduction Factor Under Bending Moment in Nonlinear Domain [4]*

The above technique for the equivalent stiffness under bending moment is based on mechanical analysis in the linear domain and assumes that the equivalent clamped body remains unchanged. However, the contact surface pressure between the clamped bodies becomes asymmetric by increasing the bending moment, and the contact pressure on the tensile side easily becomes zero, that is, it separates the contact surfaces between the clamped plates on the tensile side. Thus, the shape of the equivalent clamped body changes, resulting in nonlinear behavior. This nonlinear behavior, namely, the behavior of the contact surface separation, is shown in Fig. 3.10 as an example evaluated by finite element analysis. As the bending moment of the clamped body, M_c, increases, the contact pressure on the tensile side decreases, the contact pressure becomes zero, and the contact surfaces on the tensile side separate, as shown in Fig. 3.10b–d.

Figure 3.11 shows the bending compliance of the clamped body, which is calculated from the deformation state and the bending moment of the clamped body. The relationship between the bending compliance and the bending moment can be roughly divided into the following three regions.

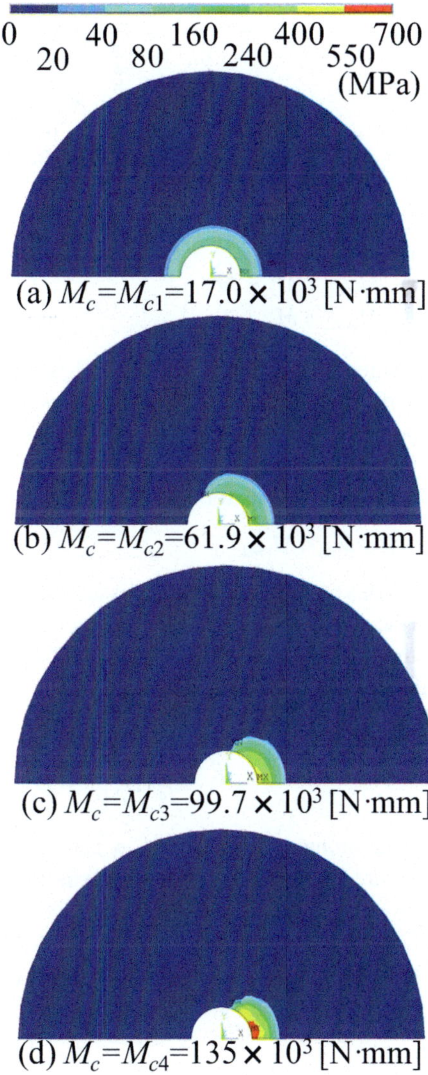

Fig. 3.10 Results of the finite element method analysis of the variation of the contact surface pressure distribution between the clamped plates (M10, t_1 = 4.5, t_2 = 6.0)

Fig. 3.11 Relationship between moment and bending compliance (M10, t 4.5, t 6.0)

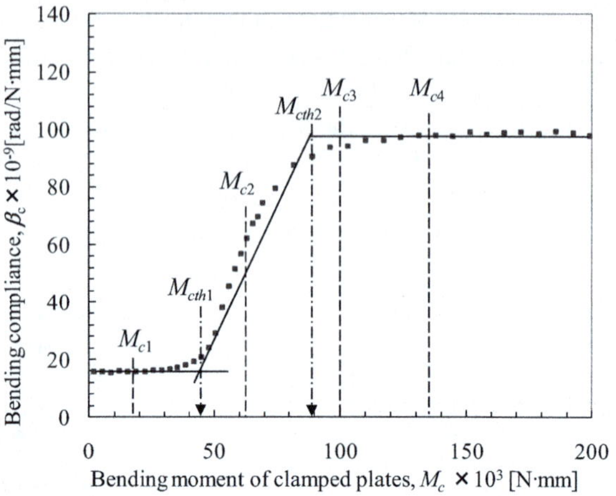

(i) $0 < M_c < M_{cth1}$.
(ii) $M_{cth1} < M_c < M_{cth2}$.
(iii) $M_{cth2} < M_c$.

Among these three regions, (ii) is considered as the transition region from (i) to (iii), in which the relationship between the bending compliance and the bending moment can be estimated along a linear relationship as shown in Fig. 3.11. Therefore, what is needed to estimate the bending stiffness of the clamped plates is the mechanical models for regions (i) and (iii).

(i) $0 < M_c < M_{cth1}$

In this region, the contact pressure between clamped plates remains and the contact surfaces do not separate. The contact pressure between the clamped plates, p, is the sum of the pressure under the preload F_f and that under the bending moment of the clamped plates, M_c, as follows.

$$p = \frac{F_f}{\frac{\pi}{4}\left(D_{A,Gr}^2 - D_i^2\right)} + \frac{M_c}{\frac{\pi}{64}\left(D_{A,Gr}^4 - D_i^4\right)}x \tag{3.31}$$

The moment threshold M_{cth1} can be obtained in the condition when the contact pressure becomes zero at the maximum diameter of the equivalent cone, namely, $x = -D_{A,Gr}/2$, in Eq. (3.31) as follows.

$$M_{cth1} = \frac{F_f\left(D_{A,Gr}^2 + D_i^2\right)}{8D_{A,Gr}} \tag{3.32}$$

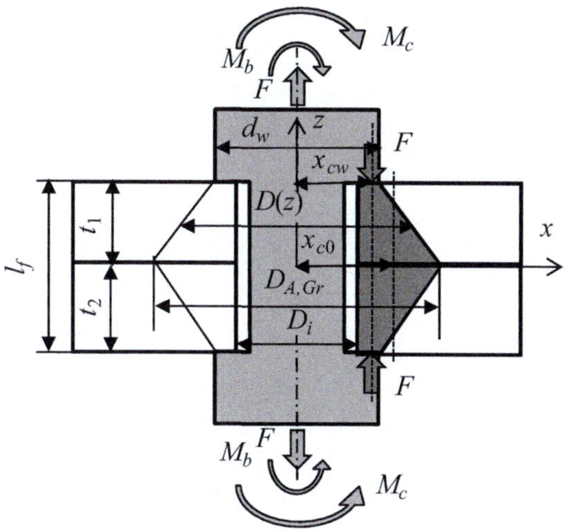

Fig. 3.12 Equivalent-solid model when $M_{cth2} < M_c$

(iii) $M_{cth2} < M_c$

In this region, as shown in Fig. 3.10c, d, the contact surfaces between the plates on the tension side open perfectly and the clamped solid on the compression side supports the bending moment. The shape of the clamped solid was assumed to be an equivalent solid, which is obtained by subtracting the bolt hole from the truncated cone with an upper diameter of $D_w/2$ and a lower diameter of $D_{A,\ Gr}/2$, as shown in Fig. 3.12. When the contact surface of the clamped plates on the tensile side is completely separated by the bending moment, the moment of the clamped body, M_c, is caused by the axial compression force F applied on the length of x_{c0} away from the bolt center axis.

$$M_c = Fx_{c0} \tag{3.33}$$

In this case, x_{c0} is thought to be the center of gravity of the contact surface of the clamped plate, namely the gravity center of the equivalent-solid section at z = 0.

Under the equilibrium of forces of the axial tension force of the bolt, F, and the axial compression force of the equivalent solid, F, the bolt extends to the amount of Δ_b and the equivalent solid shrinks by the amount of Δ_c. The bending gradient ξ is then calculated as follows.

$$\xi = \frac{\Delta_b + \Delta_c}{x_{cw}} = \frac{F(\delta_b + \delta_{c\ \mathrm{iii}})}{x_{cw}} \tag{3.34}$$

Here, x_{cw} is the gravity center of the bearing surface of the equivalent solid, δ_b is the axial compliance of the bolt, and $\delta_{c\ \mathrm{iii}}$ is the axial compliance of the equivalent solid (Fig. 3.13).

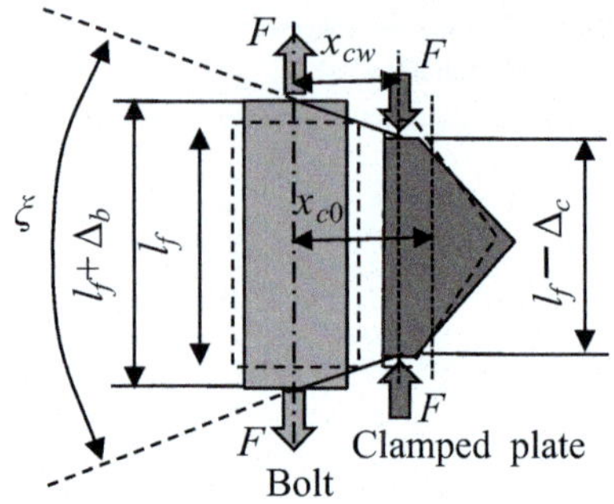

Fig. 3.13 Deformation model when $M_{cth2} < M_c$

The bending compliance of the equivalent solid, $\beta_{c\ \mathrm{iii}}$, is obtained from Eqs. (3.33) and (3.34) by the following equation.

$$\beta_{c\ \mathrm{iii}} = \frac{\xi}{M_c} = \frac{\delta_b + \delta_{c\ \mathrm{iii}}}{x_{cw} x_{c0}} \tag{3.35}$$

The center of gravity of the clamped body at the bearing surface, x_{cw}, and the center of gravity of the cross-section of the clamped body at the contact surface, x_{c0}, are obtained from the cross-section of the equivalent solid at the axial coordinate of z as shown in Fig. 3.14. Since it is difficult to obtain the axial compliance of the equivalent solid with this cross-section analytically, the section of the equivalent solid is approximately obtained by subtracting an ellipse with a short diameter of $D_i/2$ and a long diameter of $D_i\sqrt{1 - D_i^2/D^2}$ from a circle with a diameter of $D(z)/2$, as shown in Fig. 3.15.

The centers of gravity x_{cw} and x_{c0} are obtained as follows.

$$x_{cw} = \frac{1}{4} \frac{d_w^3 - 2D_i^3\sqrt{1 - \frac{D_i^2}{d_w^2}}}{d_w^2 - 2D_i^2\sqrt{1 - \frac{D_i^2}{d_w^2}}} \tag{3.36}$$

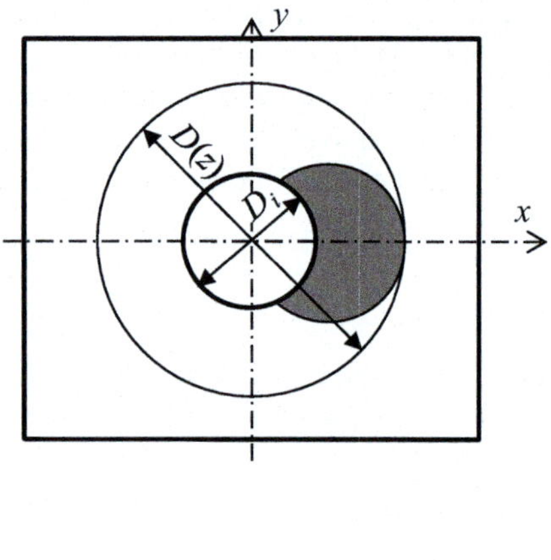

Fig. 3.14 Section of equivalent-solid model when $M_{cth2} < M_c$

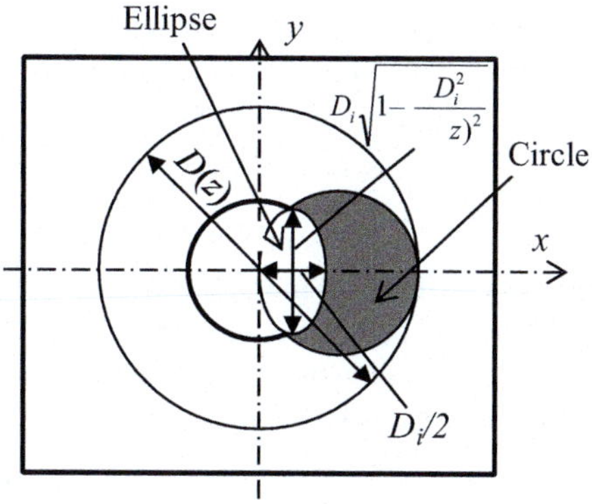

Fig. 3.15 Approximated section of equivalent-solid model when $M_{cth2} < M_c$

$$x_{c0} = \frac{1}{4}\frac{D_{A,Gr}^3 - 2D_i^3\sqrt{1-\frac{D_i^2}{D_{A,Gr}^2}}}{D_{A,Gr}^2 - 2D_i^2\sqrt{1-\frac{D_i^2}{D_{A,Gr}^2}}} \tag{3.37}$$

The stiffness of the truncated cone, K_{crc}, can be estimated from the circle area $A_{crc}(D)$ with a diameter of $D(z)/2$ as follows.

$$K_{crc} = \frac{E_c}{\int_{-t_2}^{t_1}\frac{dz}{A_{crc}(D)}} = \frac{E_c}{\int_{-t_2}^{t_1}\frac{16}{\pi D^2}dz} = \frac{\pi}{16}\frac{E_c D_{A,Gr} d_w}{l_f} \tag{3.38}$$

The stiffness of the elliptic cylinder, K_{elp}, can be calculated from the ellipse area, $A_{elp}(D)$, as follows.

$$K_{elp} = \frac{E_c}{\int_{-t_2}^{t_1}\frac{dz}{A_{elp}(D)}} = \frac{E_c}{\int_{-t_2}^{t_1}\frac{dz}{\frac{\pi}{8}D_i^2\sqrt{1-\frac{D_i^2}{D^2}}}} = \frac{\pi}{8}\frac{E_c}{l_f}\frac{D_i^2\left(D_{A,Gr} - d_w\right)}{\sqrt{D_{A,Gr}^2 - D_i^2} - \sqrt{d_w^2 - D_i^2}} \tag{3.39}$$

The stiffness of the equivalent solid is calculated by subtracting the stiffness of the elliptic cylinder, K_{elp}, from the stiffness of the truncated cone, K_{crc}. The axial compliance of the equivalent solid, $\delta_{c\,\text{iii}}$, is therefore estimated as follows.

$$\delta_{c\,\text{iii}} = \frac{1}{K_{crc} - K_{elp}} \tag{3.40}$$

The bending compliance of the clamped plates, $\beta_{c\,\text{iii}}$, can be obtained by substituting the axial compliance of the bolt, δ_b, as shown in Eq. (3.1), the axial compliance of the equivalent solid, $\delta_{c\,\text{iii}}$, which can be calculated from Eqs. (3.38) and (3.39), and the gravity centers of the equivalent-solid section, x_{cw} and x_{c0}, of Eqs. (3.36) and (3.37) for Eq. (3.35).

Moreover, the moment threshold M_{cth2} can be obtained by using Eq. (3.33) in the condition when the axial force begins to increase from the preload F_f as follows.

$$M_c = F_f x_{c0} \tag{3.41}$$

The bending compliance of the clamped body in the region of $0 < M_c < M_{cth1}$, $\beta_{c\,\mathrm{i}}$, is equal to that in the linear domain, that is β_c of Eq. (3.26).

Also, the bending compliance of the clamped body in the region of $M_{cth1} < M_c < M_{cth2}$, $\beta_{c\,\mathrm{ii}}$, is thought to be a linear interpolation of (i) and (iii), as shown in Fig. 3.11 as a function of the bending moment M_c is obtained by the following equation.

$$\beta_{c\,\mathrm{ii}} = \beta_{c\,\mathrm{i}} + \frac{M_c - M_{ch1}}{M_{ch2} - M_{ch1}}(\beta_{c\,\mathrm{iii}} - \beta_{c\,\mathrm{i}}) \tag{3.42}$$

From the bending compliance of the bolt and the clamped body determined by the above techniques, the cross-sectional quadratic moment of the equivalent beam, I_{Beq}, and its equivalent diameter d_{eq2} are obtained as follows, with the Young's modulus of the equivalent beam, E_B.

$$\frac{E_B I_{Beq}}{l_f} = \frac{1}{\beta_b} + \frac{1}{\beta_c} = \frac{\beta_b + \beta_c}{\beta_b \beta_c} = \frac{1}{\beta_b \Phi_\beta} \tag{3.43}$$

$$I_{Beq} = \frac{\pi}{64} d_{eq2}^4 = \frac{l_f}{E_b \beta_b \Phi_\beta} \tag{3.44}$$

Here, the bending compliance of the clamped body, β_c, is replaced with the bending compliances $\beta_{c\,\mathrm{ii}}$ and $\beta_{c\,\mathrm{iii}}$ in the nonlinear region.

3.4 Equivalent Stiffness and Load Factor Under Eccentric Fastening and Eccentric External Force Loading

The previous sections have mainly discussed mechanical analysis when axially symmetrical external axial forces are applied to a bolted joint with a hollow cylindrical body fastened in the center. However, in practical applications, there are many cases in which a fastened body is in a prismatic shape, and there can be eccentrically fastened cases or eccentrically loaded cases. In this section, the mechanical analysis of such cases is performed with reference to VDI2230 (1986) [5] and the equations of Yamamoto, Sawa, Maruyama, et al. [6]

3.4.1 Equivalent Stiffness and Force Factor in Linear Region

(A) Compliances of Bolt and Clamped Body, δ_b and δ_c, in the Case of an Eccentrically Fastened Square Prism

Figure 3.16 shows the model of an eccentrically fastened prism to be analyzed. The bolt compliance δ_b does not depend on the shape of the fastened body, and so Eq. (3.1) or (3.2) in Sect. 3.2.1 can be used as is.

The compliance of the clamped square prism, δ_b, is analyzed using a model in which a bolt hole with the diameter of D_i passes through two bricks with the cross-section of $a \times b$ and the length of l_f, as shown in Fig. 3.16b.

To approximate this clamped square prism to an analytical model, the cross-section of the square prism is divided into the following three regions that are sectors connecting the cylinder and cone with the same center of the bolt hole, as shown in Fig. 3.17a [5, 6].

① A prism of a fan shape with an arc EF that is tangent to the short side of the rectangular cross-section, AB, and subtends the inclination angle ψ_a.

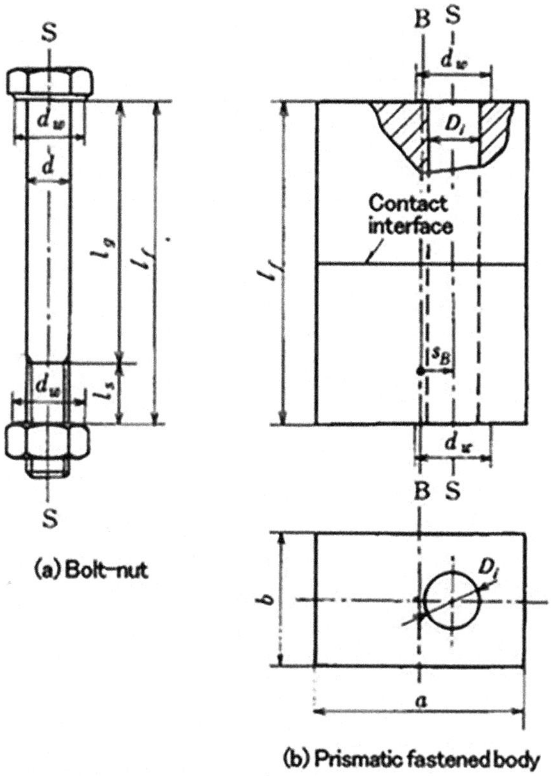

Fig. 3.16 Model of an eccentrically fastened square prism

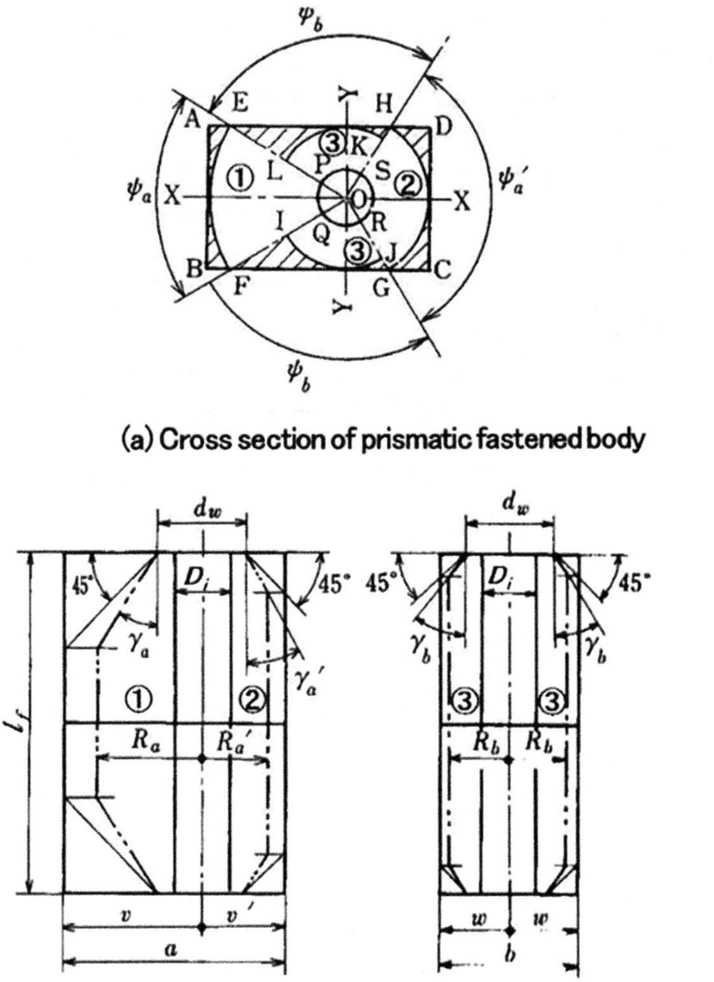

(a) Cross section of prismatic fastened body

(b) Longitudinal section X–X (c) Longitudinal section Y–Y

Fig. 3.17 Analytical model of an eccentrically fastened square prism

② A prism of a fan shape with an arc GH that is tangent to the short side of the rectangular cross-section, CD, and subtends the inclination angle ψ'_a.

③ Two prisms of fan shapes with arcs IJ and KL that are tangent to the long side of the rectangular cross-section, BC and AD, and subtend the inclination angle ψ_b.

On the hatched area shown in Fig. 3.17a, which is excluded from the above three areas, the contact pressure of plates is assumed to be zero. The longitudinal cross-sections of these three divisions are described below and shown in Fig. 3.17b, c.

① Equivalent model comprising sectors of a hollow cylinder with radius R_a and two hollow truncated cones with half apex angles γ_a above and below the cylinder.
② Equivalent model comprising sectors of a hollow cylinder with radius R'_a and two hollow truncated cones with half apex angles γ'_a above and below the cylinder.
③ Two equivalent models' sectors of a hollow cylinder with radius R_b and two hollow truncated cones with half apex angles γ_b above and below the cylinder.

These half apex angles γ_a, γ'_a, and γ_b are obtained as follows [6].

$$\begin{aligned} \tan\gamma_a &= 0.3236 + 0.032\ln\left(\frac{l_f}{2d_w}\right) + 0.153\left(\frac{2\upsilon}{2d_w}\right) \\ \tan\gamma_{a'} &= 0.3236 + 0.032\ln\left(\frac{l_f}{2d_w}\right) + 0.153\left(\frac{2\upsilon'}{2d_w}\right) \\ \tan\gamma_b &= 0.3236 + 0.032\ln\left(\frac{l_f}{2d_w}\right) + 0.153\left(\frac{2w}{2d_w}\right) \end{aligned} \quad (3.45)$$

The radii R_a, R'_a, and R_b are calculated as follows [6].

$$\begin{aligned} R_a &= \frac{d_w + (2\upsilon - d_w)\tan\gamma_a}{2} \\ R_{a'} &= \frac{d_w + (2\upsilon' - d_w)\tan\gamma_{a'}}{2} \\ R_b &= \frac{d_w + (2w - d_w)\tan\gamma_b}{2} \end{aligned} \quad (3.46)$$

The angles ψ_a, ψ'_a, and ψ_b that divide the entire circumference around the bolt hole center O shown in Fig. 3.17a are obtained as follows [6].

$$\begin{aligned} \psi_a &= 2 \times \sin^{-1}\left(\frac{b}{2\upsilon}\right) \\ \psi_{a'} &= 2 \times \sin^{-1}\left(\frac{b}{2\upsilon'}\right) \\ \psi_b &= 2\pi - (\psi_a + \psi_{a'}) \end{aligned} \quad (3.47)$$

The compliances of the prismatic bodies, δ_{ca}, δ'_{ca}, and δ_{cb}, which are shown by the three annular fans, are calculated as follows.

$$
\begin{aligned}
\delta_{ca} &= \frac{2}{\pi E_{c\tan\gamma_a}}\left\{\frac{1}{D_i}\ln\frac{(d_w+D_i)(2R_a-D_i)}{(d_w-D_i)(2R_a+D_i)}+2\frac{d_w+\ln\tan\gamma_a-2R_a}{(2R_a)^2-D_i^2}\right\}\left(\frac{2\pi}{\psi_a}\right)\\
\delta_{ca'} &= \frac{2}{\pi E_{c\tan\gamma_{a'}}}\left\{\frac{1}{D_i}\ln\frac{(d_w+D_i)(2R_{a'}-D_i)}{(d_w-D_i)(2R_{a'}+D_i)}+2\frac{d_w+\ln\tan\gamma_{a'}-2R_{a'}}{(2R_{a'})^2-D_i^2}\right\}\left(\frac{2\pi}{\psi_{a'}}\right)\\
\delta_{cb} &= \frac{2}{\pi E_{c\tan\gamma_b}}\left\{\frac{1}{D_i}\ln\frac{(d_w+D_i)(2R_b-D_i)}{(d_w-D_i)(2R_b+D_i)}+2\frac{d_w+\ln\tan\gamma_b-2R_b}{(2R_b)^2-D_i^2}\right\}\left(\frac{2\pi}{\psi_b}\right)
\end{aligned}
\tag{3.48}
$$

These three compliances are integrated into the compliance of an eccentrically fastened square prism.

$$
\frac{1}{\delta_c}=\frac{1}{\delta_{ca}}+\frac{1}{\delta_{ca'}}+\frac{2}{\delta_{cb}} \tag{3.49}
$$

(B) **Compliances of Bolt and Clamped Body in the Case of an Eccentrically Fastened Square Prism Under Eccentric Load**

(B-1) Compliance for Additional Bolt Force F_b

Figure 3.18 shows the deformation of the clamped body, focusing on the reaction force on the bearing surfaces of the clamped plate in the case of the eccentrically fastened square prism under eccentric external load. In this figure, the initial condition, in which the plates are fastened with the initial clamping force F_f, is indicated by solid lines.

If an eccentric external force W is applied to this bolted joint, internal bolt force F_b is added on the bolt, and the clamped body is loaded with a compressive force, the same as F_b on the line of S–S (on bearing surfaces). Thus, the length between the bearing surfaces of the clamped body shrinks uniformly by ε_F, as shown by the broken lines. At the same time, the bending moment M_s (=$s_B \times F_b$) causes bending deformation and the clamped body shrinks by ε_S on the line of S–S, as shown by the single-dot lines. Total deformation due to the compressive force F_b on the line of S–S is derived as follows.

$$
\varepsilon_{SF}=\varepsilon_F+\varepsilon_S \tag{3.50}
$$

ε_F on the right-hand side of Eq. (3.50) is calculated as follows, using the compliance of an eccentrically fastened square prism, δ_c, in Eq. (3.49).

$$
\varepsilon_F=\delta_c F_b \tag{3.51}
$$

ε_s on the right-hand side of Eq. (3.50) is obtained through analysis of the bending stress of the clamped body, σ_s, acting on the line of S–S as follows.

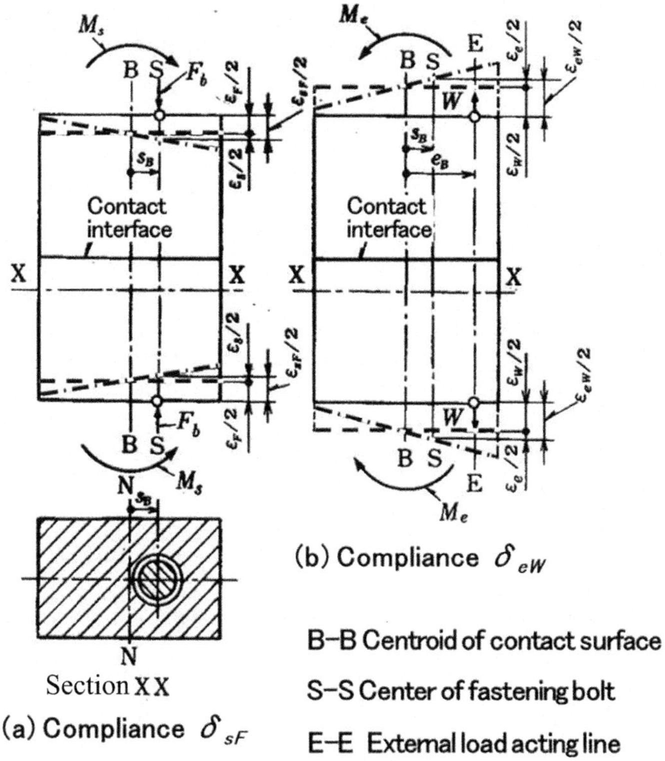

Fig. 3.18 Analytical model of an eccentrically fastened square prism under eccentric loading

$$\sigma_s = \frac{M_e}{I_B / s_B} = \frac{s_B e_B}{I_B} F_b \tag{3.52}$$

$$\varepsilon_s = \frac{\sigma_s l_f}{E_c} = \frac{s_B e_B l_f}{E_c I_B} F_b \tag{3.53}$$

I_B is the cross-sectional second moment of the square prism, which is calculated as follows.

$$I_B = \frac{ba^3}{E_c} \tag{3.54}$$

Substituting Eqs. (3.51) and (3.53) into Eq. (3.50), we can obtain the following equation.

$$\varepsilon_{sF} = \left(\delta_c + \frac{{s_B}^2 l_f}{E_c I_B} \right) F_b \tag{3.55}$$

When the compliance of an eccentrically fastened square prism for the reaction force F_b on the line of S–S is defined as δ_{sF}, it can be derived as follows.

$$\delta_{sF} = \frac{\varepsilon_{sF}}{F_b} = \delta_c + \frac{{s_B}^2 l_f}{E_c I_B} \tag{3.56}$$

(B-2) Compliance for External Force W

Figure 3.18b shows the deformation focusing on the external load W. The initially fastened deformation with the initial clamping force F_f is shown in solid lines, the same as in Fig. 3.18a. The external force W acts on both end faces of the clamped body on the line of E-E, and the length between the two end faces is uniformly elongated by ε_W as shown by the dashed lines. In addition, the clamped body is simultaneously deformed by the bending moment M_e $(= e_B \times W)$ as shown by the single point chain lines. Thus, the total elongation on the line of S- S, ε_{eW}, by the external load W is derived as follows.

$$\varepsilon_{eW} = \varepsilon_W + \varepsilon_e \tag{3.57}$$

ε_W on the right-hand side can be calculated by using the compliance of the eccentrically fastened square prism, δ_c, in Eq. (3.49) as follows.

$$\varepsilon_W = \delta_c W \tag{3.58}$$

ε_e on the right side is obtained through the analysis of the bending stress σ_e acting on the line of S–S as follows.

$$\sigma_e = \frac{M_e}{{}^{I_B}\!/_{s_B}} = \frac{s_B e_B}{I_B} W \tag{3.59}$$

$$\varepsilon_e = \frac{\sigma_e l_f}{E_c} = \frac{s_B e_B l_f}{E_c I_B} W \tag{3.60}$$

Substituting Eqs. (3.58) and (3.60) into Eq. (3.57), we can obtain the following equation.

$$\varepsilon_{eW} = \left(\delta_c + \frac{s_B e_B l_f}{E_c I_B} \right) W \tag{3.61}$$

Finally, the compliance of the eccentrically fastened square prism under eccentric load W is derived as follows.

$$\delta_{eW} = \frac{\varepsilon_{eW}}{W} = \delta_c + \frac{s_B e_B l_f}{E_c I_B} \tag{3.62}$$

(C) Load Factors Φ_e and Φ_{en}, of an Eccentrically Fastened Square Prism Under an Eccentric Load

(C-1) Eccentric External Force *W* Loaded on Both End Faces

In the above model of the eccentrically fastened square prism under the eccentric external load (shown in Fig. 3.18), our mechanical analysis applies the external load is applied on the upper and lower end faces of the clamped body. Here we calculated the load factor Φ_e, which indicates the ratio of the internal force added to the bolt, F_b, and the external load *W*.

Figure 3.19a shows the model for the calculation, which is identical to the one in Fig. 3.18. When an external force load *W* is applied, the distance between the bearing surfaces of bolt head and the nut is extended by $\delta_b \times F_b$, and the distance between the upper and lower end faces of the clamped body is extended by $\delta_{eW} \times W - \delta_{sF} \times F_b$ on the bolt center line. Since both extensions must be equal, the following formula is derived.

$$\delta_b F_b = \delta_{eW} W - \delta_{sF} F_b \tag{3.63}$$

Solving this equation, we can obtain load factor Φ_e as follows.

$$\Phi_e = \frac{F_b}{W} = \frac{\delta_{eW}}{\delta_b + \delta_{sF}} = \frac{\delta_c + \frac{s_B e_B l_f}{E_c I_B}}{\delta_b + \delta_c + \frac{s_B{}^2 l_f}{E_c I_B}} \tag{3.64}$$

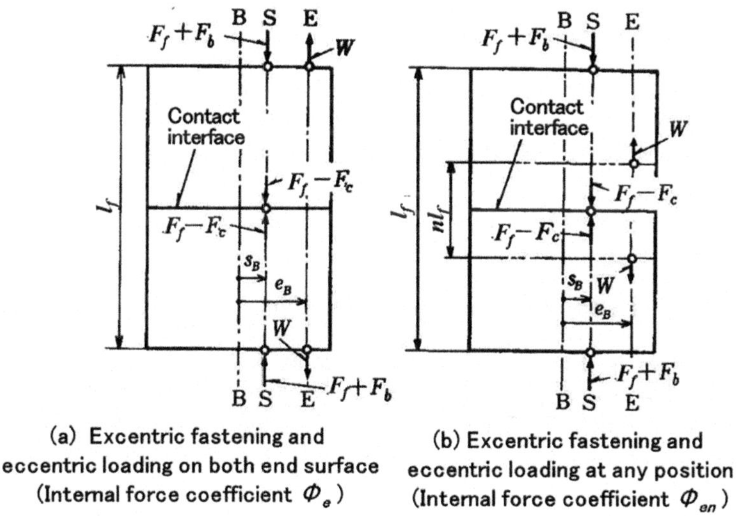

Fig. 3.19 Analytical model of an eccentrically fastened square prism under an eccentric load for calculation of load factor

(C-2) Eccentric External Force W Loaded at Any Position Between Both End Faces

Here, we set the load factor Φ_{en} when the eccentric external load W acts at the length of $n \times l_f$, that is between the upper and lower end faces of the clamped body on the line of E-E, as shown in Fig. 3.19b. In this case, the distance between the bearing surfaces of the bolt is extended by $\delta_b \times F_b$. The part of the clamped body with the length of $(1 - n)l_f$ shrinks by $(1 - n)\, \delta_{sF} \times F_b$ on the line of S–S due to the compressive internal force F_b, and the part with the length of $n \times l_f$, is extended by $n \times \delta_{eW} \times W - n \times \delta_{sF} \times F_b$.

Since the deformations of the bolt and the clamped body must be equal, the following equation is derived.

$$\delta_b F_b = -(1 - n)\delta_{sF} F_b + n\delta_{eW} W - n\delta_{sF} F_b \tag{3.65}$$

Solving this equation, we can obtain the internal coefficient Φ_{en} as follows.

$$\Phi_{en} = \frac{F_b}{W} = \frac{n\delta_{eW}}{\delta_b + \delta_{sF}} = n\frac{\delta_c + \frac{s_B e_B l_f}{E_c I_B}}{\delta_b + \delta_c + \frac{{s_B}^2 l_f}{E_c I_B}} = n\Phi_e \tag{3.66}$$

(D) **Additional Stress on Bolt, σ_{F+M}, Due to Eccentric External Load W**

It is necessary to calculate the additional applied stress on the bolt, σ_{F+M}, due to external loading W, in order to evaluate the strength of bolts in the case of an eccentrically fastened square prism under eccentric load. Specifically, the additional stress on the bolt, σ_{F+M}, is classified into an axial component (tensile stress, σ_{bF}) and a bending component (bending stress, σ_{bM}) as follows.

(D-1) Axial Additional Stress Component (Tensile Stress, σ_{bF})

As shown in Eqs. (3.62) and (3.66), the additional force on the bolt, F_b, is obtained by $F_b = (\Phi_e \text{ or } \Phi_{en}) \times W$. The bolt is subjected to additional stress σ_{bF} at the cross-sectional area of the minor diameter, A_3, as follows.

$$\sigma_{bF} = \frac{F_b}{A_3} = \frac{4\Phi_{en} W}{\pi {d_3}^2} \tag{3.67}$$

(D-2) Additional Stress Component due to Bending Moment (Bending Stress, σ_{bM})

As shown in Fig. 3.18, when an eccentric external force W is applied to the eccentrically fastened square prism, a bending moment M_s $(= - s_B \times F_b)$ is applied at the bolt center line (bolt seated position), and bending moment M_W $(=e_B \times W)$ is applied at the loading line. Then, the following total moment M_B is added to the bolted joint.

$$M_B = e_B W - s_B F_b \tag{3.68}$$

The additional bending moment on the bolt, M_b, due to this additional bending moment on the bolted joints, M_B, is derived by using the bending compliances of the bolt and the clamped body, β_b and β_c, as follows.

$$M_b = \frac{\beta_c M_B}{\beta_b + \beta_c} = \frac{\beta_c (e_B W - s_B F_b)}{\beta_b + \beta_c} = \frac{\beta_c W (e_B - \Phi_{en} s_B)}{\beta_b + \beta_c} \tag{3.69}$$

Here, we can use Eq. (3.23) in the previous section for the bending compliance of the bolt, β_b. The compliance of the clamped body, β_c, can be calculated by using the secondary moment of the fastened square prism itself, $I_c\left(= I_B - s_B^2 \times A_i = \left(b \times a^3/12\right) - s_B^2 \times \pi D_i^2/4\right)$, instead of the hollow cylindrical equivalent clamped body of Eq. (3.24) in the previous section, as follows.

$$\beta_c = \frac{l_f}{E_c I_c} \tag{3.70}$$

The bending stress σ_{bM} on the bolt at the minor diameter due to this additional bending moment M_b is obtained by using the section modulus of the bolt at the minor diameter, Z_3, as follows.

$$\sigma_{bM} = \frac{M_b}{Z_3} = \frac{32 M_b}{\pi {d_3}^3} = \frac{32 \beta_c W (e_B - \Phi_{en} s_B)}{\pi {d_3}^3 (\beta_b + \beta_c)} \tag{3.71}$$

If the load factor of bending moment, Φ_β, is substituted into $\beta_c/(\beta_b + \beta_c)$, the following equation can be obtained.

$$\sigma_{bM} = \frac{32 \Phi_\beta W (e_B - \Phi_{en} s_B)}{\pi {d_3}^3} \tag{3.72}$$

(D-3) Total Additional Stress Component (Axial Stress σ_{bF} + Bending Stress σ_{bM})

Total additional stress on the bolt, σ_{F+M}, is given as the sum of the axial stress σ_{bF} in Eq. (3.67) and the bending stress σ_{bM} in Eq. (3.72) as follows.

$$\sigma_{F+M} = \frac{4W}{\pi {d_3}^2} \left(\Phi_{en} + \Phi_\beta \frac{(e_B - \Phi_{en} s_B)}{d_3} \right) \tag{3.73}$$

3.4.2 *Equivalent Stiffness and Load Factor in Nonlinear Region*

As mentioned in Sect. 3.2, it is preferable that an axial load be smaller than the force when the contact surfaces of the clamped plates separate. However, in the case of bolted joints that are subject to bending loads or eccentric loads as described in this section, the contact surfaces of the clamped plates partially separate, and the clamped plates and the bolt deform nonlinearly. Partial separation in such a nonlinear region would often be acceptable to use, if the reliability for preventing such as container leakage, fatigue failure, or loosening is guaranteed.

Therefore, this section outlines the one-side and full separation limits and the mechanical behavior in the nonlinear region after separation under eccentrical loading.

(A) **One-side Separation Limit**

As the eccentric load W gradually increases, the contact surfaces of the prismatic joints begin to separate from one side. After this one-side separation, the deformation state changes from the linear region to the nonlinear region.

In Fig. 3.20, the one-side separation begins when the surface pressure on the contact surface, σ_c, reaches zero at $x = u$. The surface pressure distribution on the contact surface is calculated as follows.

$$\sigma_c(x) = \frac{F_f - (1 - \Phi_{en})W}{A_B} + \frac{e_B W - s_B(F_f + \Phi_{en} W)}{I_B} x \tag{3.74}$$

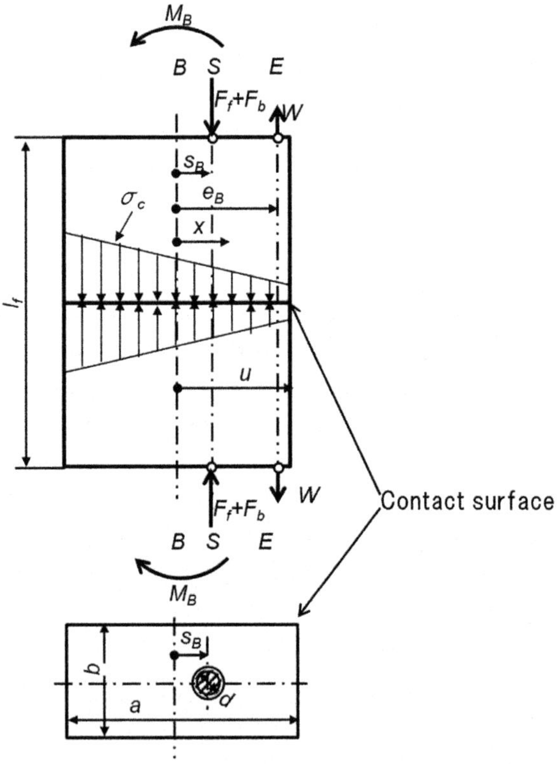

Fig. 3.20 Analytical model of an eccentrically fastened square prism under an eccentric load for calculation of one-side separation limit

When this equation is solved for W under the conditions $A_B = a \times b, I_B = b \times a^3/12$ and $\sigma_c\ (x = u) = 0$, the following equation is obtained.

$$W = \frac{F_f}{\frac{\left(1+\frac{e_B u A_B}{I_B}\right)}{\left(1+\frac{s_B u A_B}{I_B}\right)} - \Phi_{en}} = \frac{F_f}{\frac{\left(1+\frac{6e_B}{a}\right)}{\left(1+\frac{6s_B}{a}\right)} - \Phi_{en}} = \frac{F_f}{\frac{a+6e_B}{a+6s_B} - \Phi_{en}} \tag{3.75}$$

(B) **Nonlinear Behavior After one-side Separation Limit**

The bolted joint model in the progress of separation is modified by removing the length of the separation area, Δ, from the clamped part, as shown by the hatched part in Fig. 3.21. We can trace the nonlinear behavior of the bolted joint during separation by this removal process continuously. In other words, I_B, A_B, u, e_B, and s_B in the left equation of Eq. (3.75) are replaced with $I_{B\Delta} = b \times (a - \Delta)^3/12, A_{B\Delta} = (a - \Delta) \times b, u_\Delta = u - \Delta/2 = (a - \Delta)/2, e_{B\Delta} = e_B + \Delta/2$, and $s_{B\Delta} = s_B + \Delta/2$, respectively, and Φ_{en} in Eq. (3.75) is replaced with following $\Phi_{en\Delta}$.

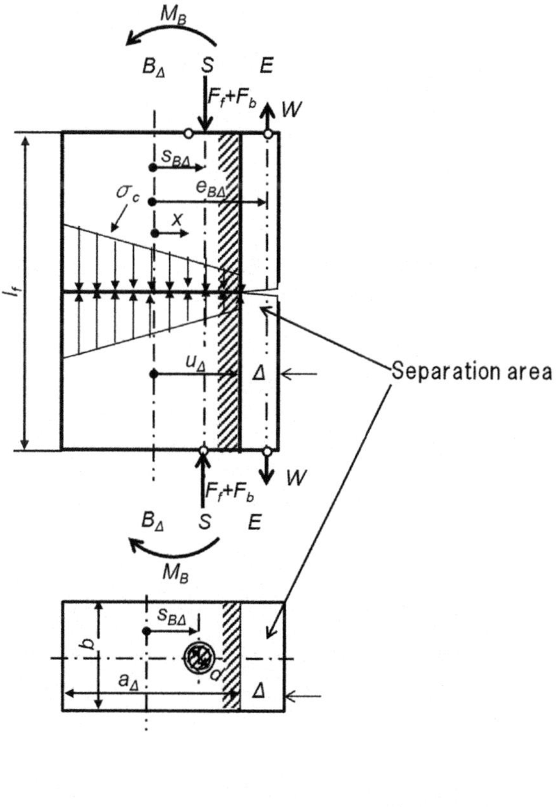

Fig. 3.21 Analytical model of an eccentrically fastened square prism under an eccentric load for calculation of nonlinear behavior after one-side separation limit

$$\Phi_{en\Delta} = \frac{F_b}{W} = n\frac{\delta_{c\Delta} + \frac{s_{B\Delta}e_{B\Delta}l_f}{E_c I_{B\Delta}}}{\delta_b + \delta_{c\Delta} + \frac{s_{B\Delta}{}^2 l_f}{E_c I_{B\Delta}}} = n\Phi_{e\Delta} \tag{3.76}$$

In this equation, $\delta_{c\Delta}$ is obtained by replacing only the cylinder and cone structure in ①–③ of the model in Fig. 3.17 as follows.

① Substitute $\nu_\Delta = \nu - \Delta$ for ν in Eq. (3.45) to obtain $\gamma_{a\Delta}$, substitute $\gamma_{a\Delta}$ in Eq. (3.46) for γ_a to obtain $R_{a\Delta}$, substitute $R_{a\Delta}$ in Eq. (3.47) for R_a to obtain $\psi_{a\Delta}$, and substitute $\psi_{a\Delta}$ in Eq. (3.48) for ψ_a to obtain $\delta_{ca\Delta}$.

②, ③ Obtain $\delta'_{ca\Delta}$ and $\delta_{cb\Delta}$ by using the same procedure.

We can then obtain the equivalent compliance of an eccentrically fastened square prism under eccentric load, $\delta_{c\Delta}$, as follows.

$$\frac{1}{\delta_{c\Delta}} = \frac{1}{\delta_{ca\Delta}} + \frac{1}{\delta'_{ca\Delta}} + \frac{2}{\delta_{cb\Delta}} \tag{3.77}$$

Finally, we can obtain the one-side separation limit W_Δ at the length of the one-side separation area, Δ, as follows.

$$\mathrm{W}_\Delta = \frac{F_f}{\frac{\left(1+\frac{e_{B\Delta}u_\Delta A_{B\Delta}}{I_{B\Delta}}\right)}{\left(1+\frac{s_{B\Delta}u_\Delta A_{B\Delta}}{I_{B\Delta}}\right)} - \Phi_{en\Delta}} \tag{3.78}$$

The additional axial force on the bolt is then obtained from the following equation.

$$F_{b\Delta} = \Phi_{en\Delta} \times W_\Delta \tag{3.79}$$

(C) Additional Stress on Bolt, $\sigma_{F+M\Delta}$, due to Eccentric External Load W over one-side Separation Limit

In 3.4.1(D), the additional stress σ_{F+M} at the minor diameter was obtained in the linear region. Here, it is similarly obtained as the sum of the bolt additional axial force component (axial stress) $\sigma_{bF\Delta}$ and the bending moment component (bending stress) $\sigma_{bM\Delta}$ under the eccentric external load over the one-side separation limit.

The axial force component $\sigma_{bF\Delta}$ is obtained as follows using the additional axial force $F_{b\Delta} = \Phi_{en\Delta} \times W_\Delta$ after one-side separation.

$$\sigma_{bF\Delta} = \frac{F_{b\Delta}}{A_3} = \frac{4\Phi_{en\Delta}W_\Delta}{\pi d_3{}^2} \tag{3.80}$$

The total additional bending moment of the bolt, M_Δ, is obtained as the sum of the bending moment $M_{s\Delta}$ $(= - s_{B\Delta} \times F_{b\Delta})$, which acts on the clamping position (bearing surface), and the bending moment $M_{W\Delta}$ $(=e_{B\Delta} \times W_\Delta)$, which acts on the external force loading position, as follows.

$$M_\Delta = M_{s\Delta} + M_{w\Delta} = e_{B\Delta}W_\Delta - s_{B\Delta}F_{b\Delta} = W_\Delta(e_{b\Delta} - s_{B\Delta}\Phi_{en}) \tag{3.81}$$

The bending moment component (bending stress) $\sigma_{bM\Delta}$ is derived as follows.

$$\sigma_{bM\Delta} = \frac{M_{b\Delta}}{Z_3} = \frac{\beta_{c\Delta}}{(\beta_b + \beta_{c\Delta})}\frac{32M_\Delta}{\pi d_3^3} = \frac{32\beta_{c\Delta}W_\Delta(e_{B\Delta} - s_{B\Delta}\Phi_{en\Delta})}{\pi d_3^3(\beta_b + \beta_{c\Delta})} \tag{3.82}$$

$\beta_{c\Delta}$ is the bending compliance of the clamped prism under the eccentric external load over the one-side separation limit and is obtained as follows.

$$\beta_{c\Delta} = \frac{l_f}{E_c I_{c\Delta}} = \frac{l_f}{E_c\left(\frac{a(b-\Delta)^3}{12} - s_{B\Delta}{}^2\frac{\pi D_i^2}{4}\right)} \tag{3.83}$$

The total additional stress $\sigma_{F+M\Delta}$, applied on the bolt section at the minor diameter under the eccentric external load over the one-side separation limit, is given as the sum of the axial stress $\sigma_{bF\Delta}$ in Eq. (3.80) and the bending stress $\sigma_{bM\Delta}$ in Eq. (3.82) as follows.

$$\sigma_{F+M\Delta} = \sigma_{bF\Delta} + \sigma_{bM\Delta} = 4W_\Delta\left\{\frac{\Phi_{en\Delta}}{\pi d_3^2} + \frac{8\beta_{c\Delta}(e_{B\Delta} - \Phi_{en\Delta}s_{B\Delta})}{\pi d_3^3(\beta_b + \beta_{c\Delta})}\right\} \tag{3.84}$$

Combined with the linear region result in 3.4.1(D), the fatigue strength of an eccentrically fastened square prism under an eccentric load is shown in Fig. 3.22. In order to ensure the safety of a bolted joints, these calculated additional bolt stresses at the minor diameter, σ_{F+M} and $\sigma_{F+M\Delta}$, must be lower than the allowable stress σ_{al} calculated from the bolt fatigue strength σ_{wk} (shown in 1.3(B), Table 1.3) and safety factor S_f, as follows.

$$\sigma_{F+M} or \sigma_{F+M\Delta} \leq \sigma_{al} = \frac{2\sigma_{wk}}{S_f} \tag{3.85}$$

(D) Leak Prevention Design for Pressure Vessel Fastened by Bolted Joints

When the flange or lid of a pressure vessel clamped with bolted joints is subjected to pressure load, each of the bolted joints is applied to an eccentric load, as shown in Fig. 3.23. If the one-side separation area spreads to the bolt hole, there is a risk of the internal fluid leakage. The relation between external load W_Δ and the length of separation area, Δ, can be calculated by Eq. (3.78), as shown in Fig. 3.24. Thus, to prevent the internal fluid leakage from the bolted flange joint of a pressure vessel, the length of separation area, Δ, should be smaller than the allowable length of the separation area, Δ_{cr}.

(E) Full Separation Limit

When the load W increases over the one-side separation limit, the length of the separation area spreads. Finally, the force between the clamped plates is transmitted only at the opposite end of the loading side. From the force balancing condition in

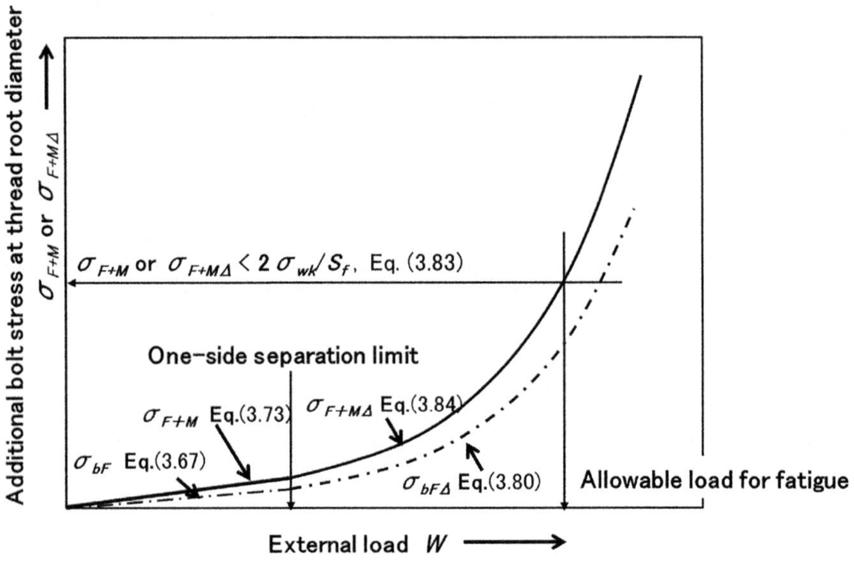

Fig. 3.22 Fatigue strength evaluation of an eccentrically fastened square prism under an eccentric load

Fig. 3.23 Flange pressure vessel clamped with a bolted joint under an eccentric load

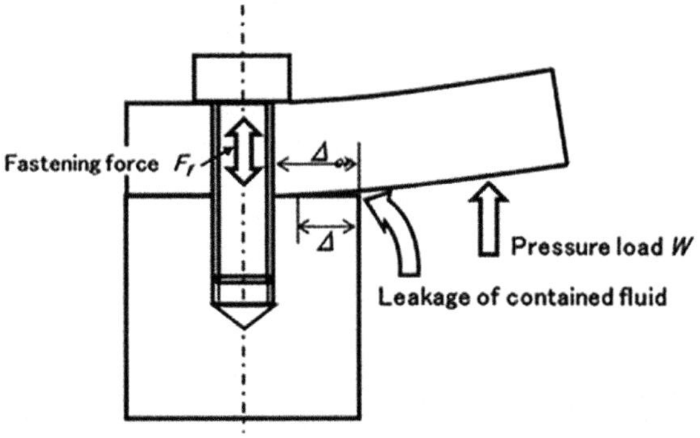

this extreme state, the relation between the total axial force of the bolt, $F_f + F_b$, and the external load W can be shown by the following equation.

$$F_f + F_b = \frac{\frac{a}{2} + e_B}{\frac{a}{2} + s_B} W \tag{3.86}$$

The relation between the axial force of the bolt and the external load is schematically shown in Fig. 3.25 by summarizing the previous part and this part. In the full-surface adherence region, the bolt axial force increases linearly with the load W from the initial clamping force F_f. When the load comes over the one-side separation

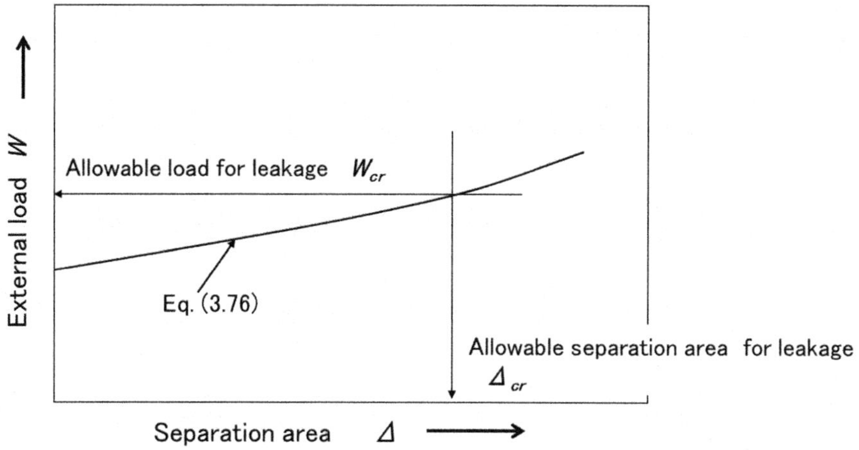

Fig. 3.24 Relation between external load W_Δ and the length of separation area Δ, and estimation of allowable load for preventing leakage

limit shown by the single-dotted line, the contact surfaces of the clamped plates gradually separate and the bolt force increases nonlinearly. Finally, the relation between the axial force of the bolt and the external load asymptotically approaching the full-surface separation limit line are as shown by the double-dotted line.

The analytical and experimental results of linear and nonlinear mechanical behavior under these eccentric external forces in specific dimensions and shapes are described in detail in Sect. 11.1, "Plastic region fastening method".

3.5 Finite Element Modeling Method for Bolted Joints

When the 3D finite element analysis of a structure with bolted joints is carried out, all of the structure and bolts can be directly modeled in a large number of elements while taking contact conditions into consideration. However, extremely high computational ability is needed for such large FE analysis, and it is too difficult to directly apply design guidelines such as VDI 2230 Part (1) to the FE result. If the bolted joints are modeled as equivalent beams, which have equivalent axial, bending and torsional stiffness to bolted joints, (as described above), the axial load and bending moment on each bolted joint can be analyzed from the result of equivalent beams, and the strength of each bolt can be evaluated by using design guidelines. Moreover, when we need to analyze a large structure with a huge number of bolted joints, it would be better to use shell elements with bolt-equivalent beam elements, and automatic meshing tools are effective. In addition, considering that the number of nodes should be reduced as much as possible, it is easier to handle the model by omitting bolt holes and dividing it into quadrilateral shapes. Figure 3.26 shows the simplified model, in

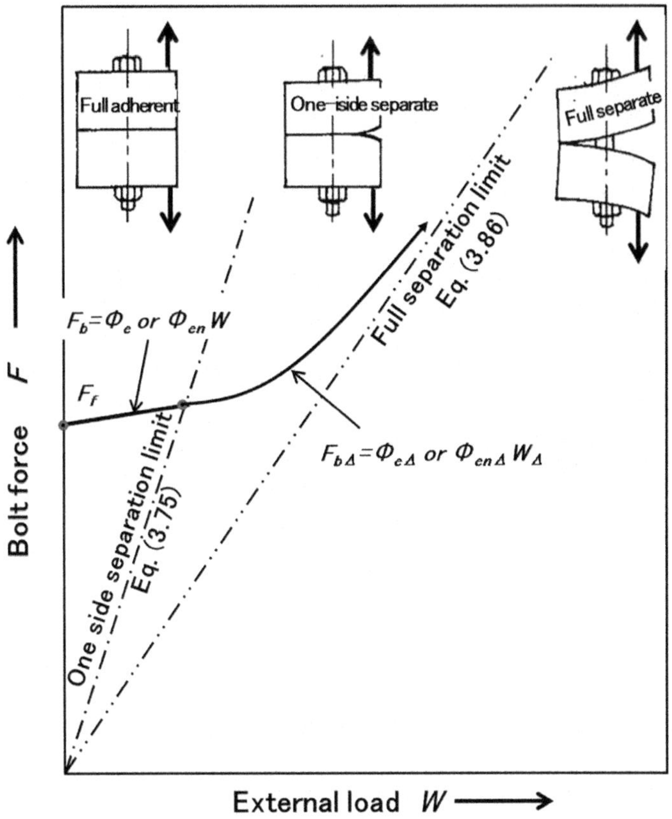

Fig. 3.25 Schematic view of the relation between bolt axial force and external load for an eccentrically fastened square prism under an eccentric load

which the plate is modeled with quadrilateral elements, the holes are omitted, and the nodes of the plates at the hole center are connected with an equivalent beam element.

The axial stiffness of the equivalent beam is determined to be the axial stiffness of the bolted joint. When the plate is modeled with a shell element, as shown in Fig. 3.27, the length of the equivalent beam representing the bolted joint is a half of the clamping length l_f, which is equal to the sum of the plate thicknesses. If the equivalent beam is subjected to an axial load W_a, the elongation of the equivalent beam is equal to one half of the elongation of the bolted joint between the bearing surfaces, ε'', as shown in Fig. 3.8a. From Eqs. (3.18) and (3.22), the following equation can be obtained.

$$\frac{1}{2}\varepsilon'' = \frac{1}{2}F\prime\prime_b\delta_b = \frac{1}{2}\Phi_n\delta_b W_a \tag{3.87}$$

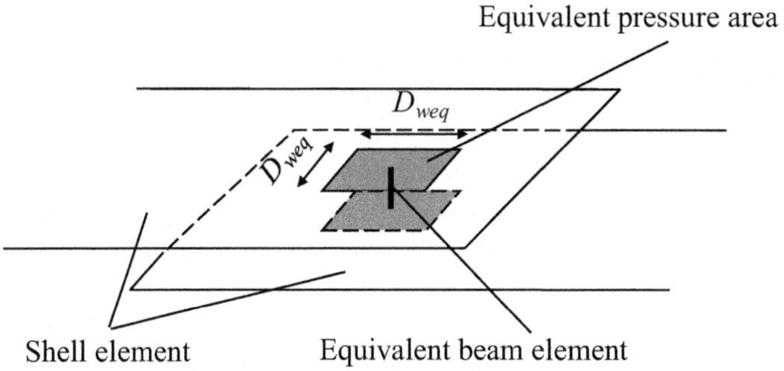

Fig. 3.26 Equivalent model of bolted joints with shell and beam elements

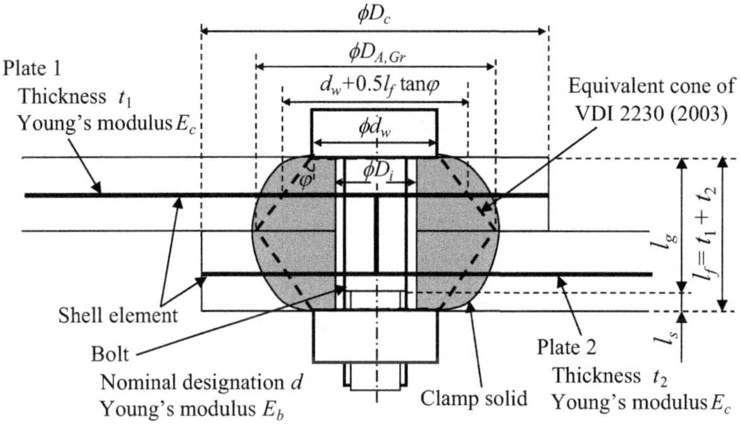

Fig. 3.27 Structure and model of bolted joints

If the Young's modulus and the cross-sectional area of the equivalent beam are E_b and A_{beq}, respectively, the elongation of the equivalent beam of $1/2l_f$ length under an axial load W_a is expressed as follows.

$$\frac{1}{2}\varepsilon'' = \frac{W_a}{E_b A_{beq}} \frac{1}{2} l_f \tag{3.88}$$

Defining the equivalent diameter d_{eq1} that gives the cross-section A_{beq} for convenience, the cross-section of the equivalent beam, A_{beq}, can be expressed from Eqs. (3.87) and (3.88) as follows.

$$\frac{\pi}{4} d_{eq1}^2 = A_{beq} = \frac{l_f}{\Phi_n \delta_b E_b} \tag{3.89}$$

The bending stiffness is considered in the same way. Let I_{beq} be the sectional second moment of the equivalent beam and define an equivalent diameter d_{eq2} that gives the sectional second moment I_{beq} for convenience. The bolt bending compliance β_b in Eq. (3.23) and the load factor $\Phi_{\beta n}$, which takes into consideration the moment loading position, in Eq. (3.30), I_{beq} is defined as follows.

$$\frac{\pi}{64} d_{eq2}^4 = I_{beq} = \frac{l_f}{\Phi_{\beta n} \beta_b E_b} \tag{3.90}$$

Torsional stiffness is considered in the same way as bending stiffness. The polar second moment, which represents the torsional stiffness of the cylinder, is defined as twice the second moment of the cylinder. Thus, the polar second moment of the equivalent beam element is equal to twice the second moment of the equivalent beam element. I_{beq}. Therefore, the equivalent diameter of the torsional stiffness is equal to the equivalent diameter of the bending stiffness, d_{eq2}, and the load factor for the torque is also equal to the load factor for the bending moment, $\Phi_{\beta n}$.

The stiffness of the equivalent beam of the bolted joint is shown in Table 3.1. The plates and bolts were made of steel materials with equal Young's modulus, and the hole diameters were class 2 of JIS B1001. The bolts were full-threaded bolts, and two plates with thicknesses of 3.2 mm and 4.5 mm (hereinafter referred to as t3.2 and t4.5) were fastened to each other. The equivalent beam diameter is larger than the nominal diameter. The thicker the plate is, the larger the equivalent beam diameter is.

In finite element analysis, if a shell element and a beam element are simply connected at the same node, the joint between the shell element and the beam element is constrained to be locally deformed. However, such local deformation does not occur in practical bolted joints because the bolt and equivalent compressive body of the plates are deformed as a single unit. Therefore, the deformation of the shell element within the equivalent pressure area must be constrained to the node of the deformation of the equivalent beam, as shown in Fig. 3.28.

The diameter of the equivalent compressive body on the shell elements should be $d_w + 0.5l_f$, as shell elements are located on the mid-face of the plates (as shown in Fig. 3.27). We defined the region within the diameter of $d_w + 0.5l_f$ as the equivalent pressure area. Then, we constrained the shell elements nodes within the equivalent pressure area to the tip node of the equivalent beam node as follows.

Figure 3.28 shows the deformed state of the equivalent pressure area. For the sake of developing a model preferable for use with an automatic meshing tool, we modeled the plates with quadrilateral elements and set the equivalent pressure area to a square with a length of $D_{weq} = d_w + 0.5l_f$. The translational displacements of the shell nodes within the equivalent pressure area, nodes 1–8, were constrained by the translational and rotational displacements of the tip node of the equivalent beam, node 0. The coordinates of node i are expressed as (x_i, y_i, z_i). The translational displacements of node i are u_{xi}, u_{yi}, and u_{zi}; and its rotational displacements are r_{xi}, r_{yi}, and r_{zi}. We determined that the z-direction translational displacement of node i, u_{zi}, within the equivalent pressure area is calculated with the degrees of freedom and coordinates of the tip node of the beam element, node 0, by the following equation.

Table 3.1 Stiffness of equivalent beam and equivalent pressure area

			M6	M8	M10	M12	M16	M20	M24
E_b		[MPa]	206000	206000	206000	206000	206000	206000	206000
E_c		[MPa]	206000	206000	206000	206000	206000	206000	206000
d_3		[mm]	4.197	6.647	8.376	10.106	13.835	17.294	20.752
d_{w}		[mm]	8.88	11.63	14.63	16.63	22.00	27.70	33.25
D_i		[mm]	6.60	9.00	11.00	13.50	17.50	22.00	26.00
D_c		[mm]	70.00	70.00	70.00	70.00	70.00	70.00	70.00
13.2 × 2	$\tan\varphi$		0.6062	0.5563	0.5139	0.4902	0.4384	0.3958	0.3620
	$D_{A,Gr}$	[mm]	12.76	15.19	17.92	19.77	24.81	30.23	35.57
	δ_b	× 10"9 [mm/N]	4225	2150	1561	1214	812.7	617.1	495.9
	δ_c	$\times 10^{-9}$ [mm/N]	594.9	430.1	286.6	278.0	167.6	112.9	78.46
	δ_{ct}	$\times 10^{-9}$ [mm/N]	165.8	132.1	98.85	94.87	63.99	45.99	33.58
	n		0.2788	0.3071	0.3449	0.3413	0.3819	0.4075	0.4280
	Φ		0.1234	0.1667	0.1552	0.1863	0.1709	0.1546	0.1366
	β_b	$\times 10^{-9}$ [rad/Nmm]	3408	700.7	318.0	169.1	60.60	29.26	16.25
	β_c	$\times 10^{-9}$ [rad/Nmm]	60.88	26.46	11.908	8.579	3.124	1.3603	0.6737
	β_{ct}	$\times 10^{-9}$ [rad/Nmm]	12.86	6.779	3.578	2.649	1.111	0.5264	0.2768
	n_β		0.2112	0.2562	0.3004	0.3088	0.3556	0.3870	0.4109
	Φ_β		0.01755	0.03639	0.03609	0.04829	0.04902	0.04443	0.03981
	A_{beq}	[mm^2]	213.7	282.3	372.0	402.5	585.6	799.0	1071.6
	I_{beq}	$\times 10^3$ [mm^4]	2.459	4.756	9.009	12.32	29.41	61.77	116.9
	d_{eq1}	[mm]	16.50	18.96	21.76	22.64	27.31	31.90	36.94
	d_{eq2}	[mm]	14.96	17.64	20.70	22.38	27.82	33.49	39.28
	D_{weq}	[mm]	10.82	13.41	16.27	18.20	23.40	28.97	34.41
14.5 × 2	$\tan\varphi$		0.6171	0.5672	0.5248	0.5011	0.4493	0.4067	0.3729
	$D_{A,Gr}$	[mm]	14.43	16.74	19.35	21.14	26.04	31.36	36.61
	δ_b	$\times 10^{-9}$ [mm/N]	5138	2514	1790	1371	896.6	670.8	533.2
	δ_c	$\times 10^{-9}$ [mm/N]	703.9	519.2	355.6	343.9	213.7	146.8	103.7
	δ_{ct}	$\times 10^{-9}$ [mm/N]	168.8	139.7	109.7	105.1	74.76	55.69	41.89
	n		0.2398	0.2691	0.3085	0.3056	0.3499	0.3794	0.4039

(continued)

Table 3.1 (continued)

		M6	M8	M10	M12	M16	M20	M24
Φ		0.1205	0.1712	0.1658	0.2005	0.1925	0.1795	0.1628
β_b	$\times 10^{-9}$ [rad/Nmm]	4237	832.4	370.3	193.7	67.62	32.13	17.64
β_c	$\times 10^{-9}$ [rad/Nmm]	66.86	29.97	14.02	10.110	3.846	1.723	0.8726
β_{ct}	$\times 10^{-9}$ [rad/Nmm]	10.72	6.192	3.542	2.673	1.215	0.6072	0.3324
n_β		0.1603	0.2066	0.2526	0.2644	0.3159	0.3523	0.3810
Φ_β		0.01554	0.03475	0.03649	0.04960	0.05382	0.05090	0.04715
A_{beq}	[mm^2]	294.3	377.3	477.4	519.9	723.6	956.2	1245.9
I_{beq}	$\times 10^3$ [mm^4]	4.140	7.310	12.80	17.20	38.00	75.82	137.9
d_{eq1}	[mm]	19.36	21.92	24.65	25.73	30.35	34.89	39.83
d_{eq2}	[mm]	17.04	19.64	22.60	24.33	29.66	35.25	40.94
D_{weq}	[mm]	11.66	14.18	16.99	18.88	24.02	29.53	34.93

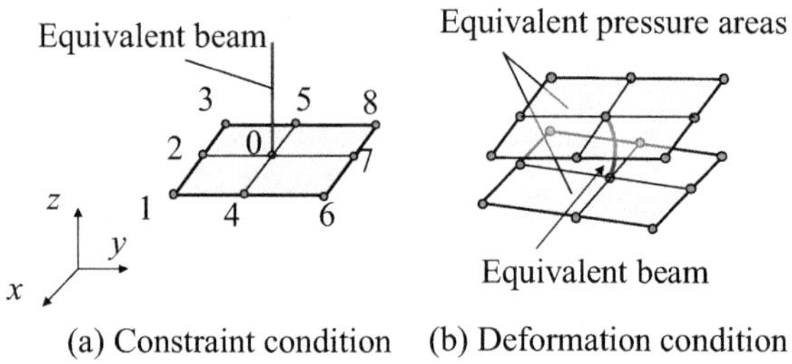

Fig. 3.28 Constraint condition of equivalent pressure areas

$$u_{zi} - u_{z0} = (y_i - y_0)r_{x0} - (x_i - x_0)r_{y0} \tag{3.91}$$

We established similar constraint equations for the translational displacements u_{xi} and u_{yi}. These constraint equations can be simply defined by using the CERIG command in Ansys Mechanical™[1] and the RBE2 element in MSC/NASTRAN®[2]. In Table 3.1, the lengths of the equivalent pressure area, D_{weq}, the diameter of the equivalent beam for axial load in Eq. (3.89), d_{eq1}, and the diameter of the equivalent beam for bending moment in Eq. (3.90), d_{eq2}, are shown as well.

[1] Ansys Mechanical™ is a trademark of Ansys, Inc.

[2] MSC/NASTRAN is a trademark or registered trademark of Hexagon AB and/or its subsidiaries in the United States and/or other countries.

3.6 Application to Eigenfrequency Analysis and Validation with Experiment Using Bolted Joint Specimens

To confirm the validity of the developed analysis model, we ran tests with bolted joint elements and compared the obtained results with the analytical results. We measured the eigenvalues of bolted joint specimens by means of an impact test, with which we could evaluate the stiffness easily. The specimen dimensions are shown in Fig. 3.29. If the plates flexibly deform, the eigenmodes of the plates become quite conspicuous and the stiffness of the bolted joints is almost impossible to evaluate. Therefore, we bent plates made of SS400 into U shapes to increase their stiffnesses and assembled two U-shaped plates with a single-bolted joint. Further, t3.2 and t4.5 specimens were prepared to evaluate the effect of the plate thickness. Class 4.8 bolts of ISO strength class were used, and the nominal diameters of bolts ensure M6 to M24 were tested. The hole diameters were Class 2 of JIS B1001. To subject sufficient clamping force, bolts were tightened by 0.7 times the yield-tightening torques, T_{fy}, which were determined with the bearing friction coefficient of $\mu_w = 0.15$ in JIS B1083.

Figure 3.30 shows the analysis model of the specimen shown in Fig. 3.29, which was built according to the modeling method shown in Figs. 3.26 through 3.28. The diameters of the equivalent beam, d_{eq1} and d_{eq2}, and the lengths of the equivalent pressure area, D_{weq}, were obtained from Table 3.1. Figure 3.31 shows the analytical eigenmodes of a t4.5 specimen tightened by M10. In Mode 1, U plates are twisted with the center of the bolt fixed. Mode 2 expresses a cantilever with the bolt center fixed. The order of Modes 1 and 2 varies from first to second respectively with the thickness of the plates and the nominal diameter of the bolt. In this test, we evaluated Mode 1, because this mode can be easily distinguished. We attached three accelerometers (points 1–3, as shown in Fig. 3.29) and measured accelerations at these three points by subjecting impact force on the other plates.

We used 353B16 accelerometers made by PCB Piezotronics, Inc., to measure the accelerations. The specimens were hit with a 086C03 impact hammer made by the same company. The measured data were processed with an OR2514 data logger made by OROS, and the accelerance and the phase were calculated by dividing the

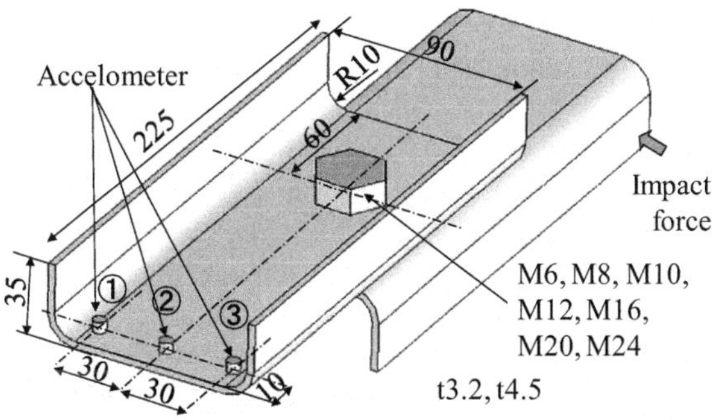

Fig. 3.29 Bolted joint specimens and testing method (unit mm)

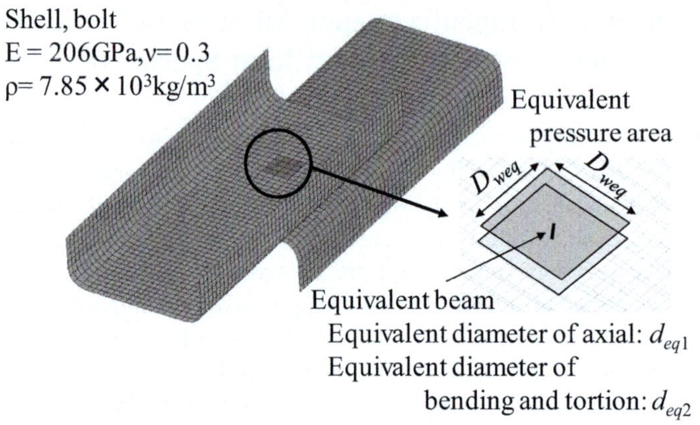

Fig. 3.30 Analysis model of test piece

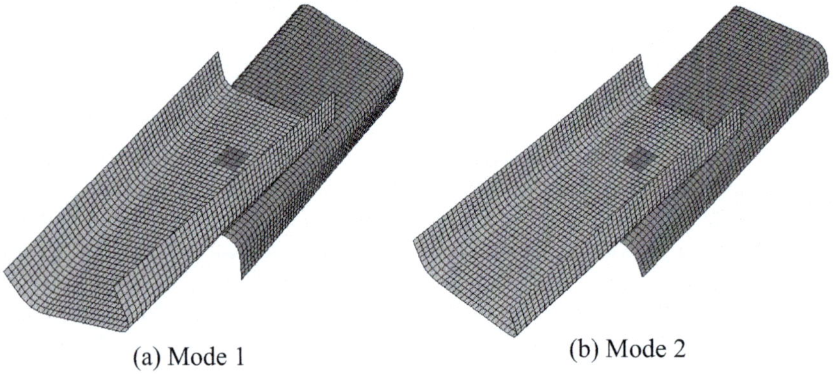

Fig. 3.31 Eigenmodes of analysis result

accelerations by the impact load. In the eigenmode of Mode 1, the phases of the accelerations of points 1 and 3 at both sides of the specimens shall be opposite, and the amplitude of the acceleration of point 2 shall be much smaller than the amplitude of the accelerators of points 1 and 3 on both sides. We evaluated the eigenvalue of Mode 1 by comparing the amplitudes and phases of the three accelerators.

Figures 3.32 through 3.37 show the test results of a t4.5 specimen with an M12 bolt. Figures 3.32, 3.34, and 3.36 show the accelerances at three points, and Figures. 3.33, 3.35, and 3.37 show the phases. The accelerances of points 1 and 3 have peaks at 350 Hz in Figs. 3.32 and 3.36. The phase at 350Hz of point 1 is opposite to that of point 3 as shown in Figs. 3.33 and 3.37. The phases of points 1 and 3 exceed +90° and –90° at 350 Hz, respectively. The accelerance of point 2 is much smaller than the accelerances of points 1 and 3 at 350 Hz as shown in Fig. 3.34. Thus, we were able to evaluate the eigenvalue of Mode 1 of this specimen as 350 Hz.

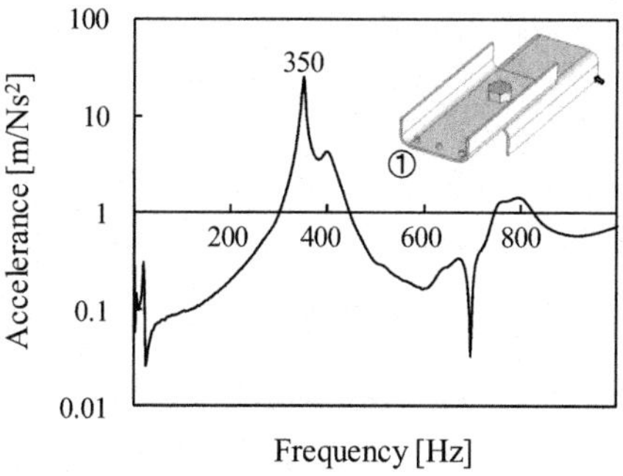

Fig. 3.32 Experiment results of accelerance for point 1 (t4.5, M12)

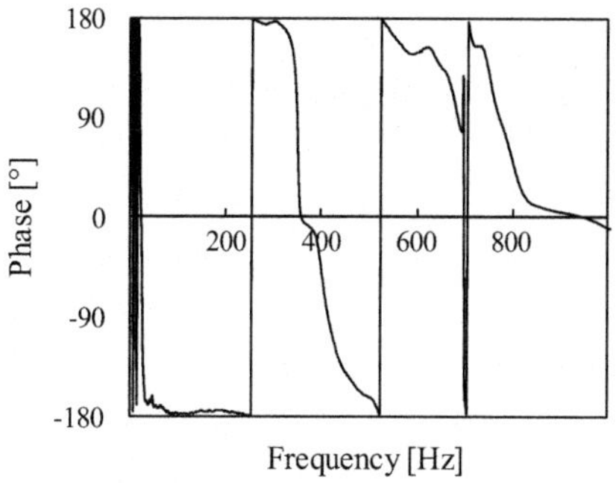

Fig. 3.33 Experiment results of phase for point 1 (t4.5, M12)

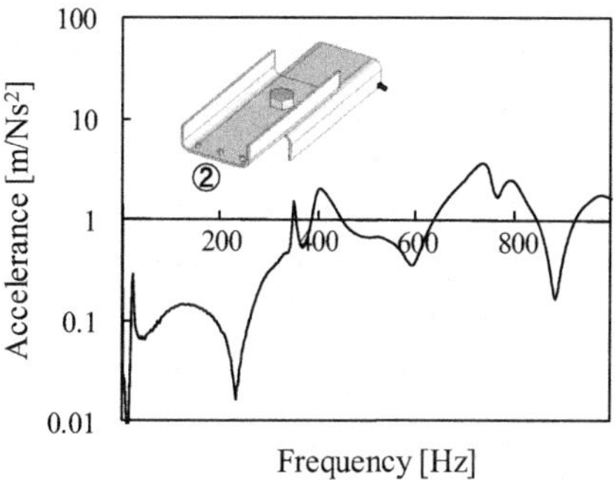

Fig. 3.34 Experiment results of accelerance for point 2 (t4.5, M12)

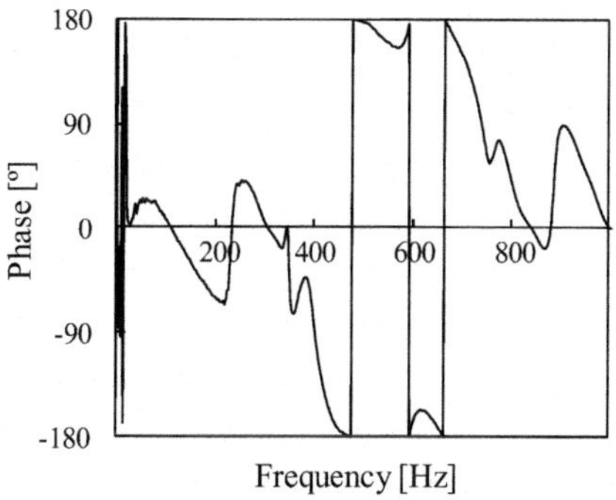

Fig. 3.35 Experiment results of phase for point 2 (t4.5, M12)

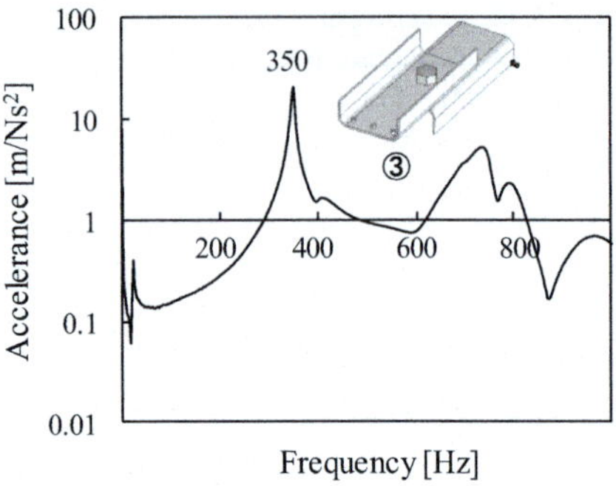

Fig. 3.36 Experiment results of accelerance for point 3 (t4.5, M12)

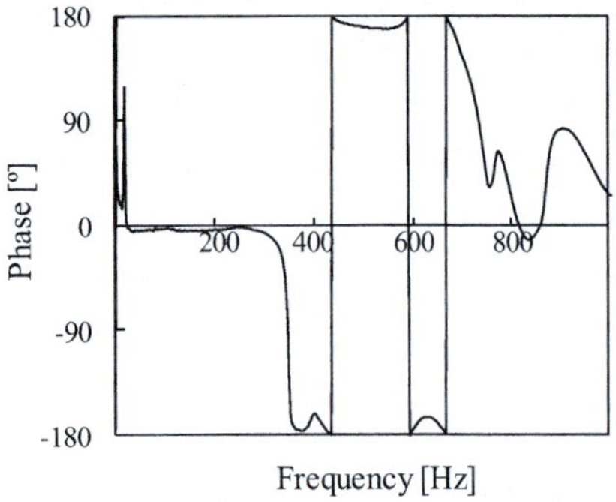

Fig. 3.37 Experiment results of phase for point 3 (t4.5, M12)

Figures 3.38 through 3.43 show the test results of a t4.5 specimen with an M20 bolt. In this case, the accelerance peaks at 380 and 428 Hz. The magnitudes of accelerance for points 1 and 3 at 380 Hz and 428 Hz are not much different in Figs. 3.38 and 3.42. However, the magnitude of accelerance for point 2 at 380 Hz is smaller than that at 428 Hz. Comparing Figs. 3.39 and 3.43 with Fig. 3.41, points 1 and 3 are in the opposite phase at 380Hz, while points 1, 2, and 3 are in the same phase at 428 Hz. Therefore, 380 Hz is considered to be eigenfrequency of Mode 1, and 428 Hz shall be a variant of Mode 2 as shown in Fig. 3.31. In this way, the eigenvalues of Mode 1 can be extracted for each test specimen (Fig. 3.40).

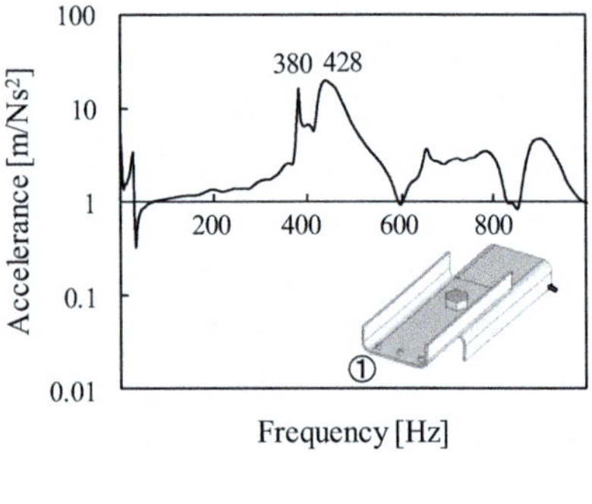

Fig. 3.38 Experiment results of accelerance for point 1 (t4.5, M20)

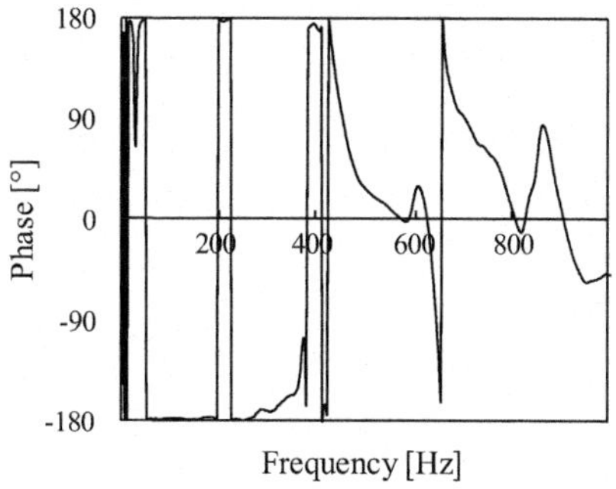

Fig. 3.39 Experiment results of phase for point 1 (t4.5, M20)

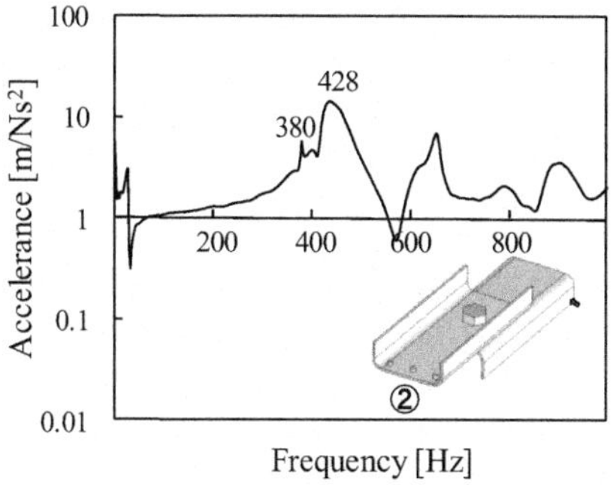

Fig. 3.40 Experiment results of accelerance for point 2 (t4.5, M20)

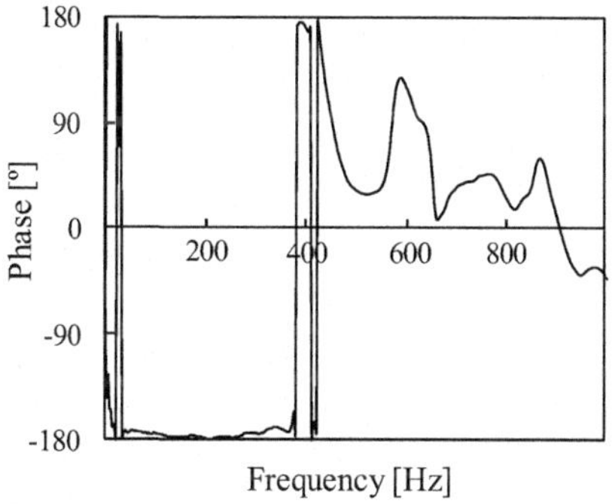

Fig. 3.41 Experiment results of phase for point 2 (t4.5, M20)

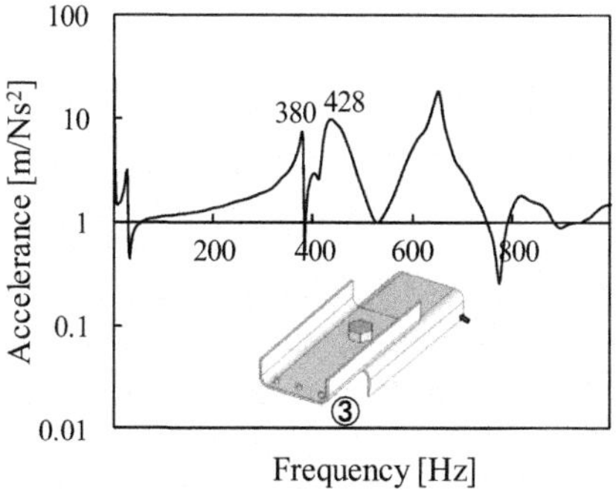

Fig. 3.42 Experiment results of accelerance for point 3 (t4.5, M20)

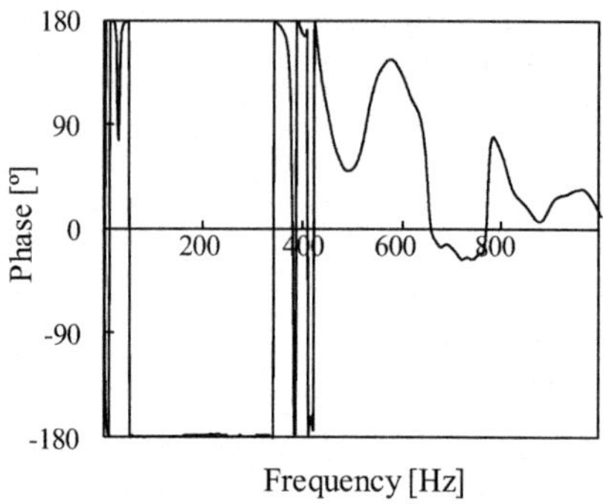

Fig. 3.43 Experiment results of phase for point 3 (t4.5, M20)

We also tested t3.2 specimens with M8, M12 and M20 bolts and t4.5 specimens with M6, M8, M10, M12, M16, M20, and M24 bolts. Figure 3.44 shows the experiment results for eigenvalues of Mode 1 for these specimens. We were able to determine that the greater the nominal diameter of the bolts and the thicker specimens were, the higher the eigenvalues became.

We carried out FE analysis of the models shown in Fig. 3.30 by using the commercial FE code Ansys Mechanical™ 11.0. The analysis results shown in Fig. 3.44 were calculated by using the diameters of the equivalent beam, d_{eq1} and d_{eq2}, and the length of the equivalent pressure area, D_{weq}, in Table 3.1. In general, the analysis results were almost the same as the test results, even though the former were slightly higher than the latter for small nominal diameters and vice versa for large nominal

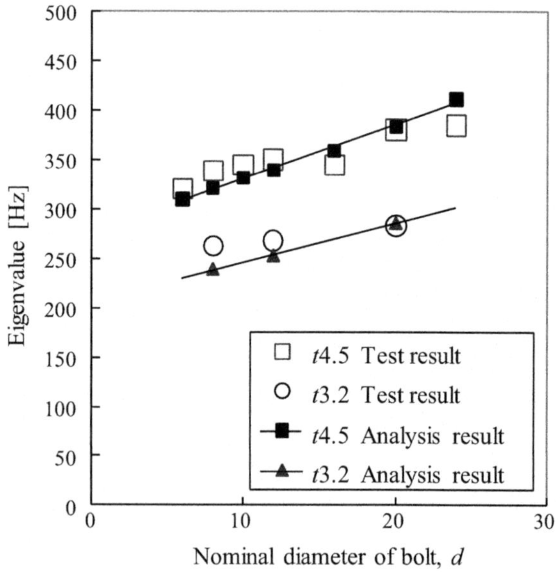

Fig. 3.44 Experiment and analysis results of eigenvalues

diameters. These findings verify the proposed modeling method of bolted joints with equivalent stiffnesses and equivalent pressure area.

3.7 Application to Strength Design of Bolted Joints

3.7.1 External Force

A bolted joint is subjected to a preload at assembly and an external force under working load. As shown in Fig. 2.3 in Chap. 2 and Fig. 3.1 in this chapter, the external force per bolt of a structure fastened with many bolts is not simply the external force divided by the number of bolts. However, the external force per bolt can be easily and accurately determined by using beam elements that take into consideration the equivalent stiffness of a bolted joint, as shown in the bolt modeling method in the previous section.

Figure 3.45 shows a schematic depiction of output forces and moments of a beam element. The output forces are calculated on the coordinate system of the beam element as shown in Fig. 3.45a. We can estimate the axial force W_a, shear force W_s, bending moment M, and torque T of the working loads in Fig. 3.45b as follows.

$$W_a = F_x \tag{3.92}$$

$$W_s = \sqrt{F_y^2 + F_z^2} \tag{3.93}$$

$$M = \max\left(\sqrt{M_{yi}^2 + M_{zi}^2}, \sqrt{M_{yj}^2 + M_{zj}^2}\right) \tag{3.94}$$

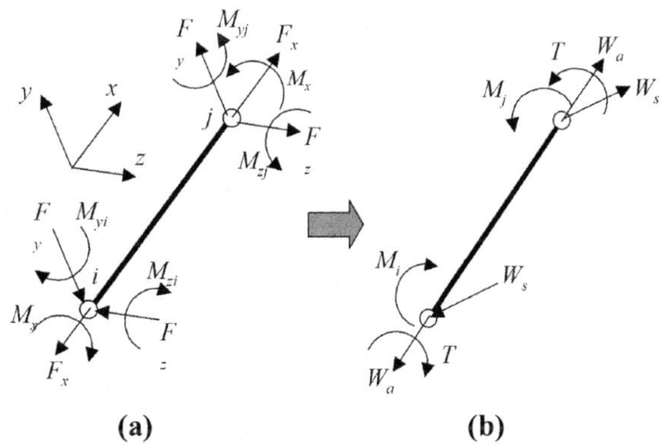

Fig. 3.45 Forces and moments of equivalent beam element of bolt

$$T = |M_x| \tag{3.95}$$

Evaluating the strength of the bolted joints mainly comprises evaluating [i] static strength, [ii] fatigue strength, and [iii] slip. In addition to these primary evaluations, we evaluate [iv] the pressure on the bearing surfaces and the preload change caused by embedding of the contact surface and thermal deformation. These four evaluations are explained in the following subsections.

3.7.2 Static Strength Evaluation

When a bolt with a nominal diameter d is tightened with a torque T_f, the pretension F_f is calculated as Eq. (3.96).

$$F_f = \frac{T_f}{Kd} \tag{3.96}$$

Here, K from Eq. (3.97) is called the "nut factor", which is calculated with the friction coefficient between the nut and bolt threads, μ_s, the friction coefficient in the head bearing area, μ_w, and the thread dimensions.

$$K = \frac{1}{2d}\left(\frac{P}{\pi} + \mu_s d_2 \sec\alpha\prime + \mu_w D_w\right) \tag{3.97}$$

Here, P is the thread pitch, d_2 is the effective diameter of the thread, α' is calculated as $\tan\alpha' = \tan\alpha + \tan\beta$ from the half-angle of the threads α and the lead angle of the threads β, and D_w is called the effective diameter of the head bearing surface, which is obtained as the following equation from the outer diameter of bearing area D_o and the hole diameter D_i.

$$D_w = \frac{2}{3}\frac{D_o^3 - D_i^3}{D_o^2 - D_i^2} \tag{3.98}$$

Since the friction coefficients vary widely and are hard to predict, the pretension scatter should be taken into consideration when bolts are tightened with the torque control method. If the scatter of the pretension is $m\%$, the maximum pretension $F_{f\max}$ and the minimum pretension $F_{f\min}$ are expressed with F_f of Eq. (3.96) as follows.

$$F_{f\max} = \left(1 + \frac{m}{100}\right)F_f \tag{3.99}$$

$$F_{f\min} = \left(1 - \frac{m}{100}\right)F_f \tag{3.100}$$

The scatter m is considered to be 15% in general.

From the above pretension values and the working forces as described in Sect. 3.7.1, the bolt stresses can be estimated. The stress on the thread section is calculated from the effective thread section A_s. The diameter of the effective thread section, d_s, is the average of the pitch diameter of the bolt thread, d_2, and the minor diameter of the bolt thread, d_3.

$$d_s = \frac{d_2 + d_3}{2} \tag{3.101}$$

In static strength evaluation, the maximum pretension $F_{f\max}$ shown in Eq. (3.99) is used. The maximum axial stress $\sigma_{f\max}$ on the effective thread section A_s under the pretension is obtained as follows.

$$\sigma_{f\max} = \frac{F_{f\max}}{A_s} \tag{3.102}$$

The maximum shear stress $\tau_{f\max}$ on the thread under the pretension is calculated by dividing the maximum torque $T_{s\max}$ under the pretension by the torsional section modulus Z_{ps} with the diameter of the effective thread section expressed as in the following equation.

$$\tau_{f\max} = \frac{T_{s\max}}{Z_{ps}} = \frac{8F_{f\max}d_2}{\pi d_s^3}\left(\tan\beta + \frac{\mu_s}{\cos\alpha\prime}\right) \tag{3.103}$$

To prevent the bolt from yielding under the pretension, the equivalent stress of the thread σ_{e_1}, which is calculated from $\sigma_{f\max}$ of Eq. (3.102) and $\tau_{f\max}$ of Eq. (3.103), should be smaller than the yield strength of the bolt material, σ_{by}, taking the safety factor a_y into consideration.

$$\sigma_{e1} = \sqrt{\sigma_{f\max}^2 + 3\tau_{f\max}^2} < \sigma_{by}/a_y \tag{3.104}$$

The yield strength of the bolt material, σ_{by}, corresponds to the lower yield point R_{eL} or 0.2% proof stress $R_{p0.2}$ in Table 3.2, with lower limits defined according to the ISO strength grade.

Table 3.2 Material property of bolt materials for ISO strength grade

Material property	ISO grade					
	4.8	6.8	8.8		10.9	12.9
			d< = 16	d > 16		
Tensile strength $R_{m,\,\min}$ [MP a]	420	600	800	830	1040	1220
Yield strength R_{eL} [MP a]	340	480				
0.2% proof strength $R_{p0\,2}$ [MPa]			640	660	940	1100

Next, we consider how to evaluate the static strength under the working forces W_a, W_s, M, and T shown in Fig. 3.45b. As previously mentioned, a bolted joint supports the working forces by dividing the forces between those of the bolt and those of the plates. In this case, the stresses caused by the shear force W_s and the bending moment M are so small that static failure caused by these working forces and moments do not need to be considered.

What needs to be evaluated, therefore, is the static strength by adding the stresses caused by the axial working force W_a and the working torque T to the stress under the pretension. When the bolt joint is subjected to the axial working force W_a, the bolt sustains the divided force $\Phi_n W_a$ and the plates support the force $(1 - \Phi_n)W_a$. Here, Φ_n is the load factor defined by Eq. (3.17). The stress σ_n on the effective thread section under the axial working force W_a is obtained by the following equation.

$$\sigma_n = \frac{\Phi_n W_a}{A_s} \tag{3.105}$$

Here, when the bolted joint is subjected to the working torque T, the bolt and the plates sustain the working torque together. The load factor of the torque is identical to the load factor of the bending moment, $\Phi_{\beta n}$, defined in Eq. (3.30). Thus, the shear stress τ_s on the effective thread section under the working torque T is calculated as follows.

$$\tau_s = \frac{\Phi_{\beta n} T}{Z_{ps}} \tag{3.106}$$

To prevent the bolt from yielding under the working forces, the equivalent stress of the thread σ_{e2}, which is calculated from $\sigma_{f\max}$ of Eq. (3.102), $\tau_{f\max}$ of Eq. (3.103), σ_n of Eq. (3.105), and τ_s of Eq. (3.106), should be smaller than the yield strength of the bolt material σ_{by}, taking the safety factor a_y into consideration.

$$\sigma_{e2} = \sqrt{\left(\sigma_{f\max} + \sigma_n\right)^2 + 3\left(\tau_{f\max} + \tau_s\right)^2} < \sigma_{by}/a_y \tag{3.107}$$

From the above static strength evaluation, we can determine the nominal diameter of the bolt by satisfying Eq. (3.104) and Eq. (3.107) under the pretension and the working forces, respectively.

3.7.3 Fatigue Strength Evaluation

Fatigue failure in a thread is caused by the cyclic stress on the axial direction under the working forces. From the stresses under the working forces W_a, W_s, M, and T shown in Fig. 3.45b, we take the axial stress σ'_n under the axial force W_a and the bending stress σ'_b under the bending moment M into consideration. The axial stress σ'_n and the bending stress σ'_b are calculated on the cross-section of the minor diameter d_3. The axial stress σ'_n under the axial force W_a is calculated by using the load factor Φ_n defined by Eq. (3.17) as follows.

$$\sigma_{n'} = \frac{\Phi_n W_a}{A_3} \tag{3.108}$$

Here, A_3 is the cross-section of the minor diameter, d_3.

The bending stress σ'_b under the bending moment M is calculated by using the load factor under bending moment, $\Phi_{\beta n}$, defined by Eq. (3.30) as follows.

$$\sigma_{b'} = \frac{\Phi_{\beta n} M}{Z_3} \tag{3.109}$$

Here, Z_3 is the section modulus of the minor diameter, d_3.

Yamamoto et al. found that the thread fatigue strength is almost constant when the pretension is sufficient. They also estimated the thread fatigue strength σ_{wk} by taking the stress concentration of the thread, the fatigue notch factor, and the fatigue strength of the bolt material into account. This estimated fatigue strength σ_{wk} is calculated with ISO strength classes and the nominal diameters, as shown in Table 3.3. To prevent the bolt from incurring fatigue failure, the thread stress amplitude σ_a, which is calculated from σ'_n of Eq. (3.108) and σ'_b of Eq. (3.109), should be smaller than this estimated fatigue strength σ_{wk} while taking the safety factor a_w into consideration.

$$\sigma_a = \max\left(\frac{|\sigma_{n'} + \sigma_{b'}|}{2}, \frac{|\sigma_{n'} - \sigma_{b'}|}{2}\right) < \sigma_{wk}/a_w \tag{3.110}$$

Table 3.3 Fatigue strength of bolt materials, σ_{wk}

Nominal designation	ISO grade				
	4.8	6.8	8.8	10.9	12.9
M6	81	56	64	78	81
M8	61	49	59	72	76
M10	52	45	55	69	72
M12	48	43	53	67	71
M16	42	39	49	62	66
M20	39	37	46	61	65
M24	37	36	45	69	63
M30	24	34	43	57	61

3.7.4 Slip Evaluation

When a bolted joint is subjected to working forces, it should be designed not to slip on the bearing surface between the bolt head and the plate, or between the nut and the plate, or on the surface between the two plates. The static friction forces on the bearing surfaces and on the surface between two plates, therefore, should be larger than the working forces. However, as shown in Subsection 2.4.5, slip-allowing design is needed in some cases. For example, when electronic devices are mounted into mechanical products with bolted joints, and when thermal load is applied to such devices by temperature change, a load perpendicular to the bolt axis is applied to the bolted joint due to a thermal expansion difference between the electronic device, which is composed of plastic, a semiconductor, a printed circuit board and so on, and the metallic mechanical base. Cases like this present design difficulties because the allowable stress of the electronic device is not so large that the bolt pretension is not sufficiently high to prevent the slip. Therefore, we adopt the slip-allowing design, in which a light amount of slip is permitted, as shown in Fig. 2.3 (allowable slip δ_{CR}). The details of the slip-allowing design are given in Chap. 4.

In this subsection, we consider non-slip design. The static friction forces are calculated by multiplying the friction coefficients by the normal forces on these surfaces. The friction coefficient on the bearing surface is μ_w, and that on the surface between the two plates is μ_c. In the slip evaluation, the nominal forces on these surfaces are estimated by adding the axial additional bolt load by the axial force W_a and the minimum pretension $F_{f\min}$ defined by Eq. (3.100).

The nominal force on the bearing surface is increased by $\Phi_n W_a$ from the minimum pretension $F_{f\min}$ under the axial force W_a. Here, Φ_n is the load factor defined by Eq. (3.17). The static friction force F_w is calculated as follows.

$$F_w = \mu_w\left(F_{f\min} + \Phi_n W_a\right) \tag{3.111}$$

The nominal force on the surface between the two plates is decreased by $(1 - \Phi_n)W_a$ from the minimum pretension $F_{f\min}$ under the axial force W_a. The static friction force F_c is calculated as follows.

$$F_c = \mu_c\left\{F_{f\min} - (1 - \Phi_n)W_a\right\} \tag{3.112}$$

We compared these static friction forces with the shear force W_s and torque T. To prevent the bolt joint from slipping under the shear force W_s, the following inequalities should be satisfied while taking the safety factor a_s into consideration.

$$W_s < F_w/a_s \tag{3.113}$$

$$W_s < F_c/a_s \tag{3.114}$$

Further, to prevent the bolt joint from slipping between the clamped plates under torque T, the following inequality should be satisfied while taking the safety factor a_s into consideration.

$$T < T_c = \frac{F_w}{a_s} \times \frac{1}{2}\frac{2}{3}\frac{D_{A,Gr}^3 - D_i^3}{D_{A,Gr}^2 - D_i^2} = \frac{F_w}{a_s} \times \frac{D_{Aw}}{2} \tag{3.115}$$

Here, T_c is the slip limit torque between the clamped plates, and D_{Aw} is the effective friction diameter of the contact surface of the clamped plate. Since slip due to torsional torque T on the bearing surface or that on the threaded surface occurs after a slip between the clamped plates, it is considered sufficient to evaluate only between the clamped plates in the linear analysis range.

3.7.5 Loosening Evaluation (Refer to Chap. 6)

There are two ways of loosening of bolted joints:

(1) Rotational loosening, in which the relative rotation between bolt and nut occurs.
(2) Non-rotational loosening, in which the relative rotation does not occur.

The mechanisms of these two types of loosening are explained by using Fig. 3.46. Rotational loosening is caused by cyclic loads such as axial direction cyclic load W_a, transverse direction cyclic load W_t, or axial rotation cyclic torque T_a. As for axial direction cyclic load W_a, loosening will not arise if the bolted joint satisfies the fatigue strength limitation (described earlier in Sect. 3.7.3). For the transverse direction cyclic load and axial rotation cyclic torque looseness can be evaluated based on the relative slip of the four contact sites (I–IV) shown in Fig. 3.46, and the loosening behaviors for the transverse direction cyclic load and the axial rotation cyclic torque will be explained in detail in Chaps. 4 and 5, respectively.

(A) **Initial Loosening due to Embedding**

As shown in Fig. 3.46, I: thread surface, II: nut bearing surface, III: bolt head bearing surface, and IV: contact surface of clamped plates are in the state of contact with compressive load in bolted joint. Each of these contact surfaces has surface roughness, undulations, and minute irregularities as shown in Fig. 3.47a. Therefore, as shown in Fig. 3.47b, the contact surfaces become flattened and permanently embedded due to cyclic contact pressure loading, and the contact surfaces reduce the distance. In this case, the clamping length (distance between the bolt head bearing surface and the nut bearing surface) will be reduced, resulting in a decrease in pretension. However, this permanent embedding phenomenon stops after it reaches the point where contact

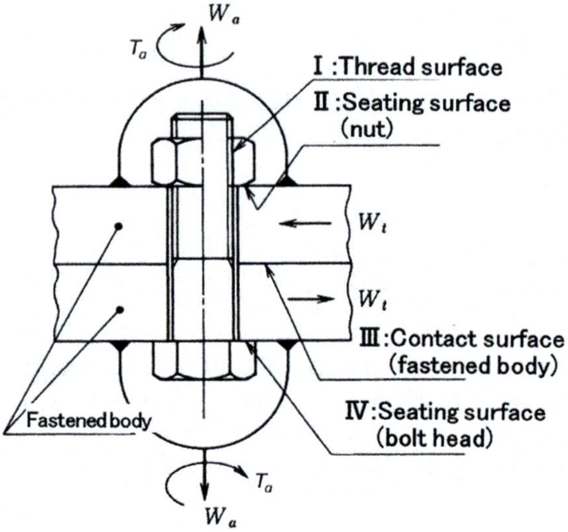

Fig. 3.46 Simplified model for the loosening mechanisms of a bolted joint

surfaces become flattened. This initial loosening of bolted joint is called (initial) embedding.

This initial loosening by embedding can be analyzed by using a clamping diagram as shown in Fig. 3.48. Here, the relation between the embedding amount s and loosening force F_s can be shown as follows.

$$s = \frac{F_s}{\tan\theta_b} + \frac{F_s}{\tan\theta_c} \tag{3.116}$$

In Eq. (3.116), $\tan\theta_b$ and $\tan\theta_c$ are the inverse of the respective compliances δ_b and δ_c, and the equation can be rewritten as follows.

$$F_s = \frac{1}{\delta_b + \delta_c} = Zs \tag{3.117}$$

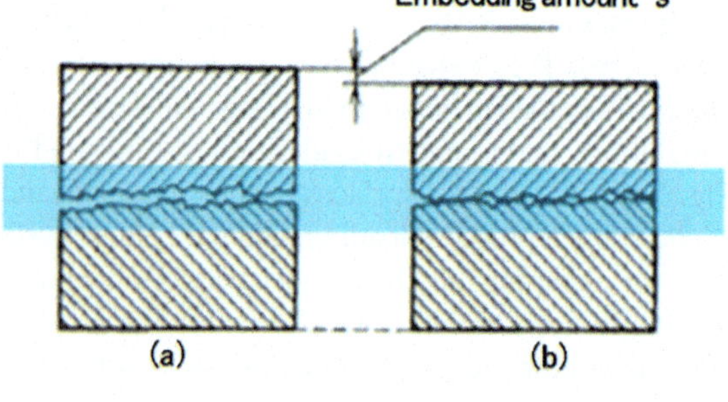

Fig. 3.47 Embedding (setting) of contact surface

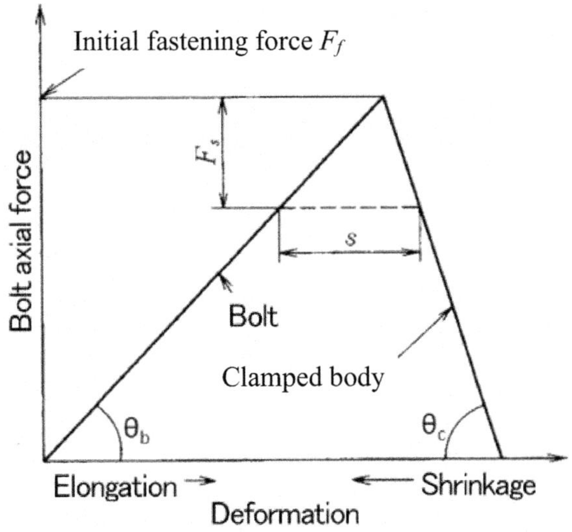

Fig. 3.48 Initial loosening by embedding

Z is called the embedding coefficient.

Figure 3.49 shows a quick calculation graph for the embedding coefficient Z of a steel bolted joint proposed by Yamamoto [6]. Here, d is the nominal diameter of the bolt and l_f is the clamping length (grip length) of the bolt.

The embedding amount s shown in Fig. 3.47 is explained as being due to microscopic surface roughness flattening, but most of this occurs during the initial fastening phase, which is excluded from embedding loosening, and only the subsequent in-service embedding amount shall be considered. Therefore, s is usually smaller than the value expected from surface roughness, and the following experimental equation

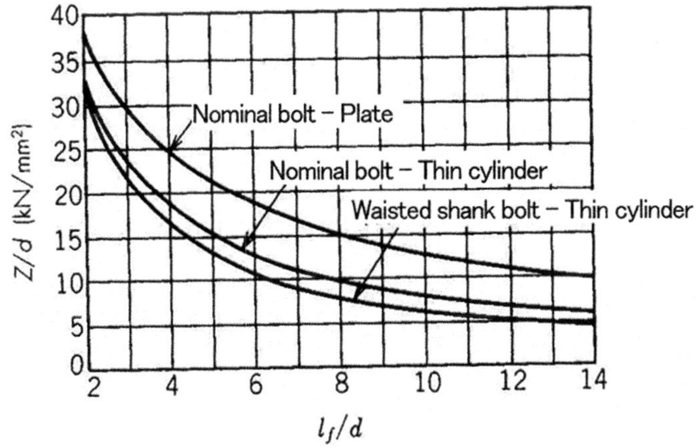

Fig. 3.49 Quick calculation graph for the embedding coefficient Z of steel bolted joint

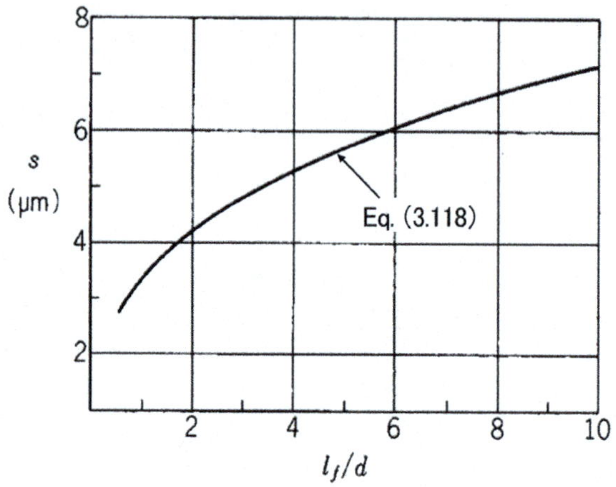

Fig. 3.50 Calculation result of the formula for the setting amount [5]

is used [5].

$$s = 3.29\left(\frac{l_f}{d}\right)^{0.34} \quad (\mu m) \tag{3.118}$$

The calculation result is shown in Fig. 3.50. However, Eq. (3.118) and Fig. 3.50 are only applicable when the surface pressure of the bearing surface does not exceed the limit bearing pressure (described below), and they cannot be used when the clamped body is a low-yield strength material such as aluminum alloy, FRP, or a thin plate structure. Please refer to the next section for information about limit bearing pressure.

(B) **Loosening due to Bearing Recession Under High Bearing Pressure**

The pressures on the bearing surfaces of I and IV in Fig. 3.46 are so high that the permanent set of bearing surfaces are large and/or the creep deformation of bearing surface can occur. Both of these can cause loosening (i.e., loosening due to bearing recession) that are more serious than the embedding loosening. To prevent large permanent set and creep deformation, the pressure on the bearing surface, p_w, should not exceed the allowable pressure of the clamped plates materials (hereafter referred to as the limit surface pressure of the clamped material), p_L, as shown in the following equation.

$$p_w = \max\left(\frac{F_{f\max}}{A_w}, \frac{F_{f\max} + \Phi_n W_{a\max}}{A_w}\right) < p_L \tag{3.119}$$

In this equation, $F_{f\max}$ is the maximum pretension in Eq. (3.99), Φ_n is the load factor shown in Eq. (3.17), $W_{a\max}$ is the maximum axial load, and A_w is the contact area of the bearing surface, which is expressed by the outer diameter of the bearing surface, D_o, and the hole diameter D_i as follows.

$$A_w = \frac{\pi}{4}\left(D_o^2 - D_i^2\right) \tag{3.120}$$

The surface pressure limit of the clamped material, p_L, has been experimentally obtained for various materials of plates in the literature [1, 2, 5] as shown in Table 3.4.

(C) Loosening due to Thermal Deformation

Next, we considered the preload change caused by thermal deformation. When the thermal expansion coefficients of the bolt material and the plate materials are different and the bolted joint temperatures are varied, or when the temperatures of the bolt and the clamped plates are different even if their thermal expansion coefficients are the same, the preload would change with thermal deformation. The modeling method of bolted joints by using shell and beam elements described in Sect. 3.7.1 can estimate the in-plane deformation by thermal expansion, but shell elements cannot evaluate the thermal deformation in the thickness direction. Thus, the preload change caused by the thermal deformation in the thickness direction should be evaluated by the following method.

The thermal expansion coefficients of the bolt and the plates are determined as α_b and α_c, respectively. The temperature rises in the bolt and the plates are represented by Δt_b and Δt_c, respectively. The preload loss F_{z2} determined by the thermal deformation

Table 3.4 Allowable pressure on bearing surface, p_L [5]

Material			Tensile strength	Allowable pressure
Type	German standard	Equivalent material in JIS		
Low carbon steel	St 37	S10C	370	260
Medium carbon steel	St 50	S30C	500	420
Low-alloy tempering steel	C 45	S45C	800	700
	42 CrMo 4	SCM440	1000	850
Austenitic CrNi steel	x5 C rNiMo 1810	SUS316	500–700	210
Cast iron	GG 15	FC150	150	600
	GG 25	FC250	250	800
	GG 35	FC350	350	900
	GG 40	–	400	1100
Magnesium alloy	GDMgAl 9	MC2	300(200)	220(140)
	GKMgAl 9	–	200(300)	140(220)
Cast aluminium alloy	GKAlSi 6 Cu 4	AC2B	–	200
Aluminium alloy	Al99	A1200	160	140
	AlZnMgCu 1.5	A7075	450	370
FRP	–	–	–	120–140

result is calculated from the load factor Φ defined by Eq. (3.16), the compliance of the plates δ_c defined by Eq. (3.12), and the clamping length l_f as follows.

$$F_{z2} = \frac{\Phi}{\delta_c}(\alpha_b \Delta t_b - \alpha_c \Delta t_c) l_f \tag{3.121}$$

We can evaluate the pretension change resulting from the thermal deformation by substituting the term of the axial force change $\Phi_n W_a$ in Sects. 3.7.2, 3.7.3, and 3.7.4 and the pressure evaluation on the bearing surface described in Sect. 3.7.5 into $(\Phi_n W_a - F_{z_2})$.

3.7.6 Evaluation Example

(A) **Evaluation Model**

We applied our strength evaluation method to the specific example shown in Fig. 3.51. Three channels were assembled into an H-shaped structure and tightened down with four M12 bolts or four M16 bolts at each of two intersections. The structure was fixed in place at the four ends and subjected to an out-of-plane load at the center. Taking the symmetry into consideration, the analysis model was a half model as shown in Fig. 3.52. The analysis was carried out with Ansys Mechanical™ 11.0. We made an FE model by using our modeling method for bolted joints described in Sect. 3.5. For modeling, it is necessary to specify the outside diameter of the bolted joint, D_c, shown in Fig. 3.28. The distance between the bolt holes and the outer edge of the plate and the bolt spacing were compared, and the outer diameter of the bolted joint, D_c, was set to 38 mm because the bolt distance of 38 mm is the minimum value for the structure shown in Fig. 3.51. The x-axis of the coordinate system corresponded to the axial directions of the bolts.

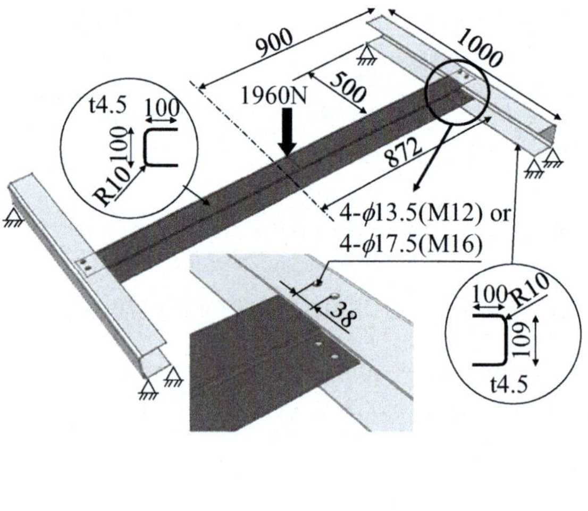

Fig. 3.51 Example of bolted joint structure

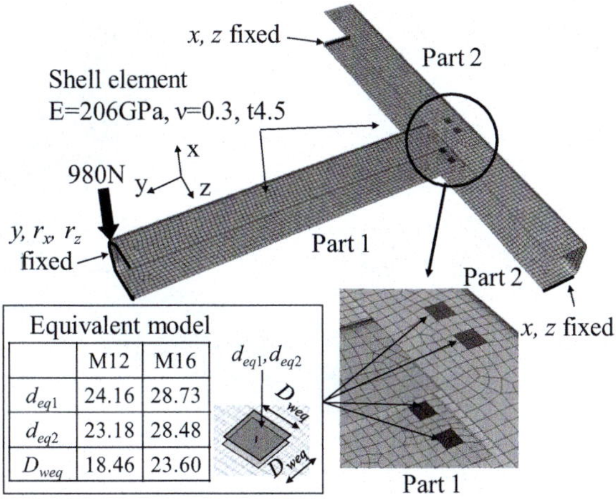

Fig. 3.52 Analysis model of example from Fig. 3.51

(B) Strength Evaluation

Figure 3.53 shows the deformation and Mises equivalent stress distribution of the M12 bolt model, where Part 1 deflects in the x-direction and rotates around the y-axis. Figure 3.54 shows the results obtained for the working forces F_x and F_y loaded to four equivalent beam elements of bolted joints 1–4 of the M12 bolt model. Working force F_z and moments M_x, M_y, and M_z were obtained as well. These forces and moments are shown in Table 3.5. Taking the force directions into consideration, axial force W_a, shear force W_s, bending moment M, and torque T calculated by using Eqs. (3.92) through (3.95) are also shown in Table 3.5. Table 3.6 shows the load sharing of the bolts in the M16 bolt model.

From Tables 3.5 and 3.6, it is clear that the loads in the x-direction of No. 1 through No. 4 bolts are not simply equal to one-fourth of the external force; rather, the four bolts intricately share the load. The load in the y-direction of No. 2 bolt is opposite to that of No. 1 bolt, and the load in the y-direction of No. 3 is opposite to that of No. 4. The total sum of the loads in the y-direction is zero. Moments M_x, M_y, and M_z also intricately share loads with No.1–4 bolts. If these external forces are manually calculated by force balance, it would be difficult to derive precisely. The simplified model using the equivalent stiffness of bolts, which was developed in this study, is quite effective in analyzing structures fastened by multiple bolted joint (see Chap. 2, Sect. 2.2).

The bolts subjected to the external forces shown in Tables 3.5 and 3.6 were examined using the strength evaluations described in Sects. 3.7.2 through 3.7.5, namely, the static strength evaluation, fatigue strength evaluation, slip evaluation, and loosening evaluation. As an example of the strength evaluation, Tables 3.7a, 3.7b, 3.7c and 3.8a, 3.8b, 3.8c show the evaluation results of No. 1 bolt, which is subjected to large axial load W_a, and that of No. 3 bolt, which is subjected to large shear load W_s. ISO strength class 4.8 and 8.8 bolts were used for the M12 and M16 bolts, respectively. The clamped plate was low-carbon streel equivalent to SS400,

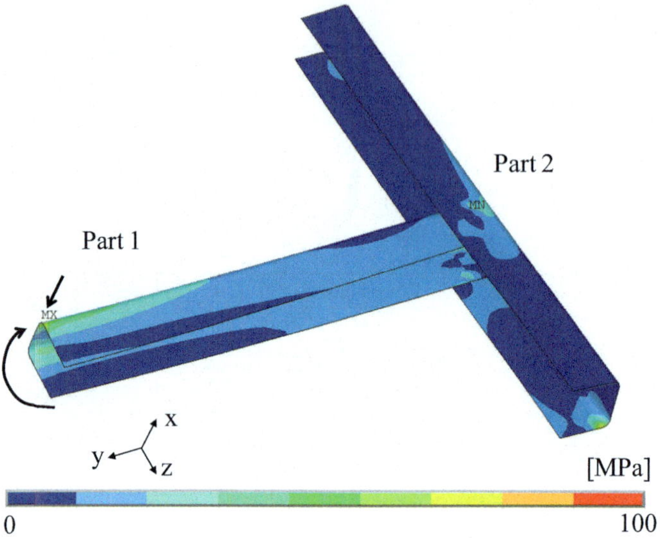

Fig. 3.53 Deformation and von Mises stress distribution of M12 bolts model

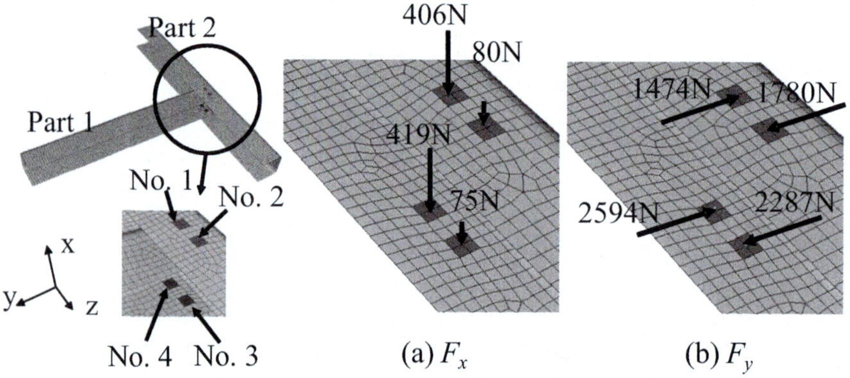

Fig. 3.54 Load distribution of analysis result of M12 bolts

and the allowable pressure on the bearing surface was that of low-carbon steel shown in Table 3.4.

The initial tightening torque was 0.5 times the yield-tightening torque T_{fy} when the coefficient of friction on the seat surface $\mu_w = 0.15$, as specified in JIS B1083. The coefficients of friction on the seat surface μ_w and on the screw surface μ_s were set to 0.15, and the nut factor K was obtained from Eq. (3.97) as $K = 0.196$ for M12 and $K = 0.194$ for M16. The pretension F_f was obtained by using Eq. (3.96), and a variation of $m = \pm 15\%$ was considered according to Eqs. (3.99) and (3.100). We estimated the preload loss obtained by embedding on the contact surface from Eqs. (3.117) and

Table 3.5 Load distribution of M12 bolts

No. of bolt	No. 1	No. 2	No. 3	No. 4
F_x [N]	–406	–80	–75	–419
F_y [N]	1,474	–1,780	2,594	–2,287
F_z [N]	–165	23	12	130
M_x [N-mm]	26,581	19,498	–31,389	–35,965
M_y [N-mm]	–4,723	1,635	1,738	–4,932
M_z [N-mm]	–10,614	8,185	10,191	–12,764
W_a [N]	406	80	–75	–419
W_s [N]	1,474	1,780	2,594	2,287
M [N-mm]	11,617	8,347	10,338	13,684
T [N-mm]	26,581	19,498	31,389	35,965

Table 3.6 Load distribution of M16 bolts

No. of bolt	No. 1	No. 2	No. 3	No. 4
F_x [N]	–410	–74	–42	–453
F_y [N]	1,390	–1,662	2,462	–2,190
F_z [N]	–127	–18	69	76
M_x [N-mm]	33,231	20,772	–36,066	–43,840
M_y [N-mm]	–4,712	1,974	2,604	–4,318
M_z [N-mm]	–12,472	8,357	10,004	–14,551
W_a [N]	410	74	–42	–453
W_s [N]	1,390	1,662	2,462	2,190
M [N-mm]	13,333	8,587	10,337	15,178
T [N-mm]	33,231	20,772	36,066	43,840

(3.118). We also estimated the safety factors of static strength, fatigue strength, slip, and bearing surface pressure as ratios of the allowable and the calculated value.

If the bolt is an ISO class 4.8 bolt, the slip evaluation safety factor is smaller than 1.0 as shown in bold italic characters in Tables 3.7a, 3.7b, 3.7c and 3.8a, 3.8b, 3.8c. This shows that the bolted joint cannot hold the shear force W_s and would slip between the two plates. If the bolt is an ISO class 8.8 bolt, the bolted joints could be tightened with higher pretension and the static friction force would be increased. The slip evaluation safety factor, therefore, would increase to greater than 1.0 with the No.1 bolt of M12 and ISO class 8.8 in Tables 3.7a, 3.7b and 3.7c.

We also evaluated fatigue strength by calculating the stress amplitude, σ_a, from the axial force W_a and the bending moment M as shown in Tables 3.7a, 3.7b, 3.7c and 3.8a, 3.8b, 3.8c. In this evaluation, the stress amplitude σ_a of the bolt is very small, as the load factor Φ_n and the load factor for the bending moment, $\Phi_{\beta n}$, are very small. We can assume that the applied load is sufficiently small for fatigue strength of

Table 3.7a Strength evaluation No. 1. (1/3)

		Namc of parameter	Symbol	Dimension	Input and output		
Input	Dimension	Nominal designation	d	[mm]	12		16
		Thread pitch diameter	d_2	[mm]	10.863		14.701
		Thread root diameter	d_3	[mm]	10.106		13.835
		Hole diameter	D_i	[mm]	13.5		17.5
		Outer bearing diameter	d_w	[mm]	16.63		22
		Thread pitch	P	[mm]	1.75		2
		Flank angle of thread	α	[°]	30		30
		Outer diameter of clamped plate	D_c	[mm]	38		38
		Friction coefficient of bearing surface	μ_w		0.15		0.15
		Friction coefficient of thread	μ_s		0.15		0.15
		Friction coefficient of clamped plates	μ_c		0.15		0.15
		Nut factor	K		0.1961		0.1924
		Youngs modulus of bolt	E_b	[MPa]	206000		206000
		ISO grade			4.8	8.8	4.8
		Yield strength of bolt material		[MPa]	340	640	340
		Fatigue strength of bolt		[MPa]	48	53	48
		Yield tightening torque (JIS B 1082)	T_{fy}	[N-mm]	52.7	99.2	131
		Ratio of tightening torque and yield tightening torque			0.5	0.5	0.5
		Tightening torque	T_f	[N-mm]	26350	49600	65500
		Preload	F_f	[N]	11200	21080	27850
		Maximum preload	$F_{f\,\max}$	[N]	12880	24240	32030
		Minimum preload	$F_{f\,\min}$	[N]	9520	17920	23670
		Youngs modulus of plate	E_c	[MPa]	206000		206000
		Thickness of plate 1	t_1	[mm]	4.5		4.5
		Thickness of plate 2	t_2	[mm]	4.5		4.5

the bolted joint. Preload changes caused by thermal deformation can be evaluated in this way as well, but we found that in this model such changes need not be evaluated because the bolt and plate materials are carbon steel types.

Table 3.7b Strength evaluation No. 1. (2/3)

		Name of parameter	Symbol	Dimension	Input and output	
Input	Load	Axial load	W_a	[N]	406	410
		Shear force	W_s	[N]	1474	1390
		Bending moment	M	[N-mm]	11617	13333
		Torque	T	[N-mm]	26581	33231
	Stiffness calculation	Tangent of helix angle of equivalent cone	$\tan\varphi$		0.4076	0.3558
		Outer diameter of equivalent cone	$D_{A,Gr}$	[mm]	20.30	25.20
		Compliance of bolt	δ_b	[mm/N]	1.371E-06	8.966E-07
		Compliance of clamped p lates	δ_c	[mm/N]	3.711E-07	2.280E-07
		Bending compliance of cylinder with $D_{A}Gr$ and $D\ i$	δ_c	[mm/N]	1.210E-07	8.456E-08
		Load factor	Φ		0.2130	0.2028
		M odified load factor	Φ_n		0.06947	0.07519
		Bending compliance of bolt	β_b	[rad/N mm]	1.937E-07	6.762E-08
		Bending compliance of clamped plates	β_c	[rad/N mm]	1.123E-08	4.203E-09
		Bending compliance of cylinder with $D_{A,Gr}$ and D_i	β_{ct}	[rad/N mm]	3.259E-09	1.437E-09
		Load factor for bending moment	Φ_β		0.05480	0.05852
		M odified load factor for bending moment	$\Phi_{\beta n}$		0.01590	0.02001
	Stiffness of equivalent beam	Area of equivalent beam	A_{beq}	[mm^2]	458.6	648.1
		Second moment of equivalent beam	I_{beq}	[mm^4]	14183	32290
		Diameter of area of equivalent beam	d_{eq1}	[mm]	24.16	28.73
		Diameter of second moment of equivalent beam	d_{eq2}	[mm]	23.18	28.48
		Equivalent pressure area	D_{weq}	[mm]	18.46	23.60
	Preload loss	Permanent deformation of bearing surface	f_z	[mm]	2.9830	2.9830
		Preload loss	F_{z1}	[N]	1712	2652

Table 3.7c Strength evaluation No. 1. (3/3)

		Name of parameter	Symbol	Dimension	Input and output		
Output	Stress calculation	Axial stress due to preload	$\sigma_{f\ \max}$	[MPa]	149	281	200
		Shear stress due to preload	$\tau_{f\ \max}$	[MPa]	69	131	89
		Axial stress due to applied load W_a	σ_n	[MPa]	0.326		0.193
		Shear stress due to applied torque T	τ_n	[MPa]	1.868		1.166
		Axial stress at thread root area due to W_a	σ'_n	[MPa]	0.351		0.205
		Axial stress at thread root area due to M	σ'_b	[MPa]	1.823		1.026
		Equivalent stress due to preload	σ_{e1}	[MPa]	192	361	253
		Equivalent stress due to preload and applied load	σ_{e2}	[MPa]	194	363	255
		Stress amplitude	σ_a	[MPa]	1.1		0.6

(continued)

Table 3.7c (continued)

		Name of parameter	Symbol	Dimension	Input and output		
	Evaluation for stress	Safety factor for σ_{e1}	σ_{by}/σ_{e1}		**1.77**	**1.78**	**1.34**
		Safety factor for σ_{e2}	σ_{by}/σ_{e2}		**1.75**	**1.76**	**1.33**
		Safety factor for fatigue strength	σ_{wk}/σ_a		**44.66**	**49.32**	**78.75**
	Evaluation for slip	Static frictional force on bearing surface	F_w	[N]	1175	2435	3157
		Static frictional force on clamped plates	F_c	[N]	1115	2375	3096
		Safety factor for slip on bearing surface by W_s	F_w/W_s		***0.80***	**1.65**	**2.27**
		Safety factor for slip on clamped plates by W_s	F_c/W_s		***0.76***	**1.61**	**2.23**
		Safety factor for slip on clamped plates by T	$F_w D_{Aw}/2/T$		***0.76***	**1.57**	**2.05**

(continued)

Table 3.7c (continued)

		Name of parameter	Symbol	Dimension	Input and output		
	Evaluation for bearing surface pressure	Pressure due to preload	$F_{f\max}/A_w$	[MPa]	174	327	229
		Pressure due to preload and applied load	$(F_{f\max} + \Phi_n W_a)/A_w$	[MPa]	174	328	230
		Allowable pressure (for low carbon steel, SS400)	p_L	[MPa]	260		260
		Safety factor for pressure due to preload	$p_L A_w/F_{f\max}$		**1.50**	***0.79***	**1.13**
		Safety factor for pressure due to preload and applied load	$p_L A_w/(F_{fmax}+\Phi_n W_a)$		**1.49**	***0.79***	**1.13**

From Tables 3.7a, 3.7b, 3.7c and 3.8a, 3.8b, 3.8c, all of the safety factors cannot be greater than 1.0 with M12 bolts of ISO strength classifications 4.8 and 8.8. By using M16 bolts of ISO strength class 4.8 with increasing the bolt diameter, all of the safety factors can be over 1.0. Therefore, we concluded that M16 bolts should be used in this evaluation example.

Table 3.8a Strength evaluation No. 3. (1/3)

		Name of parameter	Symbol	Dimension	Input and Output		
Input	Dimensio	Nominal designation	d	[mm]	12		16
		Thread pitch diameter	d_2	[mm]	10.863		14.701
		Thread root diameter	d_3	[mm]	10.106		13.835
		Hole diameter	D_i	[mm]	13.5		17.5
		Outer bearing diameter	d_w	[mm]	16.63		22
		Thread pitch	P	[mm]	1.75		2
		Flank angle of thread	α	[°]	30		30
		Outer diameter of clamped p late	D_c	[mm]	38		38
		Friction coefficient of bearing surface	μ_w		0.15		0.15
		Friction coefficient of thread	μ_s		0.15		0.15
		Friction coefficient of clamped plates	μ_c		0.15		0.15
		Nut factor	K		0.1961		0.1924
		Young's modulus of bolt	E_b	[MPa]	206000		206000
		ISO grade			4.8	8.8	4.8
		Yield strength of bolt material	σ_{by}	[MPa]	340	640	340
		Fatigue strength of bolt	σ_{wk}	[MPa]	48	53	48
		Yield tightening torque (JIS B 1082)	T_{fy}	[N-mm]	52.7	99.2	131
		Ratio of tightening torque and y ield tightening torque			0.5	0.5	0.5
		Tightening torque	T_f	[N-mm]	26350	49600	65500
		Preload	F_f	[N]	11200	21080	27850
		M aximum preload	$F_{f\,\max}$	[N]	12880	24240	32030
		Minimum preload	$F_{f\,\min}$	[N]	9520	17920	23670
		Young's modulus of plate	E_c	[MPa]	206000		206000
		Thickness of plate 1	t_1	[mm]	4.5		4.5
		Thickness of plate 2	t_2	[mm]	4.5		4.5

Table 3.8b Strength evaluation No. 3. (2/3)

		Name of parameter	Symbol	Dimension	Input and output	
Input	Load	Axial load	W_a	[N]	-75	-42
		Shear force	W_s	[N]	2594	2462
		Bending moment	M	[N*mm]	10338	10337
		Torque	T	[N*mm]	31389	36066
Output	Stiffness calculation	Tangent of helix angle of equivalent cone	$\tan\varphi$		0.4076	0.3558
		Outer diameter of equivalent cone	$D_{A,Gr}$	[mm]	20.30	25.20
		Comp liance of bolt	δ_b	[mm/N]	1.371E-06	8.966E-07
		Comp liance of clamped p lates	δ_c	[mm/N]	3.711E-07	2.280E-07
		Bending compliance of cylinder with DA,Gr and D i	δ_c	[mm/N]	1.210E-07	8.456E-08
		Load factor	Φ		0.2130	0.2028
		M odified load factor	Φ_n		0.06947	0.07519
		Bending compliance of bolt	β_b	[rad/N mm]	1.937E-07	6.762E-08
		Bending compliance of clamped plates	β_c	[rad/N mm]	1.123E-08	4.203E-09
		Bending compliance of cy linder with $D_{A,Gr}$ and D_i	β_{ct}	[rad/N mm]	3.259E-09	1.437E-09
		Load factor for bending moment	Φ_β		0.05480	0.05852
		M odified load factor for bending moment	$\Phi_{\beta n}$		0.01590	0.02001
	Stiffness of equivalent beam	Area of equivalent beam	A_{beq}	[mm^2]	458.6	648.1
		Second moment of equivalent beam	I_{beq}	[mm^4]	14183	32290
		Diameter of area of equivalent beam	d_{eq1}	[mm]	24.16	28.73
		Diameter of second moment of equivalent beam	d_{eq2}	[mm]	23.18	28.48
		Equivalent pressure area	D_{weq}	[mm]	18.46	23.60
	Preload loss	Permanent deformation of bearing surface	f_z	[mm]	2.9830	2.9830
		Preload loss	F_{z1}	[N]	1712	2652

Table 3.8c Strength evaluation No. 3. (3/3)

		Name of parameter	Symbol	Dimension	Input and output		
Output	Stress calculation	Axial stress due to preload	$\sigma_{f\,\max}$	[MPa]	149	281	200
		Shear stress due to preload	$\tau_{f\,\max}$	[MPa]	69	131	89
		Axial stress due to applied load W_a	σ_n	[MPa]	-0.060		-0.020
		Shear stress due to applied torque T	τ_s	[MPa]	2.206		1.265
		Axial stress at thread root area due to W_a	σ'_n	[MPa]	-0.065		-0.021
		Axial stress at thread root area due to M	σ'_b	[MPa]	1.622		0.796
		Equivalent stress due to p reload	σ_{e1}	[MPa]	192	361	253
		Equivalent stress due to preload and applied load	σ_{e2}	[MPa]	194	363	255
		Stress amplitude	σ_a	[MPa]	0.8		0.4
	Evaluation for stress	Safety factor for $\mathrm{cr}_{e\,1}$	σ_{by}/σ_{e1}		**1.77**	**1.78**	**1.34**
		Safety factor for $\mathrm{cr}_{e\,2}$	σ_{by}/σ_{e2}		**1.75**	**1.76**	**1.33**
		Safety factor for fatigue strength	σ_{wk}/σ_a		**57.05**	**62.99**	**117.70**

(continued)

Table 3.8c (continued)

		Name of parameter	Symbol	Dimension	Input and output		
	Evaluation for slip	Static frictional force on bearing surface	F_w	[N]	1170	2430	3152
		Static frictional force on clamped plates	F_c	[N]	1182	2442	3159
		Safety factor for slip on bearing surface by W_s	F_w/W_s		***0.45***	***0.94***	**1.28**
		Safety factor for slip on clamped plates by W_s	F_c/W_s		***0.46***	***0.94***	**1.28**
		Safety factor for slip on clamped plates by T	$F_w D_{Aw}/2/T$		***0.64***	**1.33**	**1.89**
	Evaluation for bearing surface pressure	Pressure due to preload	$F_{f\,\max}/A_w$	[MPa]	174	327	229
		Pressure due to preload and applied load	$(F_{f\,\max} + \Phi_n W_a)/A_w$	[MPa]	174	327	229
		Allowable pressure (for low carbon steel, SS400)	p_L	[MPa]	260		260

(continued)

Table 3.8c (continued)

		Name of parameter	Symbol	Dimension	Input and output		
		Safety factor for pressure due to preload	$p_L A_w / F_{f\,\max}$		**1.50**	*0.79*	**1.13**
		Safety factor for pressure due to preload and applied load	$p_L A_w/(F_{fmax}+\Phi_n W_a)$		**1.50**	*0.79*	**1.13**

References

1. Verein Deutscher Ingenieure, VDI-Richtlinien 2230 (1977) Systematic calculation of high duty bolted joints, 1977 [Translated to Japanese by the Japan Research Institute for Screw Threads and Fasteners, Systematic calculation method of high strength screw fastening, the Japan Research Institute for Screw Threads and Fasteners, 1982].
2. Verein Deutscher Ingenieure, VDI-Richtlinien Blatt 1 2230 (2003) Systematic calculation of high duty bolted joints -Joints with one cylindrical bolt-, 2003 [Translated to Japanese by Shinji Kase and Kenichi Kawai, The Japan Research Institute for Screw Threads and Fasteners, 2006].
3. Tomohiro Naruse and Yoji Shibutani, Higher Accurate Estimation of Axial and Bending Stiffnesses of Plates Clamped by Bolts, Journal of Solid Mechanics and Materials Engineering, 6-5 (2012), 397-406. https://doi.org/10.1299/jmmp.6.397
4. Tomohiro Naruse and Yoji Shibutani, Nonlinear Bending Stiffness of Plates Clamped by Bolted Joints under Bending Moment, Journal of Solid Mechanics and Materials Engineering, 6-7 (2012), 832-843. https://doi.org/10.1299/jmmp.6.832
5. Verein Deutscher Ingenieure, VDI-Richtlinien 2230 (1986) Systematic calculation of high duty bolted joints, 1986 [Translated to Japanese by the Japan Research Institute for Screw Threads and Fasteners, A Systematic Calculation Method for High Duty Bolted Joints -Cylindrical Single Bolt Fastening-, The Japan Research Institute for Screw Threads and Fasteners, 1989].
6. Akira Yamamoto, Principle and Design of Screw Fastening, Yokendo, (2004), p. 179, in Japanese
7. A. Hanau, Zum Krafteinleitungsfaktor bei der Berechnung von Schraubenvergindungen, Konstruktion, 46(1994), 99 -106(1994).
8. Masao Shibahara and Juhachi Oda, Axisymmetric deformation analysis of finite cylinder, Transactions of the JSME (in Japanese), 34-259(1968), 388-402.
9. Masao Shibahara and Juhachi Oda, Axisymmetric deformation analysis of finite cylinder (second report, the case under shear force), Transactions of the JSME (in Japanese), 36-282(1970), 168-176.
10. Masao Shibahara and Juhachi Oda, The stiffness of the clamped body of bolted joints (multi bolted joints), Transactions of the JSME (in Japanese), 37-297(1971), 1033-1039.
11. Toshiyuki Sawa, Hideaki Miyazawa and Kazuo Rikukawa, Simple method for calculating load factor of bolted joints (the case of T-flange), Transactions of the JSME C (in Japanese), 48-432(1982), 1200-1208.

12. Toshiyuki Sawa and Hiroyuki Shiraishi, Simple method for calculating load factor of bolted joints (the case for clamped body with circular flange and pipe flange), Transactions of the JSME C (in Japanese), 48-434(1982), 1598-1606.
13. Toshiyuki Sawa, Load factor of bolted joint, Transactions of JFRI, 19-11(1988), 363-371
14. Isam Yoshimoto, Kazuo Maruyama, Toshiyuki Sawa and Nobuyuki Nishiguchi, Load factor of bolted joints (the case of clamped body with hollow cylinder), Transactions of the JSME (in Japanese), 42-359(1976), 2174-2184.
15. Masaya Hagiwara, Simple method for calculating load factor of bolted joints (the case of clamped body under axi-symmetric tension), Transactions of the JSME C (in Japanese) 48-428(1982), 622-625.
16. Masaya Hagiwara and Isam Yoshimoto, The Relation between Load and Strength in Bolted Joints Subjected to Fatigue loading, Transactions of the JSME C (in Japanese), 52-475 (1986), 1072-1078.

Chapter 4
Mechanical Analysis, Looseness Analysis, and Strength Design Under Transverse Loading

4.1 Introduction

As mentioned in the previous chapters, bolted joints in mechanical structures are generally placed so that the axial direction of the bolt faces to the direction of the load. Therefore, many mechanical analyses and strength evaluations are often performed for axial loading cases. However, due to temperature differences in the structures or thermal expansion differences between members, there are many cases in which the bolted joints are subjected to loads transverse to the axis. In Chap. 3, Sect. 3.7.4, a simple treatment of such cases in the linear region where no slip occurs between the fastened parts was explained. In this section, the analyses are expanded to the nonlinear regions where slips occur between the fastened parts or between the bolt/nut/fastened parts, for loosening and failure estimations.

As an evaluation criterion for rotational loosening of bolts due to reciprocating slip in the transverse direction to the axis, an critical relative slip S_{cr} has already been proposed by many researchers [1–7]. This evaluation criterion had been derived considering the bending deformation and relative displacements between fastened elements, in which bolt was assumed to be a stepped bar, and deformations were derived considering bending compliance γ_b of the bolt shaft and the bolt head tilt rigidity k_w. However, our new experimental investigations have shown that for more accurate evaluation of this critical relative slip $S_{cr,}$ the introduction of additional mechanical considerations such as resisting bending moments from the nut M_w was indispensable. Therefore, in this chapter, we first propose analytical formulas for obtaining the more accurate tilt rigidity of the bolt head k_w and the resisting bending moment of the nut M_n. And then compare these analytical results with actual measurements of the axial stresses σ_a and bending stresses σ_b of the bolts and the loosening critical relative slips S_{cr}. And the validity of the proposed equation for the resisting bending moment M_n is verified. Furthermore, the more accurate evaluation equation of the loosening critical relative slip S_{cr} is constructed and its availability is confirmed.

© The Author(s), under exclusive license to Springer Nature Switzerland AG 2025

T. Hattori and T. Naruse, *Compendium for Design and Manufacturing of Bolted Joints*,
Solid Mechanics and Its Applications 280,
https://doi.org/10.1007/978-3-031-89628-6_4

4.2 Deformation Behavior in the Sticking State (Linear Behavior) Under Transverse Loading

As shown in Eq. (2.2) and Fig. 2.3 in Chap. 2, if the initial fastening force F_f is set sufficiently high relative to the external force W_T and satisfying following relation $F_f \mu_c \geq 1.5 \sim 2.0 W_T$, the bolted joint will not make slip at all against the external force in the direction perpendicular to the axis W_T, and there is little concern about strength or loosening behaviors, where μ_c is the coefficient of friction between the fastened body. The equivalent stiffness in this case is not specifically addressed in VDI2230, so in this chapter we analyze the bending compliances γ of bolt and fastened body as follows.

Bolt bending compliance γ_b can be derived using the both-ended rotation restrained beam model as referring same model of VDI 2230 (1977) [8] shown in Chap. 3 as follows.

$$\gamma_b = \frac{(0.4d)^3}{12E_b I_n} + \frac{l_g{}^3}{12E_b I_n} + \frac{l_s{}^3}{12E_b I_3} + \frac{(0.4d)^3}{12E_b I_n} \tag{4.1}$$

where d is the nominal diameter, E_b is the Young's modulus of the bolt, l_g is the length of the bolt cylindrical section, l_s is the thread length in the fastening section, I_n is the cross-sectional second moment at nominal diameter, and I_3 is the cross-sectional second moment at thread root diameter d_3. The first and last terms are the compliance of the bolt head and nut section, approximated as a cylinder of nominal cross-sectional area A_n with a length of $0.4d$.

In addition, in accordance with VDI 2230 (2003) [9], the following is derived.

$$\gamma_b = \frac{(0.5d)^3}{12E_b I_n} + \frac{l_g{}^3}{12E_b I_n} + \frac{l_s{}^3}{12E_b I_3} + \frac{(0.5d)^3}{12E_b I_3} + \frac{(0.4d)^3}{12E_b I_n} \tag{4.2}$$

The bending compliance γ_c of the fastened body can be derived using the equivalent hollow cylinder model with outer diameter D_{eq} and the equivalent cross-sectional second moment I_{eq} referring to VDI2230 (1977) [8] (see Fig. 3.2 in Chap. 3) as follows.

$$D_{eq} = d_w + \frac{l_f}{10} \tag{4.3}$$

$$I_{eq} = \frac{\pi}{64}\left(D_{eq}^4 - D_i^4\right) \tag{4.4}$$

$$\gamma_c = \frac{l_f^3}{12E_c I_{eq}} = \frac{16 l_f^3}{3\pi E_c \left(D_{eq}^4 - D_i^4\right)} \tag{4.5}$$

Using these, the equivalent beam's sectional second moment I_{teq} and equivalent diameter d_{teq} for transverse loading can be obtained as follows with Young's modulus of the equivalent beam as E_{beq} referring to the axial compliance model as shown in Fig. 3.2a in Chap. 3.

$$\frac{12E_{beq}I_{teq}}{l_f{}^3} = \frac{1}{\gamma_b} + \frac{1}{\gamma_c} = \frac{\gamma_b + \gamma_c}{\gamma_b \gamma_c} = \frac{1}{\gamma_b \Phi_\gamma} \tag{4.6}$$

$$I_{teq} = \frac{\pi}{64} d_{teq}^4 = \frac{l_f{}^3}{12E_{beq}\gamma_b \Phi_\gamma} \tag{4.7}$$

where Φ_γ is the load factor for transverse loading.

Using this load factor for transverse loading Φ_γ, we can calculate the transverse load W_b for a bolt as follows, using the total transverse load W_T applied to the bolted joint.

$$W_b = \Phi_\gamma \times W_T \tag{4.8}$$

The amplitude of the bending stress (axial stress on the surface) of the bolt σ_{ba} on the thread root diameter cross section) can be calculated using the section modulus Z_3 as follows.

$$Z_3 = \frac{\pi d_3{}^3}{32} \tag{4.9}$$

$$\sigma_{ba} = \frac{M_b}{2Z_3} = \frac{l_f \times W_b}{2Z_3} \tag{4.10}$$

This stress amplitude σ_{ba} acting on the bolt can be used for the fatigue strength design of bolted joint as follows. The fatigue limit σ_{wk} of the bolts are shown in Table 2.2 in Chap. 2.

$$\sigma_{ba} \leq \frac{\sigma_{wk}}{S_f} \tag{4.11}$$

4.3 Slipping Behavior Under Transverse Loading

In general, what is important in the mechanical analysis and strength design of a bolted joint under transverse loading is not the behavior in the sticking region described above, but the sliding behavior on the contact surface between the bolt/nut or the fastened bodies. Figure 4.1 shows the sliding behavior of a bolted joints under transverse loadings in chronological order.

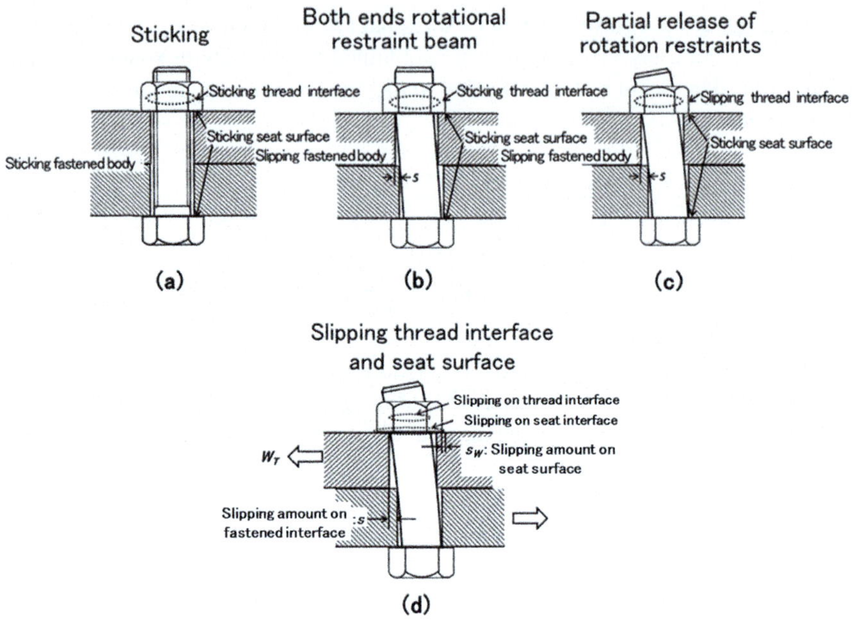

Fig. 4.1 Slipping behavior of bolted joint under transverse loading

(a) When the load is low, the bolt/nut and the fastened body (the two plates) deform as a single unit as described above Sect. 4.2 (see Fig. 4.1a).

(b) When the transverse load W_T exceeds the frictional force between the fastened bodies ($W_T = F_f\ \mu_c$), relative slippage occurs between the fastened bodies. Here μ_c is the coefficient of friction between the fastened parts. In this case, this relative slip s is absorbed by the bending deformation of the bolt and does not cause relative slip between the bolt head and the fastened body (Fig. 4.1b).

(c) Within the above range, the nut is rotationally restrained like the bolt head, and the bolt deforms as a both ends rotational restraint beam. However, as the transverse load increases further, slip occurs on the bolt/nut thread interface. Although this slip is precisely a complex three-dimensional slip, it is simplified by dividing it into transverse slip in the direction perpendicular to the axis (square screw) and bending slip around the axis perpendicular to the bolt axis (triangular screw) (Fig. 4.1c).

(c-1) Transverse slipping behavior perpendicular to the bolt axis (square screw)

Strictly speaking slipping in the direction perpendicular to the bolt axis (shown in Fig. 4.2) can't occur in screws with three-dimensional thread surfaces, such as general triangular threads, but in the case of square threads which have substantially planar thread surfaces can show this slipping behavior.

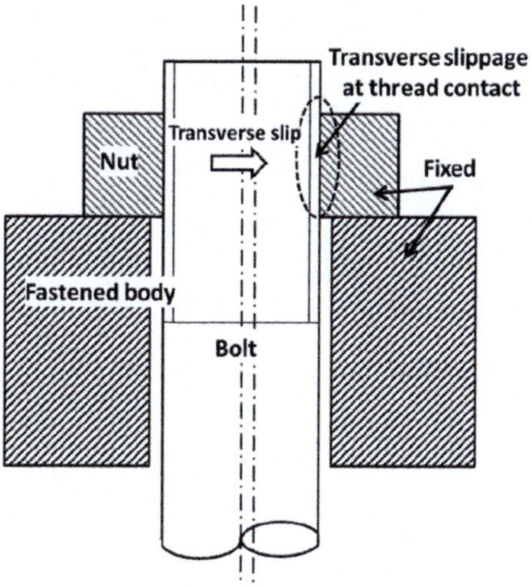

Fig. 4.2 Transverse slip at thread contact (square thread)

Figures 4.3 and 4.4 show analysis model and mechanical explanation of the conditions under which relative slipping occurs on the thread contact interface of one thread of a square screw in the direction perpendicular to the axis. When relative slipping occurs on the nut thread surface in the direction shown in the figure, the upper half of the figure shows sliding down along the lead angle, and the lower half shows sliding up along the lead angle. In this case, the reaction force on the bolt thread surface in the sliding direction is $N\mu_s \cos\beta' - N \sin\beta'$ at the coordinate *(r, θ)* in the upper half, and $N\mu_s \cos\beta' + N \sin\beta'$ works in the lower half.

where N and β' are denoted by

$$N = \frac{F_f}{\cos\beta} \tag{4.12}$$

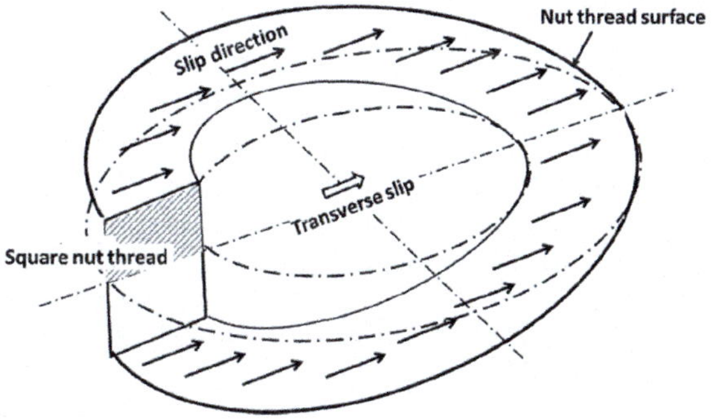

Fig. 4.3 Transverse slipping behavior of one thread of square screw

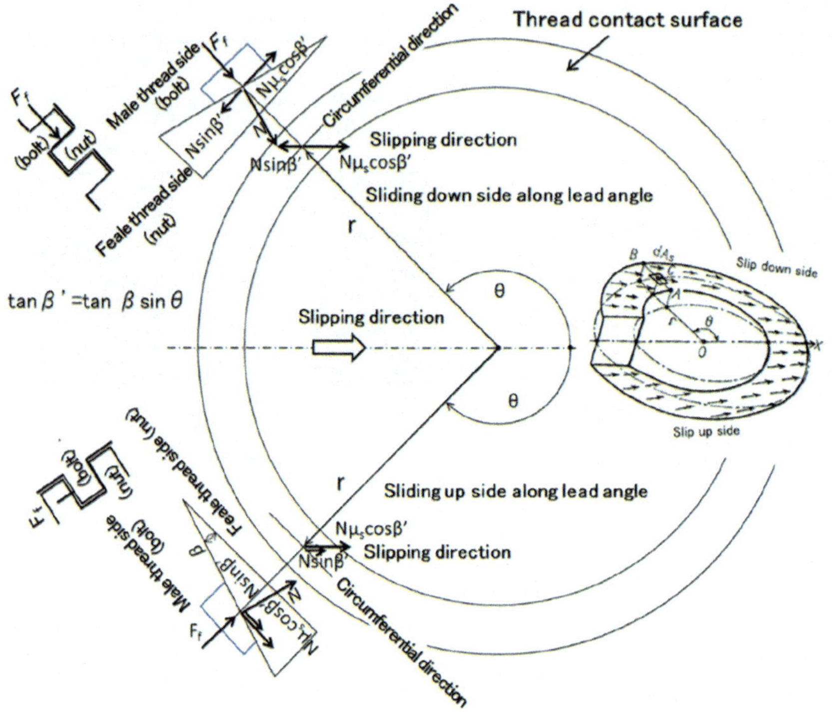

Fig. 4.4 Transverse slip behavior and mechanical explanation of a single thread of square screw

$$tan\beta' = tan\beta sin\theta \tag{4.13}$$

F_f Bolt fastening force.
β Lead angle

$$\tan\beta = p/\pi d_2$$

p Thread pitch
μ_s Friction coefficient of thread contact surface.
d_s Thread effective diameter.

Summing these over the entire circumference of one strip, the total transverse reaction force W_s is shown below.

$$W_s = \int_{r=r_1}^{r=r_3}\int_{\theta=-\pi}^{\theta=0} \frac{F_f}{A_s \cos\beta}\left(\mu_s \cos\beta' + \sin\beta'\right)dA_s$$
$$+ \int_{r=r_1}^{r=r_3}\int_{\theta=-\pi}^{\theta=0} \frac{F_f}{A_s \cos\beta}\left(\mu_s \cos\beta' - \sin\beta'\right)dA_s$$
$$= 2\int_{r=r_1}^{r=r_3}\int_{\theta=-\pi}^{\theta=0} \frac{F_f}{A_s \cos\beta}\left(\mu_s \cos\beta'\right)dA_s \tag{4.14}$$

whereas

$$A_s = \pi\left({r_3}^2 - {r_1}^2\right)/\cos\alpha$$

$$\mathrm{d}A_s = rd\theta dr/\cos\alpha$$

r_3 Outer radius of thread contact surface
r_1 Inner radius of thread contact surface.

Also, for simplicity, let us approximate $\cos\beta \approx \cos\beta' \approx 1,\ \cos\alpha' \approx \cos\alpha \simeq 1$,

$$W_s = F_f \mu_s \tag{4.15}$$

(c-2) Bending slipping behavior around the axis perpendicular to the bolt axis (triangular screw)

Strictly speaking this bending slipping (shown in Figs. 4.5 and 4.6) can't occur in spiral surface-to-spiral surface. So, in this section, we approximate this slipping behavior by a cone surface-to-cone surface slipping model as shown in Fig. 4.7a. And the critical slip moment M_c can be estimated by summing local tangential slipping force (sharing stress) along all contact surface, as shown in Fig. 4.7b and Eq. (4.15).

$$M_c = \frac{N\mu_s}{2\pi r_s}\int_0^{2\pi} r_s\sqrt{\cos^2\theta + 3}r_s d\theta$$
$$= \frac{2N\mu_s r_s}{\pi}\int_0^{\pi/2}\sqrt{\cos^2\theta + 3}d\theta = \frac{F_f\mu_s d_s}{\pi\cos\alpha}2.93$$
$$= 1.08F_f\mu_s d_s \tag{4.16}$$

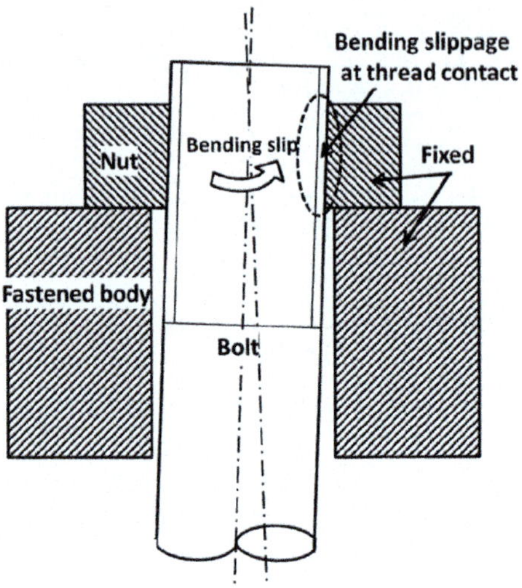

Fig. 4.5 Bending slip around the axis perpendicular to the bolt axis (triangular screw)

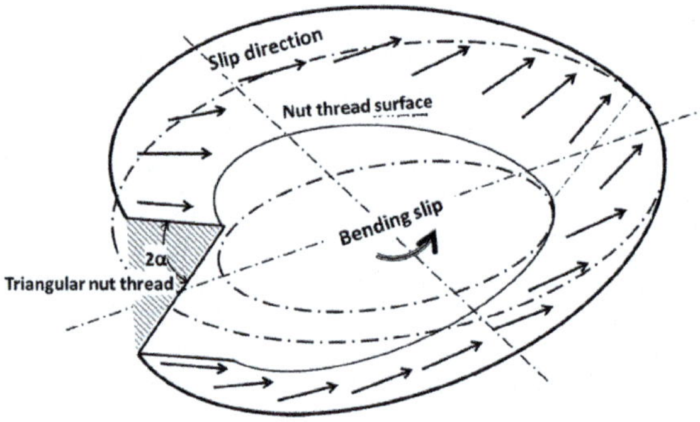

Fig. 4.6 Bending slipping behavior of one thread of triangular screw

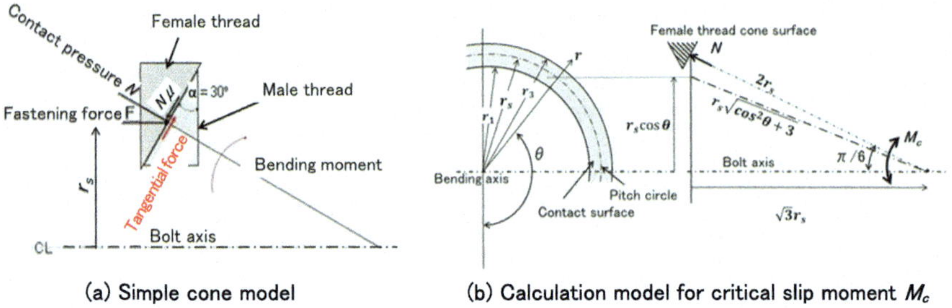

Fig. 4.7 Analytical models for calculation of critical slip moment M_c. **a** Simple cone model **b** Calculation model for critical slip moment M_c

Before this bending slip occurs, the bolt behaves as a both ends rotational restraint beam, as shown in Fig. 4.1b, so the bending moment M_n acting on the nut thread contact surface is expressed by the following equation.

$$M_n = \frac{W_T l_f}{2} \tag{4.17}$$

where l_f is the fastening length of the bolt. Using this relationship, the conditions under which this bending slip occurs are

$$W_b = 2.16 F_f \mu_s \frac{d_2}{l_f}$$

For ISO standard threads, from $d_s \approx 0.9d$,

$$W_b = 1.94 F_f \mu_s \frac{d}{l_f} \tag{4.18}$$

(d) When the transverse load W_T reaches to the total frictional force between the whole fastening bodies ($W_3 = F_f \mu_c + F_f \mu_w$), relative slip between the seat interface between bolt head and the fastened object occurs (see Fig. 4.1d), and after which the bolt remains fixed in deformation and the perpendicular load W_T keep constant as follows.

$$W_T = W_3 = F_f \mu_c + F_f \mu_w \tag{4.19}$$

If the frictional coefficient on seat interface μ_w is same as frictional coefficient on thread surface μ_s, this condition is just same as (c-1) and Eq. (4.15) in the case of square screw. However, in general, the majority of threads are triangular threads, so we will use the (c-2) condition for triangular threads and illustrate this series of behaviors (a)–(d) as shown in Fig. 4.8.

(e) Slipping behavior under cyclic transverse loading.

This section examines how the slipping behavior progress under the cyclic transverse loading. Under cyclic transverse loading, the process of releasing the bending moment of the bolt that was accumulated during the simple incremental loading process make it slightly complicated, but the load/displacement hysteresis under the displacement constraint condition can be shown schematically as Fig. 4.9. As shown later, it is known that nut rotation loosening will occur if seat surface slipping process (see above section (d)) is involved in this cyclic loading. Therefore, the critical amount of slip δ_{cr} for loosening under cyclic transverse loading is just twice the amount of deformation difference between on loading W_1, where slip between the fastened bodies occurs, and on W_3, where seat slippage occurs. This analysis is discussed in detail in Sects. 5.3 and 5.4.

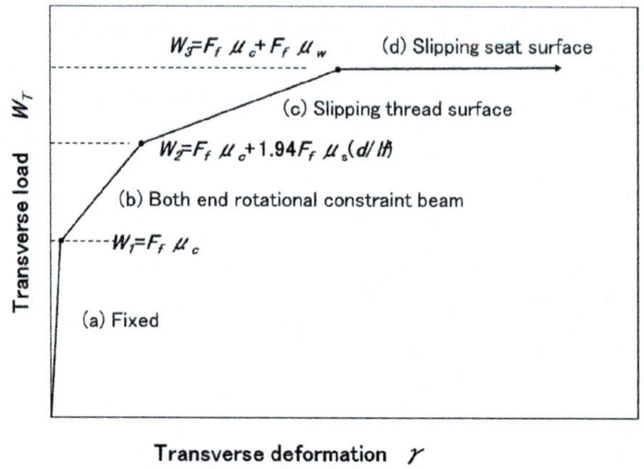

Fig. 4.8 Deformation behavior under transverse loading

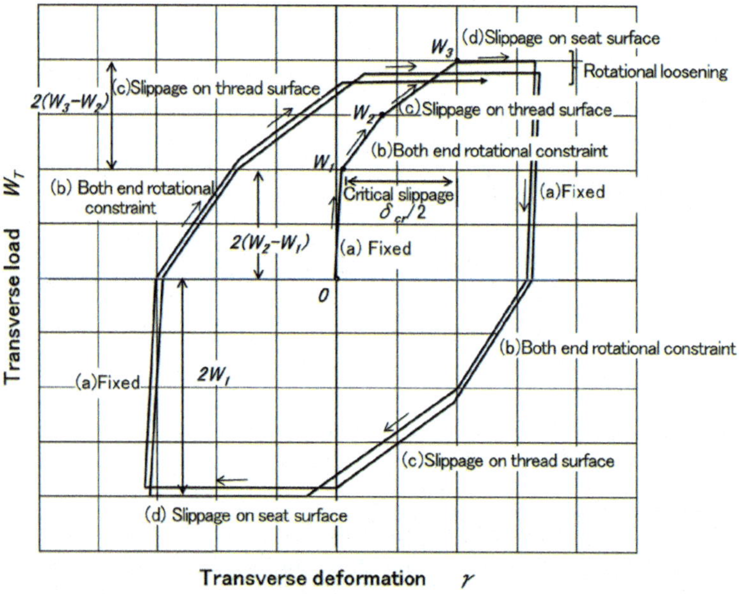

Fig. 4.9 Hysteresis behavior under cyclic transverse loading

4.4 Slipping Behavior and Rotational Loosening Mechanism Under Cyclic Transverse Loading

In the previous section, it was suggested that when relative slippage occurs between the bolt head or nut and the fastened body, rotation is allowed between the nut or bolt head and the fastened body and will lead the loosening of the bolt. In this case,

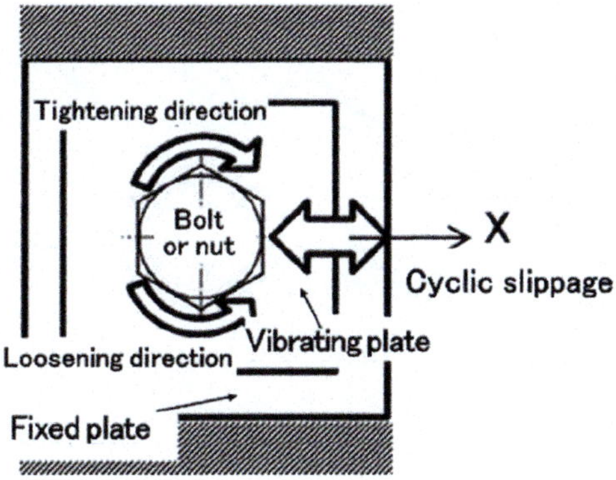

Fig. 4.10 Schematic illustration of cyclic relative slippage and rotational loosening of bolt or nut

however, as shown in Fig. 4.10, it is not clear whether the cyclic relative slippage between bolt head or nut and the fastened body leads to the rotation toward loosening direction or tightening direction of the bolt. Here, we will explain the mechanism of this rotational loosening by focusing on the slippage on the contact surface of thread fitting.

4.4.1 In the Case of Square Threads

Figure 4.11 shows the state of slippage and direction of force on the thread surface of a square thread [10].

Let O is a point on the bolt shaft centerline and X is the coordinate axis in the slip direction. And A, B, and C are positions on the slip surface at an angle θ from the X axis. And dA_s is a small area at point C on the slip surface.

The helical slope, viewed from the axial and radial directions, is shown in Fig. 4.12.

The lower half of Fig. 4.12 shows slip up behavior along the lead angle, and the upper half shows slip down behavior along the lead angle. Even if the bending

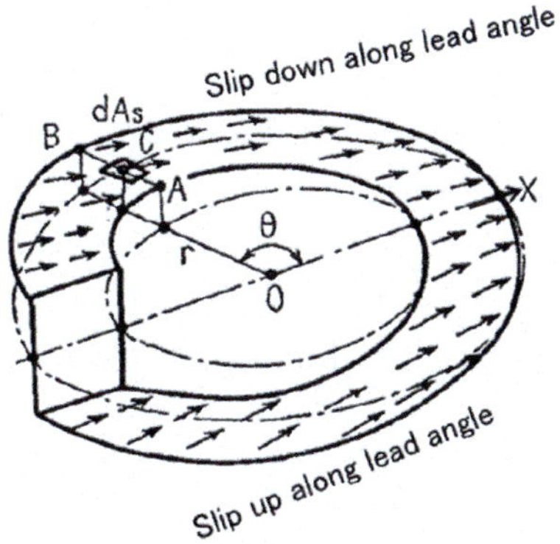

Fig. 4.11 Slipping conditions and force directions on thread surface of square thread under transverse loading

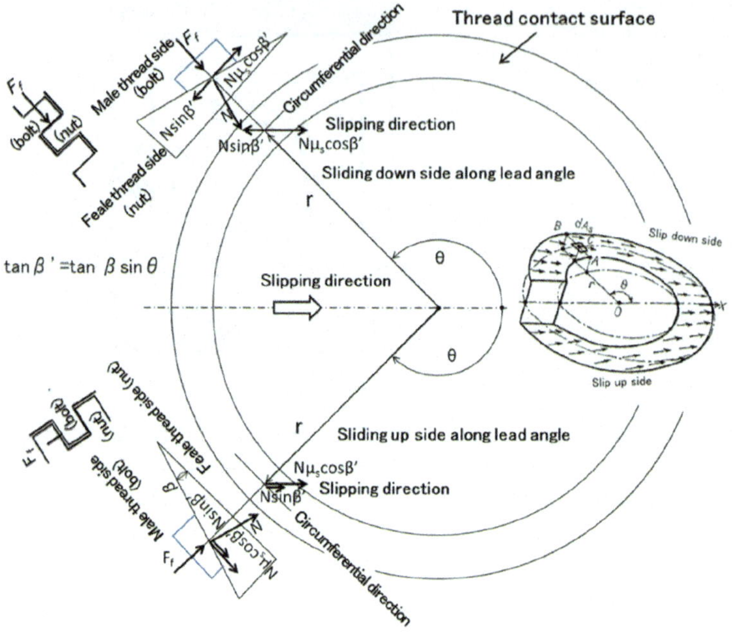

Fig. 4.12 Mechanical explanation of a single thread of square screw under transverse loading

deformation of the bolt is symmetrical about the neutral axis of bending as shown in Fig. 4.1, the mechanical asymmetry between the lower and upper halves of nut slip surface will receive some torque from the bolt thread surface by this mechanical asymmetry. The magnitude and direction of this torque (tightening direction or loosening direction) are mechanically analyzed as follows.

The relationship between the tightening force F_f and the contact pressure N acting on thread contact surface is calculated as follows,

On the sliding up side

$$N = F_f \frac{1}{\cos\beta - \mu_s \sin\beta} \tag{4.20}$$

On the sliding down side,

$$N = F_f \frac{1}{\cos\beta + \mu_s \sin\beta} \tag{4.21}$$

where β is the lead angle of the thread and μ is the coefficient of friction on the thread interface. As $\mu\sin\beta$ is sufficiently small compared to $\cos\beta$, both Eqs. (4.20) and (4.21) can be approximated as follows.

$$N = F_f / \cos\beta \tag{4.22}$$

The transverse component of the frictional force on both sides can be calculated with reference to Fig. 4.12 as follows.

On the sliding up side

$$\frac{F_f}{\cos\beta}\left(\mu_s\cos\beta' + \sin\beta'\right)$$

On the sliding down side,

$$\frac{F_f}{\cos\beta}\left(\mu_s\cos\beta' - \sin\beta'\right)$$

By multiplying these transverse component of frictional forces by the distance *rsinθ* from the bolt axis, the torque T_{ss} around the bolt axis can be calculated as follows.

$$T_{ss} = \int_{r=r_1}^{r=r_3}\int_{\theta=-\pi}^{\theta=0} \frac{F_f}{A_s\cos\beta}\left(\mu_s\cos\beta' + \sin\beta'\right)\cdot r\sin\theta dA_s - \int_{r=r_1}^{r=r_3}\int_{\theta=0}^{\theta=\pi} \frac{F_f}{A_s\cos\beta}\left(\mu_s\cos\beta' - \sin\beta'\right)\cdot r\sin\theta dA_s \tag{4.23}$$

The first term of this equation is the slide up side component and the second term is the slide down component. Here each parameter is shown below.

$$\tan\beta' = \tan\beta\sin\theta \tag{4.24}$$

$$\tan\beta = p/\pi d_2$$

β Lead angle

$$A_s = \pi\left(r_2{}^2 - r_1{}^2\right)$$

$$\mathrm{d}A_s = rd\theta dr$$

F_f Bolt fastening force.
P Thread pitch.
r_3 Outer radius of thread contact surface.
r_1 Inner radius of thread contact surface.
μ_s Friction coefficient of thread contact surface.

The both frictional force terms on the slide up side (first term) and the slide downside (second term) in Eq. (4.23) are canceled out, and resulting in the following

$$T_{ss} = \int_{r=r_1}^{r=r_2} \int_{\theta=-\pi}^{\theta=0} \frac{2F_f \sin\beta\prime}{A_s \cos\beta} r^2 \sin\theta dr d\theta \tag{4.25}$$

If β is sufficiently small, Eq. (4.24) can be approximated by

$$sin\beta' = sin\beta \sin\theta \tag{4.26}$$

$$\begin{aligned} T_{ss} &= \frac{2F_f \tan\beta}{A_s} \int_{r_1}^{r_3} r^2 dr \int_{-\pi}^{0} sin^2\theta d\theta \\ &= \frac{2F_f \tan\beta}{A_s} \left\{ \frac{1}{3}\left(r_3^3 - r_1^3\right)\frac{\pi}{2} \right\} \\ &= \frac{\pi F_f \tan\beta}{3A_s}\left(r_3^3 - r_1^3\right) \\ &= \frac{F_f \tan\beta\left(r_3^2 + r_3 r_1 + r_1^2\right)}{3(r_3 + r_1)} \end{aligned} \tag{4.27}$$

And if the following formula for the screw thread shape of JIS triangular threads

$$\left.\begin{aligned} r_1 &= \frac{d_2}{2} - \frac{0.866}{4}p \\ r_3 &= \frac{d_2}{2} + \frac{0.866 \times 3}{8}p \end{aligned}\right\} \tag{4.28}$$

Hear, d_2 is effective diameter of thread.

are applied to the square threads here, Eq. (4.25) can be approximated by the following equation.

$$T_{ss} \approx \frac{F_f \tan\beta}{3}\left(\frac{3}{4}d_2 + 0.082p\right) \cong \frac{F_f d_2 \tan\beta}{4} \tag{4.29}$$

On the other hand, the loosening torque T_l required to rotate and loose a bolt that is fastened by the axial force F_f can be derived by referring to the following equation for the tightening torque T_f (same with (2.3) (2.4) and (2.5) in Chap. 2),

$$T_f = \frac{F_f}{2}\left(\frac{p}{\pi} + \mu_s d_2 \sec\alpha' + \mu_w D_w\right) \tag{4.30}$$

and changing the sign of the term for the lead angle to minus (–) as the follows.

$$T_l = \frac{F_f}{2}\left(-\frac{p}{\pi} + \mu_s d_2 \sec\alpha' + \mu_w D_w\right) \tag{4.31}$$

In these equations, each symbols are shown below.

μ_s Friction coefficient of thread contact surface.
μ_w Friction coefficient of seat contact surface.
D_w Equivalent seat diameter

$$D_w = \frac{2\left(d_w{}^3 - D_i{}^3\right)}{3\left(d_w{}^2 - D_i{}^2\right)}$$

d_w Outer diameter of seat contact surface (width across flats of nut)
D_i Seat contact surface inner diameter (bolt hole diameter).

Therefore, the following equation must be satisfied in order to rotate and loosen bolt by only this torque T_{ss} alone.

$$\frac{F_f}{2}\left(-\frac{p}{\pi} + \frac{\mu_s d_2}{\cos\alpha'} + \mu_w D_w\right) < \frac{F_f d_2 \tan\beta}{4} \tag{4.32}$$

whereas, if we approximate each parameter as follows,

$$\frac{p}{\pi d_2} = \tan\beta \approx \beta \tag{4.33}$$

$$\alpha\prime \approx \alpha \tag{4.34}$$

P Pitch of bolt thread.
β Lead angle of thread.
d_2 Effective diameter of thread.
α Flank angle of thread.
α' Flank angle of cross section perpendicular to thread.

Equation (4.32) can be approximated as follows.

$$\frac{\mu_s}{\cos\alpha} + \frac{D_w \mu_w}{d_2} < 1.5\beta \tag{4.35}$$

And if we assume that the thread surface friction coefficient μ_s and the seat surface friction coefficient μ_w are equal, this equation can be shown more simple as follows.

$$\mu_s = \mu_w < 0.025 \tag{4.36}$$

Such a low coefficient of friction is unrealistic, and at first glance it seems that this loosening will not occur. However, this condition is a condition for rotational slippage to occur from a stationary contact state, but in reality, as shown in Fig. 4.9, slipping has already occurred on the thread surface (c) under this seat surface slipping load T_{ss} (d), so rotational slipping from this state is probably quite possible. with a normal friction coefficient of about 0.15. Details will be explained in Sect. 4.8, Fig. 4.26.

4.4.2 In the Case of Triangular Threads

In the case of a triangular thread, Fig. 4.13 can be used as the mechanical model. In this mechanical model, the contact surface pressure N and the slope tanβ' in the sliding direction are modified and expressed as follows instead of Eqs. (4.21) and (4.24).

$$N = F_f / \cos\beta \cos\alpha' \tag{4.37}$$

$$\tan\beta' = \tan\alpha\cos\theta + \tan\beta\sin\theta \tag{4.38}$$

As a result, Eq. (4.29) for square thread case can be modified as follows.

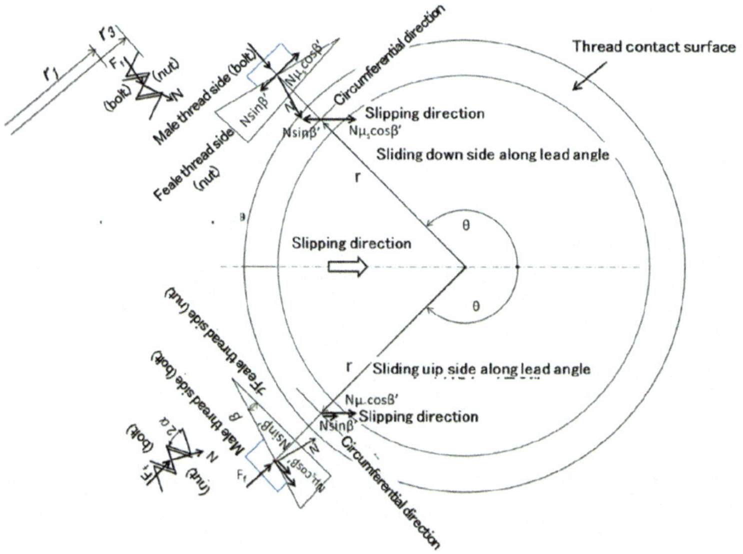

Fig. 4.13 Mechanical explanation of a single thread of triangular screw under transverse loading

$$T_{ss} \cong \frac{F_f d_2 \tan\beta}{4\cos\alpha'} \tag{4.39}$$

And the loosening condition can be expressed as follows.

$$\frac{F_f}{2}\left(-\frac{p}{\pi} + \frac{\mu_s d_2}{\cos\alpha\prime} + \mu_w D_w\right) < \frac{F_f d_2 \tan\beta}{4\cos\alpha'} \tag{4.40}$$

and, using the same approximation as for the square thread case, Eq. (4.40) becomes

$$\frac{\mu_s}{\cos\alpha} + \frac{D_w \mu_w}{d_2} < \tan\beta\left(1 + \frac{1}{2\cos\alpha'}\right) \tag{4.41}$$

and assuming that the thread surface friction coefficient μ_s and the seat surface friction coefficient μ_w are equal; this equation can be shown more simple as follows.

$$\mu_s = \mu_w < 0.032 \tag{4.42}$$

As explained in Sect. 4.4.1, this condition is a condition for rotational slippage to occur from a stationary contact state, but in reality, as shown in Fig. 4.9, slipping has already occurred on the thread surface (c) under this seat surface slipping load T_{ss} (d), so rotational slipping from this state is probably quite possible. with a normal friction coefficient of about 0.15. Details will be explained in Sect. 4.8, Fig. 4.26.

Please refer to Sect. 4.3 for the slip limit criteria when transverse slip and rotational slip perpendicular to the axis are superimposed.

4.5 Critical Relative Slippage S_{cr}

In Sect. 4.4, the mechanism of rotation and loosening of bolt due to slipping of the seat surface under transverse loading was explained. Yamamoto and Kase et al. [4, 7] proposed an evaluation equation for the critical relative slipping amount for the rotational loosening of a bolt under transverse loading, using experimental results when the thread contact between bolt and nut is in metal contact and the fastened bodies are metal material. In this section, we modified these equations by introducing the bending compliance β_{wh} and β_{wn} that take into account the local deformation of the contact area between the bolt head or nut and the fastened body, and slip resistance moment M_n at the thread fitting part of bolt and nut.

Figure 4.14 shows a simple calculation model for bending deformation of bolts. Yamamoto and Kase et al. [4, 7] derived critical relative slippage S_{cr} by assuming that the bolt is a stepped round cantilever bar with the unthreaded part of the bolt nominal diameter and the threaded part of the effective thread diameter, and thread fitting part of bolt and nut is rotation free constraint.

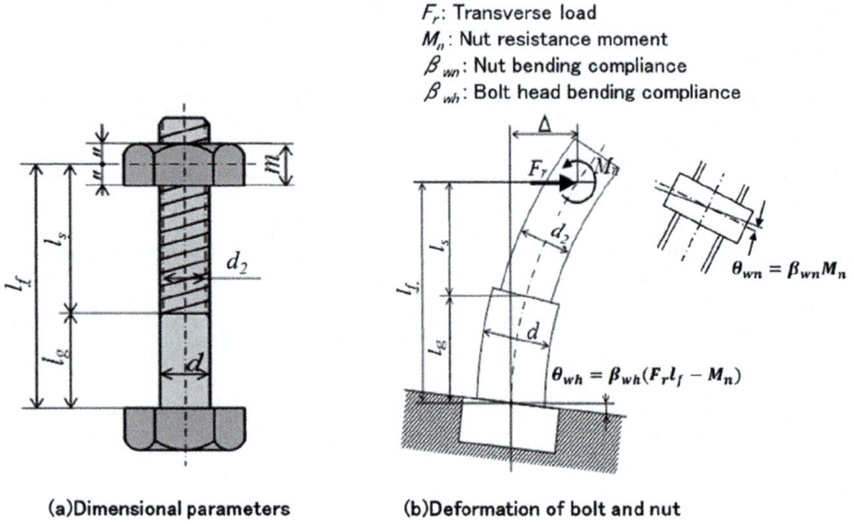

Fig. 4.14 Dimensional parameters and calculation model for bending deformation of bolt and nut under transverse loading

When a bolted joint which is fastened with a bolt axial force F_f experiences slip under the transverse loading, the frictional force (transverse load) F_r, and a resistance moment M_n due to the nut will act on thread fitting part of bolt and nut. As shown in Fig. 4.14, considering the transverse load $F_r = F_f \, \mathrm{X}\mu_w$, the nut resistance moment M_n, and the bending compliance β_{wh} and β_{wn} of the contact area between the bolt head or nut and the fastened body, the bending deformation Δ of the bolt under this load is expressed as follows.

$$\Delta = F_f \mu_w \left(\frac{l_g^3}{3EI_g} + \frac{l_s^3}{3EI_s} + \frac{l_g l_s l_f}{EI_g} + \beta_{wh} l_f^2 \right) - M_n \left(\frac{l_g^2}{2EI_g} + \frac{l_s^2}{2EI_s} + \frac{l_g l_s}{EI_g} + k_{wh} l_f \right) + M_n \beta_{wn} l_f \tag{4.43}$$

And the critical relative slippage S_{cr}, which is the whole slip amount under reciprocal repetitive load cycle, is expressed as follows.

$$S_{cr} = 2\Delta = 2 \left\{ F_f \mu_w \left(\frac{l_g^3}{3EI_g} + \frac{l_s^3}{3EI_p} + \frac{l_g l_s l_f}{EI_g} + \beta_{wh} l_f^2 \right) - M_n \left(\frac{l_g^2}{2EI_g} + \frac{l_s^2}{2EI_p} + \frac{l_g l_s}{EI_g} + k_{wh} l_f \right) + M_n \beta_{wn} l_f \right\} \tag{4.44}$$

where each symbol represents the following:

- E Modulus of longitudinal elasticity of bolt.
- I_g Cross-sectional secondary moment of unthreaded section.
- I_s Sectional secondary moment of the thread (effective diameter d_2).
- l_g Length of unthreaded part.
- l_s Length of threaded part.
- l_f Tightening length ($l_g + l_s$).
- μ_w Coefficient of friction on the seat surface.
- β_{wn} Nut bending compliance.
- β_{wh} Bolt head bending compliance.

4.6 Bending Moments Acting on Bolts

In addition to the aforementioned slippage and loosening, there is also bolt bending fatigue to consider in the strength design of bolted joints that are subjected to cyclic transverse loads. So, bending fatigue strength evaluation is performed here using the bolt bending moment analysis formula shown earlier. The relationship between the transverse load W_t and the bending moments M_h, M_n acting just below the bolt head (below the neck) and the fit edge with nut, respectively, are shown in Fig. 4.15.

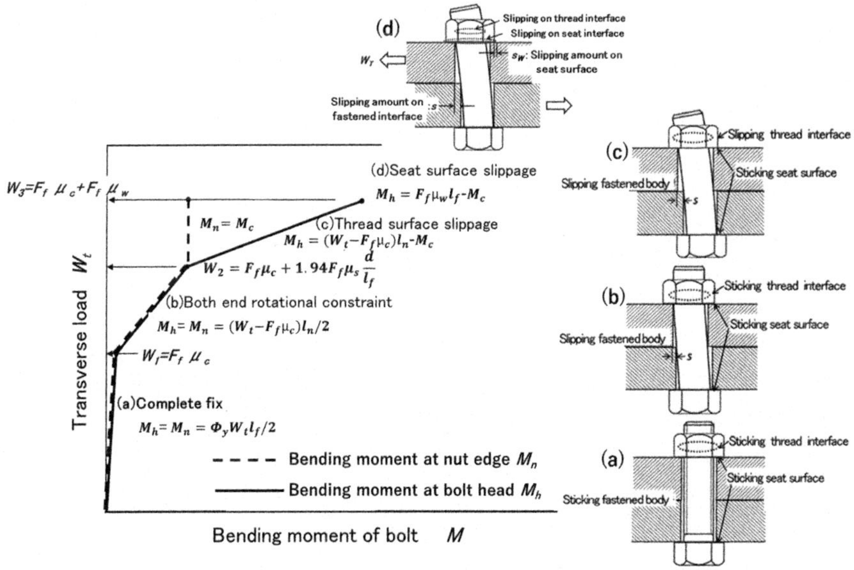

Fig. 4.15 Bending moments in bolt under transverse loading

4.6.1 Complete Fixed Area

As described in Sect. 4.3, if the transverse load is less than the frictional force between the fastened body ($F_f \mu_c$), show the simple linear elastic deformation behavior and the bending moments M_h, M_n acting on bolt can be derived using the bending compliance β_b of the bolt, the bending compliance β_c of the fastened body and the load factor Φ_β as follows.

$$M_h = M_n = \Phi_\beta W_t l_f/2 = W_t l_f \beta_c/2(\beta_b + \beta_c) \tag{4.45}$$

4.6.2 Both Ends Rotationally Constrained Area

When the transverse load W_t become greater than $W_I = F_f \mu_c$, the contact surface between the fastened bodies slips, while causing an increase in bending moment to the bolt. On this transverse loading, the bolt shaft is deformed in a state where both ends are restrained from rotating, and the moments M_n, M_h acting on the nut fit edge and the bottom of the bolt neck are derived as follows.

$$M_h = M_n = (W_t - F_f \mu_c) l_n/2 \tag{4.46}$$

4.6.3 Thread Surface Slip Area

When the transverse load is greater than

$$W_2 = F_f \mu_c + F_f \mu_s \frac{d}{l_f},$$

the bending moment generated in the bolt increase and if bending moment at nut fit edge exceeds the nut's critical moment of resistance M_c, partially release of the rotational restraint occur on the nut side. So, the bending moment at nut side is shown as follows.

$$M_n = M_c \tag{4.47}$$

And bending moment at bolt head side is shown as follows.

$$M_h = \left(W_t - F_f \mu_c\right) l_f - M_c \tag{4.48}$$

4.6.4 Seat Slip Area

When the transverse load W_t become greater than $W_3 = F_f \mu_c + F_f \mu_w$, both seating surfaces of the bolt head and nut slip, and the transverse load cannot increase any further and remains constant, as shown in Fig. 4.15. So, the bending moment at nut side is shown as follows.

$$M_n = M_c \tag{4.49}$$

And bending moment at bolt head side is shown as follows.

$$M_h = F_f \mu_w l_f - M_c \tag{4.50}$$

4.7 Bending Compliances of Bolt Head β_{wh} and Nut Seat Surface β_{wn}

Since a simplified solution for the bending compliance (reciprocal of stiffness) of the bolt head and nut seating surface is difficult to obtain, a finite element method analysis is used here. Figure 4.16a shows the FEM model of the contact seating area between the bolt head and the fastened body. Figure 4.16b shows the calculated result of bending deformation at contact seat. The analytical results of bending compliance at contact seat for M6, M10, and M20 are shown in Fig. 4.17 with marks (○). symbol, and the following equation was obtained as their regression equation.

$$\beta_{wh} = 0.168d^{-3}\,(\mathrm{kN\,mm})^{-1} \tag{4.51}$$

This equation is shown as a solid line in Fig. 4.17. The tilt coefficient at the nut seating surface, β_{wn}, is also the same as in Eq. (4.51).

4.8 Experimental Verifications of Critical Relative Slippage S_{cr} and Bending Moment of Bolt M_b, M_n

4.8.1 Experimental Apparatus and Methods

Experiments were conducted to verify the validity of the prediction equations for critical slip S_{cr} and bending moment of bolt M_b, M_n under transverse loading. Figure 4.18 shows a schematic diagram of the experimental model. Two SS400 steel plates (each 9 mm thick) were tightened with M10 or M16 bolts and nuts, and subjected to cyclic

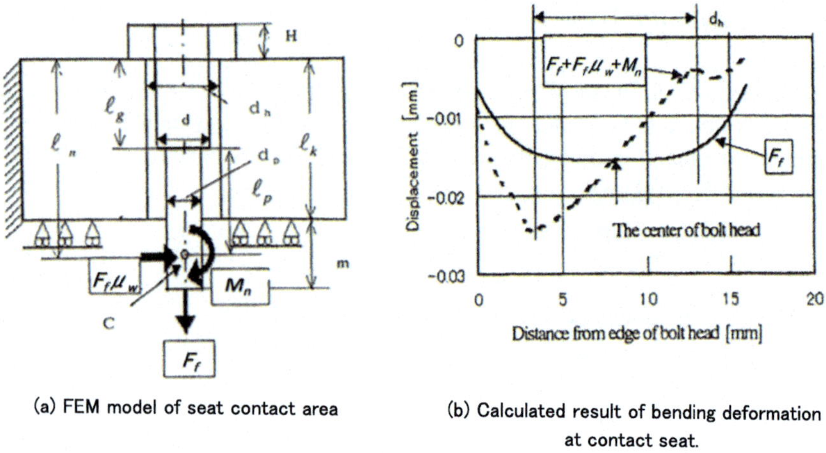

Fig. 4.16 FEM model and deformation result at contact seat

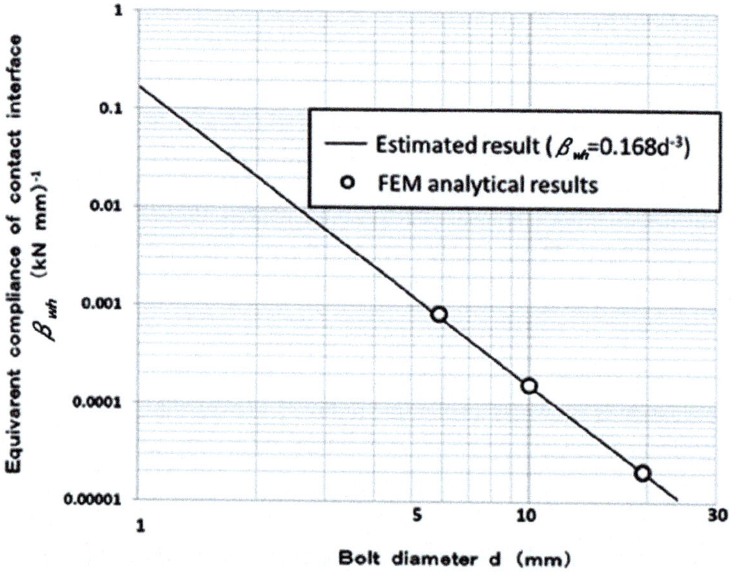

Fig. 4.17 Calculation results and regression line of bending compliance of bolt seating surface

loading in the direction perpendicular to the bolt axis (cycle sped 1 Hz) using an electro-hydraulic servo fatigue testing machine. Load was measured by a load cell, relative slippage was measured by a strain gage displacement sensor, bolt axial force (tightening force) was measured by a bolt gage, and bending moment was measured by a strain gage pasted to both sides of the bolt head (see Fig. 4.19).

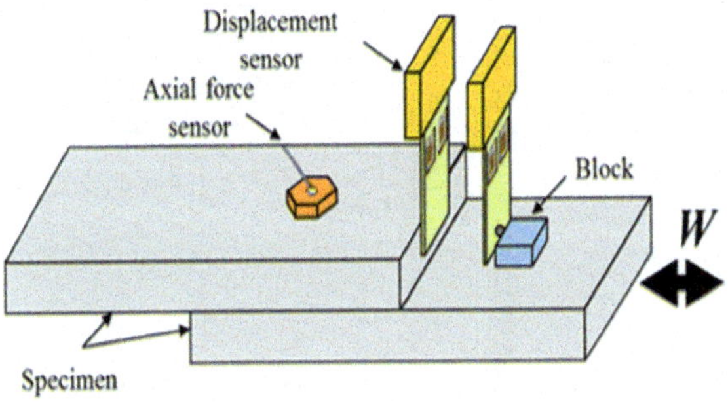

Fig. 4.18 Schematic diagram of the experimental model

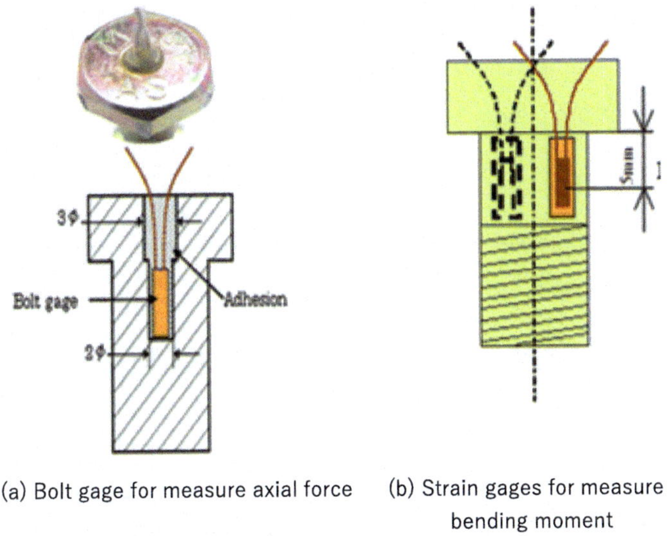

Fig. 4.19 Strain gages for measurement of bolt axial force and bending moment

4.8.2 Experimental Results

An example of the decrease in fastening force during cyclic loading is shown in Fig. 4.20. This is the result for an M16 bolt with an initial fastening force of 15 kN. The relationship between the loosening velocity *dF/dN* and the relative slip amplitude *S* at 1000 cycles for various bolts are shown in Fig. 4.21. The "reduced shank bolt" in this figure will be explained in detail in Sect. 4.10. It was found that when the relative slip amplitude exceeds a certain value, the loosening speed increases rapidly and the nut begins to rotate. This relative slip amplitude is the measured critical slip of loosening, S_{cr}.

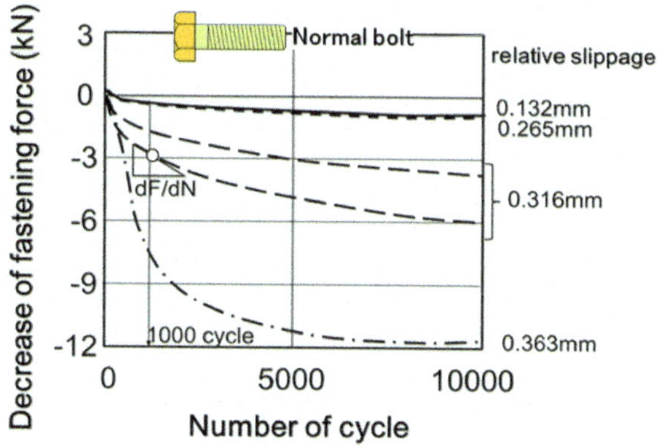

Fig. 4.20 Experimental results of loosening behavior under various cyclic transverse loadings (M16)

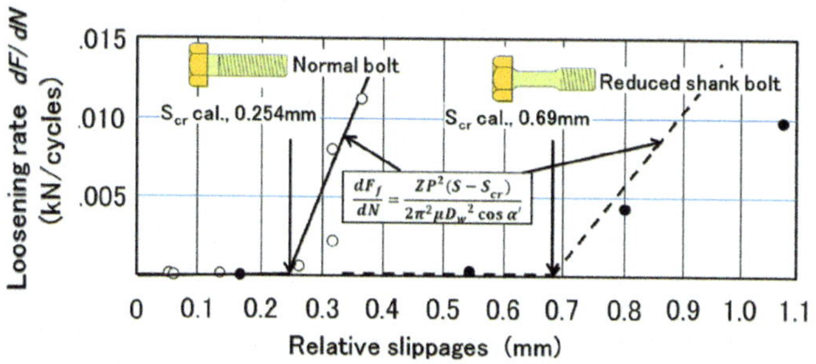

Fig. 4.21 Critical slippages S_{cr} determined from experimental results and comparison with analytical results (M16)

The measured results of this critical slip amount S_{cr} under various bolt sizes and initial fastening forces are shown in Figs. 4.22 and 4.23 with marks (●). The analytical results using Eq. (4.41) in Sect. 4.5 are also shown as solid lines. Both results are in relatively good agreement with each other, so we can confirm the validity of the prediction method for the critical slip amount. The measured bending moment at the bottom of the bolt neck under cyclic loading in the direction perpendicular to the axis is also shown in Figs. 4.24 and 4.25 with marks (●). The analytical results based on Eqs. (4.42) and (4.46) in Sect. 4.6 are shown as dashed lines and marks (□).

4.8.3 *Mechanical Analysis of Rotational Loosening Speed Under Transverse Slippage*

In Sect. 4.4, we explained that the nut loosening drive torque T_{ss} under transverse slipping is small, and in order for the threaded surface to slip from a static state, the coefficient of friction on thread surface must be $\mu_s \leqq 0.032$. Therefore, rotational loosening seemed impossible on real structures.

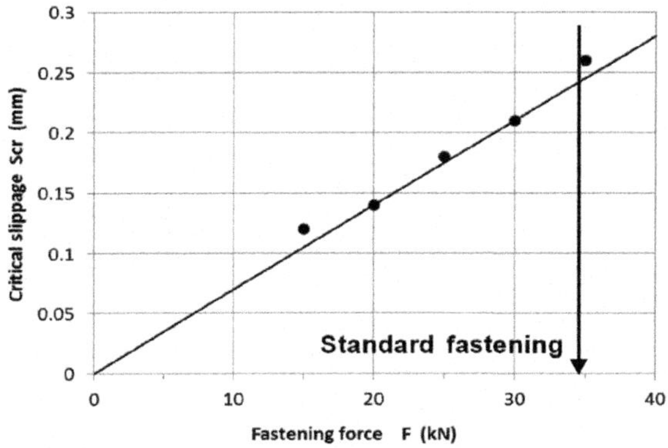

Fig. 4.22 Experimental results of critical slippages S_{cr} and comparison with analytical result (M16, $l_f = 18$ mm, normal bolt)

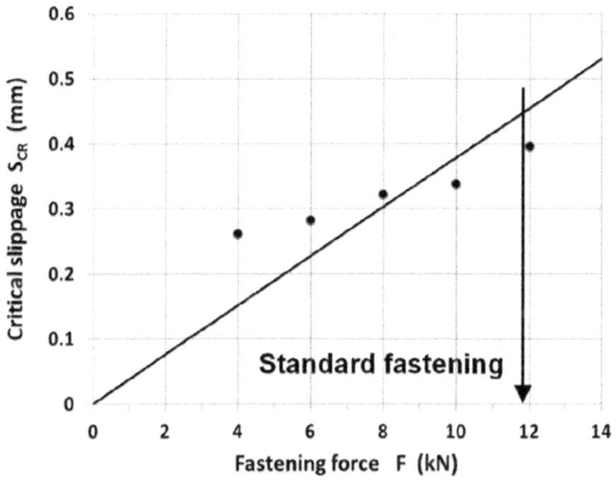

Fig. 4.23 Experimental results of critical slippages S_{cr} and comparison with analytical result (M10, $l_f = 18$ mm, normal bolt)

However, under the loading of this seat surface slippage (Fig. 4.8d), a slip in the direction perpendicular to the axis has already occurred on the thread surface (Fig. 4.8c), and if the circumferential (loosening) torque T_{ss} acts in conjunction with this transverse slip, it is thought that circumferential slippage also occurs on the screw surface and rotational loosening can occurs even if T_{ss} is small. This conjunctional loosening mechanism will be explained using figure. Figure 4.26 shows the slipping behavior on the thread surface and seat surface when the relative slipping amount between fastened bodies S is above S_{cr}. Here, the slippage amount S_w on the seat surface is given by the following equation.

$$S_w = S - S_{cr} \tag{4.52}$$

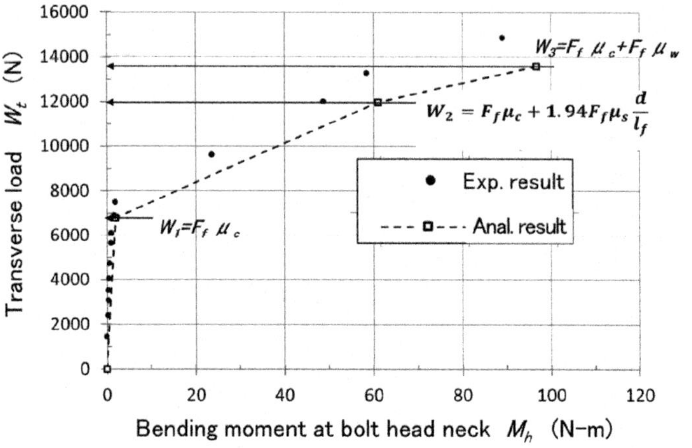

Fig. 4.24 Experimental results of bending moment at bolt head neck and comparison with analytical result (M16, $l_f = 18$ mm, normal bolt)

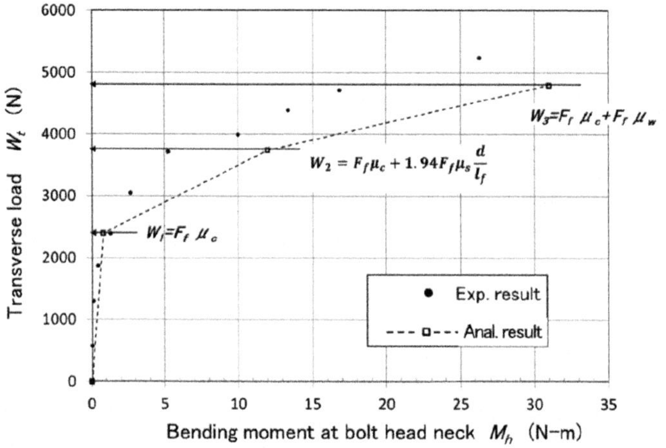

Fig. 4.25 Experimental results of bending moment at bolt head neck and comparison with analytical result (M10, $l_f = 18$ mm, normal bolt)

A mechanical analysis of the local slip condition at the seat surface under these conditions is shown in Fig. 4.27. A fastening force F_f acts on the seating surface in the axial direction, a friction force $F_f \times \mu$ acts in the direction perpendicular to the axis, and T_{ss}/r_w acts in the circumferential direction. Here, r_w is half of the effective diameter D_w of the seat surface. Assuming that the combined direction of the frictional force $F_f \times \mu$ and the circumferential force acting on the seating surface T_{ss}/r_w coincide with the slipping direction, the circumferential slip component S_T is expressed by the following equation.

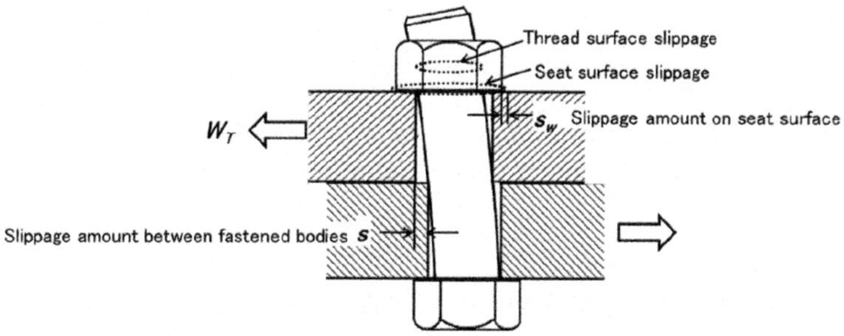

Fig. 4.26 Slipping behavior on the thread surface and seat surface when the relative slipping amount between fastened bodies S is over S_{cr}

$$\begin{aligned} S_T &= S_w \times \frac{T_{ss}/r_w}{F_f \mu_w} = S_w \times \frac{2T_{ss}}{F_f \mu_w D_w} = S_w \frac{\frac{F_f d_2 \tan \beta}{2\cos \alpha'}}{F_f \mu_w D_w} \\ &= S_w \frac{d_2 \tan \beta}{\mu_w D_w 2\cos \alpha'} = \frac{S_w P}{\pi \mu_w D_w 2\cos \alpha'} \end{aligned} \tag{4.53}$$

If this amount of circumferential slip occurs at the effective diameter of the seat surface, the slip angle $d\theta/dN$ of the nut or bolt sea surface per cycle of transverse loading can be expressed as follows.

$$\frac{d\theta}{dN} = \frac{S_T}{r_w} = \frac{PS_w}{\pi \mu_w D_w{}^2 \cos\alpha'} \tag{4.54}$$

Therefore, the loosening rate dF_f/dN per cycle of transverse loading is given by using the embedment coefficient $Z = 1/(\delta_b + \delta_c)$, as follows.

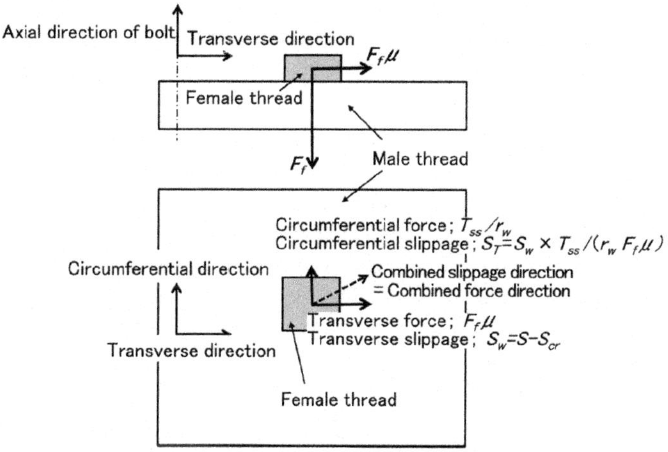

Fig. 4.27 Mechanical analysis of sliding direction on thread surface under combined load

$$\frac{dF_f}{dN} = Z\frac{P}{2\pi}\frac{PS_w}{\pi\mu_w D_w{}^2\cos\alpha'} = \frac{ZP^2S_w}{2\pi^2\mu_w D_w{}^2\cos\alpha'} or = \frac{ZP^2(S - S_{cr})}{2\pi^2\mu_w D_w{}^2\cos\alpha'} \tag{4.55}$$

In Fig. 4.21, a comparison of this predicted result (solid line) with the experimental results (○) is shown and relatively good agreement can be observed.

4.9 Optimal Design of Bolted Joints Considering Loosening and Fatigue Behaviors

Using the analytical results for the loosening limit slip S_{cr} and the bending moment M applied to the bolt, for which the validity of the analysis was confirmed in the previous chapter, we will try an optimal design study that takes both loosening and fatigue. Assuming that the fatigue limit at the bolt neck is σ_{wb} and the fatigue limit at the fitting area with the nut is σ_{wn}, the respective fatigue-limiting bending moments of both areas M_{bc} and M_{nc} can be obtained as follows

$$M_{bc} = Z_b\sigma_{wb} \tag{4.56}$$

$$M_{nc} = Z_n\sigma_{wn} \tag{4.57}$$

Inserting these fatigue-limiting bending moments into Fig. 4.28, the respective fatigue-limiting external forces W_{bc} and W_{nc} are obtained from the intersection with the dashed line. The relative slip S under these conditions can be obtained by inserting $(W_{bc} - F_f\mu_c)$ or $(W_{nc} - F_f\mu_c)$ instead of $F_f\mu_w$ in Eq. (4.44). These relative slippages represent the critical slippages S_{cr} taking into account the fatigue failure, and applied examples on M10, $l_f = 18$ mm bolt and M10, $l_f = 18$ mm bolt are shown in Figs. 4.29 and 4.30 as single-pointed and dashed lines. The design allowable area considering loosening and fatigue failure is shown by the hatched area in Figs. 4.29 and 4.30.

In these two example cases (M10, $l_f = 18$ mm bolt and M10, $l_f = 18$ mm bolt), the tightening length of the bolts is relatively short compared to the bolt diameter, so in these cases the critical slippages are determined based on fatigue failure criteria rather than loosening criteria. On the other hand, considering the fact that the transverse cyclic loadings are generally experienced in low-cycle regions such as thermal cycling, and therefore low-cycle fatigue strength criteria of about $N = 10^4$ cycles are often used. In such cases, the design based on loosening criterion will be considered to be necessary. In the actual design of bolted joint structures, this should be taken into consideration.

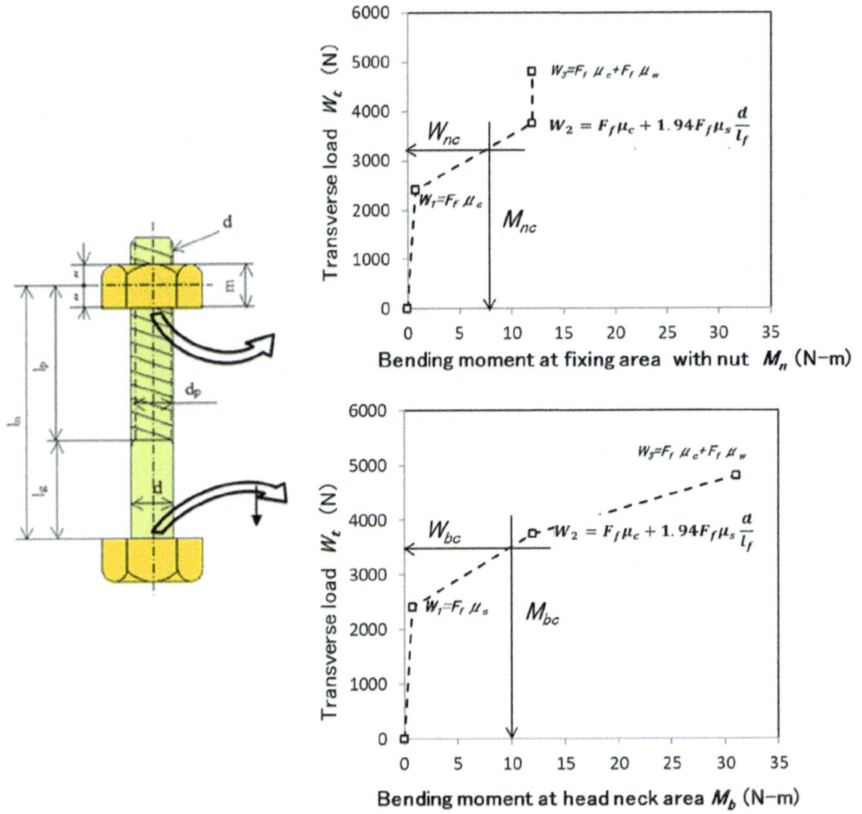

Fig. 4.28 Critical bending moments M_{bc} and M_{nc} and critical transverse loads W_{bc} and W_{nc} based on fatigue limit

4.9.1 Increase of Critical Slip S_{cr} and Reduction of Loosening Speed by Using Reduced Shank Bolts

In Chap. 3, Sect. 3.3.3, we proposed an reduced shank bolt (with a thinner bolt body diameter) as a measure to improve fatigue strength under axial loading. This measure to reduce bolt rigidity is also effective for transverse loading, and here we present effect results confirmed through experiments and analysis. First, regarding the critical slip S_{cr}, it is found that it increases as the cross-sectional secondary moment I_g of the bolt shank is reduced in Eq. (4.44). As for the loosening velocity, it is found that it decreases in Eq. (4.55) in accordance with the embedment coefficient Z becomes smaller. These calculation results and experimental results are shown below. Figure 4.31 shows a comparison between the reduced shank bolts used in the analysis and experiments, and the ordinary bolts used in previous experiments and analyses. Figure 4.32 shows the measured results of the loosening behavior of the reduced

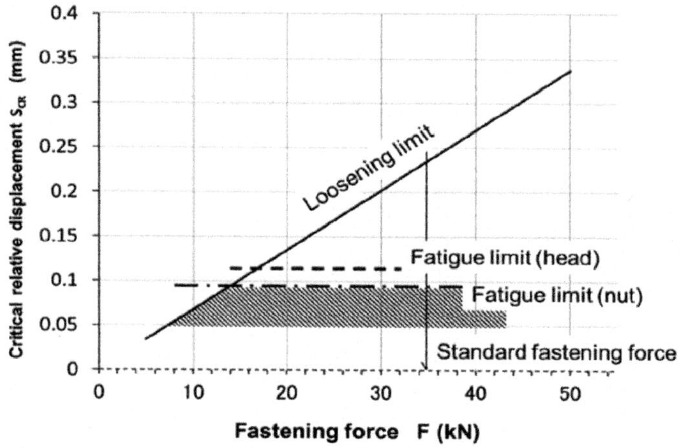

Fig. 4.29 Critical slippage S_{cr} considering loosening and fatigue behaviors (M16, $l_f = 18$ mm Normal bolt)

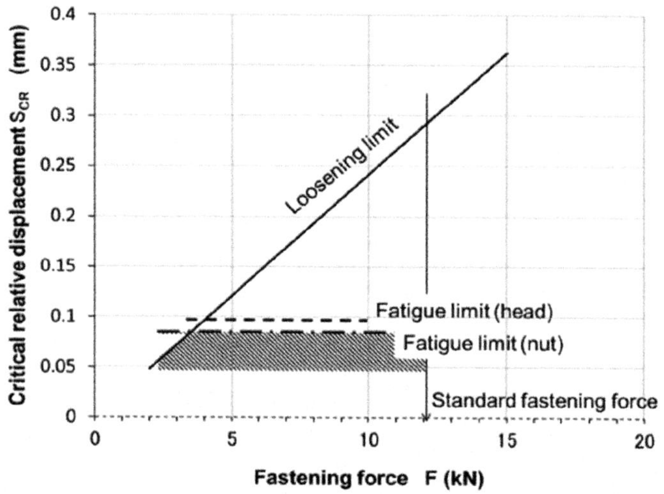

Fig. 4.30 Critical slippage S_{cr} considering loosening and fatigue behaviors (M10, $l_f = 18$ mm Normal bolt)

shank bolt. The relationship between the loosening velocity dF_f/dN and the relative slipping amount S at 1000 cycles is shown in Fig. 4.21 and it was experimentally verified that the critical slipping amount S_{cr} can be increased to about twice by using an reduced shank bolt. The predicted results of the relationship between loosening rate dF_f/dN and relative slip amount S and S_w are in good agreement with the experimental results marked with ●, as shown by the dashed line in Fig. 4.21.

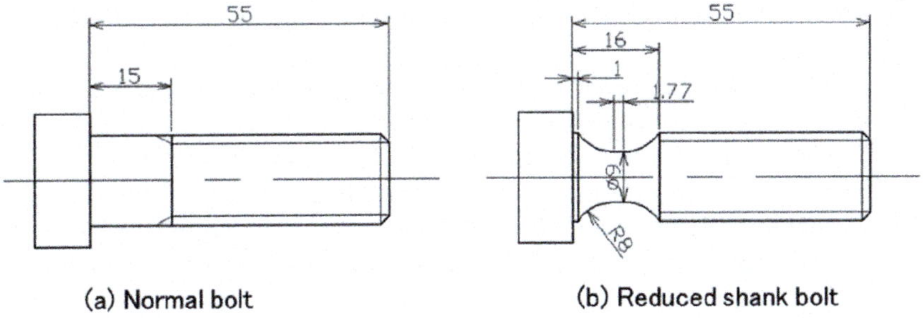

Fig. 4.31 Comparison of shapes of normal bolt and reduced shank bolt for experimental use

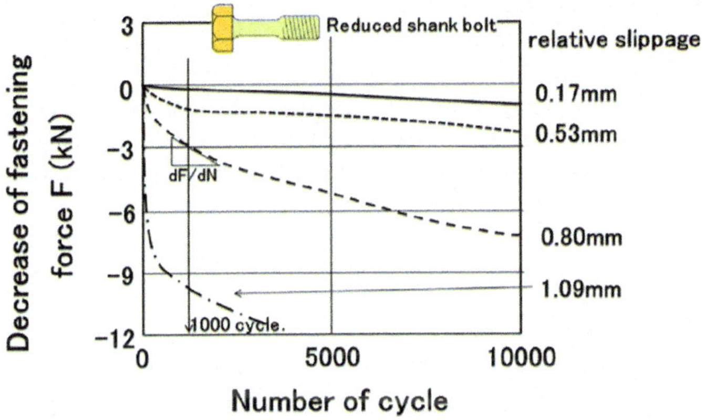

Fig. 4.32 Experimental results of loosening behavior under various cyclic relative slippages (M16, reduced shank bolt, $l_f = 18\,\text{mm}$)

4.9.2 Conclusion

(1) The relationship between the transverse load W_t with the bending moment on bolt M_b or M_b, and the amount of relative slippage S is analyzed divided into four regions: (a) complete sticking, (b) slipping on the contact surface of the fastened body, (c) slip on the nut thread surface, and (d) slip on the seat surface.

(2) The equivalent compliance of the screw fastener in the complete sticking region, where the transverse load is less than $W_1 = F_f \mu_c$, is obtained. And by using this equivalent compliance, enabling the equivalent beam setting during FEM analysis.

(3) When transverse load W_t exceeds following limit

$$W_2 = F_f \mu_c + 1.94 F_f \mu_s \frac{d}{l_f}$$

bending slip occur on the thread surface around the axis perpendicular to the bolt axis.

The critical relative slipδ_{cr}, and bolt bending moments M_h and M_n in this nut thread slipping condition were derived using the resisting moment of the nut, bolt head, and frictional coefficients μ_w on bolt head or nut seating.

(4) Experimental validation of the analytical equations for the loosening critical slip δ_{cr} and the bolt bending moment M_h was performed. This enabled the setting of equivalent beams in the FEM analysis for transverse loading higher than F_f μ_c.
(5) The optimum design method that satisfies both loosening and fatigue conditions under transverse loading was studied using analytical expressions for the relative slip at the loosening limit δ_{cr}, and bolt bending stresses M_h and M_n.
(6) We proposed a method to analyze the loosening velocity under cyclic transverse loading at the loosening critical slippage δ_{cr} or higher, and compared it with the actual measurement and confirmed a good correlation.
(7) It was confirmed analytically and experimentally that the reduced shank bolts used to improve fatigue strength are also effective in increasing the critical slippage S_{cr} and reducing the loosening rate under transverse loading.

References

1. Junker, G. H., SAE Trans. Vol. 78 (1969), pp. 314–335.
2. Pai, N. G. and Hess, D. P., J. Sound Vibration. Vol. 253-3 (2002), pp. 585–602.
3. Yamamoto, A., Principle and design of thread joint (Yokendo Ltd., Tokyo 1995) (in Japanese).
4. Yamamoto, A. and Kasei, S., Jpn. Soc. Precision Eng. Vol. 43 (1977), p. 470–475 (in Japanese).
5. Izumi, S., Yokoyama, T., Iwasaki, A. and Sakai, S., Eng. Fail. Analyis. Vol. 12 (2005), pp. 604–615.
6. Nakamura, M., Hattori, T., Sato, S. and Umeki, K., Trans.
7. Kasei, S., A Study of Self-Loosening of Bolted Joints Due to Repetition of Small Amount of Slippage at Bearing Surface, Journal of Advanced Mechanical Design, Systems, and Manufacturing, Vol. 1, No. 3, (2007),pp. 358–367.
8. Verein Deutscher Ingenieure, VDI-Richtlinien 2230 (1977) Systematic Calculation of high duty bolted joints, 1977.
9. Verein Deutscher Ingenieure, VDI-Richtlinien Blatt 1 2230 (2003) Systematic calculation of high duty bolted joints -Joints with one cylindrical bolt-, 2003.
10. Tomotugu Sakai, Loosening of bolts (1st Report, when subjected to loads perpendicular to the axis), Transactions of the Japan Society of Mechanical Engineers (Part 3), Vol. 44, No. 377, (Shou 53–1), pp. 279–287.

Chapter 5
Mechanical Analysis, Loosening Analysis, and Strength Design Under Rotational Loading

Generally, bolted joints of mechanical structures are rarely composed of only one bolt, and most of them are composed of multiple arrangement as shown in Fig. 5.1. In such cases, even if the load torque T around the axial direction is applied to the multiple bolted structure, it is replaced by the transverse load to the individual bolts (described in the previous chapter). Therefore, the rotational load acting on each individual bolt fastener is negligible in many cases, and was therefore we omitted from the overall explanation in Chap. 3. However, in limited cases, such as mounting the impeller of a turbomachinery on an axle shaft, as shown in Fig. 5.2, a single bolt fastening may be used for each shaft.

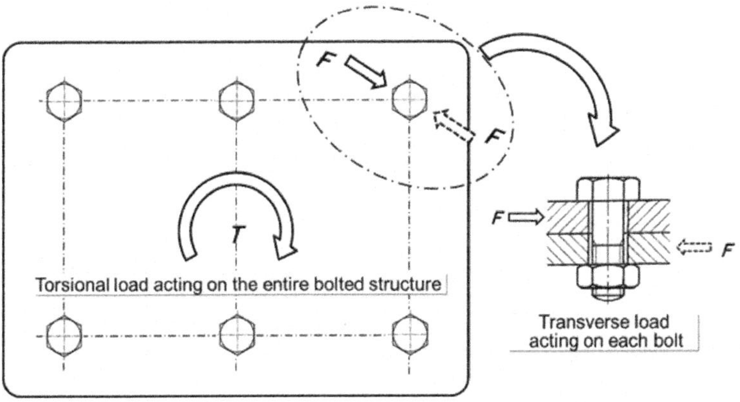

Fig. 5.1 Rotational load T acting on a multibolted structure and transverse load F acting on the individual bolted joint

© The Author(s), under exclusive license to Springer Nature Switzerland AG 2025

T. Hattori and T. Naruse, *Compendium for Design and Manufacturing of Bolted Joints*,
Solid Mechanics and Its Applications 280,
https://doi.org/10.1007/978-3-031-89628-6_5

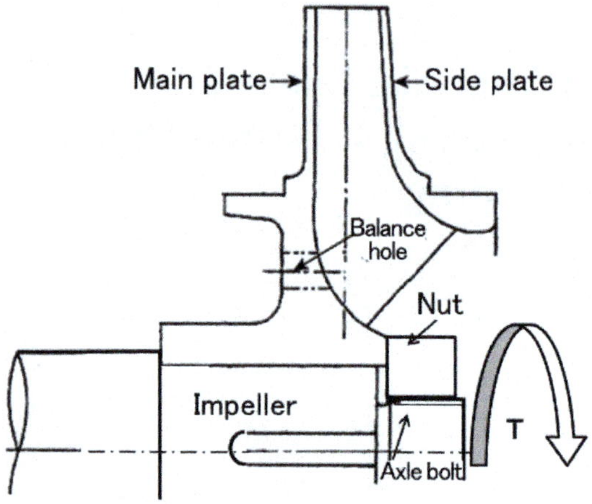

Fig. 5.2 Rotational load T acting on a single screw fastener

In such a case, a rotational load T (in this case, the torque applied to the impeller) is applied to the single bolted fastener, and mechanical analysis, loosening analysis, and strength design for such a rotational load case should be considered.

5.1 Deformation and Slipping Behavior Under Rotational Loading

The deformation and sliding behavior of a single bolt fastener subjected to a rotational load T is described for each load level using a simple model in Fig. 5.3.

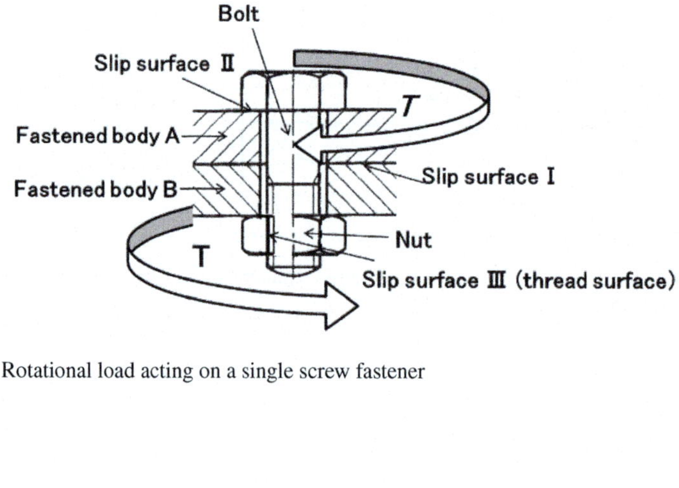

Fig. 5.3 Rotational load acting on a single screw fastener

5.1.1 Deformation Behavior in the Sticking (Linear) Region Under Rotational Loading

When a rotational load T is applied between the fastened bodies A and B, the fastened bodies A, B, bolts and nuts are elastically deformed as a unit when the load T is lower than the slip limit torque T_c at the slip surface I shown below.

$$T_c = \frac{\frac{2}{3}\frac{(D_m{}^3 - D_i{}^3)}{(D_m{}^2 - D_i{}^2)} F_f \mu_c}{2} \tag{5.1}$$

where D_m is the equivalent cylindrical body diameter shown in Fig. 5.4a, D_i is the bolt hole diameter, F_f is the fastening force, and μ_c is the coefficient of friction between the fastened bodies.

The torsional equivalent compliance (reciprocal of stiffness) of this joint is not specifically addressed in VDI 2230, but when analyzed as the torsional compliance of cylinder and hollow cylinder as in Chaps. 3 and 4, it can be derived as follows.

The torsional compliance of a bolt φ_b is obtained by applying and expanding VDI 2230 (1977) [1] as follows.

$$\varphi_b = \frac{0.4d}{G_b \frac{\pi}{32} d^4} + \frac{l_g}{G_b \frac{\pi}{32} d_g{}^4} + \frac{l_s}{G_b \frac{\pi}{32} d_3{}^4} + \frac{0.4d}{G_b \frac{\pi}{32} d^4} \tag{5.2}$$

Alternatively, VDI 2230 (2003) [2] can be applied and expanded to obtain the following.

$$\varphi_b = \frac{0.5d}{G_b \frac{\pi}{32} d^4} + \frac{l_g}{G_b \frac{\pi}{32} d_g{}^4} + \frac{l_s}{G_b \frac{\pi}{32} d_3{}^4} + \frac{0.5d}{G_b \frac{\pi}{32} d_3{}^4} + \frac{0.4d}{G_b \frac{\pi}{32} d^4} \tag{5.3}$$

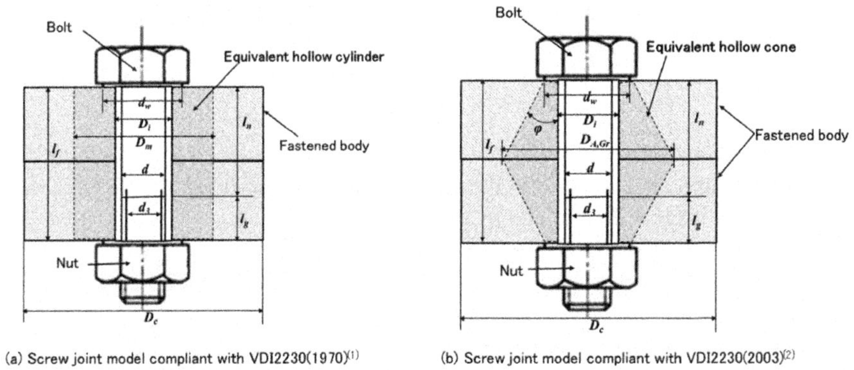

Fig. 5.4 Bolted joint models compliant with VDI2230 [1, 2]

where G_b is the rigidity of the bolt material, d is the nominal diameter of bolt, l_g is the length of the bolt cylindrical section shown in Fig. 5.4, l_s is the thread length of the fastening range, and d_3 is the thread valley diameter.

Also, the torsional compliance of the fastened body φ_c can be derived using the equivalent hollow cylinder model proposed by VDI2230 (1977) [1] as Eq. (5.5), where D_m is the equivalent outer diameter and D_i is the inner diameter (see Fig. 5.4a),

$$D_m = d_w + \frac{l_f}{10} \tag{5.4}$$

$$\varphi_c = \frac{32 l_f}{\pi \left(D_m^4 - D_i^4\right) G} \tag{5.5}$$

Also, using the equivalent hollow cone model proposed in VDI2230 (2003) [2], (see Fig. 5.4b).

The sectional second-order polar moment I_{peq} of the equivalent hollow cone is obtained from the following equation,

$$I_{peq} = \frac{3\pi}{32} \frac{(D_c - D_w) {d_w}^3 {D_c}^3}{{D_c}^3 - {d_w}^3} - \frac{\pi}{32} {D_i}^4 \tag{5.6}$$

When the fastened body is sufficiently wide $D_c > D_{A,Gr}$, Eq. (5.4) uses $D_{A,Gr}$ instead of D_c.

Using this, the torsional compliance of the fastened body φ_c is expressed as follows by applying and expanding VDI 2230 (1977) [1].

$$\varphi_c = \frac{l_f}{G_c I_{peq}} \tag{5.7}$$

Using these and other factors, the load factor for axial torsional loading Φ_φ is derived as follows.

$$\Phi_\varphi = \frac{\varphi_c}{\varphi_b + \varphi_c} \tag{5.8}$$

Using this load factor Φ_φ, which is the ratio of the shared load of the bolt itself T_b to the torsional load T applied to the entire screw fastened body, the torsional load T_b of the bolt itself is determined by the following equation,

$$T_b = \Phi_\varphi \times T \tag{5.9}$$

Then the amplitude of shear stress in the bolt, τ_a, is calculated as the following equation using the polar section modulus at thread root diameter of the bolt, Z_{p3}

$$Z_{p3} = \frac{\pi {d_3}^3}{16} \tag{5.10}$$

$$\tau_a = \frac{T_b}{2Z_{p3}} \tag{5.11}$$

The stress amplitude τ_a acting on the bolt is to be designed to be lower than the shear fatigue strength of the bolt. The shear fatigue limit τ_{wk} of the bolt should be about half of the vertical stress fatigue limit σ_{wk} of various bolts shown in Table 2.2 in Chap. 2. Furthermore, considering the safety factor S_f, the allowable shear stress amplitude τ_a should be lower than the following allowable shear stress amplitude.

$$\tau_a \leq \frac{\tau_{wk}}{S_f} \tag{5.12}$$

5.1.2 Slipping Behavior Under Rotational Loading

In general, what is important in the mechanical analysis and strength design of bolted joints under rotational loading is not the behavior in the sticking region (linear region) described above, but the sliding behavior on the contact surface between the bolt and the fastened body.

When the torque around the bolt axis indicated by T in Fig. 5.3 acts between the fastened bodies, first slip occurs on the sliding surface I, and when the torque increases further, slippage may also occur on the seat surface between the nut or bolt head and the fastened body on the sliding surface II. In this case, the sliding surface III (threaded surface) between the nut and bolt may slipping, causing the nut to return and rotate, resulting in loosening [3].

Figure 5.5 illustrates a typical return rotation loosening mechanism. For the sake of convenience, let us assume that one fastened body is fixed (fixed plate) and the other is in relative sliding vibration (vibrate plate), and that the bolt head is in contact with the vibrating plate. In the case of a right-hand thread, when the vibrating plate turns to the right, the bolt head slips on the seat surface and bolt maintains the same relative position without rotating to the right, as shown in Fig. 5.5a, and then when the vibrate plate turns to the left, the bolt does not slip on the seat surface but on the threaded part as shown in Fig. 5.5b, causing relative sliding between the bolt and nut, this causes a loosening rotation.

Next, quantitative looseness evaluation is performed by clarifying the sliding torque between the seat surface (sliding surface II) and the threaded surface between the bolt and nut (sliding surface III). If the fastening force of the bolt is F_f, the torque required to slip the threaded surface T_s, and the torque required to slip the seat surface T_w, can be expressed as follows.

$$T_s = \frac{F_f}{2}\left(\frac{d_2\mu_s}{\cos\alpha} \pm \frac{P}{\pi}\right) \tag{5.13}$$

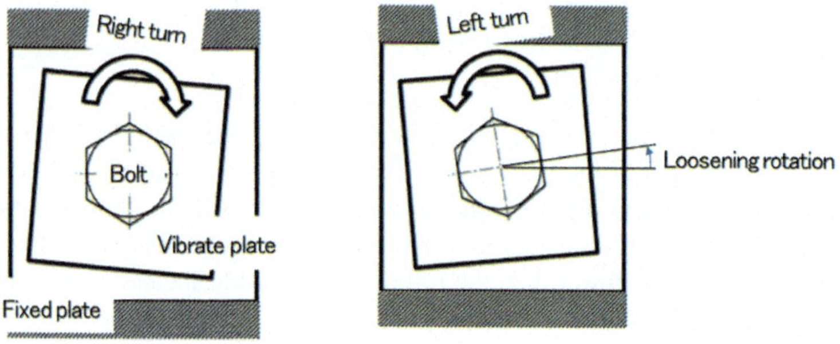

Fig. 5.5 Loosening mechanisms under cyclic rotational loading

$$T_w = \frac{F_f \mu_w D_w}{2} \tag{5.14}$$

where d_2 is the effective diameter of the screw, μ_s is the coefficient of friction of the thread surface, P *is the* pitch of the screw, α is the flank angle of the screw thread (thread half angle), μ_w is the coefficient of friction of the seating surface, and D_w is the effective diameter of the seating surface. The + in Eq. 5.13 represents the torque T_{st} on the tightening side and the—represents the torque T_{sl} on the loosening side. In order for the return rotation loosening described in Figs. 5.3 and 5.5 to occur, when the vibrate plate rotates to the right, it must slip on the seat face and the thread surface must stick, i.e., $T_w < T_{st}$, and when the vibrate plate rotates to the left, it must slip on the thread surface and the seat face must stick, i.e., $T_{sl} < T_w$. By summarizing these

$$T_{sl} < T_w < T_{st}$$

and the bolt will return and rotate loose only if the following equation is satisfied

$$\frac{d_2 \mu_s}{\cos\alpha} - \frac{P}{\pi} < \mu_w D_w < \frac{d_2 \mu_s}{\cos\alpha} + \frac{P}{\pi} \tag{5.15}$$

This mechanism is outlined in the figures below.

Figure 5.6 shows a schematic of the deformation and sliding behavior under cyclic torsional loading when $T_w < T_{sl} < T_{st}$. In this case, when a torsional load T is applied between the fastened bodies A and B shown in Fig. 5.3, the fasteners A, B, bolt, and nut are elastically deformed as a unit when the load T is lower than the slip limit torque T_c on the slip surface I shown in Eq. (5.1).

When a torsional load T exceeds T_c, torsional deformation proceeds for the bolt only and when the rotational load T reach to $T_c + T_w$, the seating surface slips. Thereafter, under unloading and left-rotational load processes, similarly, slippage at the slip surface I and slippage of the seat surface continue, and the threaded surface remains fixing and no loosening occurs.

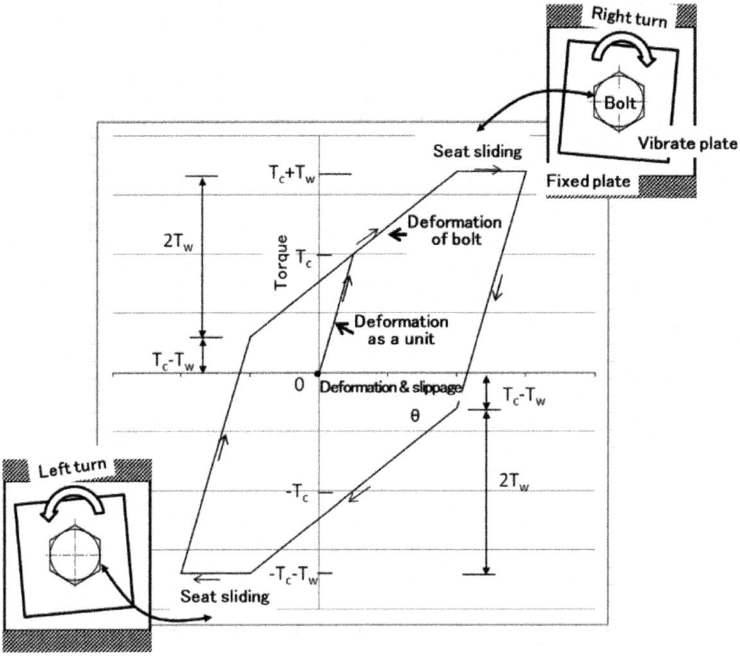

Fig. 5.6 Deformation and slipping behavior under cyclic torsional loading ($T_w < T_{sl} < T_{st}$, No looseness)

Figure 5.7 shows a schematic of the deformation and sliding behavior under cyclic torsional loading when $T_{sl} < T_w < T_{st}$, In this case, slipping occurs at the seat surface during right rotation and slipping occurs on the thread surface during left rotation and this repetitions resulting in progressive loosening.

Figure 5.8 shows a schematic of the deformation and slipping behavior under cyclic torsional loading when $T_{sl} < T_{st} < T_w$. In this case, slipping is repeated on the thread surface during both right and left rotation processes. So, in this case, tightening and loosening will repeat and the loosening will not progress.

Now, let's calculate which of the above three conditions applies to bolts according to the actual JIS or SI standards, assuming a friction coefficient of the seat surface and thread surface by setting $\mu_w = \mu_s = 0.15$. The results are shown in Fig. 5.9 [3]. The horizontal axis is the ratio of the average friction diameter of the bolt or nut seat surface D_w to the effective diameter of the thread d_2, and the vertical axis is the lead angle β of the thread. And three conditions areas of no loosening ($T_w < T_{sl} < T_{st}$), progressive loosening ($T_{sl} < T_w < T_{st}$) and repeated loosening and tightening ($T_{sl} < T_{st} < T_w$) are shown. For reference, the positions for some JIS or IS standard nuts are indicated by each mark. Strictly speaking, the slope of the line changes with the value of each coefficient of friction, μ_w and μ_s, and the range of each region also varies, but it can be seen that many practical nuts are in the danger zone of loosening. However, it can be seen that a nut with a larger seating surface, such as a flanged

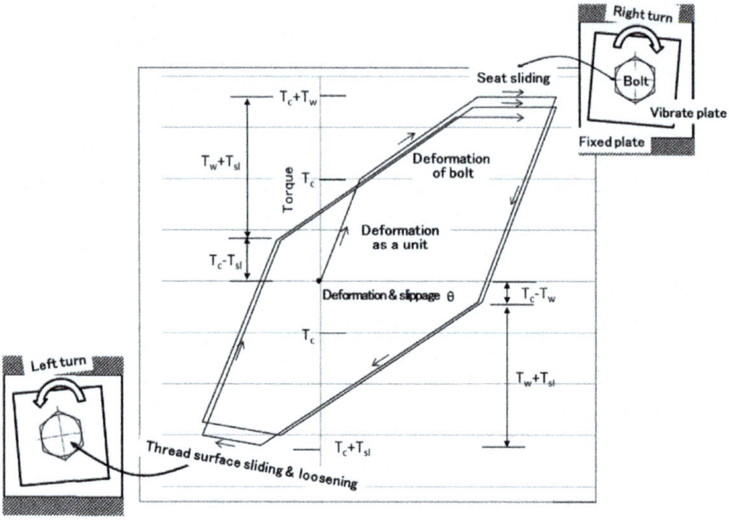

Fig. 5.7 Deformation and sliding behavior under cyclic torsional loading ($T_{sl} < T_W < T_{st}$, Loosening progress)

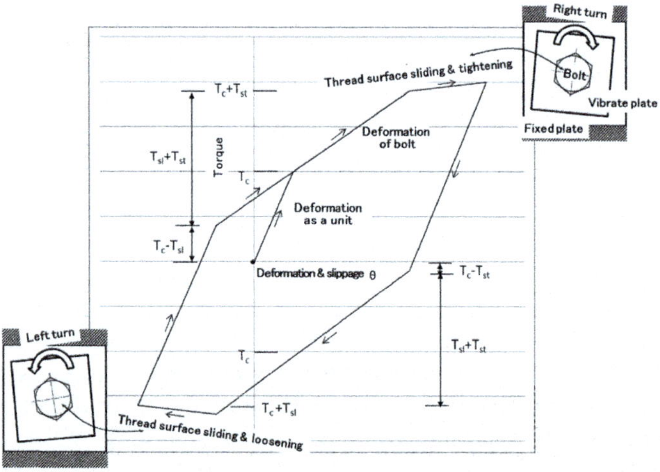

Fig. 5.8 Deformation and slipping behavior under cyclic torsional loading ($T_{sl} < T_{st} < T_W$, Repetition of loosening and tightening)

nut, is effective in preventing rotational loosening under cyclic torsional load. This figure should be utilized in the evaluation and design of actual threaded fasteners.

Sakai et al. [4] confirmed this deformation and slip behavior analyses under cyclic torsional loading in the following experiments. Figure 5.10 shows a schematic of the experimental and measurement apparatus. The test bolts are M10, P1.25, with an

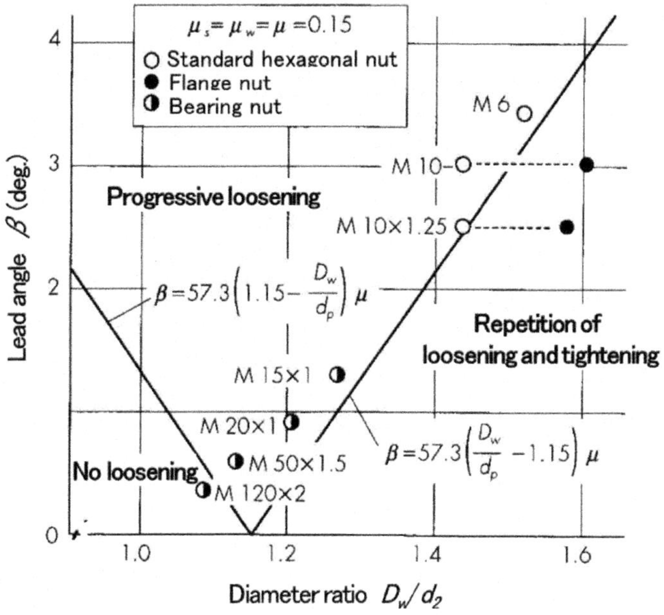

Fig. 5.9 Looseness occurrence evaluation diagram under cyclic rotational loading

initial fastening torque of 59 N–m and a repeated relative rotation angle of ±2.3. The results of the experiments are shown on Fig. 5.11. The upper part of this figure shows the loosening process of this fastened bolt under cyclic loading. The lower part of Fig. 5.11 shows the schematic diagram of changes of the respective sliding limit torques T_{sl}, T_{st}, T_w and increase in friction coefficients μ_w, μ_s and the transition of the sliding surface during this cyclic torsional loading process. These graphs conceptually represent the situation in which each slip surface is transferred between seat surface and thread surface due to the change in the friction coefficient during the cyclic torsional loading cycle.

5.1.3 Loosening Limit Rotation Angle Under Rotational Loading

In Sects. 5.2.1 and 5.2.2, it was shown that when the rotational load indicated by T in Fig. 5.3 acts between the fastened bodies, the slip surface I first slips, and as the torque increases, slip also occurs on the seat surface between the nut or bolt head and the fastened bodies (slip surface II). In this case, the slipping surface (threaded surface) between the nut and bolt may slip and resulting in loosening [3, 4]. Therefore, here we calculate the critical relative torsional angle θ_{cr}, which is the limit of slippage initiation on the slip surface II and initiation loosening.

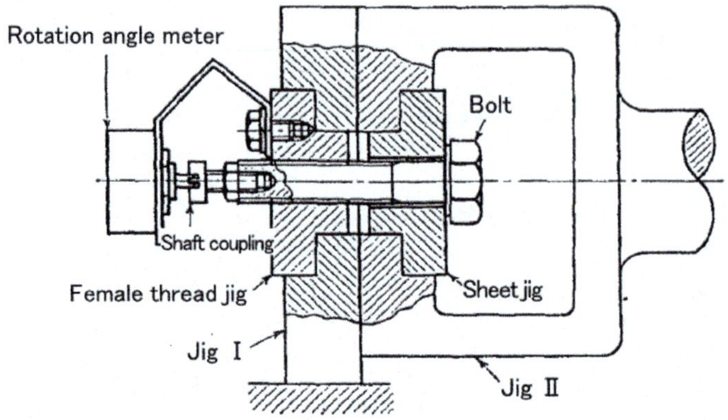

Fig. 5.10 Schematic of the experimental and measurement apparatus for cyclic rotational loading [4]

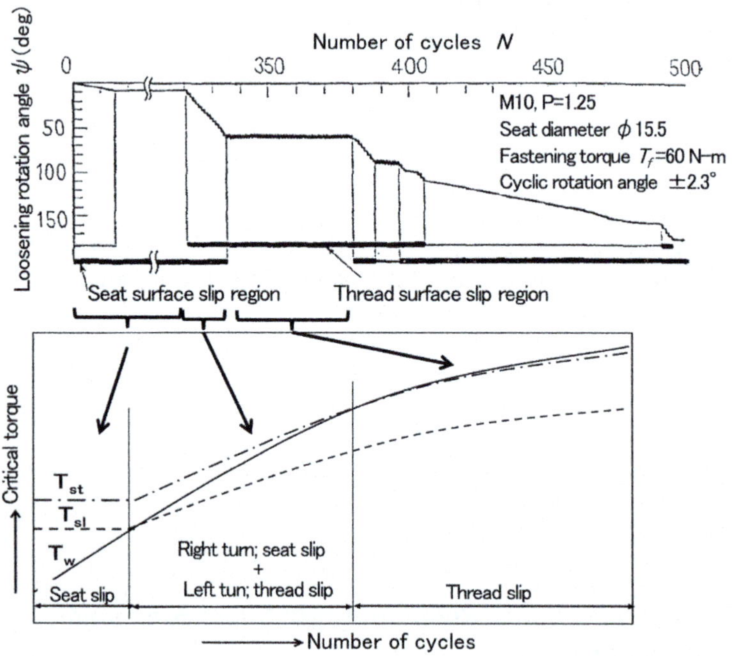

Fig. 5.11 Experimental results of slipping and loosening behavior under cyclic rotational loadings [4]

Referring to the deformation process in Fig. 5.7, the critical relative torsional angle θ_{cr} for loosening can be found as the sum of the following two components.

① The critical torsion angle θ_{cr1} of the bolt in the linear region without slipping is determined by considering the deformation of the entire screw fastened body under the torque $2T_c$ and the internal force coefficient Φ_b.
② The critical torsion angle θ_{cr2} of the bolt in the slip region is found as the deformation of the bolt itself under torque T_{sl}.

And total critical relative torsional angle θ_{cr} can be calculated as a sum of θ_{cr1} and θ_{cr1} as follows.

$$\begin{aligned}\theta_{cr} &= \theta_{cr1} + \theta_{cr1} = \Phi_{\psi}\psi_b(2T_c) + \psi_b(T_{sl}) \\ &= F_f\left(\frac{0.4d}{G_b\frac{\pi}{32}d^4} + \frac{l_g}{G_b\frac{\pi}{32}d_g^4} + \frac{l_s}{G_b\frac{\pi}{32}d_3^4} + \frac{0.4d}{G_b\frac{\pi}{32}d^4}\right) \\ &\cdot\left(\frac{2\left(D_m^3 - D_i^3\right)}{3\left(D_m^2 - D_i^2\right)}\Phi_{\psi}\mu_w + \frac{d_2\mu_s}{\cos\alpha} - \frac{P}{\pi}\right)\end{aligned} \tag{5.16}$$

where G_b is the rigidity of the bolt material, d is the nominal diameter, l_g is the length of the bolt cylindrical section shown in Fig. 5.4, l_s is the thread length in the fastening range, d_3 is the thread valley diameter, P *is the* thread pitch, α is the flank angle (thread half angle), μ_w is the friction coefficient of the seating surface, μ_s is the friction coefficient of the thread surface, and D_w is the effective diameter of the seating surface.

5.2 Design of Bolted Joints Under Rotational Loadings Considering for Both Loosening and Fatigue

The previous sections have described the analysis of slipping, loosening, etc. This section describes a design method for threaded fasteners under rotational loading that also takes torsional fatigue of bolts into account.

In Sect. 5.2.2, it was shown that the only condition under which loosening occurs is when the sliding limit torques of the thread and seat surface satisfy the relationship $T_{sl} < T_w < T_{st}$, or Eq. (5.15). This section classifies this relationship into the following three areas ((a)–(c)) and describes the design methods that take loosening and fatigue into account in each area. And in this section, we will first define the fatigue-limiting relative torsion angle θ_{cr} just on which condition the shear stress τ_b at the bolt valley bottom diameter d_3 section reaches to the allowable shear stress amplitude τ_{ba} for the bolt. This allowable shear stress amplitudeτ_{ba} can be estimated as the half of the allowable nominal stress amplitude σ_a. So, by referring Sect. 4.2.2, Eq. (4.15) and Table 2.2, we can show as follows.

$$\tau_{ba} \leq \frac{\sigma_{wk}}{2S_f} \tag{5.17}$$

where S_f is the safety factor, usually 2 or greater.

And we define the critical torque T_{cr}, just on which the shear stress at the bolt valley bottom diameter d_3 section reaches to the allowable shear stress amplitudeτ_{ba} for the bolt.

T_{cr} and θ_{cr} can be shown as follows.

$$T_{cr} = Z_{P3}\tau_{ba} \tag{5.18}$$

$$\theta_{cr} = 4\varphi_b T_{cr} = 4\varphi_b Z_{P3}\tau_{ba} \tag{5.19}$$

where φ_b is the torsional compliance of the bolt shown in Eqs. (5.2), (5.3), and Z_{P3} is the polar section modulus at the bolt valley bottom diameter.

(a) $T_w < T_{sl} < T_{st}$

Since the bolt will not loosen under these conditions, only the fatigue-limiting relative torsion angle θ_{cr} should be considered.

(a-1) In the case of fatigue limit critical torque $T_{cr} < T_w$.

In this case, the initial fastening force F_f should satisfy following equation,

$$F_f > 4Z_{p3}\tau_{ba}/\mu_w D_w \tag{5.20}$$

And as shown in Fig. 5.12, if the critical cyclic torque that reaches the bolt fatigue strength is lower than the seating surface slip torque T_w, the fatigue-limiting relative torsion angle θ_{cr} is same as above-mentioned Eq. (5.19).

In this case, the relationship between the allowable limit relative torsional angle θ_{cr} and the initial fastening force F_f is shown in Fig. 5.13 as a dashed line, and the allowable design range is shown in hatching.

In this case, the initial fastening force F_f should satisfy following equation,

$$F_f \leqq 4Z_{p3}\tau_{ba}/\mu_w D_w \tag{5.21}$$

And in this case, the bolt will slip on the seating surface before the torque reach fatigue failure, i.e., $T_{cr} < Z_{P3}\tau_{ba}$ and there is no risk of fatigue failure as shown in Fig. 5.13.

(b) $T_{sl} < T_w < T_{st}$

Since loosening occurs under these conditions, the limiting relative torsion angle θ_{cr} is considered for both loosening and fatigue.

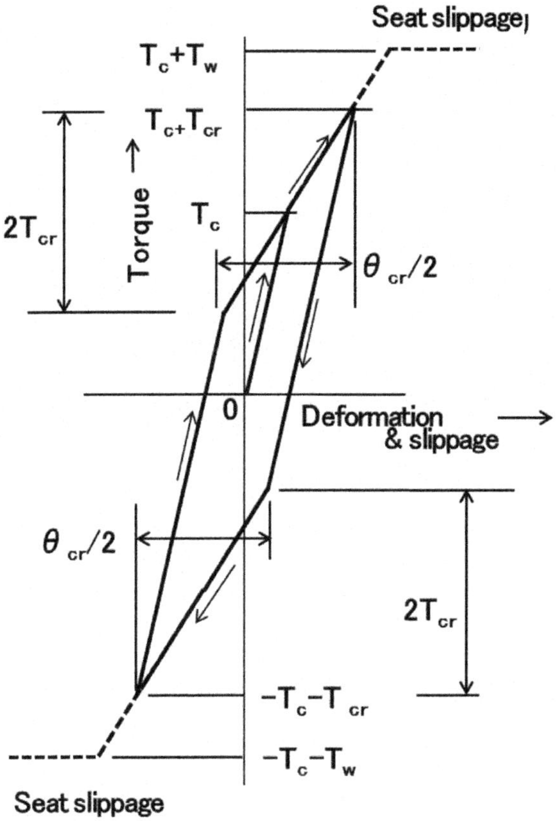

Fig. 5.12 Fatigue strength evaluation for the case of (a) $T_w < T_{sl} < T_{st}$, (a-1) $T_{cr} < T_w$ (no loosening)

(b-1) Critical relative torsional angle θ_{cr} for loosening

As shown in Fig. 5.14 (b-1), the limiting relative torsion angle for loosening, θ_{cr}, is the relative torsion angle when the load torque reaches the sliding torque T_w on the seat face for clockwise rotation and the sliding torque T_{sl} on the screw face for counterclockwise rotation, as already shown in Eq. (5.16). This is shown in Fig. 5.15 as a solid line.

(b-2) Critical relative torsional angle θ_{cr} for fatigue

(b-2-1) Repeated torque $2T_{cr} < T_w + T_{sl}$

Figure 5.15, as well as (a-1), is shown in Eq. (5.16). This relationship is shown by the dashed line in Fig. 16. The allowable design range, taking into account both this limiting relative torsional angle of fatigue θ_{cr} and the limiting relative torsional angle of looseness θ_{cr} shown in (b-1), is also shown by hatching.

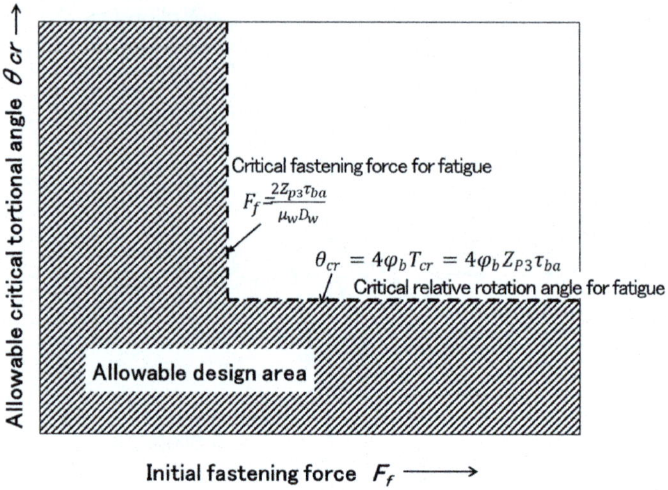

Fig. 5.13 Relationships between allowable critical torsional angle θ_{cr} and the initial fastening force F_f for the case of (a) $T_w < T_{sl} < T_{st}$ (a-2) In the case of fatigue limit critical torque $T_{cr} \geqq T_w$

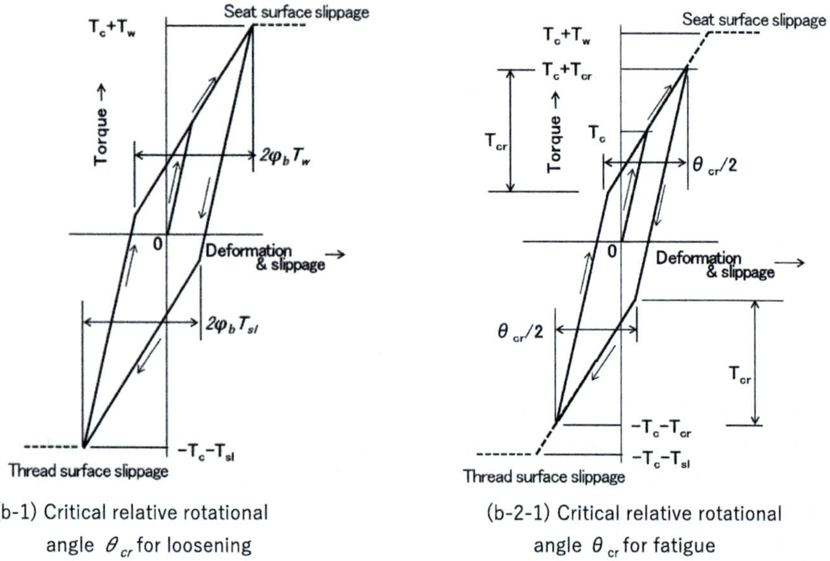

Fig. 5.14 Loosening and fatigue strength evaluation for the case of (b) $T_{sl} < T_w < T_{st}$

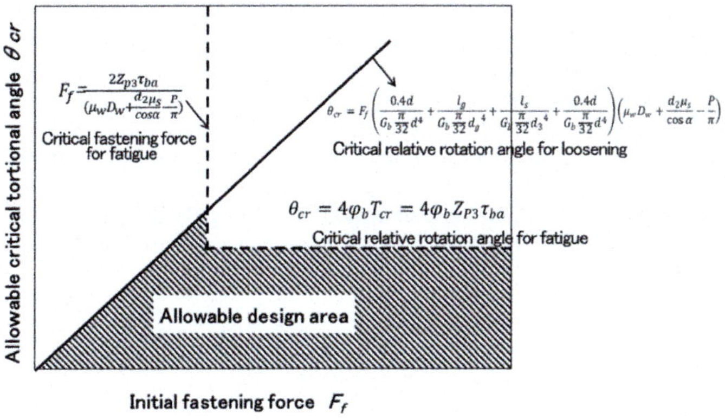

Fig. 5.15 Relationships between allowable critical torsional angle θ_{cr} and the initial fastening force F_f for the case of (b) $T_{sl} < T_w < T_{st}$

(b-2-2) Repeated torque $2T_{cr} \geq T_w + T_{sl}$

In this case, the bolt will slip on the seat and threaded surface before fatigue failure torque is applied to the bolt, i.e., $T_{cr} < Z_{P3}\tau_{ba}$ and there is no risk of fatigue failure.

(c) $T_{sl} < T_{st} < T_w$

Under these conditions, tightening and loosening are considered to be repeated and loosening does not progress. Therefore, the fatigue-limiting relative torsion angle θ_{cr} is examined.

(c-1) Repeat torque $T_{cr} < (T_{sl} + T_{st})/2$

As shown in Fig. 17, if the cyclic torque that reaches the bolt fatigue strength is T_{cr} and this is lower than the average of the sliding torques on both thread faces $(T_{sl} + T_{st})/2$, the fatigue-limiting relative twist angle θ_{cr} is the same as in Eq. (5.16) in (a-1) above and can be obtained as follows.

$$\theta_{cr} = 4\varphi_b T_{cr} = 4\varphi_b Z_{P3}\tau_{ba} \tag{5.16}$$

In this case, the relationship between the allowable limit relative twist angle θ_{cr} and the initial tightening force is also the same as in Fig. 14.

(c-2) Repeat torque $T_{cr} \geq (T_{sl} + T_{st})/2$

In this case, the bolt will slip on both threaded surfaces before fatigue failure torque is applied to the bolt, i.e., $T_{cr} < Z_{P3}\tau_{ba}$ and there is no risk of fatigue failure.

References

1. Junker, G. H., SAE Trans. Vol. 78 (1969), pp. 314–335.
2. Pai, N. G. and Hess, D. P., J. Sound Vibration. Vol. 253-3 (2002), pp. 585–602.
3. Isamu Yoshimoto, Key Points for Designing Screw Fasteners, Japanese Standards Association 2002, pp. 228.
4. Tomoji Sakai, Loosening of Bolts (2nd Report, For Bolts Subjected to Rotational Load), Transactions of the Japan Society of Mechanical Engineers (Part 3) **44–377** (Shou 53–1), pp. 288–292.

Chapter 6
Rotational and Non-rotational Loosening and Anti-loosening Technology

6.1 Classifications and Characteristics of Loosening [1, 2]

All of the mechanical analyses, deformation, and slip evaluations of threaded fasteners up to the previous chapters were conducted on the assumption that the threaded fastener is united by the pretension force (initial fastening force) generated in the bolt shaft by the tightening of the bolt and nut and the compression force generated in the fastened bodies.

However, this pretension and compression force (initial fastening force) may decrease for some reason during use of the machine or component. This decrease in initial fastening force is called "loosening of threads". Loosening can lead to many accidents, such as the rigidity of the fastener deteriorating, parts falling off, and fatigue fracture of bolts, as described in Chap. 1. Therefore, this section describes the mechanism and mechanical analysis of this loosening.

As shown in Table 6.1, loosening can be broadly classified into two categories: loosening that occurs without return rotation of the screw and loosening that occurs with return rotation. In general, loosening not caused by return rotation is less progressive, and there is less risk of loss of preload. On the other hand, loosening caused by return rotation is progressive and requires more attention.

Table 6.1 shows the classification of these types of loosening, and the following sections describe the mechanisms and countermeasures for each of these types of loosening.

6.2 Non-rotational Loosening

In Table 6.1, items (1) through (6) show non-rotational loosening, all of which are caused by permanent deformation, wear, and thermal expansion differences in each part of the bolted joint under external loads or over time, resulting in a decrease

© The Author(s), under exclusive license to Springer Nature Switzerland AG 2025

T. Hattori and T. Naruse, *Compendium for Design and Manufacturing of Bolted Joints*, Solid Mechanics and Its Applications 280,
https://doi.org/10.1007/978-3-031-89628-6_6

Table 6.1 Classification of many types of loosening

Non-rotational loosening	(1) Short-term relaxation (embedment, settling)
	(2) Loosening by seat sink
	(3) Loosening by cyclic micro slippage
	(4) Loosening by thermal deformation
	(5) Loosening by high temperature creep
	(6) Loosening by deformation of seal or gasket
Rotational loosening	(7) Loosening by cyclic transverse loading
	(8) Loosening by cyclic rotational (twisting) loading
	(9) Loosening by cyclic axial loading

in the fastening allowance (elastic expansion of the bolt and elastic contraction of the fastened body) given during initial fastening process. Considering that materials such as aluminum alloys, titanium alloys, and FRP have been increasingly used in recent years to reduce weight, and as that the operating environments become increasingly harsh, this type of non-rotational loosening is becoming increasingly important. Therefore, we will explain each item in detail below.

6.2.1 Short-Term Loosening

Microscopically, thread contact surfaces, seat surfaces, and contact surfaces between fastened bodies have surface roughness, waviness, or burrs as shown in Fig. 6.1, and if these minute unevennes cause embedding or smoothing due to surface pressure action during fastening process, the initial fastening force will decrease over a short period of time.

The fastening diagrams in Fig. 6.2 are used to quantitatively explain the initial loosening due to this denting. Let s be the amount of shrinkage of the contact surface shown in Fig. 6.1 after applying surface pressure, and F_s is the amount of reduction in initial fastening force due to these embedding can expressed as follows using the axial compliance δ_b and δ_c of the bolt and the fastened body.

$$F_s = \frac{1}{\delta_b + \delta_c} s = Zs \tag{6.1}$$

where $\mathbf{Z} = \frac{1}{\delta_b + \delta_c}$ is called the embedment coefficient.

A quick calculation of this embedment coefficient Z using normal screw standard dimensions is shown in Fig. 6.3.

The embedment coefficient Z_s of stud or tap bolted fasteners are similarly expressed as follows, using the fastener compliance δ_{sc} of the stud or tap bolted fastener as shown in Eq. (3.12) in Chap. 3, instead of δ_c in Eq. (6.1).

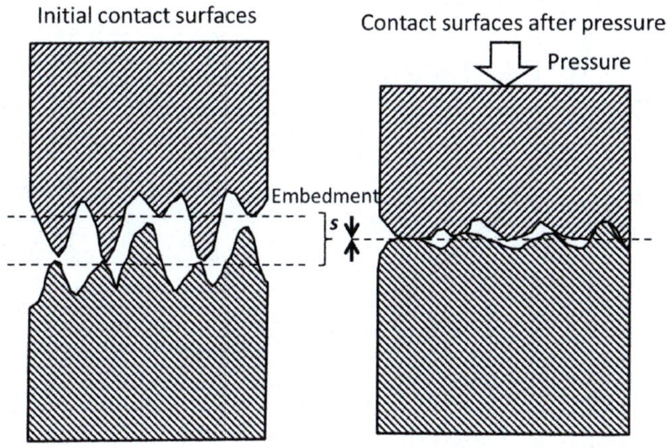

Fig. 6.1 Microscopical surface roughness and embedment by acting of pressure

Fig. 6.2 Fastening diagrams and degradation of fastening force by embedment

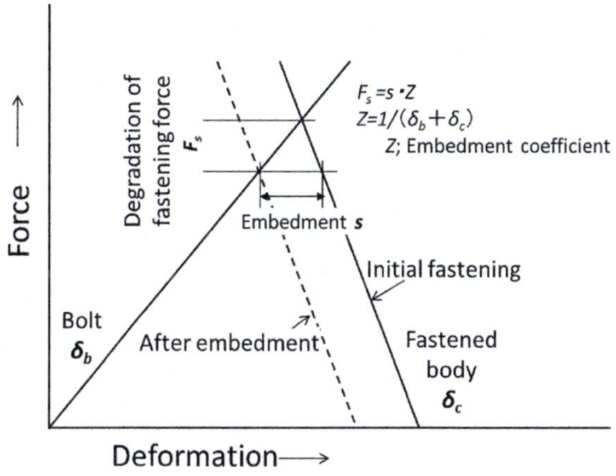

$$F_s = \frac{1}{\delta_b + \delta_{sc}} s = Z_s s \tag{6.2}$$

Similarly, the embedment coefficient Z_s calculated using normal screw standard dimensions for stud or tap bolted joint is shown in Fig. 6.4.

Although the embedment amount s may be affected by surface roughness, material, etc., VDI 2230 [3] provides the following empirical equation for a bolted joints consisting of a fastened body with sufficient thickness.

$$s \fallingdotseq 3.29 \left({}^{l_f}/_{d} \right)^{0.34} (\mu m) \tag{6.3}$$

Figure 6.5 shows a graph of the calculation results of Eq. (6.3) [1, 3].

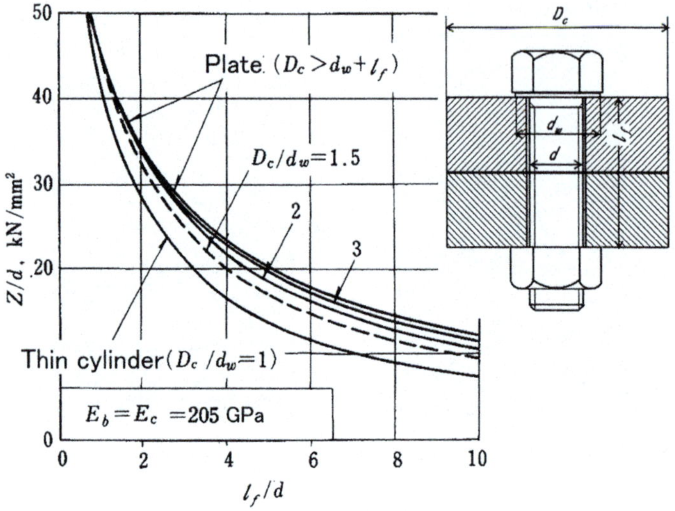

Fig. 6.3 Quick calculation diagram of embedment coefficient Z using normal screw standard dimensions (bolt/nut joint)

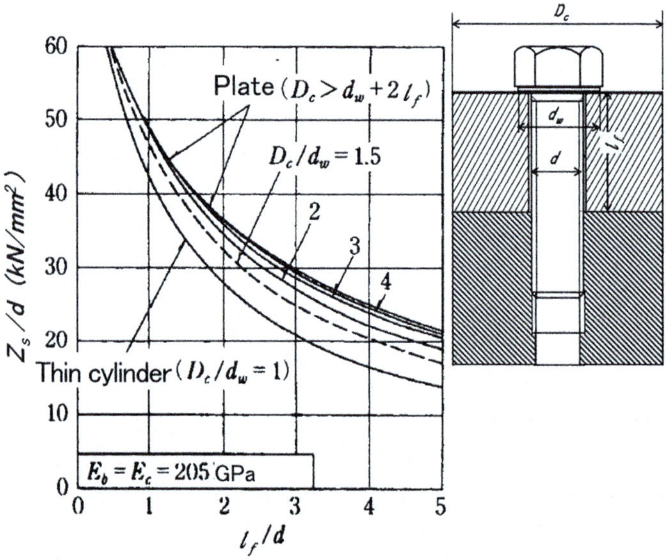

Fig. 6.4 Quick calculation diagram of embedment coefficient Z using normal screw standard dimensions (stud or tap joint)

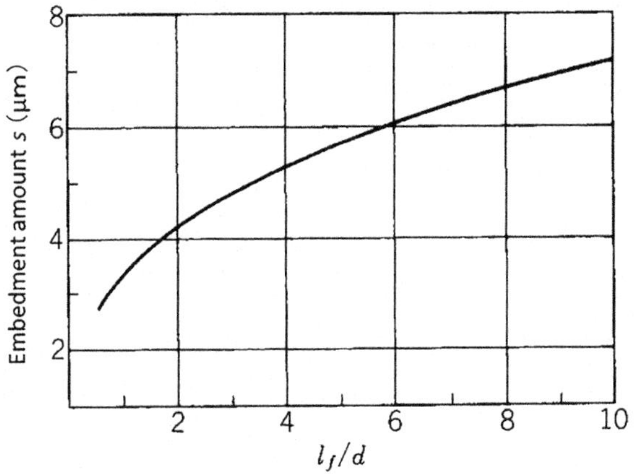

Fig. 6.5 Calculation result of embedment amount *s* (VDI2230 Blatt 1 (1986) [3])

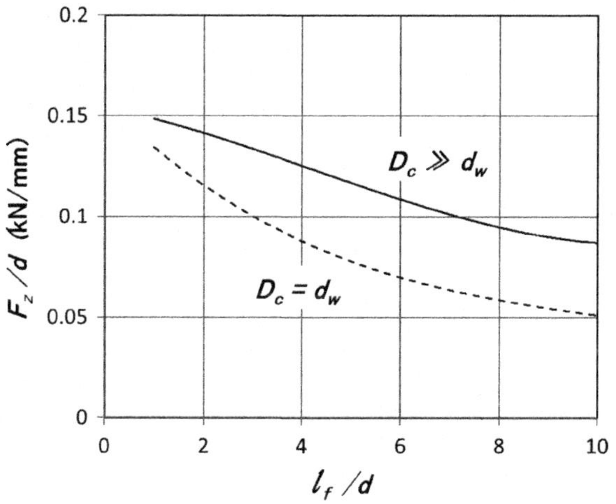

Fig. 6.6 Quick calculation diagram of short-term loosening (bolt/ nut joint)

Amount of reduction in initial fastening force F_S can be calculated using the embedment coefficient Z shown in Figs. 6.3 and 6.4, and embedment amount s shown in Fig. 6.5 or Eq. (6.3), and the calculated results are shown in Figs. 6.6 and 6.7 for bolt/nut fasteners, and stud or tap bolt fasteners, respectively.

6.2.2 Bearing Surface Sinking and Loosening

Recently, softer materials such as aluminum alloys and FRP are increasingly used for fastening bodies of mechanical structures to reduce weight. In such cases, the initial deformation caused by the smoothing of the micro-surface roughness and unevenness of the contact surface described above is not sufficient, and the bearing surface is plastically deformed in a rolling manner, causing the bearing surface to

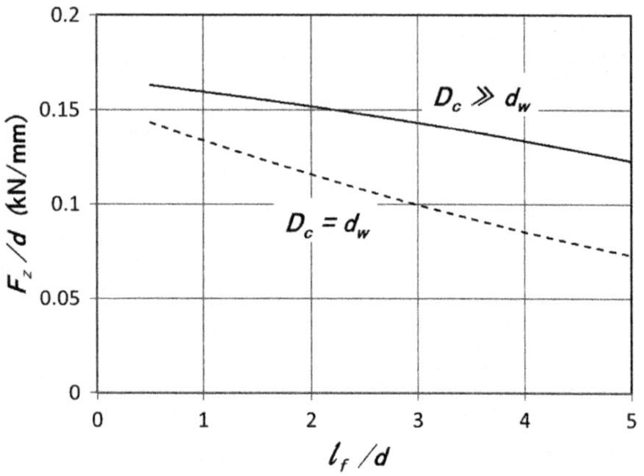

Fig. 6.7 Quick calculation diagram of short-term loosening (stud or tap bolt joint)

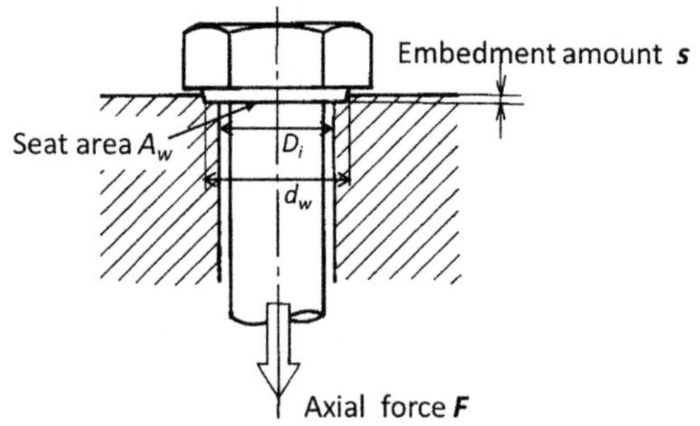

Fig. 6.8 Seat surface sinking on bolt head

sink in as shown in Fig. 6.8. If this occurs only during initial fastening, it will not be a particular problem, but if it progresses due to repeated loading or creep after fastening, it becomes a loosening problem, and this type of loosening is called sunken seat surface loosening.

This loosening will not occur if the surface pressure p_w of the seating surface, calculated from the maximum axial force of the bolt in operation, is less than the allowable surface pressure value (limit surface pressure) p_L of the body to be fastened. In other words, in order to prevent the bearing surface from sinking and loosening, the added bolt axial force φW_A, which is added by the external force W_A to the initial fastening axial force F_f, must be taken into account,

$$p_w = \frac{F_f + \Phi W_A}{A_w} \leq p_L \tag{6.4}$$

Φ Internal force coefficient (Eq. (3.26) in Chap. 3)

A_w Seating area $= \frac{\pi\left(d_w{}^2 - D_i{}^2\right)}{4}$

d_w Seat outer diameter.

D_i Inside diameter of seating surface (diameter of bolt hole).

Table 6.2 Critical pressure p_L for each material [2]

Material			Tensile strength σ_B(Mpa)	Critical pressure P_L(Mpa)
Maerial types	DIN	JIS		
Mild steel	St 37	S10C	370	260
Middle carbon steel	St 50	S30C	500	420
Carbon steel (quench and temp.)	C 45	S45C	800	700
Cr–Mo steel	42CrMo4	SCM440	1000	850
Stainless steel	X5CrNiMo1810	SUS316	500–700	210
Cast steel	GG15	FC15	150	600
	GG25	FC25	250	800
	GG35	FC35	350	900
	GG40	–	400	1100
Mg alloy casting	GD MgA 19	MC2	300	220
			200	140
	GK MGA 19		200	140
			300	220
Al alloy casting	GK AlSi6Cu4	AC2B	–	200
Al alloy	Al 99	A1200	160	140
	AlZnMgCu 1.5	A7075	450	370
FRP	–	–	–	120–140

The critical surface pressure p_L of various fastened materials is shown in Table 6.2 [3].

Special care should be taken when fastening in the plastic range, which will be discussed later in Chap. 10, Sect. 10.1. If necessary, a washer with a larger diameter and sufficient thickness may be used.

6.2.3 Loosening Due to Cyclic Micro-Slippage Wear

When the seat surface, thread surface, etc. undergoes minute relative slippage due to cyclic loading during use, it is necessary to consider loosening due to micro-slippage wear. The following Archard's equation [4] is used as a simple prediction method for wear due to slippage, and is applied to the analysis of loosening behavior of threaded fasteners.

$$\mathrm{w} = k_w p S \tag{6.5}$$

w Wear depth (mm)
k_w Wear coefficient (mm^2/N).
p Surface pressure (N/mm)^2
S Relative slippage (mm).

Equation (6.5) can be rewritten using the contact area A, bolt fastening force F, number of load cycles N, and cyclic relative slip amplitude s_a as follows

$$\mathrm{w} = k_w \frac{F}{A} s_a N \tag{6.6}$$

Differentiating this by the number of load cycle N, we get,

$$\frac{dw}{dN} = \frac{k_w s_a}{A}\left(\frac{dF}{dN}N + F\right) \tag{6.7}$$

The relationship between this amount of wear and the decrease in fastening force can be obtained by using the embedment coefficient Z in Eq. (6.1),

$$\frac{dF}{dN} = -Z\frac{dw}{dN} \tag{6.8}$$

Substituting Eqs. (6.7) into (6.8), the following equation is obtained

$$\frac{dF}{dN} = -Z\frac{k_w s_a}{A}\left(\frac{dF}{dN}N + F\right) \tag{6.9}$$

By transforming,

$$\frac{dF}{dN} = -\frac{F}{\frac{A}{Zk_w s_a} + N} \tag{6.10}$$

Solving this differential equation by variable separation method,

$$\int \frac{1}{\frac{A}{Zk_w s_a} + N} dN = \int -\frac{1}{F} dF + C \tag{6.11}$$

$$log\left(\frac{A}{Zk_w s_a} + N\right) = -logF + C \tag{6.12}$$

$$logF\left(\frac{A}{Zk_w s_a} + N\right) = C$$

By applying initial condition, at $N = 0$, $F = F_f$

$$C = logF_f\left(\frac{A}{Zk_w s_a}\right) \tag{6.13}$$

$$\frac{F}{F_f} = \frac{\frac{A}{Zk_w s_a}}{\frac{A}{Zk_w s_a} + N} \tag{6.14}$$

This formula is applied to the actual model case, M10, $l_f/d = 10$ for a coarse thread hex bolt/nut fastener. There is not much database for wear coefficients for bolted joint materials, but referring to the wear test case [5] of S45C steel we can refer $K_w = 4.5 \times 10^{-8}$ (mm^2/N). By applying relative slip amplitudes s_a with 1, and 5 μm, the calculated results are shown in Fig. 6.9.

This method of predicting wear loosening was compared with Sakai's experimental results of relative slip loosening under cyclic torsional loading [6]. The experimental specimens are shown in Fig. 6.10. The bolts are made of M14, p1.5, S45C steel, and the shaft side, which serves as the female thread, is made of FCD70 cast iron. The relative torsional angle was measured by observing the relative misalignment between the marking line on the outer circumference of the pulley and the tip of the shaft guide, and a rotation angle control test was conducted to set the torsional angle amplitude. The bolt shaft force is measured using an ultrasonic shaft force measuring device.

The experimental results for various relative slip angles are shown in Fig. 6.11. The predicted results of wear loosening corresponding to these experimental results are also shown in Fig. 6.11 for reference. The experimental and predicted results

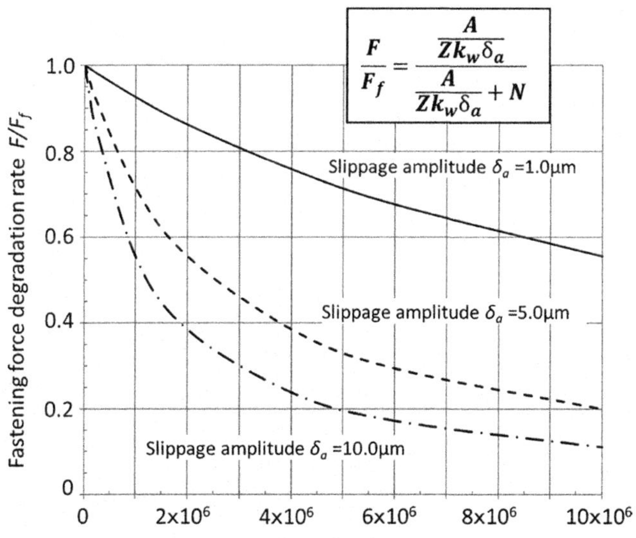

Fig. 6.9 Calculated examples for loosening due to cyclic micro-slippage wear

Fig. 6.10 Experimental specimen for measuring loosening due to rotational cyclic micro-slippage wear [6]

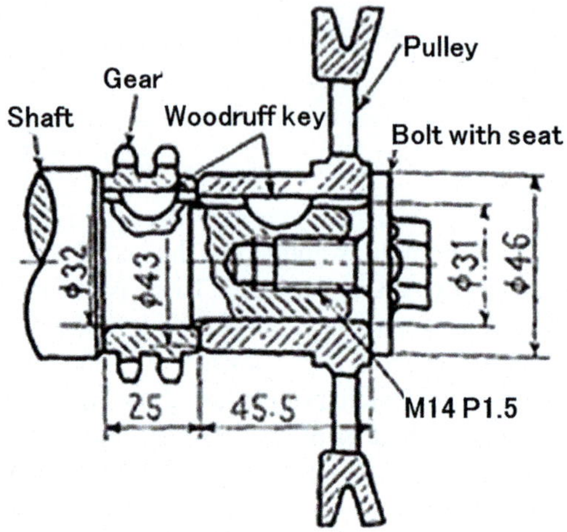

are in relatively good agreement, even considering that the wear coefficient k_w is not exact because the female threads are made of cast iron, which has little data. The results of Sakai's cyclic torsional loading experiment [6] may be questioned whether this is the reverse sliding rotation loosening described in Chap. 5, but in this experiment, repeated slipping was confirmed, and it was not unidirectional slippage loosening, and that loosening is due to wear. In other words, the sliding torques T_{sl} and T_{sf} at the threads were lower than the sliding torque T_w at the seat surface, and the conditions for rotational loosening ($T_{sl} < T_w < T_{sf}$) were not met in this test case. The reason for this may be that FCD70 material was used for the female threads in the experimental specimens, and the coefficient of friction μ_s at the threads was small and did not satisfy the Eq. (5.12) in Chap. 5.

To reduce loosening due to this micro-wear, following measures may be effective.

(i) Use materials that are resistant to wear (materials with small k_w, generally hard materials).

Fig. 6.11 Comparison of calculated results with experimental results for loosening due to cyclic micro-slippage wear (M14, *p*1.5, S45C steel bolt, FCD70 cast iron female thread)

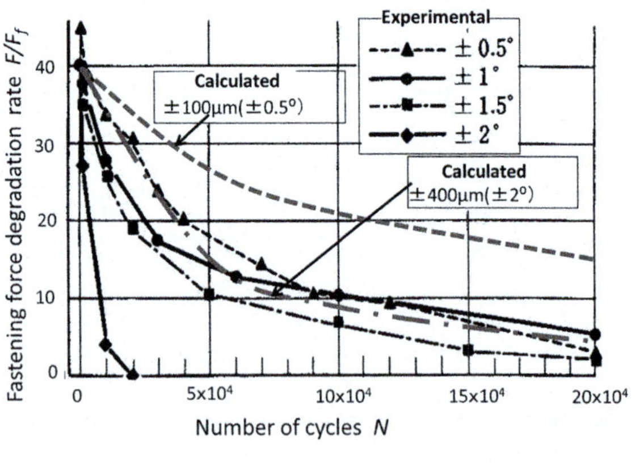

Table 6.3 Linear expansion coefficients α for several materials

			10^{-6}°C
Iron	11.7	Stainless steel	17–18
Alluminum alloy	23	10%Ni steel	13
Cupper	16.6	20%ni steel	20
Steel	10–11	Duralumin	23
Cast iron	10–12	Aluminum bronze	16

Mechanical Engineering Hanndobook, Japan Society of Mechanical Engineers

(ii) Increase the surface pressure contact area A and decrease the surface pressure p.
(iii) Reduce the embedment coefficient Z (set longer clamping length l_f and make bolts thinner (e.g., reduced shank bolts).

6.2.4 Loosening Due to Thermal Deformation

When the coefficients of linear expansion of the bolt and the fastened body differs and the temperature of the bolted joint structure changes, the difference in thermal deformations of both materials in the bolt axial direction causes the fastening force to change. Let α_b, α_c be the linear expansion coefficients of the bolt and the fastened body, and let Δt_b, Δt_c be the temperature changes of the bolt and the fastened body, respectively. In this case, the change in fastening force ΔF_t is expressed by the following equation using the embedment coefficient **Z** in Eq. (6.1) and the clamping length l_f as described above.

$$\Delta F_t = Z(\alpha_b \Delta t_b - \alpha_c \Delta t_c) l_f \tag{6.15}$$

This change, F_t if negative, is needed to evaluate slippage and loosening, and if positive, to evaluate seat sinking, etc. Linear expansion coefficients for several materials are shown in Table 6.3.

6.2.5 Deformation and Loosening of Seals and Gasket Materials

(A) **Compliance of seals, gaskets, bolts, and flanges**

Gaskets are used to prevent leakage in pressure vessels and pressure piping that are subject to internal pressure. It is important to maintain a high surface pressure to ensure that the gasket maintains a seal over a long period of time. However, it is inevitable that the gasket will gradually lose surface pressure due to stress relaxation and creep caused by time and temperature. Although retightening of bolts is an

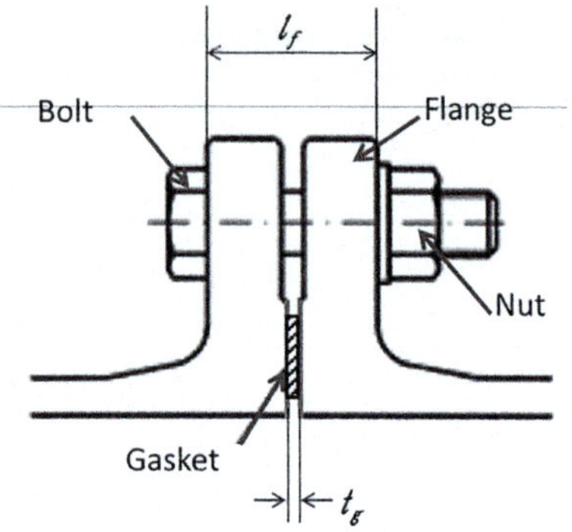

Fig. 6.12 Bolted fastener through the gasket

effective means of recovering the reduced surface pressure, the timing and method of retightening are not specified in standards and are left to the discretion of the user.

In this section, to grasp the basics of this complex phenomenon, a simple analysis of loosening is performed using the basic mechanism of the aforementioned embedment coefficient Z, linear thermal expansion coefficient, etc. The effects of re-fastening and other factors are discussed with examples.

The bolted fastener through the gasket is generally as shown in Fig. 6.12. In this case, the bolt compliance δ_b, will be used the following Eq. (2.10) or Eq. (2.11) in Chap. 2 using total bolt tightening length $l_f = l_a + l_s$..

$$\delta_b = \frac{0.4d}{E_b \frac{\pi}{4} d^2} + \frac{l_a}{E_b \frac{\pi}{4} {d_g}^2} + \frac{l_s}{E_b \frac{\pi}{4} {d_3}^2} + \frac{0.4d}{E_b \frac{\pi}{4} d^2}$$

The compliance of the flange of the fastened body is considered to be smaller than that of the gasket, so it is ignored, and only the compliance of the gasket is assumed, as follows.

$$\delta_c = \frac{t_g}{A_g E_g} \tag{6.16}$$

Here, t_g is the thickness of the gasket, A_g is the contact area of the gasket, and E_g is the elastic modulus of the gasket (the nonlinearities are ignored and considered equivalently linear). Using these, the fastening diagram (tightening triangle) of a bolted fastener through the gasket is shown in Fig. 6.13.

(B) Mechanical behavior due to temperature change

Let F_c be the initial fastening force when the bolt is initially tightened at room temperature. When this bolt/gasket joint is in operation and becomes hot, the gasket's rigidity decreases, while the bolt's rigidity does not change due to temperature, so the fastening force of this bolt/gasket joint will loose at high temperatures. However,

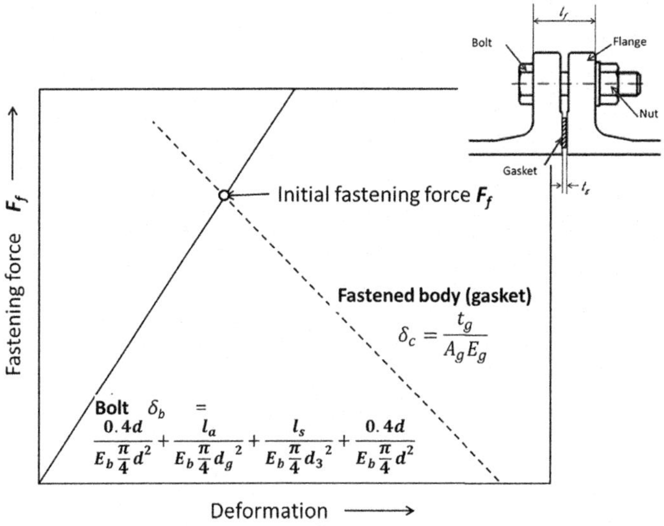

Fig. 6.13 Fastening diagram (tightening triangle) of a bolted fastener through the gasket

at the same time, the tightening force increases at high temperatures due to the difference in thermal expansion between the bolt and gasket.

These processes are explained quantitatively using Fig. 6.14. First, the fastening force at high temperature F_h', which only considers high-temperature softening, is obtained by the following equation using the embedment coefficients Z_c and Z_h at room temperature and high temperature, and the compliance δ_{cc}, and δ_{ch}, of the fastened body at room temperature and high temperature.

$$F_{h'} = \frac{Z_h}{Z_c} F_c \tag{6.17}$$

$$Z_c = \frac{1}{\delta_b + \delta_{cc}} \tag{6.18}$$

$$Z_h = \frac{1}{\delta_b + \delta_{ch}} \tag{6.19}$$

Next, the amount of increase in fastening force due to thermal expansion difference, ΔF_t, will be derived based on Eq. (6.15) as follows.

$$\Delta F_t = Z_h \left(\alpha_b \Delta t_b - \alpha_{gh} \Delta t_g \right) t_g \tag{6.20}$$

where α_b, α_{gh} is the coefficient of thermal expansion of the bolt and gasket at high temperature, and Δt_b, Δt_g indicate the temperature change of the bolt and gasket.

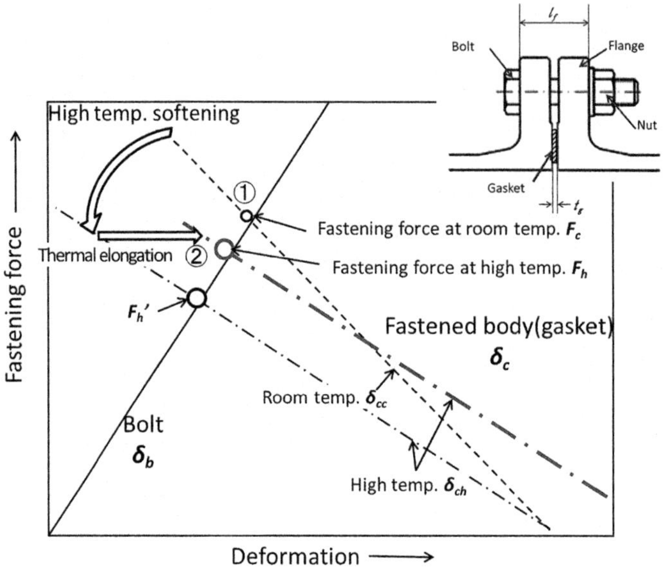

Fig. 6.14 Deformation and loosening process of a bolted fastener through the gasket during elevating temp

Finally, the fastening force $\boldsymbol{F_h}$ at high temperature of the bolted fastener through the gasket is given by

$$F_h = F_{h'} + \Delta F_t = \frac{Z_h}{Z_c}F_c + Z_h\left(\alpha_b \Delta t_b - \alpha_{gh}\Delta t_g\right)t_g \tag{6.21}$$

Next, from this situation at high temperature, consider what happens when the temperature is lowered to room temperature. Figure 6.15 shows the stiffness (compliance) of a single gasket material at room temperature and high temperature, and the deformation process during temperature rise and fall. The gasket material is treated as a linear elastic body for the sake of simplicity in order to elucidate its basic behavior. When the temperature rises to a high temperature while the clamping allowance is maintained from the room temperature clamping point shown in Fig. 6.15, point ①, the gasket softens at the high temperature as described above and shifts to the stiffness (compliance δ_{ch}) of the broken line, and clamping point moves to point ②. Next, consider the behavior when the temperature is lowered from this high-temperature state to room temperature. In this case, the deformation caused by the softening at high temperature includes plastic deformation, and this plastic deformation is not recovered during the temperature reduction and unloading process, so the path ② to ③ is followed instead of ② to ① in Fig. 6.15. Figure 6.16 shows this process on a fastening diagram.

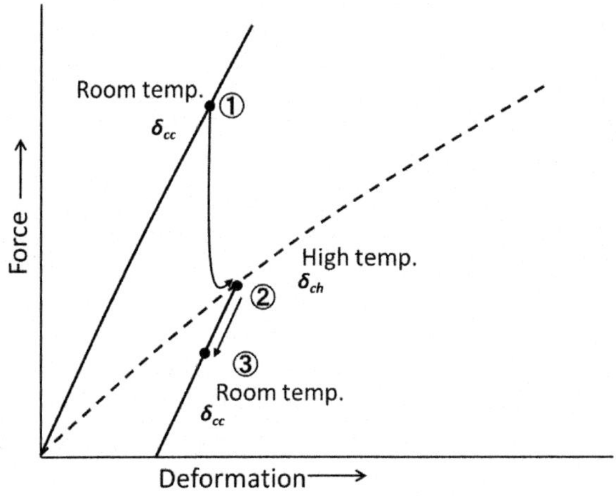

Fig. 6.15 Deformation and loosening process of gasket during elevating or cooling temp

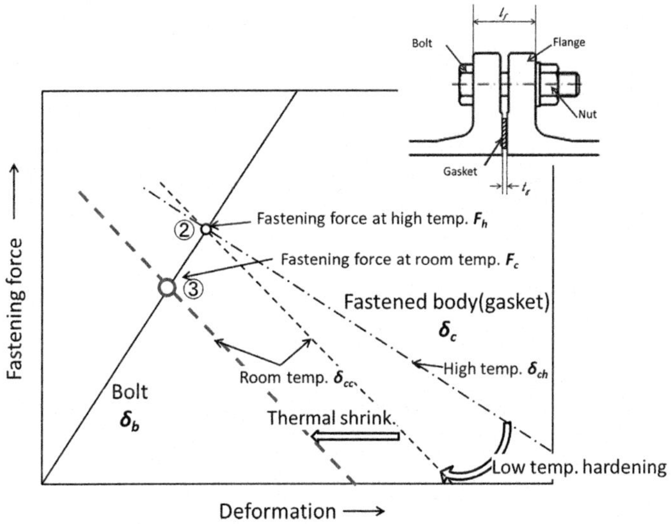

Fig. 6.16 Deformation and loosening process of a bolted fastener through the gasket during cooling temp

(C) Experimental confirmations

The response of the fastening force during these temperature raising and lowering processes was tested on an actual bolt/gasket fastener and the results are shown below [7]. Figure 6.17 shows a schematic diagram of the test apparatus. The gasket to be used as a sample is fastened to the test flange with bolts to which strain gages are mounted. The gasket surface contact pressure is calculated from the bolt axial force measured by the strain gage and the area of the gasket contact surface. This allows continuous evaluation of changes in gasket surface pressure in all processes after fastening. The strain gages are heat-resistant, and the fastened body is heated by an electric furnace or band heater to enable evaluation under conditions that simulate the operating environment at high temperatures. The sample gasket is a high-performance sheet gasket No. GF300 manufactured by VALQUA Ltd. Table 6.4 shows the test conditions. Using this test apparatus, changes in surface pressure were measured at 200 °C after fastening at room temperature and under temperature changing conditions simulating an operating cycle of room temperature ⇔ 200 °C. The results are shown in Fig. 6.18, which clearly shows how the behavior at elevated temperatures and under thermal cycling well reproduces the behavior from ① to ② shown in Fig. 6.14 and from ② to ③ and ③to ② shown in Fig. 6.16, respectively. In other words, when the temperature rises, the rigidity of the gasket decreases, plastic deformation progresses, and the clamping force decreases, resulting in a total surface pressure decrease even if the increase in clamping force due to the difference in thermal expansion is included (① to ②). Then, when the temperature decrease, the surface pressure decrease by the difference in thermal expansion, And the plastic deformation that occurs when the temperature rises cannot be recovered when the temperature falls, and it behaves as an elastic body. So, further reducing the surface pressure (② to ③). However, during the next heating process, the plastic deformation has already disappeared and the material behaves almost as an elastic body, so by the difference in thermal expansion surface pressure increase (③ to ②). As mentioned above, the process in ② to ③ to ② assumes that the gasket is a linear elastic body, so creep and relaxation behavior over a long period of time are not included. If this is necessary, there are FEM analysis examples that also take into account viscoelastic properties [8–11], I hope this is also helpful.

(D) **Confirmation of mechanical behavior and effect of retightening**

As shown in the previous section, when gaskets are fastened at room temperature and become hot, the rigidity of the gasket decreases (plastic deformation), resulting in a decrease in fastening force (decrease in surface pressure), and over time there

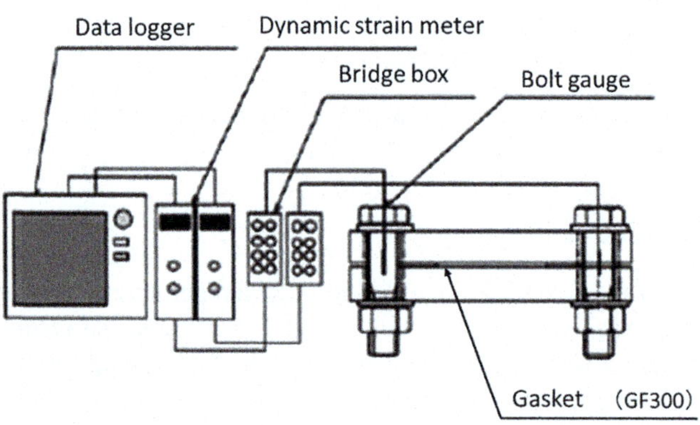

Fig. 6.17 Schematic diagram of the test apparatus of gasket/bolt joint

Table 6.4 Thermal cycle test specifications

Test temp	Room temp. and 200 °C, Roomtemp. ⇔200 °C cycle (24 h)
Gasket	VALQUA Ltd. No. GF300, thickness 1.5 mm
Flange sheet dimension	JIS 10 K 50A RF
Flange specification	Material; SS400, Steel surface loughness; (Rz) 14.4 μm (Av)
Bolt (with strain gauge)	Material; SS400, 4XM16

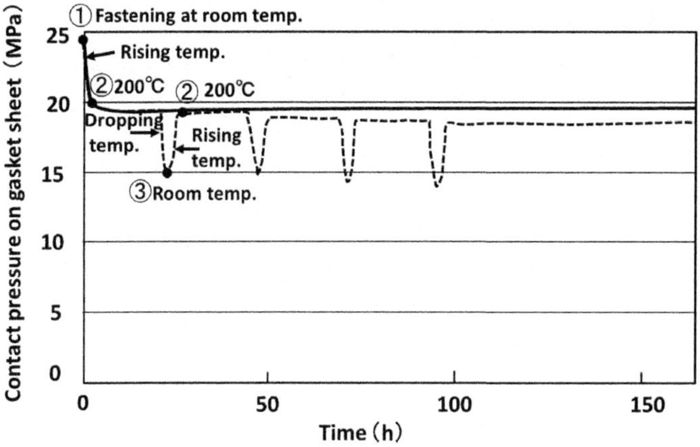

Fig. 6.18 Thermal heat cycle test results for gasket/bolt fastened joints

will be further decrease in fastening force due to creep of the gasket. In the case of piping equipment in power plants, etc., where internal pressure is applied and it was concern about leakage due to internal pressure, so retightening was performed as a preventative measure. This section presents analytical considerations and experimental confirmation of the deformation behavior and effects of retightening. There are two types of retightening: hot retightening, which is tightening at high temperature, and cold retightening, which is tightening after the temperature has dropped to room temperature. Figure 6.19 shows the mechanical and deformation behavior of hot retightening using a clamping diagram. In this case, since there is no change in material properties or thermal expansion, the fastening force that has decreased to point ② in Fig. 6.14 is simply returned to point ①, and the retightening effect is big. However, retightening at this high temperature is not easy, and if the temperature drops during retightening operation due to plant shutdown, etc., the internal pressure will increase from a very loose (low seat surface pressure) state as shown in Fig. 6.16 or Fig. 6.18 at point ③ when the plant starts up again, so there is a risk of leakage at this restart. To avoid this risk, cold retightening is also used, which has a small retightening effect but maintains safety.

Figure 6.20 shows the mechanical and deformation behavior using the clamping diagram during cold retightening. When the temperature drops from point ② where

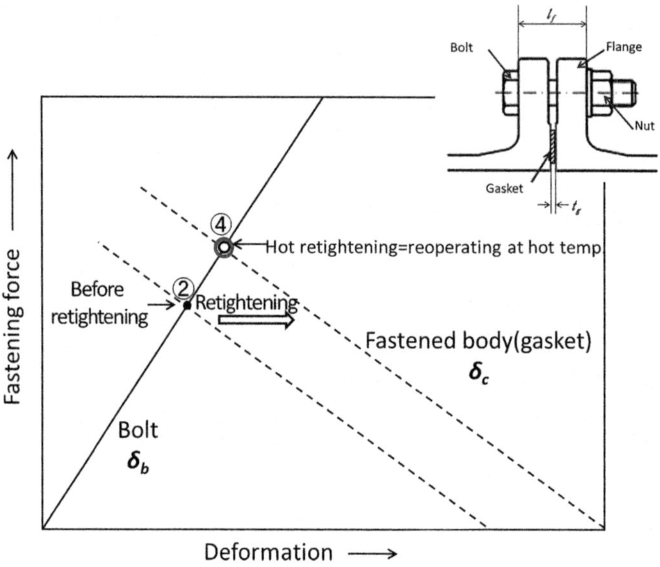

Fig. 6.19 Deformation and loosening process of a bolted fastener through the gasket during hot retightening process

high-temperature operation before retightening, the thermal contraction of the gasket causes the clamping force to drop to point ③. If it is restarted and pressure is increased from this state, there is a risk of leakage. So cold retightening must be performed in this low-temperature state. This is point ⑤. This is basically the same as the fastening force at the initial fastening point ①. Therefore, when restarting from this cold retightening point, both behaviors of gasket softens + thermal expansion occur in combination, and eventually the system returns to point ②, as shown in Fig. 6.14, when restarting at high temperatures.

Experiments were conducted to confirm the behaviors of the gasket pressure after both of these retightenings. The experimental conditions are shown in Table 6.5. The experimental results are shown in Fig. 6.21. It can be seen that the behavior during hot retightening and cold retightening is in good agreement with the mechanical processes shown in Figs. 6.19 and 6.20, except for the viscoelasticity and creep behavior over a long period of time.

6.3 Rotational Loosening

Items for rotational loosening are shown in (6) to (8) in Table 6.1 in Sect. 6.1.

Here, we will further explain each item in detail below.

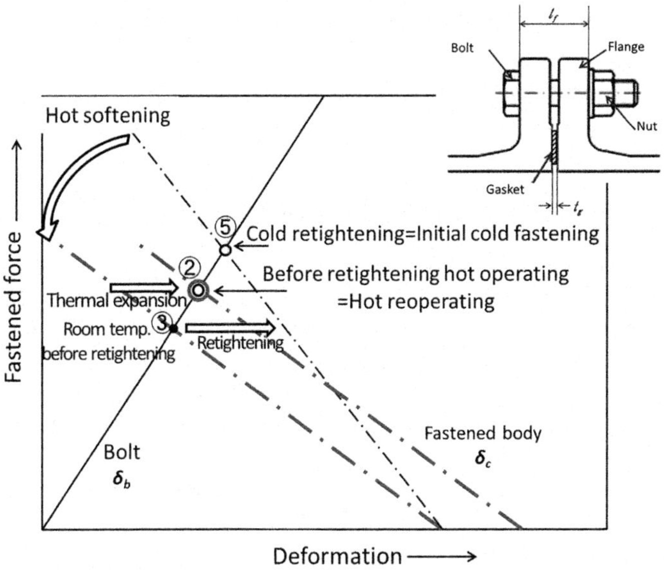

Fig. 6.20 Deformation and loosening process of a bolted fastener through the gasket during cold retightening process

Table 6.5 Retightening conditions

Types of retightning	Timing of retightning
Hot retightning	24 h heating (200 °C)
Cold retightning	When cooling after 24 h heating (room Temp.)

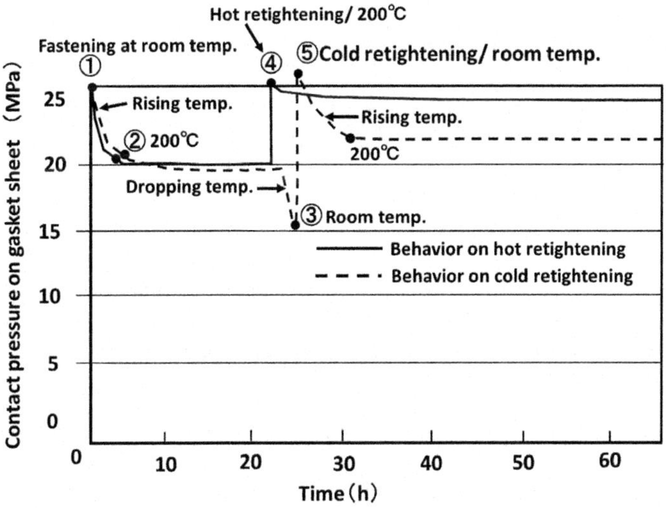

Fig. 6.21 Thermal heat cycle test results for gasket/bolt fastened joints with retightening

6.3.1 Loosening Due to Cyclic Transverse Loading

The basic considerations based on mechanical analysis were explained in Chap. 5. This section provides a summary of the mechanical behaviors under transverse loading and explains some know-hows to prevent transverse loosening.

(A) **Loosening conditions based on mechanical analysis**

Even if the external force W acts in the direction perpendicular to the bolt axis (shear external force), if it is less than $F_f \mu_c$, no relative slip occurs between the fastened bodies, and no slip occurs on other contact surfaces either. However, if W is greater than $F_f \mu_c$, slip occurs between the fastened bodies, inducing nonlinearity and asymmetry in the inclination of the bolt and the surface pressure of the thread and seat surfaces, which in turn induces rotational slip in the bolt/nut contact thread surface and may lead to loosening.

The contact surface of the bolt/nut has a three-dimensional helical shape, and if relative slip perpendicular to the shaft occurs under a fastening force $\boldsymbol{F_f}$, loosening torque $\boldsymbol{T_{ss}}$ is generated on this thread surface.

$$\mathrm{T}_{ss} \cong \frac{F_f d_2 \tan\beta}{4\cos\alpha'} \qquad \text{Chapter4(4.40)}$$

and if this torque is greater than the bolt-loosening torque T_l as shown below equation, this torque T_{ss} will cause the bolt and nut to rotate relative to each other and loosen.

$$\frac{F_f}{2}\left(-\frac{p}{\pi} + \frac{\mu_s d_2}{\cos\alpha'} + \mu_w D_w\right) < \frac{F_f d_2 \tan\beta}{4\cos\alpha'} \qquad \text{Chapter 4(4.41)}$$

If we apply the ISO standard triangular thread specifications to this condition, above equation is simplified as follows.

$$\mu_s = \mu_w < 0.032 \qquad \text{Chapter 4(4.43)}$$

However, this is a condition where the bolt and nut are loosened by relative rotation only with the above T_{ss} under the condition that there is no relative slip on the thread surface and the seat surface. Under the condition where the relative slip is already occurring on the thread surface and seating surfaces, it is quite possible by the above torque T_{ss} to gradually show rotational component to the main transverse slippage component. The conditions for the occurrence of relative slip at these surfaces are shown below, respectively.

(A-1) **Slip condition at the seating surface**

As already shown in Sect. 4.3 in Chap. 4, the condition in which the external transverse load W_3 causes the two friction surfaces to slip, one between the fastened bodies and the other between the seating surfaces,

$$W_3 = F_f \mu_c + F_f \mu_w \qquad \text{Chapter4(4.20)}$$

The amount of relative slippage between the fastened bodies reaches the limit amount of slip S_{cr} just at this load condition W_3.

$$S_{cr} = 2\Delta = 2\left\{ F_f \mu_w \left(\frac{l_g^3}{3EI_g} + \frac{l_p^3}{3EI_p} + \frac{l_g l_p l_n}{EI_g} + k_{wh} l_n^2 \right) \right.$$
$$\left. -M_n \left(\frac{l_g^2}{2EI_g} + \frac{l_p^2}{2EI_p} + \frac{l_g l_p}{EI_g} + k_{wh} l_n \right) + M_n k_{wn} l_n \right\} \quad \text{Chapter 4 (4.45)}$$

(A-2) **Slip condition at thread surface**

In the case of a square thread, as shown in Sect. 4.3, Eq. (4.14), Figs. 4.2, and 4.3, a transverse load $W = F_f \mu_s$ on the thread surface causes a slip in the direction perpendicular to the axis.

$$W_s = F_f \mu_s \qquad \text{Chapter 4(4.14)}$$

In the case of triangular threads, the thread contact surface is a cone-shaped helix, which does cannot cause planar slip perpendicular to the bolt axis, but rather a condition for rotational slip around the axis perpendicular to the bolt. For simplification, we approximate the helical thread surface to a cone-shaped surface and the slip-generating bending moment M_c between concentric cones can be derived as follows.

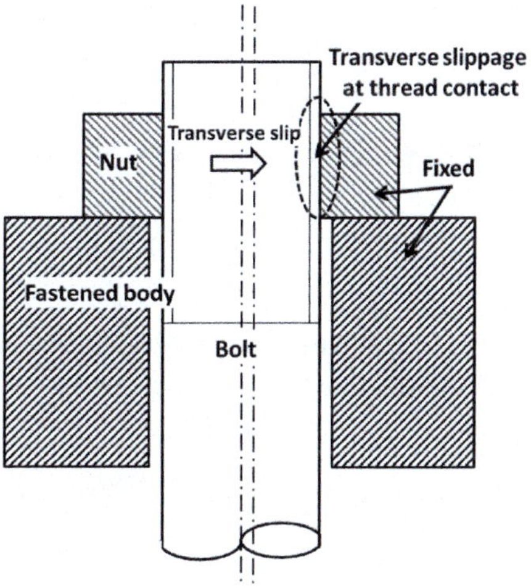

Fig. 4.2 Transverse slip at thread

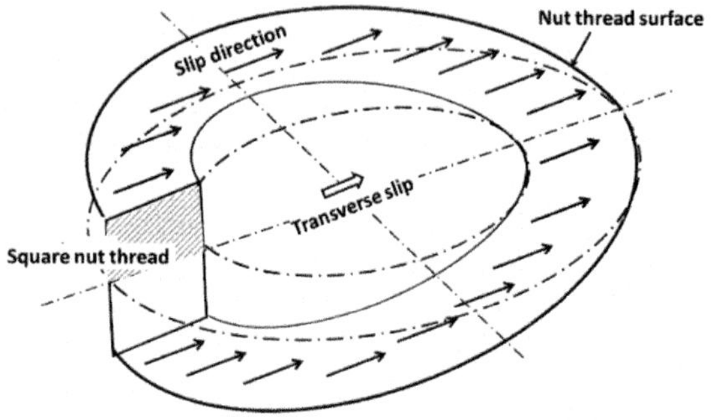

Fig. 4.3 Transverse slipping behavior of (square thread) one thread of square screw

$$M_c = \frac{N\mu_s}{2\pi r_2}\int_0^{2\pi} r_2\sqrt{\cos^2\theta + 3}r_2 d\theta = \frac{2N\mu_s r_2}{\pi}\int_0^{\pi/2}\sqrt{\cos^2\theta + 3}d\theta = \frac{F_f\mu_s d_2}{\pi\cos\alpha}2.93$$
$$= 1.08F_f\mu_s d_2 \quad \text{Chapter 4(4.15)}$$

Before this bending slip occurs, the bolt behaves as a both ends rotational restraint beam, as shown in Sect. 4.3, Fig. 4.1b, so the bending moment M_n acting on the nut thread contact surface is expressed by the following equation.

$$M_n = \frac{W_T l_f}{2} \quad \text{Chapter4(4.16)}$$

where l_f is the fastening length of the bolt. Using this relationship, the conditions under which this bending slip occurs are

$$W_b = 2.16F_f\mu_s\frac{d_2}{l_f}$$

For ISO standard threads, from $d_s \approx 0.9d$,

$$W_b = 1.94F_f\mu_s\frac{d}{l_f} \quad \text{Chapter 4(4.17)}$$

So, by comparing with Eq. (4.20) if seat surface frictional coefficient μ_w and thread surface frictional coefficient μ_w are same, this thread slip can occur when

$$1.08d_2 < \frac{l_f}{2}, \left(l_f > 2.16d_2\right)$$

It can also be seen that the tightening length l_f must be approximately twice or more than the nominal diameter d of the bolt to prevent the screw from sliding and loosening against an external force in the perpendicular direction to the axis.

(B) **Technology to prevent loosening under transverse loading**

(B-1) **Effects of each specification on loosening characteristics and methods for preventing loosening**

Basically, the key is to increase the critical slip amount S_{cr} according to Eq. (4.45).

(i) The higher the fastening force F_f, the less likely it is to loosen.
⇒ High fastening force recommended.
(ii) The longer the fastening length l_f (thickness of the fastened body), the less likely it is to loosen.
⇒ Long bolt or spacer will be recommended.
(iii) The smaller the bolt's cross-sectional secondary moment I_g, the less likely it is to loosen.
⇒ Reduced shank bolt will be recommended.
(iv) The larger the coefficient of friction μ_w on the bearing surface, the less likely it is to loosen.
⇒ Surface treatment of bearing surface or washer surface will be recommended.
(v) The smaller the screw fitting length, the less likely it is to loosen (k_n smaller).
⇒ Thin nuts are recommended.
(vi) The larger the clearance perpendicular to the axis of the male thread and female thread, the more difficult it is to loosen (k_n smaller).
⇒ Do not increase thread accuracy.
(vii) The larger the clearance between the threaded hole and the outside diameter of the bolt, the less likely it is to loosen.
⇒ Reamer bolts will be not recommended.

(B-2) **Verification of the benefits and harms of various washers and double nuts based on experiments**

For evaluation of rotational loosening of bolted joints under transverse loading, a testing machine invented by Junker has long been used [12, 13]. Figure 6.22 shows a schematic of the machine [14]. The movable plate (fastened body) is connected to the eccentric disk via a load cell, and the rotation of the eccentric disk gives the movable plate a cyclic repetitive displacement and transverse load. Cyclic relative displacement is measured by a differential transformer type displacement meter (LVDT), and cyclic transverse load is measured by load cell. Yamamoto et al. used a testing machine shown in Fig. 6.23, which is basically the same as Junker's testing machine, but with modifications such as eliminating the plate spring to give a precise relative displacement between the movable plate and the fixed plate. In these testing machines, the contact surfaces between the fixed and movable plates are roller bearings without frictional force in order to accurately control the cyclic relative displacement. So, these are different from the mechanical analysis model described in Chap. 4. However, when comparing the results of this test with the results of the mechanical analysis in Chap. 4, we can simply add the friction force $F_f \mu$ between the fastened bodies to the transverse load W_T.

The results of evaluating the loosening performance of various bolted joints under cyclic transverse loading using such a testing machine [15, 24] are shown [16] bellow. Figure 6.24 shows the change of fastening force F_f under the cyclic transverse loading with total amplitude of displacement S. If the loosening rate is defined as the amount of axial force decrease per load cycle, d F_f/dN, then in this test case (M10, hexagonal bolt/nut) dF_f/dN was obtained as 0.224 kN/cycle. The relationship between S and dF_f/dN obtained experimentally for the hexagonal bolt/nuts and various loosening devices shown in Table 6.6 is shown in Fig. 6.25. In this figure, the S value at which dF_f/dN increase from 0 is called as S_{cr} as mentioned in Chap. 4. The measured values S_{cr} are shown in the bottom column of Table 6.6. These are the results of anti-loosening performance for various locking devices. The performance and characteristics for each locking device are shown below.

(i) The spring washer and conical spring washer are easier loosening than the normal hexagon nut alone

Figure 6.26 shows the fastening diagram for the case with spring washers. As the stiffness C_{sw} of the spring washer is sufficiently lower than the stiffness C_c of the fastened body, it is almost impossible for the spring washer to become involved in an axial force below point C in fastening diagram.

Therefore, if the spring washer is properly fastened, there is no mechanical contribution from the spring washer.

Rather, as described in the FEM analysis below (see Section (B-3), impairing the uniformity of such circumferential surface pressure bring the disadvantage of lowering the critical slippage S_{crr} and making it easier to loosen.

(ii) The effect of the flange nut is recognized, which can be verified analytically by the fact that the larger the effective diameter D_w of the nut seating surface, the larger the loosening torque T_l becomes, as shown in Eq. (4.29) in Chap. 4, the less likely it is to loosen.

Table 6.6 Hexagonal nut/bolt fasteners with various locking devices and experimental results of critical slippage amplitude S_{cr} for each locking devices [17]

Type	Normal nut	Spring washer	Conical spring washer	Double nut	Locking adhesive
Figure					
S_{cr} (mm)	0.6	0.35	0.4	0.85	1.1

(iii) Double nuts are also effective, but they are not used much these days. This may be due to the fact that, recently the true mechanical basis was not understood and the correct fastening method was forgotten, so the wrong fastening method was used, and the effect of double nut was no longer confirmed. The correct fastening method is shown in Fig. 6.27. First, tighten the lower nut with torque T_1 (fastening force F_1) (Fig. 6.27a). Next, tighten the upper nut with the normal target torque T_f (target fastening force F_f). As shown in Fig. 6.27b, the threaded surface of the lower nut is not crimped with the threaded surface of the bolt and no effect can be obtained. Then the correct fastening method can be expressed as follows. After the lower nut was fastened, the upper nut is fixed and the lower nut is rotated in the direction of loosening until the threaded surfaces of both the upper nut and the lower nut are crimped with the threaded surface of the bolt. The presence of this crimping force (locking force F_{loc}) causes the lower nut, upper nut, and bolt to lock together and restrain both rightward and leftward rotation (Fig. 6.27c). This is called locking, and since the two wrenches are subjected to torque in opposite directions in this process, this locking operation is called a inter-locking. The effect of the double nut is realized only after this entire process. Therefore, if the mechanical mechanism of the double nut is not shared between the design site and the production site, workers at the production site will finish their work in the state shown in Fig. 6.27b, it will have the opposite effect and become easier to loosen, and this negative image seems to have hindered the use of double-nut methods.

Ordinary the arrangement of the upper and lower nuts in a double nut are as shown in Fig. 6.28a, the upper nut a normal nut and the lower one a low nu. In this arrangement, however, there is a problem that a standard wrench cannot fit into the lower nut when fastening by inter-locking. So sometimes thin-type wrench was used as shown in Fig. 6.29. However, this thin-type wrench does not have sufficient strength, so it is effective to use a special jig for double nuts that has sufficient thickness, rigidity, and strength with a socket insertion hole in a torque wrench for the upper nut as shown in Fig. 6.30.

(iv) Locking adhesive (anaerobic adhesive) has the best results among the experimental results in Table 6.6. An anaerobic adhesive is an adhesive that begins to solidify when contact with air is discontinued after application. When filled this locking adhesive to the gap between the male thread and female thread during bolt fastening process, the adhesive completely solidifies after several hours. During the fastening operation, the adhesive is in a liquid state, so there is no particular concern about the coefficient of friction. Figure 6.31 shows an example of a product in which the anaerobic adhesive is trapped in small capsules in advance to improve workability, and each capsule is pre-applied with a normal adhesive, so that the capsule breaks when the screw is tightened and the anaerobic adhesive leaches out and solidifies. These effects are thought to be due not only by the adhesive strength, but also by the fact that the clearance between the male screw and female screw is made of a solid material which prevents relative slippage of the contacting surfaces. Dismantling or loosening of this adhered screw fastener can be easily performed with a wrench, but replacement or reapplication is necessary for reuse.

(B-3) **Verification of the benefits and harms of various washers and double nuts based on fem analysis**

(i) Fundamental analysis for normal nut/bolt fastening model

FEM analysis of sliding and loosening behavior of a bolted joint under transverse loading has not been widely performed due to the complexity of the three-dimensional contact problem, but Yokoyama et al. [17, 18] have performed sliding and loosening analysis using a structure and loading model that reproduces the Junker-type testing machine described in (i) to verify the results of the above simple mechanical analysis and experiments. The geometry of the target mechanical model is shown in Fig. 6.32 and the finite element method analysis model is shown in Fig. 6.33. This object is composed of a metric coarse thread with nominal diameter d of M16, a fastening length l_f of 30 mm, and a bolt hole diameter of 18 mm, and three-dimensional solid elements are used for the bolt and nut, taking into account the spiral of the threads. Nuts are modeled as round nuts with a diameter of two face widths, omitting the hexagonal shape. The movable and fixed plates are modeled using three-dimensional shell elements. Contact elements that consider isotropic Coulomb friction are used for the thread surface, bolt, and nut seating surfaces.

The boundary conditions of the fixed plate constrain displacement in all directions, and the boundary conditions of the movable plate are such that the side in the x-axis direction constrains displacement in the z-direction, and the side in the z-axis direction is forced to displacement or external force in the x-direction.

The circumferential displacement is constrained at the outer surface of the nut.

The initial fastening force $F_f = 8000$ N is given by forced displacement of the movable plate in the axial direction.

The material properties of the bolt, nut, fixing plate and movable plate were set as Young's modulus $E = 205$ GPa and Poisson's ratio $\nu = 0.3$. The friction coefficient is set to $\mu = 0.1$. For the analysis, a quasi-static linear analysis was performed for the inside of the solid object, and a nonlinear contact problem analysis was performed for the contact surface.

(ii) Loosening in bearing surface slippage (total surface slippage)

Figure 6.34 shows the analytical results of the cyclic load/displacement (4 cycles) relationship for the case of complete seat slip. This corresponds to the load/displacement hysteresis relationship using the simplified mechanical model shown in Fig. 6.4.5 in Chap. 4, where the interval between the load $\pm W_1$ is cut off, because the load model of the Junker-type testing machine has roller bearings between the fastened bodies and so the sliding friction force $W_1 = F_f\,\mu_c$ between the fastened bodies shown in Chap. 4 is ignored. Considering this, the results are in good agreement with those of Chap. 4 and this load/displacement hysteresis relationship consists of three different gradient regions. In other words, the steep slope area (O-A, B'-C, D'-E) corresponds to part (b) in Fig. 6.4.5 of Chap. 4, and both the thread and seat surfaces are adhered, i.e., (a) in the Fig. 6.35. The gradual slope (A-B, C-D, E–F) corresponds to part (c) in Fig. 6.4.5 of Chap. 4, where the thread surface has shifted to the sliding state and the seat surface is still in the adhering area. In other words, it corresponds to Fig. 6.35b. The flat areas (B-B', D-D') correspond to (d) in Fig. 5. 5 of Chap. 5, indicating that the seat surface has also shifted to the sliding state. In other words, it corresponds to Fig. 6.35b. When the seat surface shifts in a sliding state, the thread surface and the seat surface can slide and rotate, indicating that the screw will be loosening after each cycle.

In the simple mechanical analysis in Chap. 4, we also explained that loosening occurs when seat surface slipping is repeated in this way, but we have not mentioned the loosening behavior caused by repeated thread surface slipping.

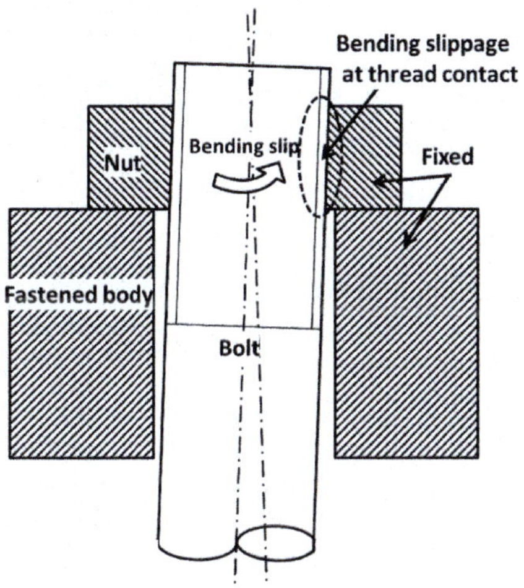

Fig. 4.5 Bending slip at thread

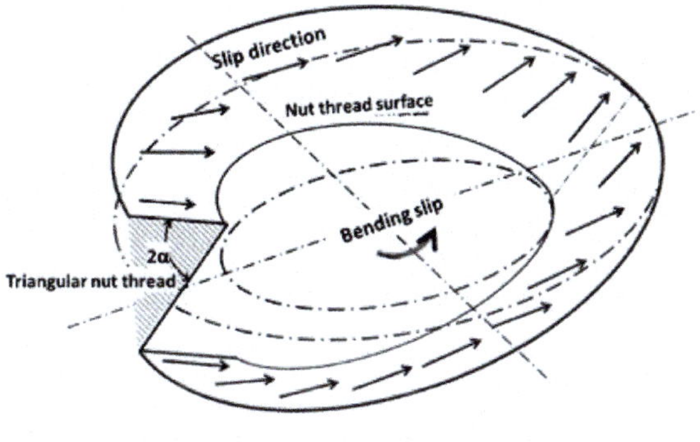

Fig. 4.6 Bending slipping behavior of (square thread) one thread of triangular screw

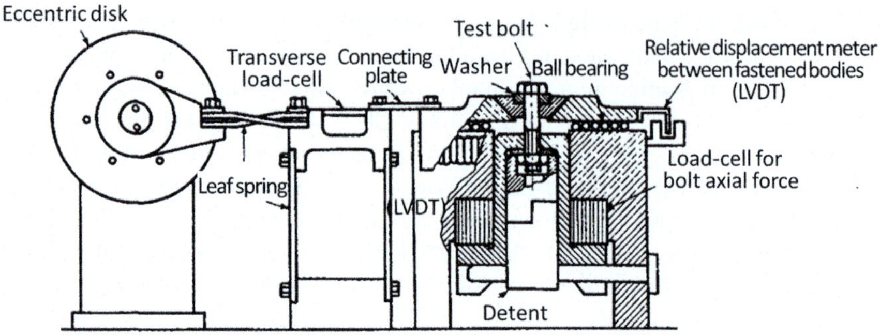

Fig. 6.22 Schematic diagram of the Junker's bolt-loosening test machine [14]

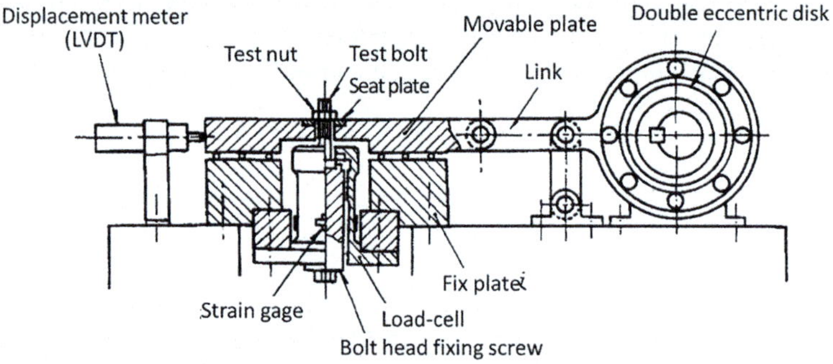

Fig. 6.23 Schematic diagram of Yamamoto et al.'s improved bolt-loosening test machine [15, 16]

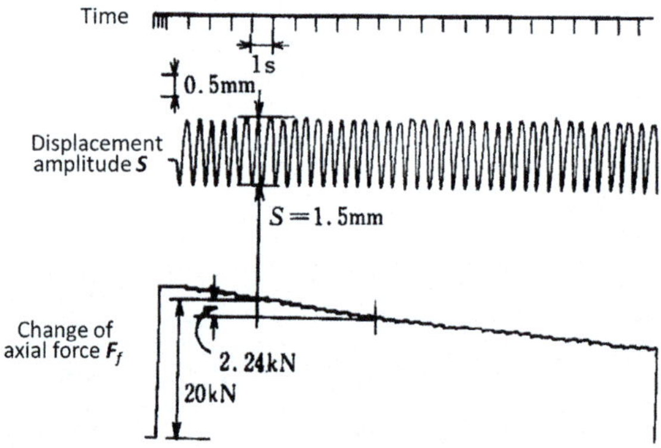

Fig. 6.24 Loosening test result under cyclic transverse loading with constant displacement amplitude [16]

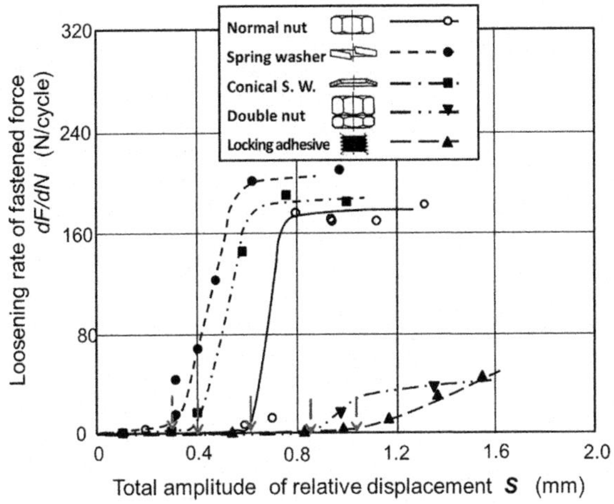

Fig. 6.25 Relationship between S and dF_f/dN obtained experimentally for hexagonal bolt/nuts with various loosening devices [17]

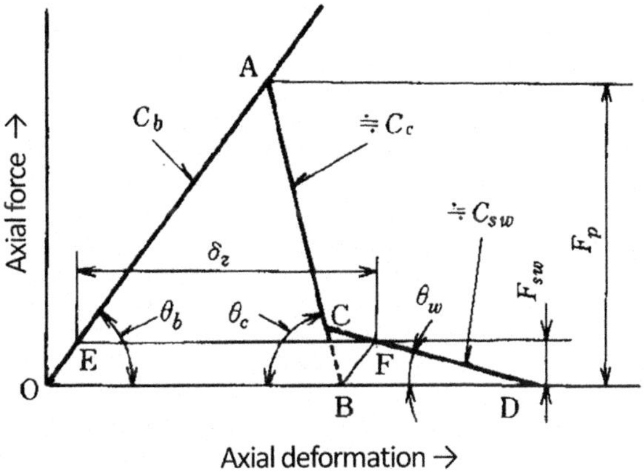

Fig. 6.26 Tightening diagram for the case with spring washers

(iii) Loosening in thread surface slippage

Figure 6.36 shows the analytical results of load/displacement under cyclic lateral loads in a load range of 90% of the seat sliding limit load (F_f 8000N Xμ 0.1 = 800 N). It is stated that under this load cycle, the thread surface slips, and loosens slightly with each cycle. In this region, as shown in Fig. 6.35b, there is a sticking area on the seat surface in the bolt bending side, so although the bolt does not slip macroscopically on the seat surface, and also there is a slip zone on the opposite side. In this region, the mismatch in local elastic deformation between the nut side

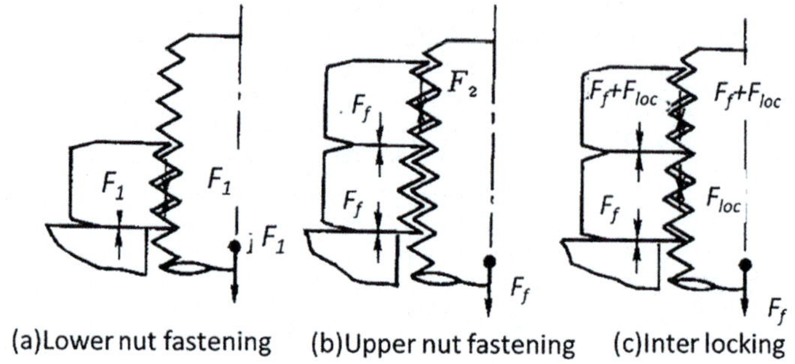

Fig. 6.27 Fastening processes for double-nut fastening method and mechanical conditions on thread and seat surfaces

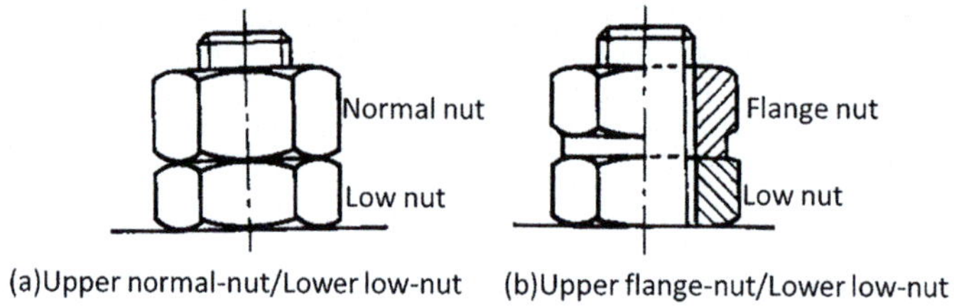

Fig. 6.28 Arrangements of upper and lower nuts in double-nut fastening

Fig. 6.29 Thin-type wrenches for inter-locking in double-nut fastening

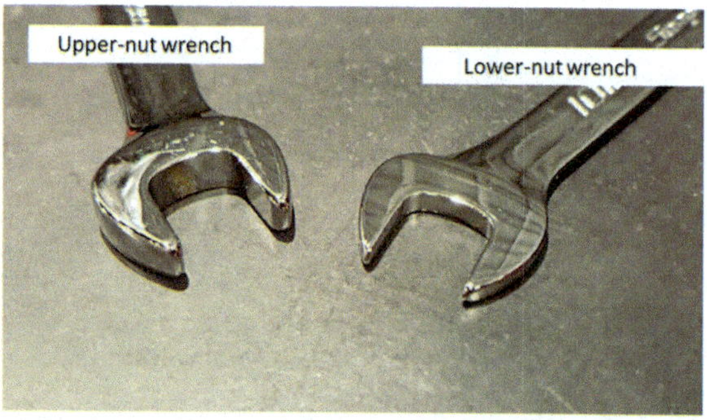

Fig. 6.30 Auxiliary wrench for inter-locking in double-nut fastening

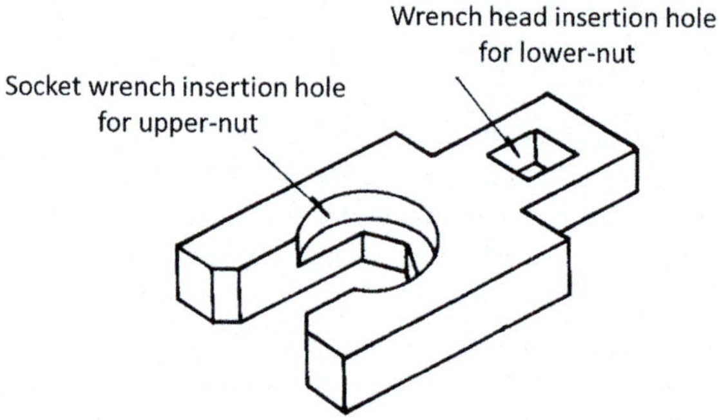

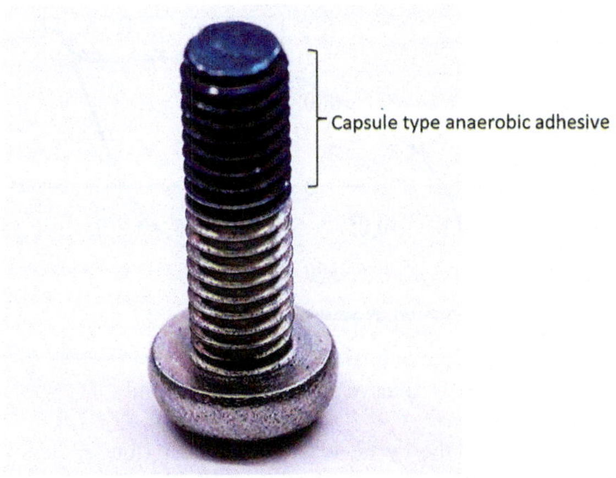

Fig. 6.31 Applied example of capsule type locking adhesive (Anaerobic adhesive)

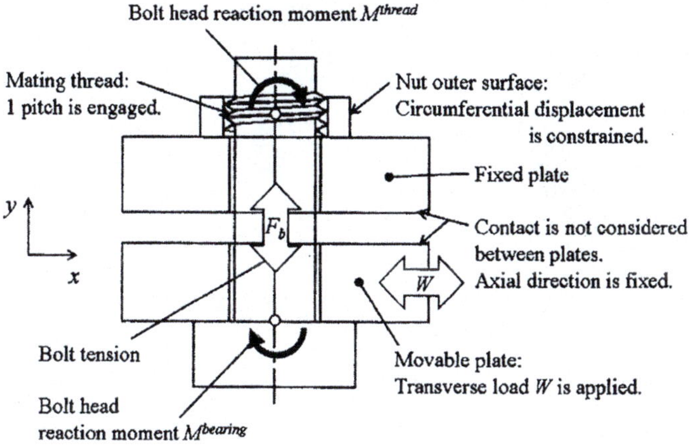

Fig. 6.32 Geometry of the mechanical mode for analysis [18–21]

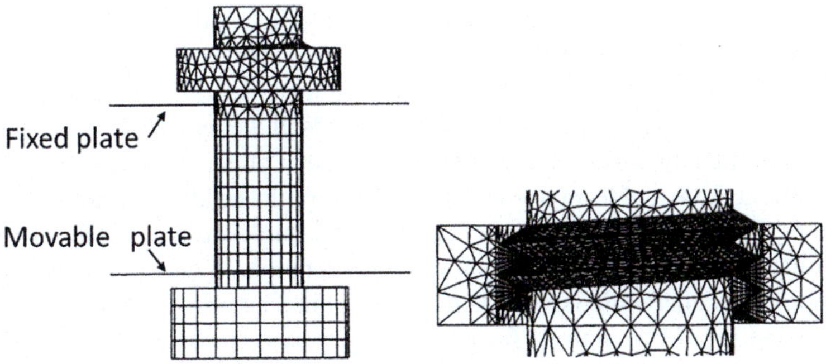

Fig. 6.33 Finite element method analysis model for bolted joints under transverse loading [18–21]

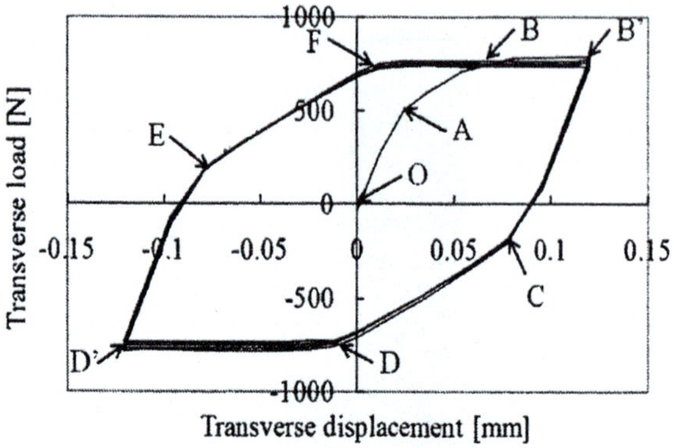

Fig. 6.34 Analytical results of the cyclic load/displacement (four cycles) relationship under transverse loading in bearing surface slippage condition [18–21]

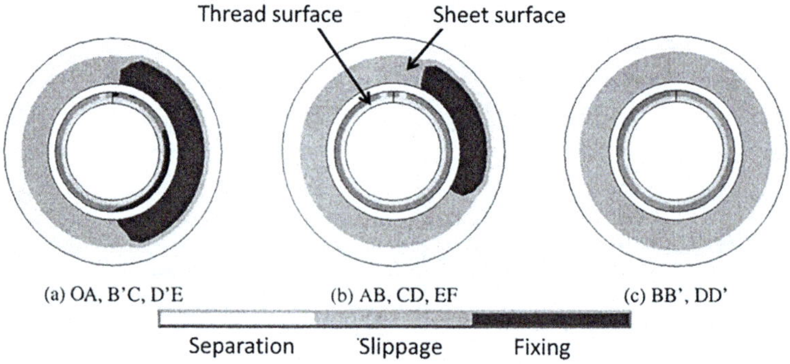

Fig. 6.35 Contact conditions of thread surface and seat surface on each load situation [18, 19]

and the fixed plate (fastened body) is allowed to local slip. In this case, as long as there are no sticking points throughout the entire cycle, this local mismatch in elastic deformation will move just like the movement of the inchworm as shown in Fig. 6.37. In Fig. 6.37a, the rear end is fixed and the worm body is extended, then as (b) and (c), the front end is fixed and the body is shrinked and then the rear end is fixed again and the body moves by δ after one cycle. In other words, if we replace the expansion and contraction of the inchworm body with the mismatch of elastic deformation and the fixation of the front and rear ends with the fixation of the right side of the seat surface in Fig. 6.35b and the fixation of the left side after half a cycle, we can see that the accumulation of these cycles allows for gradual rotation. In the mechanical analysis in Chap. 4, this micro-slip is not taken into account since elastic deformation is not so large and should be ignored for practical purposes. However, the fact that such micro-slip can cause micro-rotational slip and small loosening is a valuable result of this finite element analysis.

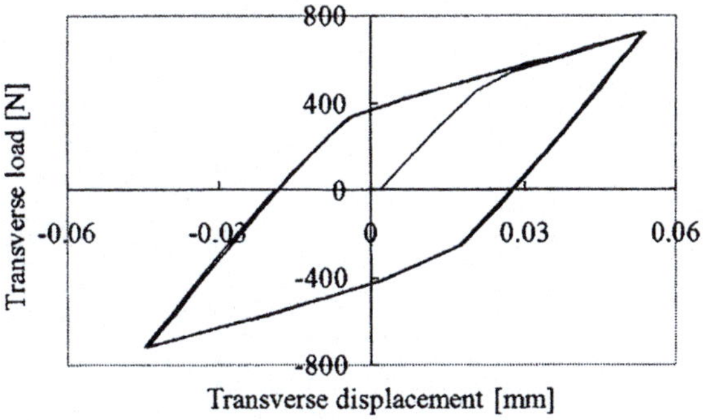

Fig. 6.36 Analytical results of the cyclic load/displacement (four cycles) relationship under transverse loading in thread surface slipping condition [18–21]

Fig. 6.37 Moving cycle of inchworm

(iv) Analysis of bolt bending moment

In Fig. 4.10 of Chap. 4, the response of the bending moment of a bolt under transverse loading was shown by simple mechanical analysis, and the relationship between transverse load/bolt bending stress was analyzed using FEM in the case where the transverse load exceeds the frictional resistance between fastened bodies $W_1 = F_f \mu_c$ as shown in Fig. 6.38. In this figure, the simplified analytical results from Chap. 4, Sect. 4.6 are shown as solid lines (bolt neck lower bending moment M_h) and dashed lines (nut section M_n). In the simplified analytical solution, the bending rigidity k_{wh} of the bolt head neck area and the bending rigidity k_{wn} of the nut area are equal for now (see Sect. 4.7 in Chap. 4), so the relations between transverse load and bolt bending moment under thread surface slipping load didn't match well under

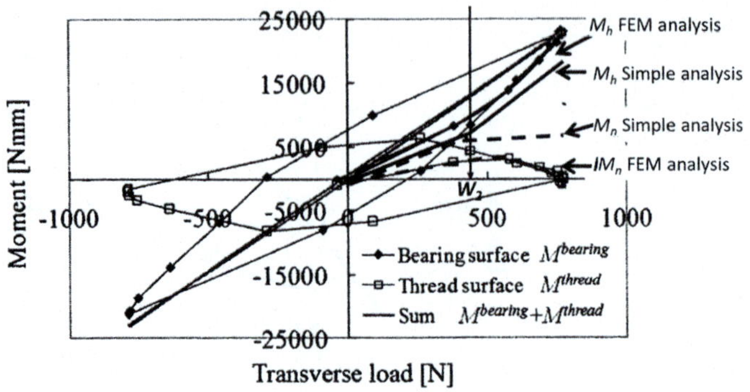

Fig. 6.38 Comparison of FEM analysis and simple analysis of bolted joint under transverse loadings [18–21]

the thread surface slipping load W_2 shown in Eq. (6.22). This will be a good fit if the bending rigidity of the nut contact area k_{wn} is estimated to be about 70% of the bending rigidity of the bolt head contact area k_{wn}.

$$W_2 = 1.94 F_f \mu_s \frac{l_f}{d} \tag{6.22}$$

This should be considered as a future improvement of the accuracy of the estimation.

The resisting moment M_c of the nut is assumed to be constant as shown in Sect. 4. 3 of Chap. 4, but the FEM analysis results show that it decreases after W_2. This was found to be due to the fact that when thread face slip occurs, the bolt thread face rises up the nut thread face, generating a moment in the opposite direction as shown in Fig. 6.39b. This is another valuable result of the FEM analysis that has been applied to this type of screw fastening. We hope that this will lead to higher precision in the future.

(v) Analysis for nut/ bolt fastening models with washer and double nut

Here, we cite the paper by Yokoyama et al. [19]. They use the FEM model shown in Fig. 6.40, and software ANSYS10.0, which is a general-purpose finite element method software. The model reproduces a helical profile of the internal and external threads, but detailed shapes such as the curvature of the bottom of the thread are not taken into account since a detailed evaluation of stress distribution is beyond the scope of the present study. The nominal size of the bolt and nut is M10, and the grade and position of the crossover is 6H/6 g (expressed in terms of the Japanese Industrial Standards). Because the friction between the movable top plate and the fixed plate can be ignored in the loosening apparatus, only part of the movable plate is modeled, and the displacement of the bottom of the movable plate in the axial direction is fixed. Contact elements are used to incorporate contact behavior at the interfaces

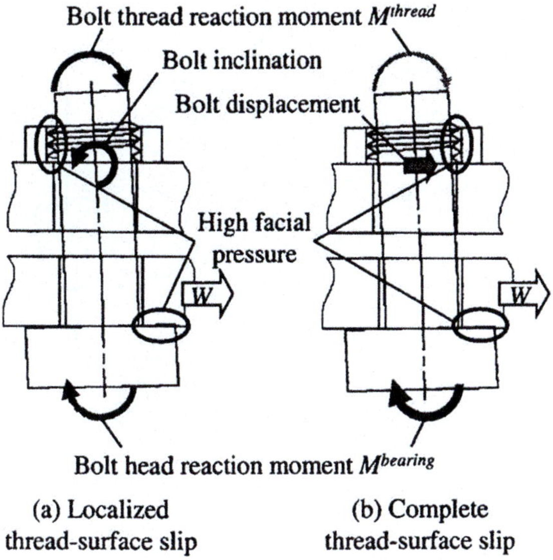

Fig. 6.39 Analytical results of reaction moments on bolt heads and bolt threads [18–21]

such as the mating threads, the nut bearing surface, and so on. In this model created using ANSYS, the contact element pair TARGE170 and CONTA174, which realizes surface-surface contact between three-dimensional objects and which can deal with the Coulomb friction, is used. The pure penalty method is employed as a contact algorithm. To reduce calculation costs, axial force is generated by allowing initial interference between the movable plate and the lower nut (in the case of a double-nut joint) or the washer (in the case of spring washer and conical spring washer joints). The transverse load is applied to the end surface of the movable plate as constrained displacement or force. A displacement-constrained boundary is employed in order to observe the loosening due to complete contact surface slip, while a force-constrained boundary is employed to observe the loosening due to micro-contact surface slip. The Young's modulus and Poisson's ratio for all the components (bolt, nut, spring washer, conical spring washer, and movable top plate) are 205 GPa and 0.3, respectively. The friction coefficient of the contact surfaces is set to 0.15. The performed FEM analysis is quasi-static and elastic. Geometric nonlinearities are taken into account. Detailed analysis conditions for each case of a double-nut joint, spring washer joint, and conical spring washer joint are explained in the sections below. For each case, loosening progress is compared to the case of a conventional joint. The progress of loosening is observed from two viewpoints. One is the loosening rotation angle of the nut in relation to the torsion angle of the bolt (loosening rotation angle). The other is the decrease in axial force. Here, axial force means bolt tension.

(v-1) In the case of spring washer/Analytical model of spring washer joint

The detailed modeling process is described according to Fig. 6.41. As shown in step (a), while one end of the SW is fixed, the other end is displaced by 2.5 mm along the

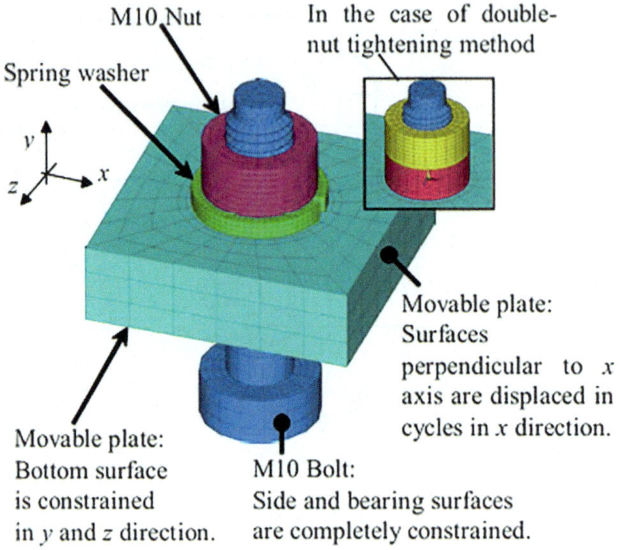

(a) Schematic examples of FEM models for loosening analysis

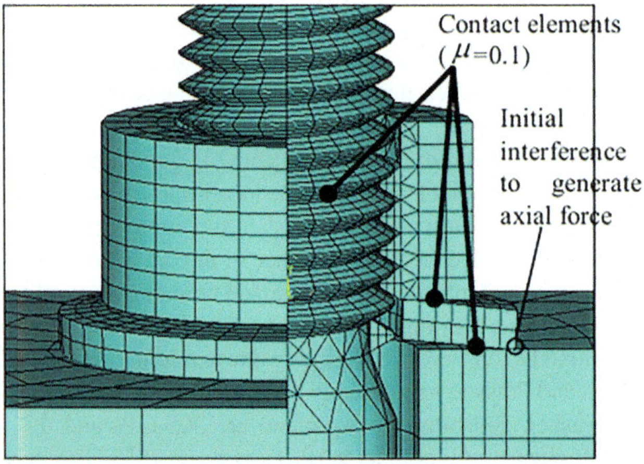

(b) Detail of thread part of FE model

Fig. 6.40 FEM models for loosening analysis under transverse loadings with few loosening resistance component [18–21]

normal direction. In the next step (b), the shape is updated to the deformed shape and the developed stress is cleared. In step (c), the resulting updated shape is compressed so that its deformed shape becomes flat like the initial shape shown in step (a). In step (d), the developed stress is output as initial stress. In step (e), the recorded initial stress is given to the initial flat C-shaped volume. The spring force becomes about 1 kN. Bolt tension is set to 10 kN. Both complete and micro-bearing surface slips

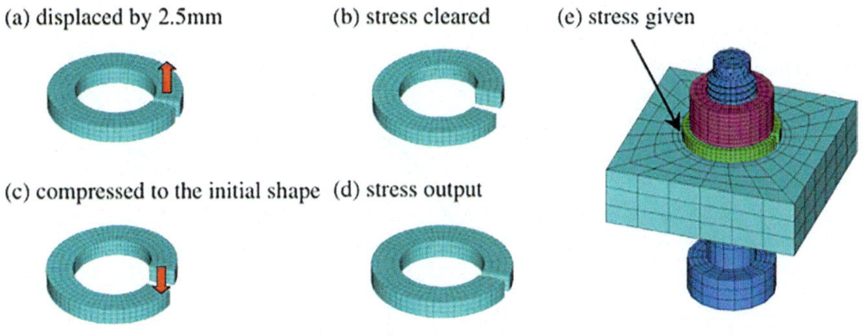

Fig. 6.41 Modeling procedure for spring washer [18–21]/Analytical results for spring washer joint

are investigated. In the former case, the amplitude of the movable plates' transverse displacement is set to 0.4 mm. Two vibration directions (x- and z-directions) are investigated.

The loosening rotation angles during three vibration cycles are shown in Fig. 6.42. It is observed that the loosening rotation of the spring washer joint proceeds faster than that of the standard joint. It can be also said that there is no significant dependence on the vibration directions (x and z directions). Two effects are considered to increase loosening rotation. One is that the torsion of the bolt axis increases because the grip length of the joint is increased by adding a spring washer under the nut. The other is that the pressure distribution of a spring washer is concentrated on the corner edges. The latter effect is discussed in detail below.

The relationship of the applied load and the transverse displacement (x-direction) is shown in Fig. 6.43 accompanied by the case of standard joint. It is widely accepted that a load–displacement curve consists of three kinds of slopes: a steep slope (A), a gradual slope (B), and a flat slope (D). However, it can be seen that an extremely gradual slope (C) is added in the spring washer joint.

Schematics of the contact state during x-direction vibration are shown in Fig. 6.44. Symbols of A-D correspond to those denoted in Fig. 6.43. First, in slope A, partial slip

Fig. 6.42 Loosening rotation angle of the nut [18–21]

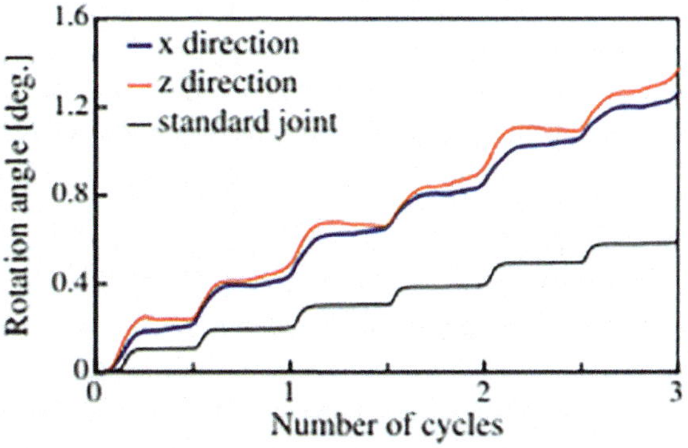

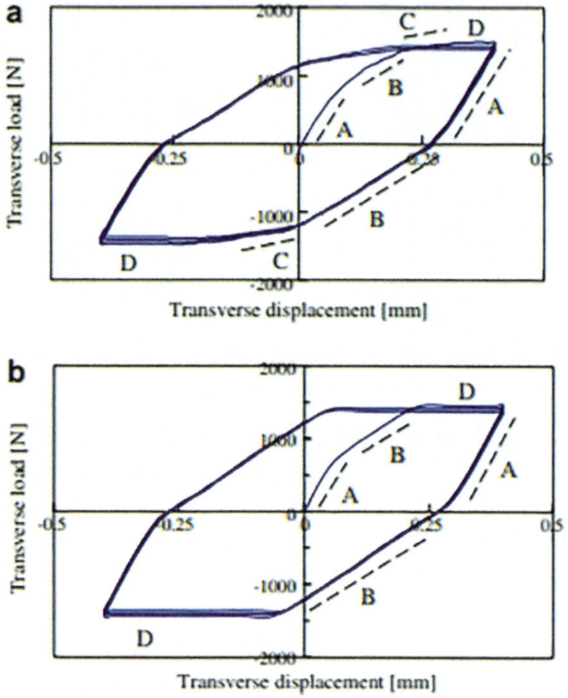

Fig. 6.43 Relation between transverse load and displacement: (a) spring washer joint and (b) standard joint [18–21]

occurs both on the thread and the spring washer surfaces. Secondly, in slope B, the entire thread surface undergoes complete slip. At the end of slope B, the stuck regions on the spring washer surfaces are reduced around three or four edge points. Next, in slope C, the stuck regions are further limited to only two edge points. The rotation of the spring washer around these stuck edge points occurs at this stage. Details of the mechanism of this rotation are discussed later. Finally, in the former part of slope D, complete slip occurs both on the thread and the spring washer surfaces. In the latter part of slope D, the thread surface regains a stuck state.

The progress in the rotation angles of the nut and the spring washer in the case of x vibration are shown in Fig. 6.45. It is found that the rotation direction depends on the vibration direction. The spring washer rotates in the loosening direction during stage C of + x displacement while it rotates slightly in the tightening direction during that of x displacement. This reverse rotation is caused by rotation around two edge points as illustrated in Fig. 6.46. Depending on the vibration direction, the rotation direction of the spring washer changes. Due to its effect, nut rotation during + x displacement (around cycle 0, 1, and 2) becomes larger than that in the x-direction (around cycles 0.5, 1.5, and 2.5). Particularly, the nut rotation proceeds mainly during stage C while it does not proceed as much during stage D in which complete bearing surface slip occurs. This indicated that the spring washer joint could loosen more than the standard joint even prior to the load of complete bearing surface slip.

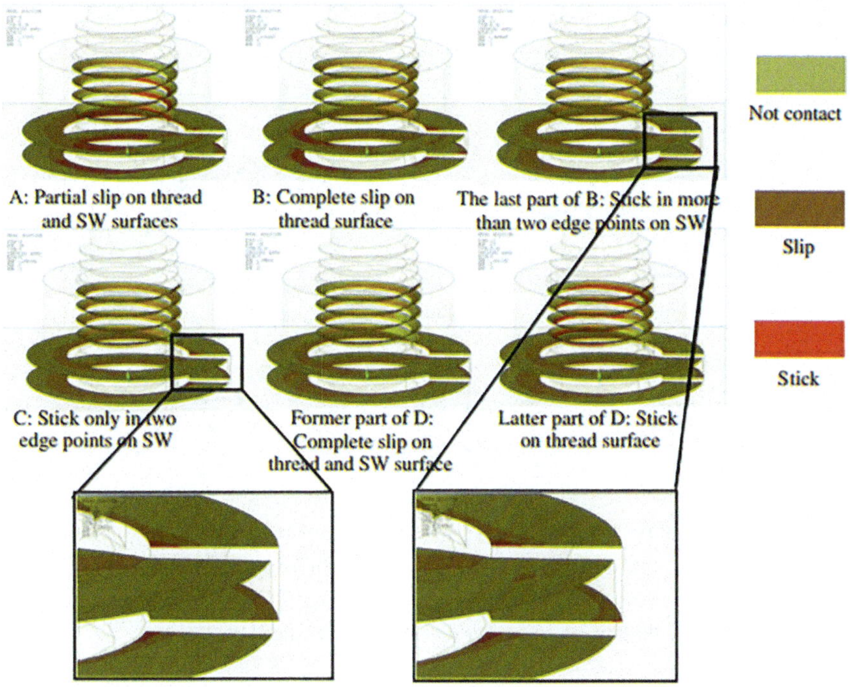

Fig. 6.44 Classification in contact states [18–21]

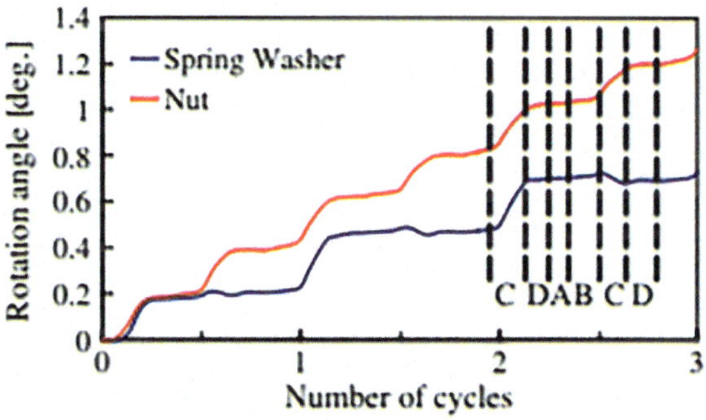

Fig. 6.45 Loosening rotation angle of the spring washer [18–21]

The progresses in loosening rotation angle caused by the 1000 N transverse load are shown in Fig. 6.47. It is observed that the loosening rotation of the spring washer joint proceeds faster than that of the standard joint as the case of complete bearing surface slip. However, the vibration in z-direction causes faster loosening rotation in this case than that in x-direction does. In particular, spring washer joint shows significant loosening rotation at the beginning of the vibration. The rates of the loosening rotation angle under various transverse loads, which are less than critical for complete bearing surface slip, are plotted in Fig. 6.48. Here, the transverse load W

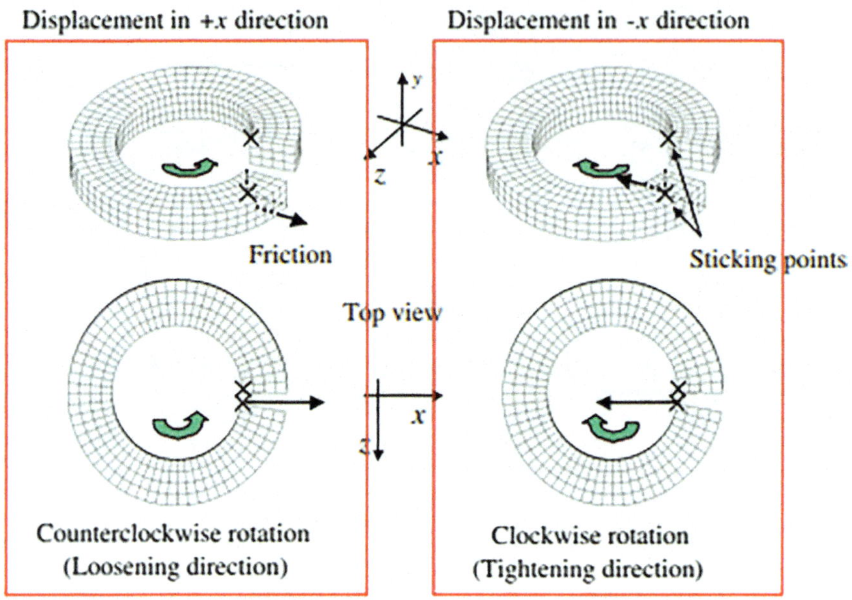

Fig. 6.46 Schematic illustrations of washer rotation during C state [18–21]

is normalized by the critical value W_{cr}. It is found the loosening rotations of the SW joint are larger than that of the standard joint. The rate increases drastically around $W/W_{cr} = 0.80$ ($W = 1200$ N) in the case of the spring washer joint. The value of 1200 N corresponds to the shifting point from stage B to C in Fig. 6.43a. Therefore, it is supposed that the increase in loosening rate is caused by the rotation around the two stuck edge points, as described in Fig. 6.44.

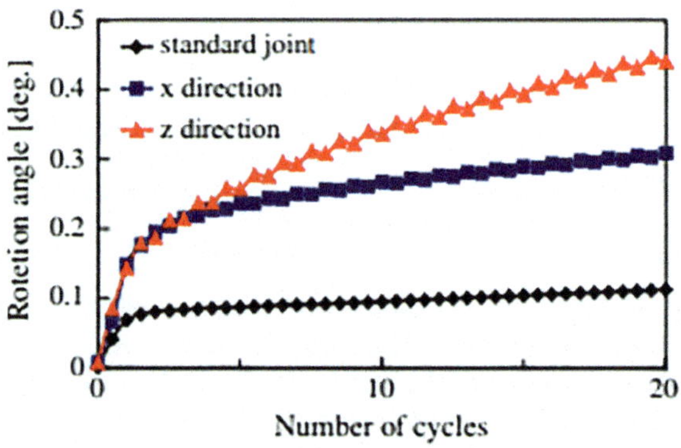

Fig. 6.47 Loosening rotation angle under $W = 1000$ N loading [18–21]

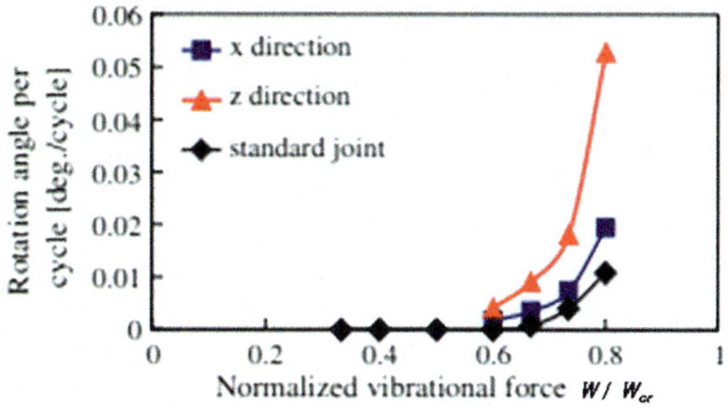

Fig. 6.48 Dependences of loosening rate on normalized vibration force W/W_{cr} [18–21]

(v-2) In the case of conical spring washer/Analytical model of conical spring washer joint

The FEM model of conical spring washer joint was shown above in Fig. 6.40b. The modeled conical spring washer is classified as "1L" in JIS B1251. The axial force–deflection relationship is shown in Fig. 6.49. Its total deflection is 0.5 mm. Two kinds of axial force are applied; one is a low axial force (10 kN) in which the conical spring washer is not fully compressed, and the other is a high axial force (20 kN) in which the CSW is fully compressed. Both complete and micro-contact surface slips are investigated. In the former case, the amplitude of the movable plate is set to 0.3 mm in the case of low axial force (10 kN) and 0.4 mm in the case of high axial force (20 kN). For details, readers are referred to (20).

The loosening rotation angle and the decrease in axial force during three vibration cycles are shown in Fig. 6.50. In the case of low axial force (10 kN) (Fig. 6.50a), it can be seen that the loosening rotation of the conical spring washer joint proceeds faster than that of the standard joint. The loosening rotation angles during the third cycle of the conical spring washer and standard joints are 0.26 and 0.21 deg, respectively. On the other hand, the decrease in axial force of the conical spring washer joint proceeds slower than that of the standard joint. The decrease in axial force during the third cycle is 81 N for the conical spring washer joint while it is 241 N for the standard joint. Larger loosening rotation might be caused by changes in the distribution of

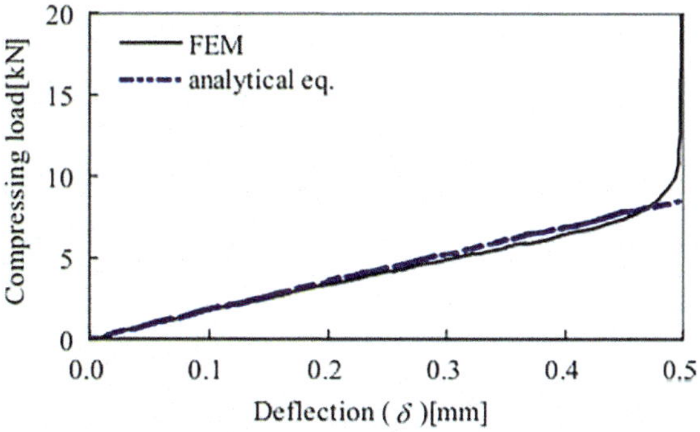

Fig. 6.49 Compressing property of conical spring washer [18–21]/Loosening due to complete contact surface slippage [18–21]

contact pressure. The contact pressure distribution between the nut and the conical spring washer, shown in the upper part of Fig. 6.50, is apparently concentrated around the inner diameter of the conical spring washer, as compared with the standard joint shown in the lower part. The equivalent diameter of friction torque based on the pressure distribution of the conical spring washer joint is estimated as 12.4 mm while that of the standard joint is 14.6 mm (assuming uniform contact pressure distribution). To sum up, it can be said that loosening rotation in the conical spring washer joint proceeds more quickly because of the smaller equivalent diameter of friction torque. On the other hand, it is considered that the decrease in axial force depends on the compressing spring constant of the cramped component. When a bolted joint is tightened and the grip length is shortened by ε_c the generated axial force F_s is expressed as below using the compressing spring constant of fastened bodies C_c,

$$F_s = C_c \varepsilon_c \tag{6.23}$$

When the fastened bodies consist of two components, conical spring washer and fastened body, each compressing spring constants are C_1 and C_2., C_c is expressed as

$$C_c = \frac{C_1 C_2}{C_1 + C_2} \tag{6.24}$$

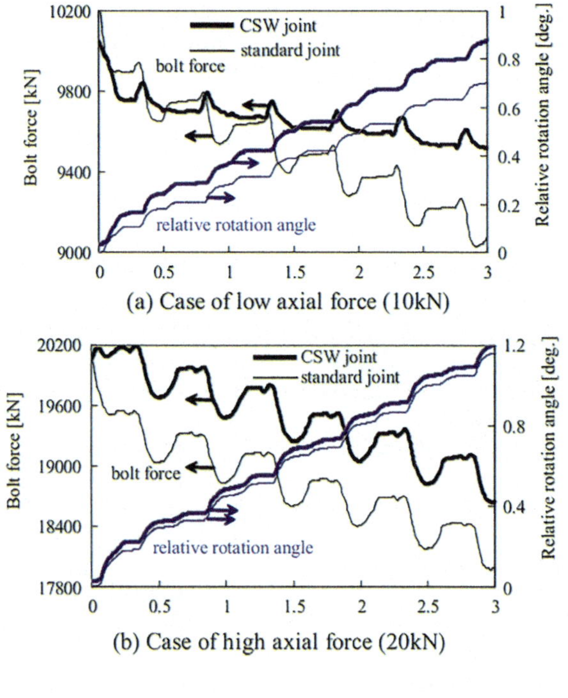

Fig. 6.50 Variation in loosening rotation angle and bolt force in the case of **a** low axial force (10 kN) and **b** high axial force (20 kN) [18–21]

If C_1 is sufficiently smaller than C_2, C_c can be approximated as

$$C_c \approx C_1 \tag{6.25}$$

As can be seen from Fig. 6.49, the deformation-force relation can be divided into two regions. One is the region below approximately 10 kN, where the axial force is sustained mainly by the reaction force of the conical spring washer. The other is over approximately 10 kN, where the conical spring washer is fully compressed. The compressing spring constant of the conical spring washer at 10 kN estimated from the curve in Fig. 6.49 is 1.29×10^5 N/mm while that of the fastened body, which is calculated using the expression proposed by LORI et al. [21], is estimated as 1.94×10^6 N/mm. Thus, the total compressing spring constant C_c is approximated as the spring constant of the conical spring washer using Eq. (6.23). Because the smaller spring constant of the fastened body induces a slower decrease in axial force as can be seen in Eq. (6.23), the decrease in axial force proceeds slower in the conical spring washer joint. In the case of high axial force (20 kN) (Fig. 6.50b), both the loosening rotation angle and the decrease in axial force show similar behaviors, except for the initial drop in axial force in the case of the standard joint. These values are the same (0.36 deg) in the third cycle. According to the deformation-force relation of the conical spring washer shown in Fig. 6.49, when the axial force reaches 20 kN, the macroscopic deflection of the conical spring washer has already occurred and the entire interface is in a compression state. Thus, the contact pressure distribution between the nut and the conical spring washer becomes more uniform and similar to that of the standard joint, as compared with the case of 10 kN shown in Fig. 6.51. This leads to the equal loosening behavior of the conical spring washer joint and the standard joint in the case of high axial force (20 kN).

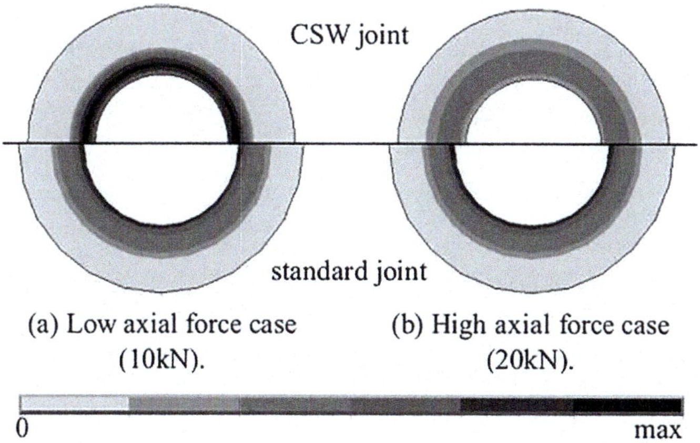

Fig. 6.51 Pressure distribution on nut bearing surface after axial force is applied [18–21] (max is 200 MPa for (**a**), 400 MPa for (**b**))/Loosening due to micro-contact surface slippage [18–21]

The relationships between a transverse load and the loosening rotation angle and that between a transverse load and a decrease in axial force are shown in Fig. 6.52a, b in the case of low axial force (10 kN), and in Fig. 6.52c, d in the case of high axial force (20 kN), respectively. Here, the transverse load W is normalized by the critical value W_{cr}. It is found that in the case of low axial force, the loosening rotation angle of the conical spring washer joint becomes larger than that of the standard joint, as well as in the case of complete contact surface slip. However, the difference in the decrease in axial force becomes small. In the case of high axial force, those differences are negligibly small. The normalized critical transverse load required to induce small loosening (W/W_{cr}) ranges from 0.4 to 0.5 for all cases. It can be said that the conical spring washer does not affect loosening.

(v-3) In the case of double nut [18–21]/Overview of double-nut fastening method and FEM analytical method

Two methods are well-known to set up the double-nut joint, that is, the upper nut rotation method and the lower nut reverse rotation method. This study is intended for the latter and detail is shown in Fig. 6.53. After the processes of lower nut reverse rotation method, thread of the lower nut comes in contact on the surface opposite to the one that is in contact in the standard tightening process. This state is called as "locking state", and the force produced on the thread of lower nut is called as

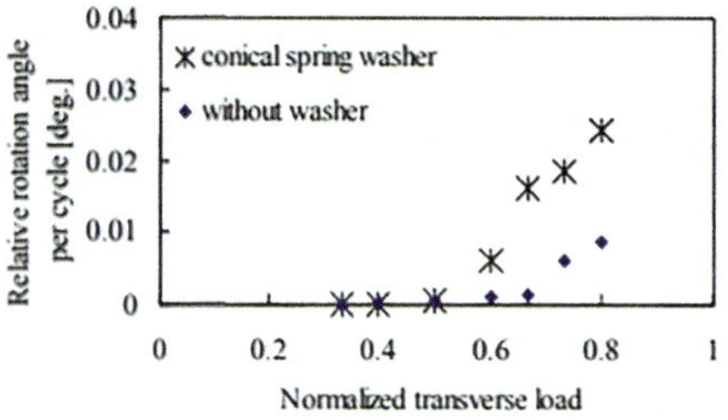

(a) Relation between loosening rotation angle and transverse load in the case of low axial force (10kN).

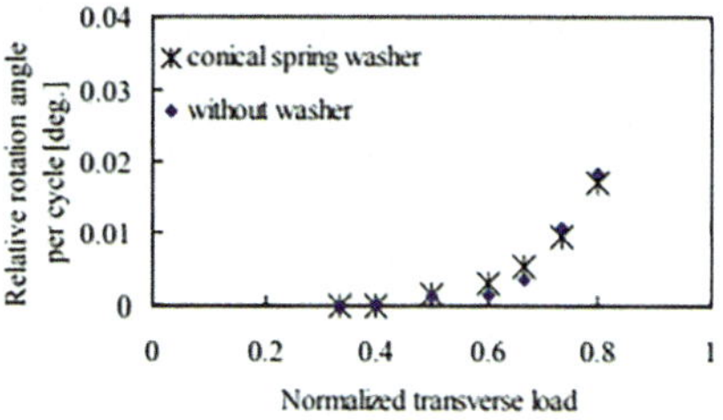

(c) Relation between loosening rotation angle and transverse load in the case of high axial force (20kN).

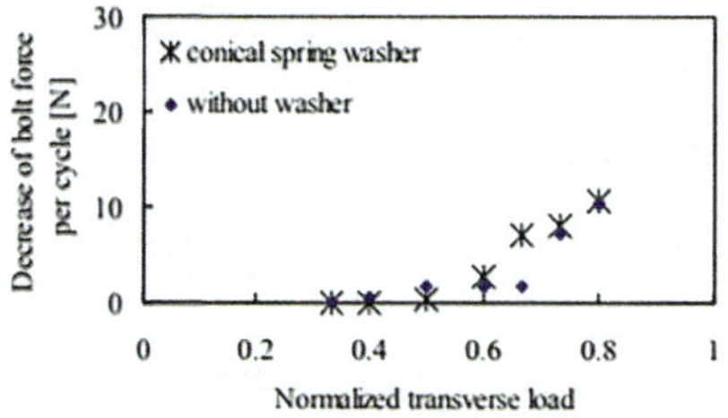

(b) Relation between decrease of bolt force and transverse load in the case of low axial force (10kN).

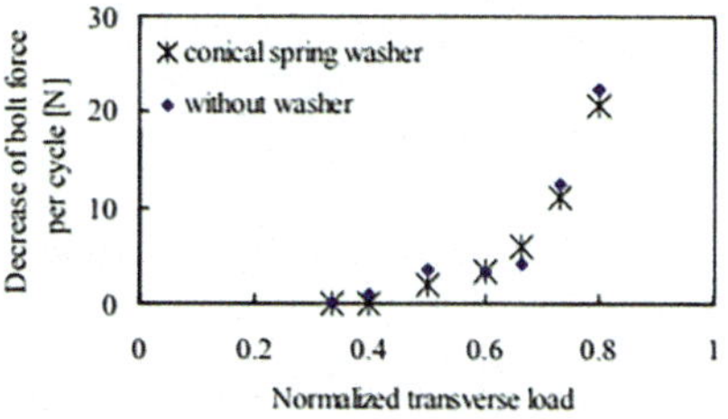

(d) Relation between decrease of bolt force and transverse load in the case of high axial force (20kN).

Fig. 6.52 Dependence of loosening rate on normalized transverse load W/W_{cr} [18–21] locking force. The details of FEM analysis for the tightening process are shown by Yokoyama et. al. [20]. Here, in order to reduce calculation costs, bolt tension and locking force are generated by using an initial interference scheme. The bolt tension and the locking force are generated by the interference between the movable plate and the lower nut and that between the upper nut and the lower nut, respectively. At the locking state, the lower surface of the lower nut thread is in contact with the upper surface of the bolt thread. And the thread of the upper nut supports both bolt tension and locking force. Bolt tension is adjusted to 10 kN. Four kinds of locking states are employed. Case 1 has a high locking force of 8.8 kN, while Case 2 has a low locking force of 1.1 kN. Although Case 3 is in the locking state, its locking force is close to zero. Case 4 is not in the locking state, which means that there is a gap between the thread of the lower nut and that of the bolt. Therefore, the thread of the lower nut is not loaded at all and the thread of the upper nut supports only the bolt tension. Only the loosening due to complete bearing surface slip is investigated since no loosening occurs in the case of micro-bearing surface slip. The amplitude of the movable plates' transverse displacement is set to 0.3 mm.

The variations in axial force during three transverse vibration cycles are shown in Fig. 6.54 (Cases 1–3) and Fig. 6.55 (Case 4). And the variations in double-nut locking force relative to the initial value are shown in Figs. 6.56 and 6.57, respectively. It is found that the variation in axial force shows a half-cycle period. In addition, it can be seen that the variation becomes large as the double-nut locking force becomes small. Judging from Figs. 6.56 and 6.57, it can be considered that the variation in axial force originates from the variation in double-nut locking force. In Case 4, although there is a gap between the thread of the lower nut and the bolt in its tightening process, it is

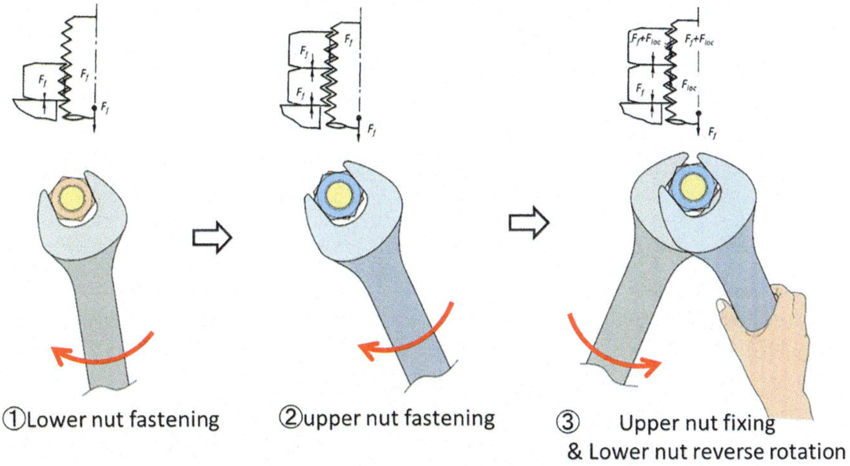

Fig. 6.53 Inter-rocking process for double-nut fastening (lower nut reverse rotation method)/ Analytical results for double-nut joint [18–21]

found that double-nut locking force appears due to the applied vibration. A significant decrease in axial force occurs only in Case 4. Even though complete contact surface slip occurs, there is no significant decrease in Cases 1 to 3. The amounts of the decrease in axial force during the third cycle are 0.65, 2.7, 4.7, and 220 N/cycle, respectively. These results indicate that if the locking state is properly achieved, the double-nut tightening method significantly prevents loosening regardless of the magnitude of the locking force. However, if locking state is not properly achieved, the loosening resistance performance completely vanishes and loosening is increased to about twice as much as a conventional joint. This result agrees with Yamamoto's and Kasei's experiments [23]·

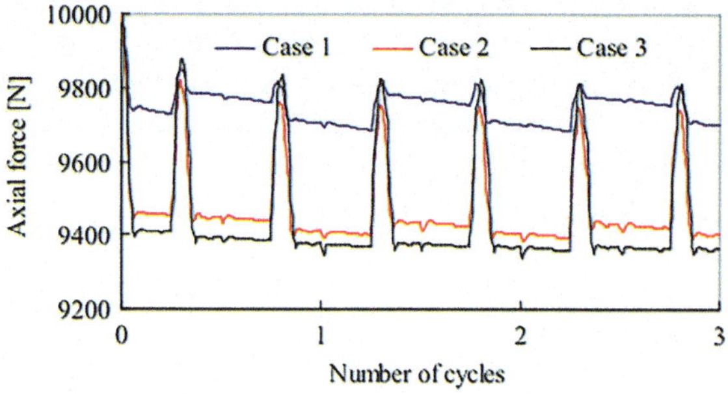

Fig. 6.54 Variations in axial forces (Case1 to Case3) [18–21]

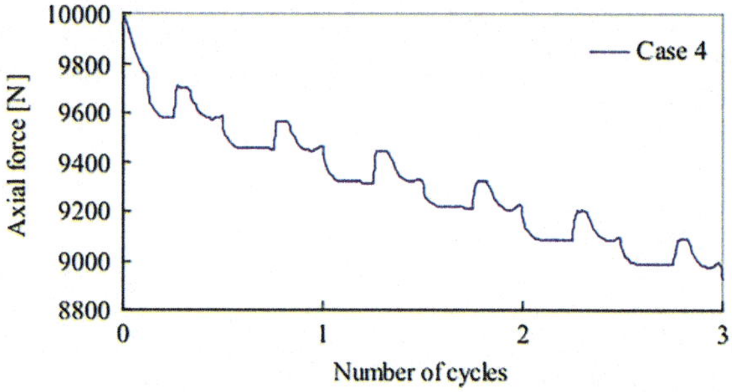

Fig. 6.55 Variation in axial force (Case4) [18–21]

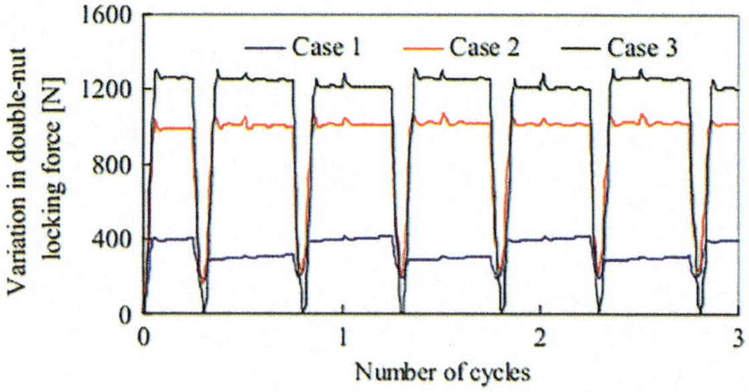

Fig. 6.56 Variations in double-nut locking forces (Case1 to Case3) [18–21]

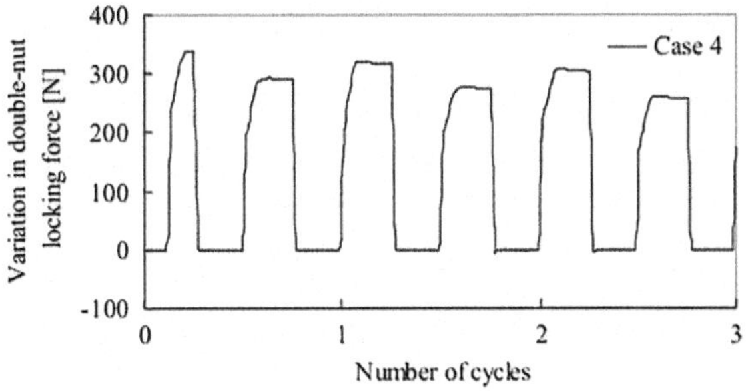

Fig. 6.57 Variation in double-nut locking force (Case4) [18–21]

(v-4) Comparison of spring washer, conical spring washer and double nut

On the basis of the results shown above, we discuss the comparison of loosening resistance components. It is clear that double nut has the best loosening resistance performance among the three components. Even if the transverse load enough to cause complete contact surface slip is applied, loosening would be drastically suppressed by the double nut. However, in actual tightening operation, the problem is a reliable control of locking state. To set up the locking state is not easy because serious control of the tightening torque is necessary in three tightening processes; tightening of the lower nut, tightening of the upper nut, and rotating of the lower nut in the loosening direction. On the contrary, such a troublesome problem does not take place in the spring washer and conical spring washer tightening. However, as compared with double nut, spring washer and conical spring washer are not able to sufficiently suppress the loosening due to both seat surface and thread surface slip. In particular, it should be noted that the spring washer induces significant loosening rotation prior to the seat surface slip. Thus we suggest that spring washer should not be used as a loosening resistance component without sufficient discussion. On the other hand, we can see that conical spring washer provides a certain effect on the loosening resistance since it gives a higher spring force than spring washer and its spring force contributes to suppress the decrease in axial force during loosening. Therefore, it is expected that conical spring washer shows better loosening resistance performance than spring washer.

(vi) Conclusion

Three-dimensional finite element analyses of M10 bolted joints using the double-nut tightening method, spring washer, and conical spring washer subjected to a transverse load were performed and the loosening resistance performance of those components were investigated. First, in regard to double nut, the effect of the locking state on preventing loosening was clarified. We have found that double shows significant loosening resistance if locking is properly achieved in its tightening process. However, if locking is not properly performed, its ability completely disappears. Secondly, spring washer shows negative loosening resistance because the spring washer rotation occurs around two stuck edge points. Thus, the loosening rotation of the spring washer joint is increased drastically under the condition of micro-seat surface slip. It can be concluded that using a spring washer is problematic from the point of view of loosening resistance. Finally, regarding conical spring washer, in the case of low axial force, under which a conical spring washer is not fully compressed, two conflicting effects on loosening were seen. One is a negative aspect; that is, a conical spring washer increases the loosening rotation angle due to the concentration of contact pressure around the inner diameter of bearing surface. The other is a positive one; that is, it prevents a decrease in axial force due to the smaller compressing spring constant compared to that of fastened body. The effect of a conical spring washer on loosening is determined by the trade-off of these two effects. In the case of high axial force under which a conical spring washer is fully compressed, the loosening rotation angle and the decrease in axial force do not depend on the use of a conical spring washer.

6.3.2 Loosening Due to Cyclic Rotational Load (Torsion)

Basic considerations based on mechanical analysis were explained already in Chap. 5. This section summarizes the results and describes know-how such as practical methods for preventing loosening.

(A) **Loosening conditions based on mechanical analysis**

Even if a torsional load T around the bolt axis is applied on the fastened body, if it is less than T_c shown in the following equation, relative slip occurs between the fastened bodies.

$$T_c = \frac{\frac{2}{3}\frac{(D_m{}^3 - D_i{}^3)}{(D_m{}^2 - D_i{}^2)} F_f \mu_c}{2} \quad \text{Chapter5(5.1)}$$

Under this critical torque, no slip occurs on other seat surface and thread surface, and the threaded fastener deforms elastically as a single unit, so loosening is not possible. However, when T become greater than T_c, the fastener slips between the fastened bodies, inducing nonlinearity and asymmetry in the tilt of the bolt and the thread surface and seat surfaces, which in turn induces rotational slippage in the bolt/nut contact area and may cause loosening.

The sliding torques of the thread surface and bearing surfaces are shown below, Thread surface slip torque in tightening direction T_{st} is

$$T_{st} = \frac{F_f}{2}\left(\frac{d_2 \mu_s}{\cos\alpha} + \frac{P}{\pi}\right) \quad \text{Chapter5(5.13)}$$

Thread surface slip torque in loosening direction T_{sl} is

$$T_{sl} = \frac{F_f}{2}\left(\frac{d_2 \mu_s}{\cos\alpha} - \frac{P}{\pi}\right) \quad \text{Chapter5(5.13)}$$

The seat sliding torque T_w is

$$T_w = \frac{F_f \mu_w D_w}{2} \quad \text{Chapter5(5.14)}$$

Rotational loosening occurs only when these large and small relationships are $T_{sl} < T_w < T_{st}$. In other words.

When $T_w < T_{sl} < T_{st}$, the seat surface only slides when turning to the right or to the left, and the thread surface does not slide and does not loosen. Also,

When $T_{sl} < T_{st} < T_w$, the thread surface only repeats clockwise and counterclockwise rotation, and the nut does not loosen because the bolt itself cannot rotate. Then, the conditions of friction coefficient, pitch angle, etc. that satisfy $T_{sl} < T_w < T_{st}$ can be formulated as follows

$$\frac{d_2 \mu_s}{\cos\alpha} - \frac{P}{\pi} < \mu_w D_w < \frac{d_2 \mu_s}{\cos\alpha} + \frac{P}{\pi} \quad \text{Chapter5(5.15)}$$

This is the condition for loosening due to repeated loading (torsion) around the axis.

The above describes the loosening condition from the load side, but consider the loosening condition from the rotational displacement side, which is that the torque must not loosen up to the rotational displacement angle between the leftward torque T_{sl} and the rightward torque T_w as shown in the following equation.

$$\theta_{cr} = 2\varphi_b(T_w + T_{sl}) = F_f \left(\frac{0.4d}{G_b \frac{\pi}{32} d^4} + \frac{l_g}{G_b \frac{\pi}{32} d_g^4} + \frac{l_s}{G_b \frac{\pi}{32} d_3^4} + \frac{0.4d}{G_b \frac{\pi}{32} d^4} \right)$$
$$\left(\mu_w D_w + \frac{d_2 \mu_s}{\cos\alpha} - \frac{P}{\pi} \right) \quad \text{Chapter5(5.16)}$$

Therefore, the above θ_{cr} is the loosening limit rotation angle.

(B) **Technology to prevent loosening under cyclic rotational load (torsion)**

To avoid satisfying Eq. (5.15) in Chap. 5 above,

(i) Reduce the difference between the first and third terms in Eq. (5.15), i.e., to reduce the pitch. Recommendation of fine threads.

(ii) Increase the seat diameter D_w to prevent seat slippage. The flange nut is recommended.

It is also effective to increase the looseness limit rotation angle θ_{cr} in Eq. (5.15) of Chap. 5,

(iii) The tightening force F_f should be large and high tightening. Plastic area fastening will be recommended.

(iv) Take a long fastening length l_f. Use spacers when fastening thin plates.

(v) Reduce the torsional stiffness I_{pb} of the bolt. Recommendation of reduced shank bolts with smaller bolt shank diameter.

6.3.3 Loosening Due to Cyclic Axial Loading

In conclusion, this axial cyclic loading loosening rarely occurs in practice. This is because, unlike in the case of cyclic transverse loading and cyclic torsional loading, there is no macroscopic slippage at the fastened bodies contact surface and thread contact surface. For example, as described in the mechanical analysis of the thread contact surface under transverse loading in Chap. 5, due to the presence of a pitch angle β, the screw will experience a torque component in the loosening direction

$$\mathrm{T}_{ss} \cong \frac{F_f d_2 \tan\beta}{4\cos\alpha'} \quad \text{Chapter 4(4.40)}$$

However, in order to slip with this loosening directional torque, in a situation where there is no macroscopic sliding, the following low friction coefficient is a necessary condition.

$$\mu_s = \mu_w < 0.032 \quad \text{Chapter4(4.43)}$$

In the case of axial cyclic loading, the same loosening direction torque T_{ss} works as described above, but unlike in the case of transverse loading and rotation loading, macro-slip (full-surface sliding) cannot occur in this axial direction loading, so in reality, no loosening occurs. However, strictly speaking, micro-slip (local slip with a sticking point) may occur on the contact surface due to a mismatch in the elastic deformation of the two contacting bodies. The results of the finite element analysis shown in Figs. 6.35b and 6.36 indicate that even the repetition of micro-slip with a sticking point rotates slightly due to the accumulation of micro-slip in the manner of the movement of an inchworm, and that it is not negligible in terms of mechanism. Here, we will confirm whether the mismatch amount (micro-slip amount) due to elastic deformation between the contact surface is meaningful or negligible compared to the aforementioned macro-slip amount.

(a) **Analysis of relative slip and loosening velocity under axial cyclic loading**

When subjected to axial load variation $\Delta F = \Phi W$, contact surface pressure ΔN and radial force ΔR act on the thread contact surface, as shown in Fig. 6.58.

$$\Delta R = \Phi W \frac{\sin\alpha - \mu_s \cos\alpha}{\cos\alpha + \mu_s \sin\alpha} \tag{6.26}$$

where α is the half angle of the thread

This radial force ΔR is the force that widens the nut and makes thinner the bolt, and when the nut is solved as a thick-walled cylindrical model with internal pressure and the bolt as a cylindrical model with external pressure, as shown in Fig. 6.59, the internal pressure p_o acting on the nut and the external pressure p_i acting on the bolt are,

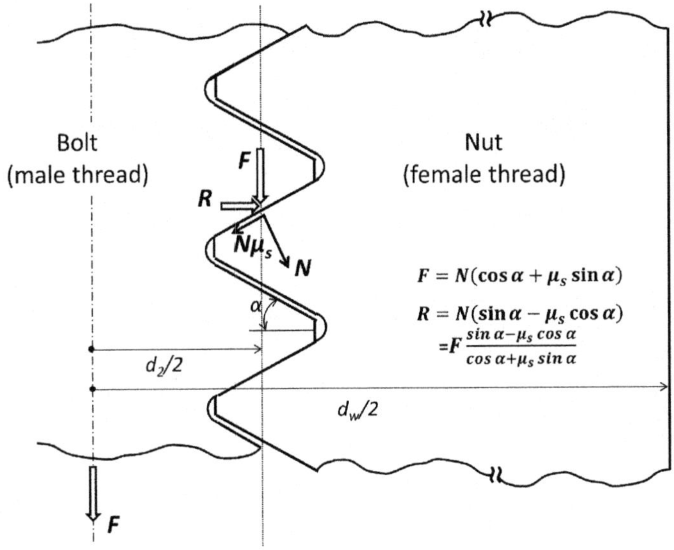

Fig. 6.58 Mechanical analysis model for bolt/nut joint under axial loading

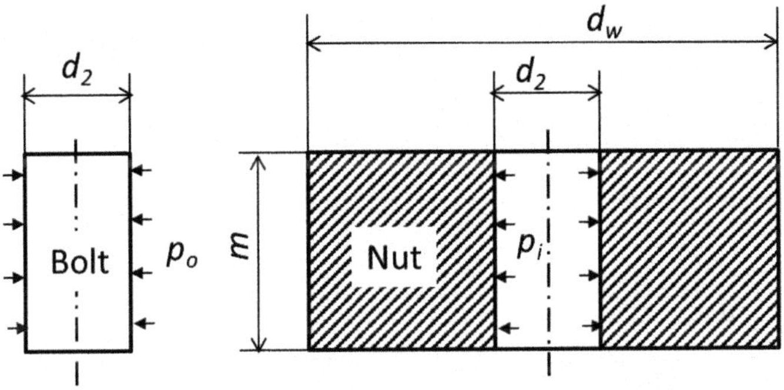

Fig. 6.59 Simple analytical model of bolt/nut joint with internal and external loadings

$$p_o = p_i = \frac{\Delta R}{\pi d_2 m} \tag{6.27}$$

The radial displacement of the nut inner diameter u_{rn}, the radial displacement of the bolt outer diameter u_{rb} is,

$$u_{rn} = \frac{p_i d_2/2}{E}\left(\frac{d_w/2^2 + d_2/2^2}{d_w/2^2 - d_2/2^2} + \nu\right). \tag{6.28}$$

$$u_{rb} = \frac{p_o d_2/2}{E}(1+\nu) \tag{6.29}$$

The relative radial slip amount S of the bolt and nut at the thread contact surface can be expressed as follows,

$$S = \frac{u_{rn} + u_{rb}}{\cos\alpha} = \frac{\Phi W}{\pi E m \cos\alpha} \frac{\sin\alpha - \mu_s \cos\alpha}{\cos\alpha + \mu_s \sin\alpha} \left(\frac{d_w^2}{d_w^2 - d_2^2} + \nu \right) \tag{6.30}$$

where

d_2 Screw effective diameter.
d_w Nut outer diameter (two side widths).
E Young's modulus of bolts and nuts.
ν Poisson's ratio of bolt and nut.
Φ Load factor.
W Axial load.
m Nut height.

Using this relative slip amount S, the amount of wear per load cycle w can be obtained using Archard's equation, shown in Sect. 6.2.3 of this chapter, as follows.

$$\begin{aligned} \mathrm{w} &= k_w \times \mathrm{S} \times \mathrm{P} \\ &= k_w \times \frac{F}{(\cos\alpha + \mu_s \sin\alpha)} \frac{4}{\pi\left(d^2 - d_3^2\right)} \\ &\frac{\Phi W}{\pi E m} \frac{(\sin\alpha - \mu_s \cos\alpha)}{(\cos\alpha + \mu_s \sin\alpha)} \left(\frac{d_w^2}{d_w^2 - d_2^2} + \nu \right) N \end{aligned} \tag{6.31}$$

The amount of wear per cycle (wear rate) dw/dN, is,

$$\begin{aligned} \frac{dw}{dN} &= \frac{k_w}{(\cos\alpha + \mu_s \sin\alpha)} \frac{4}{\pi\left(d^2 - d_3^2\right)} \\ &\frac{\Phi W}{\pi E m} \frac{(\sin\alpha - \mu_s \cos\alpha)}{(\cos\alpha + \mu_s \sin\alpha)} \left(\frac{d_w^2}{d_w^2 - d_2^2} + \nu \right) \left(F + \frac{dF}{dN} N \right) \end{aligned} \tag{6.32}$$

Also, using the embedment coefficient $Z = 1/(\delta_b + \delta_c)$, the rate of decrease in clamping force dF/dN is.

$$\begin{aligned} \frac{dF}{dN} &= -Z\frac{dw}{dN} = -Z\frac{dw}{dN} \\ &= -Z\frac{k_w}{(\cos\alpha + \mu_s \sin\alpha)} \frac{4}{\pi\left(d^2 - d_3^2\right)} \\ &\frac{\Phi W}{\pi E m} \frac{(\sin\alpha - \mu_s \cos\alpha)}{(\cos\alpha + \mu_s \sin\alpha)} \left(\frac{d_w^2}{d_w^2 - d_2^2} + \nu \right) \left(F + \frac{dF}{dN} N \right) \end{aligned} \tag{6.33}$$

By rearranging this equation,

$$\frac{dF}{dN} = \frac{-Z\frac{k_w}{(\cos\alpha+\mu_s\sin\alpha)}\frac{4}{\pi\left(d^2-d_3^2\right)}\frac{\Phi W}{\pi Em}\frac{(\sin\alpha-\mu_s\cos\alpha)}{(\cos\alpha+\mu_s\sin\alpha)}\left(\frac{d_w^2}{d_w^2-d_2^2}+\nu\right)\mathrm{F}^\circ}{\left(1+ZN\frac{k_w}{(\cos\alpha+\mu_s\sin\alpha)}\frac{4}{\pi\left(d^2-d_3^2\right)}\frac{\Phi W}{\pi Em}\frac{(\sin\alpha-\mu_s\cos\alpha)}{(\cos\alpha+\mu_s\sin\alpha)}\left(\frac{d_w^2}{d_w^2-d_2^2}+\nu\right)\right)} \tag{6.34}$$

Solving using the variable separation method,

$$\int\frac{}{-Z\frac{k_w}{(\cos\alpha+\mu_s\sin\alpha)}\frac{4}{\pi\left(d^2-d_3^2\right)}\frac{\Phi W}{\pi Em}\frac{(\sin\alpha-\mu_s\cos\alpha)}{(\cos\alpha+\mu_s\sin\alpha)}\left(\frac{d_w^2}{d_w^2-d_2^2}+\nu\right)F}\mathrm{dF} =$$
$$\int\frac{1}{\left(1+ZN\frac{k_w}{(\cos\alpha+\mu_s\sin\alpha)}\frac{4}{\pi\left(d^2-d_3^2\right)}\frac{\Phi W}{\pi Em}\frac{(\sin\alpha-\mu_s\cos\alpha)}{(\cos\alpha+\mu_s\sin\alpha)}\left(\frac{d_w^2}{d_w^2-d_2^2}+\nu\right)\right)}\mathrm{dN}+\mathrm{C} \tag{6.35}$$

$$\frac{1}{-Z\frac{k_w}{(\cos\alpha+\mu_s\sin\alpha)}\frac{4}{\pi\left(d^2-d_3^2\right)}\frac{\Phi W}{\pi Em}\frac{(\sin\alpha-\mu_s\cos\alpha)}{(\cos\alpha+\mu_s\sin\alpha)}\left(\frac{d_w^2}{d_w^2-d_2^2}+\nu\right)}$$
$$\log\left\{-Z\frac{k_w}{(\cos\alpha+\mu_s\sin\alpha)}\frac{4}{\pi\left(d^2-d_3^2\right)}\frac{\Phi W}{\pi Em}\frac{(\sin\alpha-\mu_s\cos\alpha)}{(\cos\alpha+\mu_s\sin\alpha)}\left(\frac{d_w^2}{d_w^2-d_2^2}+\nu\right)F\right\}$$
$$=\frac{1}{Z\frac{k_w}{(\cos\alpha+\mu_s\sin\alpha)}\frac{4}{\pi\left(d^2-d_3^2\right)}\frac{\Phi W}{\pi Em}\frac{(\sin\alpha-\mu_s\cos\alpha)}{(\cos\alpha+\mu_s\sin\alpha)}\left(\frac{d_w^2}{d_w^2-d_2^2}+\nu\right)}$$
$$\log\left\{\left(1+ZN\frac{k_w}{(\cos\alpha+\mu_s\sin\alpha)}\frac{4}{\pi\left(d^2-d_3^2\right)}\frac{\Phi W}{\pi Em}\frac{(\sin\alpha-\mu_s\cos\alpha)}{(\cos\alpha+\mu_s\sin\alpha)}\left(\frac{d_w^2}{d_w^2-d_2^2}+\nu\right)\right)\right\}+C \tag{6.36}$$

Using boundary condition $F = F_f$ at $N = 0$,

$$C = \frac{1}{Z\frac{k_w}{(\cos\alpha+\mu_s\sin\alpha)}\frac{4}{\pi\left(d^2-{d_3}^2\right)}\frac{\Phi W}{\pi Em}\frac{(\sin\alpha-\mu_s\cos\alpha)}{(\cos\alpha+\mu_s\sin\alpha)}\left(\frac{{d_w}^2}{{d_w}^2-{d_2}^2}+\nu\right)}\bullet$$
$$\log\left\{-Z\frac{k_w}{(\cos\alpha+\mu_s\sin\alpha)}\frac{4}{\pi\left(d^2-{d_3}^2\right)}\frac{\Phi W}{\pi Em}\frac{(\sin\alpha-\mu_s\cos\alpha)}{(\cos\alpha+\mu_s\sin\alpha)}\left(\frac{{d_w}^2}{{d_w}^2-{d_2}^2}+\nu\right)F_f\right\} \tag{6.37}$$

By rearranging this equation, we can get following,

$$\frac{F}{F_f} = \frac{1}{1+ZN\frac{4k_w}{\pi^2\left(d^2-{d_3}^2\right)}\frac{\Phi W}{Em}\frac{(\sin\alpha-\mu_s\cos\alpha)}{(\cos\alpha+\mu_s\sin\alpha)^2}\left(\frac{{d_w}^2}{{d_w}^2-{d_2}^2}+\nu\right)} \tag{6.38}$$

(b) **Analysis of loosening rate of bolted joints under cyclic axial loading and comparison with experimental results** [22, 23]

Loosening behavior of crank rod joints under cyclic axial loads is experimented by Sakai [22, 23]. Figure 6.60 shows the test bolt ($M10, p1.25$), test connecting rod and test equipment.

Figure 6.61 shows the measurement results of loosening behavior when the initial fastening force F_f are at 30, 20, and 10 KN, respectively, and the cyclic axial load ΔW is applied at 0–30 kN. Although initial loosening by setting wear is observed, wear loosening is rarely observed except in the case of low fastening force, and it can be seen that in these axial loading case, fatigue fracture of the bolt dominates rather than loosening.

Then, we will compare these experimental results [22, 23] with the analysis results using the following initial embedding loosening analysis formula Eq. (6.2) and the above-mentioned (Sect. 6.3.3 (A)) micro-slip wear loosening analysis Eq. 6.39.

The amount of looseness F_s due to the initial settling is obtained as follows using the initial embedment amount s and the embedment coefficient Z.

$$Fs = 1 / (\delta_b + \delta_c)s = Zs \tag{6.39}$$

Here, the embedment amount is empirically calculated by the following formula.

$$s \doteq 3.29\left(l_f/d\right)^{0.34}(\mu\text{m}) \tag{6.40}$$

Here, l_f is the fastening length of the bolt, and d is the nominal diameter of the bolt.

The calculation of fastened body's compliance δ_c and embedment coefficient Z (Eq. (6.40) is based on the behavior of the linear range with a sufficient initial tightening force, but since the experiment here includes the behavior of the nonlinear range with a low tightening force, Sakai [22, 23] actually measured the nonlinear

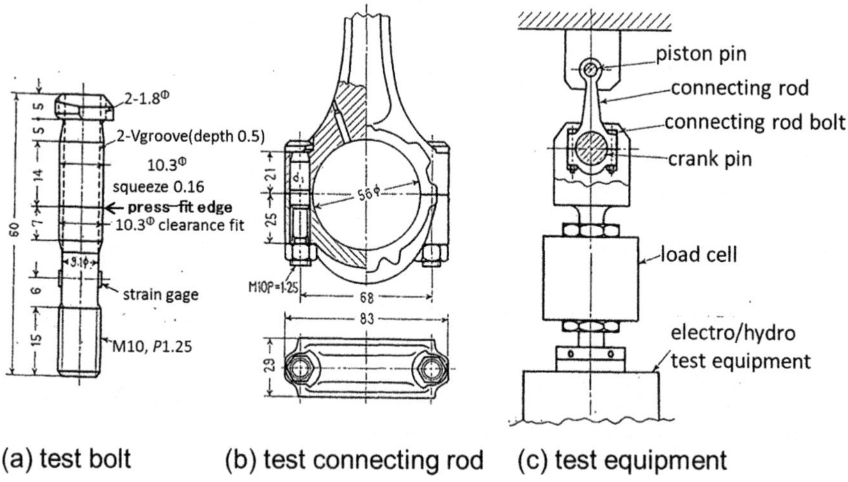

Fig. 6.60 Loosening behavior test specimen and equipment under cyclic axial loads [22, 23]

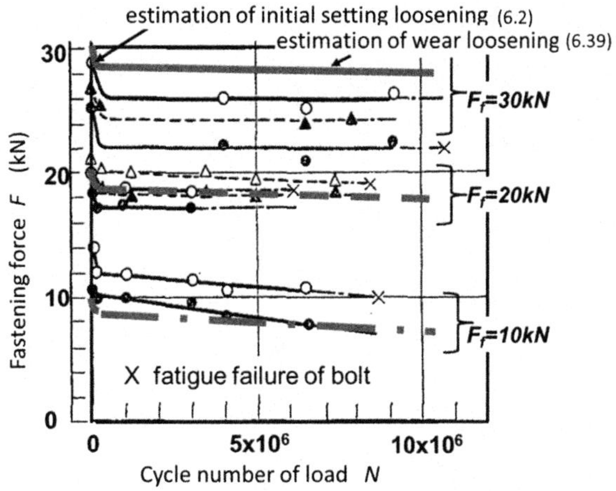

Fig. 6.61 Experimental and analytical results of loosening behavior under cyclic axial loadings, cyclic axial load W (0–30 kN), ($M10, p1.25$) [22, 23]

compliance and setting coefficient as shown in Fig. 6.62. Results are shown in Table 6.7.

For the calculation of micro-slip wear looseness, the analysis formula for the micro-slip wear looseness of Archard's equation was used. The data base of wear coefficient k_w for SCM4 is not sufficient, so in this case we applied $k_w = 1.15 \times 10^{-10}$ mm^2/N referring to the data of Ni–Cr–Mo steel [23–26], which is considered to be close to bolt material SCM4. The slip wear and looseness were analyzed using Eq. (6.39).

These analysis results are shown by a thick solid line ($F_f = 30$ kN), a thick broken line ($F_f = 20$ kN), and a thick alternate long, and short dash line ($F_f = 10$ kN) on the experimental result [24] in Fig. 6.61. By comparing these experimental results and analytical results, there was a difference that there was almost no loosening in the actual measurement in the high fastening state, unlike analytical result. The reason of this will be that, in the calculation of relative slip amount, only the radial elastic deformation was considered, but this relative slip will be reduced due to the shear

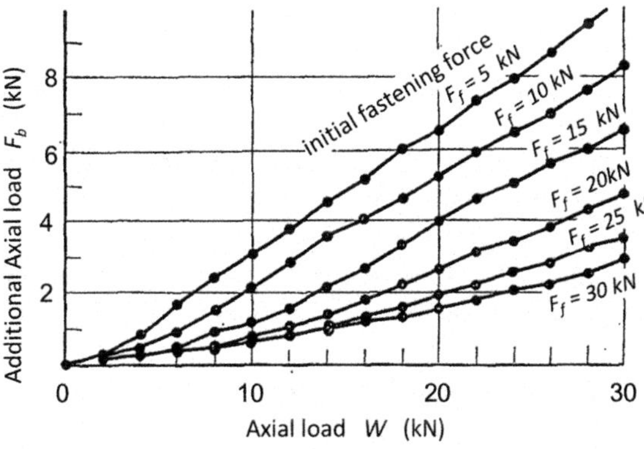

Fig. 6.62 Actual measurement result of bolt shaft additional load under axial load repeated axial load (0–30 kN), (M10, p 1.25) [22, 23]

Table 6.7 Specifications and results of experiment and analyses

Exp. and anal. conditions	Initial fastening force F_f		
	10 kN	20 kN	30 kN
Axial load W	30 kN		
Nominal diameter d	10		
Effective diameter d_2	9.188 mm		
Seat diameter d_w	14.63 mm		
Height of nut m	8.4 mm		
Fastening length l	46 mm		
Compliance of bolt δ_b	2.79X10^{-6} mm/N		
Load factor $\Phi = \delta_c/(\delta_b + \delta_c)$	0.27	0.17	0.1
Embedment coefficient $Z = 1/(\delta_b + \delta_c)$	0.262×10^6 N/mm	0.298×10^6 N/mm	0.323X10^6 N/mm
Embedding amount s	5.5×10^{-3} mm		
Embedding loosening force F_s	1.44 kN	1.6 kN	1.8 kN
Wear coefficient k_w	1.15×10^{-10} mm^2/N		

elastic deformation at the contact surface. On the contrary, under the low fastening force condition, the measured result tends to be larger, the reason of this will be the separation of the contact surfaces and repeated struck and worn under cyclic dynamic loadings. For more highly accurate predictions, it is necessary to analyze the shear elastic deformation of the contact surface and predict the repeated tapping wear accompanied by the contact surface separation, and we look forward to future research.

Figure 6.63 shows a comparison between the actual measurement result of the initial embedment loosening amount F_s [22, 23] and the prediction result by Eq. (6.2). The actual measurement results vary considerably, but it seems that they are almost in line with the predicted results.

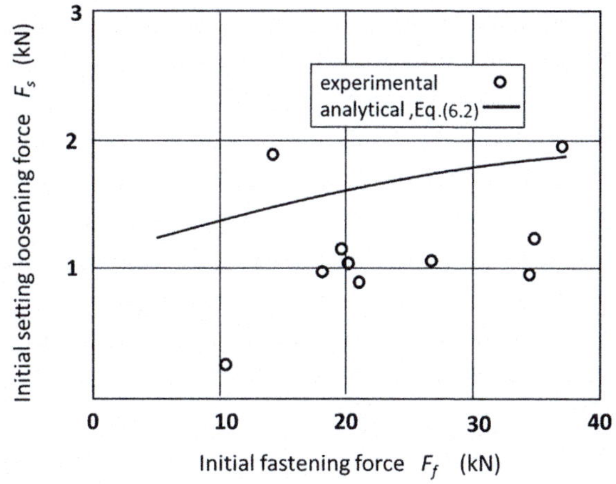

Fig. 6.63 Comparison of experimental results and analytical result of initial embedding loosening force [22, 23] (M10, p 1.25)

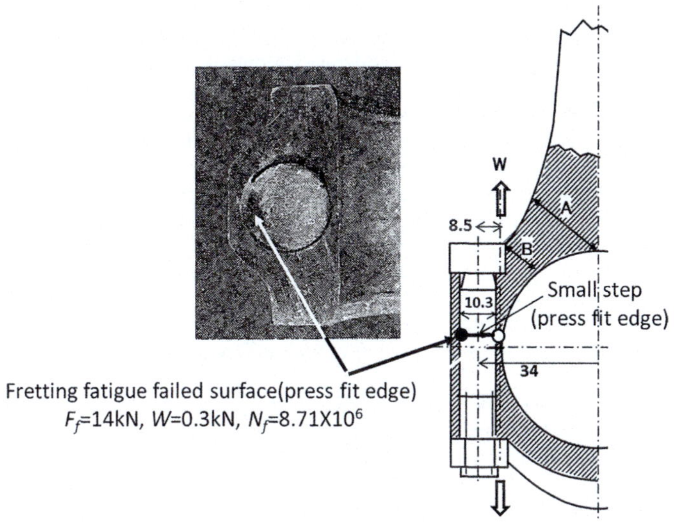

Fig. 6.64 Fatigue failure surface of connecting rod bolt under cyclic axial loadings axial load W (0–30 kN), bolt size M10, p 1.25

In these loosening tests [22, 23], bolt fatigue fracture occurred as a result, and it was confirmed that there is almost no concern about loosening if the fatigue strength of the bolt is designed.

The state of fatigue fracture of the bolt is shown in Fig. 6.64. The fatigue fracture position is at the press-fitting edge shown in Fig. 6.60a, which is the boundary between the press-fitting region and the clearance region. From this situation, it is considered that fretting fatigue failure occurs at this contact edge [21–26]. For the mechanism and countermeasures of such bolt fretting fatigue, refer to Chap. 9.

References

1. Isamu Yoshimoto, Key Points for Designing Screw Fasteners, Japanese Standards Association 2002, pp. 217, 221
2. Akira Yamamoto, Principle and Design of Screw Fastening, Yokendo, 2004, pp102
3. Kazuo Maruyama, Shinji Gase, Toshiyuki Sawa Translation (1989): VDi2230 Blatt 1(1986) Systematic Calculation Method for High Strength Screw Fastening, pp.55, Japanese Society for Screw Research
4. J. F. Archard, J. Appl. Phys. **24**,981 (1953).
5. T. H. Childs, The sliding wear mechanisms of materials, Tribology International, 13 (1980) pp. 285-293
6. Tomoji Sakai, Loosening of Bolts (2nd Report, For Bolts Subjected to Rotational Load), Transactions of the Japan Society of Mechanical Engineers (Part 3) **44–377** (Shou 53–1), pp. 288–292
7. K. Kurihara, Surface pressure behavior of gasket fasteners and effect of retightening, Journal of Bulker Technology, No. 29, Summer 2015, pp. 7–9
8. Tomohiro Takagi, Norihiro Nago, Hirotsugu Sato, Sachi Yamanaka, Evaluation of Mechanical Characteristics of Bolted Fasteners Considering Viscoplastic Behavior of Gasket, Transactions of the Japan Society of Mechanical Engineers, Vol.73, No.728,pp.1245-1252 (2007).
9. Norihiro Nago, Tomohiro Takagi, Hirotsugu Sato, Visco-elastic-plastic properties of gaskets and prediction of axial force relaxation of bolted fasteners, VALQUA Technical Journal, Vol.16, pp.11-19 (2009).
10. Masahiro Watanabe, Atsushi Yamaguchi, Hisashi Honda, Yuichi Tsuji, Evaluation of High Temperature Viscoelastic Properties of Gasket for Flange Joint, Proc. of the JSME Yamanashi Conference, pp.145–146 (2008).
11. Hirotsugu Sato, Shinya Kurokawa, Tomohiro Takagi, Study of Prediction Method for Long-Term Leakage Characteristics of Fasteners Using Gaskets, Proc. of the JSME Yamanashi Conference, pp.139–140 (2008).
12. G.H.Junker, New criteria for self-loosening of fasteners under vibration, SAE Paper No.690055 (1969–1)
13. N.G. Pai, D.P. Hes, "Experimental Study of Loosening of Threaded Fasteners Due to Dynamic Shear Loads," Jornal of Sound and vibration,253,pp.585–602 (2002)
14. Tomotsugu Sakai, Enlarged and supplemented Introduction to Screw Fastening, Yokendo, 2003, pp134
15. Akira Yamamoto and Toshihito Miyake, Prototype and Study of Two Types of Screw Fastening Performance Test Trees, Bulletin of the Faculty of Science and Technology, Tokyo Denki University, No. 5 (1983–6) pp. 35
16. Akira Yamamoto, Shinji Kase, Teruyoshi Kubo, A Study on Loosening of Screws by Right Angle Vibration of the Axis: Characteristics of Various Loosening Prevention Devices, Preprints of the 1977 Spring Meeting of the Japan Society for Precision Engineering, 323
17. Akira Yamamoto, Shinji Kase, Loosening behaviors of bolted joints under transverse loadings, Evaluation and comparison of the performance of various anti-loosening devices, Precision Machinery, 42(6), (1976), pp 507
18. T. Yokoyama, Construction of a mechanical model of bolted fastener behavior subjected to external force perpendicular to the axis, Doctoral dissertation, University of Tokyo, 2010.
19. Satoshi Izumi, Takashi Yokoyama, Atsushi Iwasaki, Shinsuke Sakai, Three-Dimensional Finite Element Analysis of Tightening and Loosening Mechanisms of Bolted Fasteners, Transactions of the Japan Society of Mechanical Engineers A, vol.71, No.702, pp.204-212, (2005).

20. Satoshi Izumi, Takashi Yokoyama, Masatake Kimura, Shinsuke Sakai, Loosening-resistance evaluation of double-nut tightening method and spring washer by three-dimensional finite element analysis, Engineering Failure Analysis, vol. 16, 1510–1519 (2009)
21. M. Kimura, S. Izumi, S. Sakai, Self-loosening Behavior of a Spring Washer: Three-dimensional Finite Element Method Study (in Japanese), Transactions of the Japan Society of Mechanical Engineers, Series A, 73, 734, (2007), pp. 1105-1110.
22. Tomotsugu Sakai, Study of loosening characteristics of connecting rod cap bolts, Transactions of the Japan Society of Mechanical Engineers (Part 3) **43–368** (2000–4), pp. 1454–1461
23. Hattori, T., Naruse, T., Bolted joints technology system learned from accident cases, NTS, (2021)
24. T. Hattori, M. Yamashita and N. Nishimura, Fretting fatigue strength and life estimation considering the fretting wear process, ,Proc. the 2nd JSME/ ASME Int. Conf. on Materials and Processing 2005 pp.(ICS-01)-1–6 (2005).
25. T. Hattori, M. Yamashita and N. Nishimura, Fretting fatigue strength and life estimation in ultra high cycle region considering the fretting wear process JSME Int. J., **48–4**, pp. 246–250 (2005).
26. . T. Hattori and T. Watanabe, Fretting fatigue strength estimation considering the fretting wear process Tribology International, **39**, pp.1100-1105 (2006).

Chapter 7
Mechanical Analysis and Experimental Verification for Threaded Joint Pipe

7.1 Introduction

Piping is used in various fields such as social infrastructure, general industry, transportation equipment, etc., and each piping equipment suitable for each application is required. In order to meet various requirements, piping equipment needs to connect piping elements such as elbows and valves, as shown in Fig. 7.1.

Recently, with the diversification of equipment, higher pressures, and higher flow rates, piping joints are required not only to be safe and lightweight, but also to have functions that meet the needs of piping workability and maintenance-free operation. Piping joints are the most important piping element in terms of workability and maintenance-free requirements. There are various types of pipe joints, for example, the pipe screw coupling, union joint type that connects pipes by sandwiching a union nut and union screw, the butt-welding type that connects the ends butt, the flange type connects pipe flanges with bolts and nuts, the mechanical type joints, and the bite-type joints by inserting a metal sleeve into the end of the pipes.

Therefore, in this chapter, we will focus on IHARA SCIENCE CORPORATION's bite-type joint for NE-shaped steel pipes [1, 2], derive the fastening mechanism, mechanical behavior of the joint against external load, and looseness evaluation criteria, and confirm these analytical results through experiments. We believe that this mechanical and experimental handling method focused on biting joints can be used as a reference for other piping joints.

7.2 Outline of Bite-Type Pipe Joints

The bite joint consists of three parts: the main body, the nut, and the NE-type sleeve, and when tightened according to the prescribed procedure, the bite-in state shown in Fig. 7.2 is obtained.

© The Author(s), under exclusive license to Springer Nature Switzerland AG 2025

T. Hattori and T. Naruse, *Compendium for Design and Manufacturing of Bolted Joints*, Solid Mechanics and Its Applications 280,
https://doi.org/10.1007/978-3-031-89628-6_7

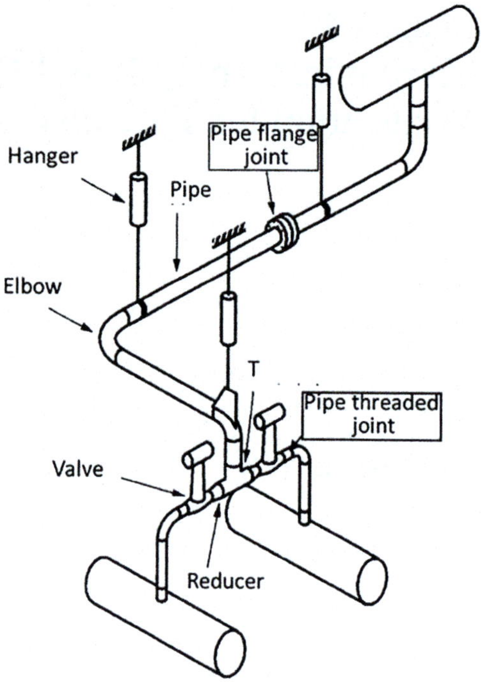

Fig. 7.1 General piping equipment

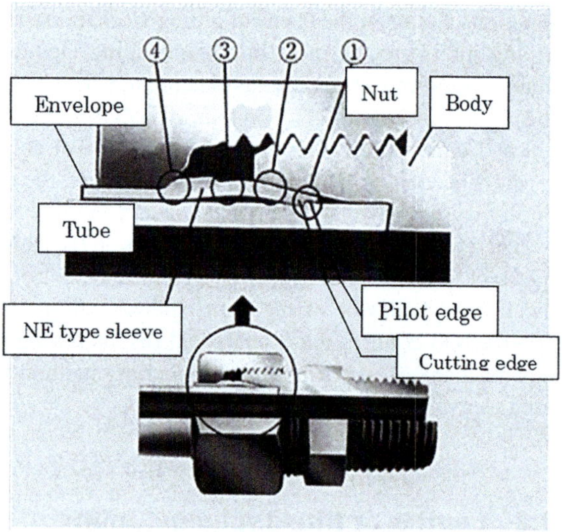

Fig. 7.2 Cross-sectional view of bite joint for NE type for steel pipe [1, 2] (Manufactured by Ihara Science Co., Ltd., Tube diameter 12 mm, thickness 1.4 mm)

This joint has the advantage of reducing piping man-hours because it does not require thread cutting, welding, flaring, etc., for piping.

The mechanism of this joint will be described below.

(i) The cutting edge of the sleeve bites into the tube, holding the tube and creating a seal between the sleeve and the tube.
(ii) The outer circumferential surface of the sleeve is crimped onto the tapered surface of the joint body, creating a seal between the sleeve and the joint body.
(iii) The curved deformation of the sleeve acts as a strong spring action, maintaining sealing performance over a long period of time and preventing the nut from loosening due to vibrations, etc.
(iv) The rear part of the sleeve (envelope part) is crimped to the outer periphery of the pipe to hold the pipe and also alleviates stress concentration on the biting part due to vibrations and improves fatigue strength.

7.3 Fastening Procedure of Bite-Type Pipe Joints

7.3.1 Pre-fastening

Follow the steps below to tighten this bite-type joint.

(i) Insert the pipe with the nut and sleeve into the fitting body.
(ii) Tighten the nut by wrench.
(iii) If you tighten the nut while turning the tube lightly, there will be a point where the nut will no longer turn; this point is called the "grip point". At this time, the sleeve begins to sink into the outer surface of pipe. Mark the fit to confirm the next final tightening amount.
(iv) Tighten the nut 1–1/4 turns from the above fit mark (grip point). The sleeve is now completely embedded in the pipe.
(v) Loosen the nut and check the biting condition of the sleeve into tube.

7.3.2 Final Fastening

Insert the pre-fastened tube set into the joint body and tighten the nut with a wrench. When turning the nut with wrench, there is a point where the response suddenly increases. This point is called the "sharp torque rise point (STP)", and from this point, turn the nut an additional 1/4 turn. (Converted to tightening torque and force, approximately 38 N-m and approximately 10.7 kN.)

7.3.3 *Disassembly and Refastening*

Biting joints can be disassembled by simply loosening the nut, and re-fastening after disassembly is the same as the final fastening described in above (Sect. 7.3.2). It can be disassembled and re-tightened more than 8 times by following the same procedure.

7.4 Experimental and Mechanical Analyses of Slipping and Loosening Behaviors of Bite-Type Pipe Joints

Below, we will sequentially describe loosening experiments under internal pressure, bending, torsion, and combined bending/torsion loads. Then slip analysis is performed, and comparison of the both results is conducted. Slip analysis was performed using simple mechanical analysis of the slip at the nut/body screw fitting part or between the pipe and sleeve. In addition, the loosening experiment was conducted by measuring the decrease in tightening force under various cyclic loads using a strain gauge attached to the inner circumference of the tube at the biting position where the tightening force was greatest. The pasting situation is shown in Fig. 7.3.

7.4.1 *Slipping and Loosening Behavior Under Cyclic Inner Pressure Loadings*

The cyclic internal pressure test was conducted using a cyclic impact pressure tester shown in Fig. 7.4. The impact internal pressure was set at 31 MPa, and the loosening limit was determined by varying the initial fastening force. A strain gauge was pasted on the inner surface of the test pipe, and the looseness was evaluated from the variation in the strain gauge.

Figure 7.5 shows the loosening behavior under cyclic impact internal pressure of 31 MPa under rated initial fastening force conditions. Slight loosening

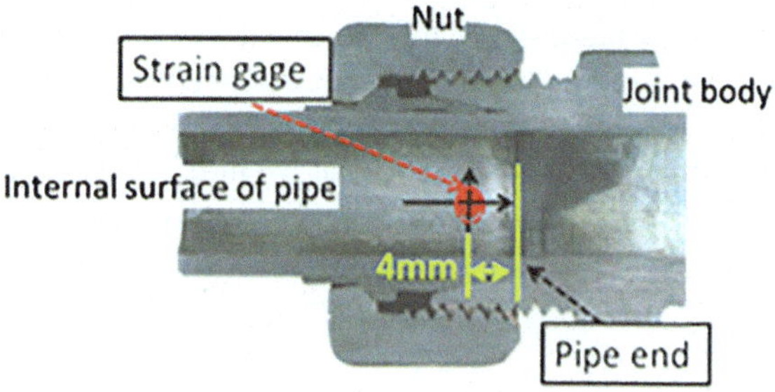

Fig. 7.3 Situation of strain gauge attachment for loosening measurement [1, 2]

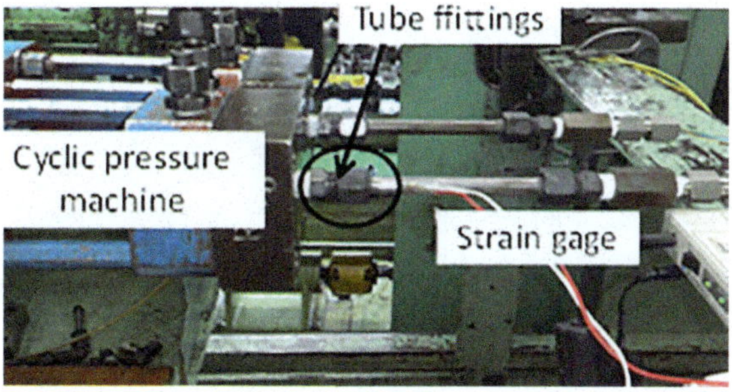

Fig. 7.4 Cyclic internal impact pressure test machine (impact internal pressure 31 MPa)

(circumferential strain of approximately 430 × 10^{-6}) was observed after 20,000 cycles.

Figure 7.6 shows the loosening behavior after 20,000 cycles under cyclic impact internal pressure of 31 MPa for various initial fastening forces. It can be seen that the loosening speed increases rapidly when the initial fastening force is less than about 80% of rated fastening force.

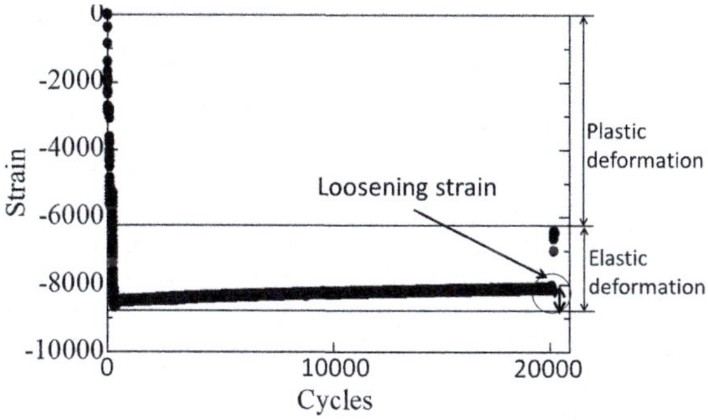

Fig. 7.5 Loosening behavior by cyclic internal impact pressure test (impact internal pressure 31 MPa)

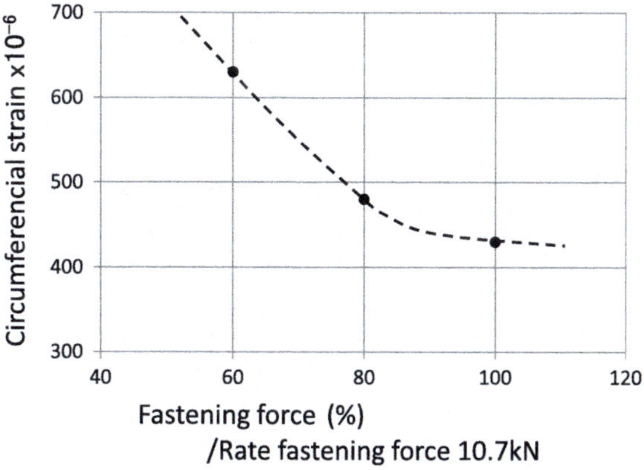

Fig. 7.6 Loosening amounts under each initial fastening force (impact internal pressure 31 MPa, number of loading cycles 2000)

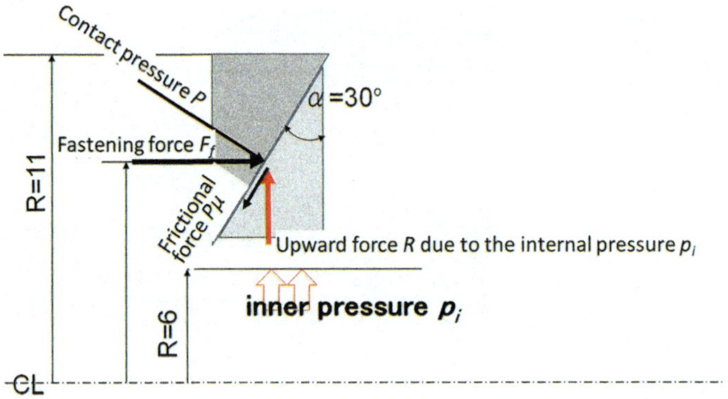

Fig. 7.7 Mechanical model of nut/joint-body thread contact under internal pressure

The loosening conditions under cyclic internal pressure loads was mechanically analyzed below as conditions for slippage of the nut/body screw fitting due to internal pressure loads. This fitting surface is actually a spiral, but this is simplified to a rotationally symmetrical cone shape and mechanical model is constructed as shown in Fig. 7.7. When internal pressure is applied to the thick-walled cylindrical model in which the joint body and nut are completely fixed, the radial compressive stress in the effective diameter portion of the threaded fit is derived as follows.

$$\sigma_r = \frac{p_i a^2}{b^2 - a^2} - \frac{p_i a^2 b^2}{b^2 - a^2}\frac{1}{r^2} \tag{7.1}$$

In this section, we estimate loosening condition under cyclic internal pressure loading is just when the slippage occurs on the thread contact surface shown in Fig. 7.7.

This slip condition is estimated such that the upward force R due to the internal pressure p_i in Fig. 7.7 matches to the radial stress in Eq. (7.1).

$$R = F(\sin\theta + \mu\cos\theta)/(\cos\theta - \mu\sin\theta) \tag{7.2}$$

From Formulas (7.1) and (7.2), the slip limit internal pressure p_c at rated tightening is 52.9 MPa.

It is predicted that the loosening limit internal pressure at rated tightening is converted from the experimental loosening limit cyclic impact internal pressure 31 MPa with initial fastening force about 80% of rated fastening force as follows.

$$31.0 \text{ MPa}/0.8 = 39 \text{ MPa}$$

This was slightly lower than the analytical predicted value, but this is thought to be due to dynamic behavior due to the impact loading, effect of uneven contact on actual thread surface, and a decrease in the coefficient of friction.

7.4.2 Slipping and Loosening Behavior Under Cyclic Bending Loadings

For the cyclic bending test, a pipe bending load jig shown in Fig. 7.8 was manufactured, and it was attached to a hydraulic fatigue testing machine, and the loosening and damage of the test jointed pipe specimen were observed by applying cyclic bending loads.

Figure 7.9 shows the measurement results of loosening behavior due to changes in strain in the circumferential direction of the inner surface of the pipe under cyclic bending moment loads of ±17.7 and ±45.1 N-m, respectively, under rated fastening (fastening torque 38 N-m, fastening force 10.7 kN). In this state, there was no change in strain and no loosening was observed, but fatigue failure occurred in the pipe near the fastened part when a load of ±45.1 N-m was applied. Therefore, a similar cyclic bending test was conducted with the fastening force reduced to 75% of the rated value (tightening torque 28.5 N-m, tightening force 8 kN). As a result, as shown in Fig. 7.10, loosening was observed under a bending moment load of ±33.8 N-m.

For slipping and loosening analyses under cyclic bending loads, the conditions for slip and loosening occurrence at the nut/joint-body screw fitting are considered same as those for cyclic internal pressure loading. And if the thread surface is modeled approximately rotationally symmetrical shape, the basic mechanical model for bending loading can be shown as Fig. 7.11. The slipping occurrence condition M_c on the thread surface is derived as follows using the three-dimensional development as Fig. 7.12.

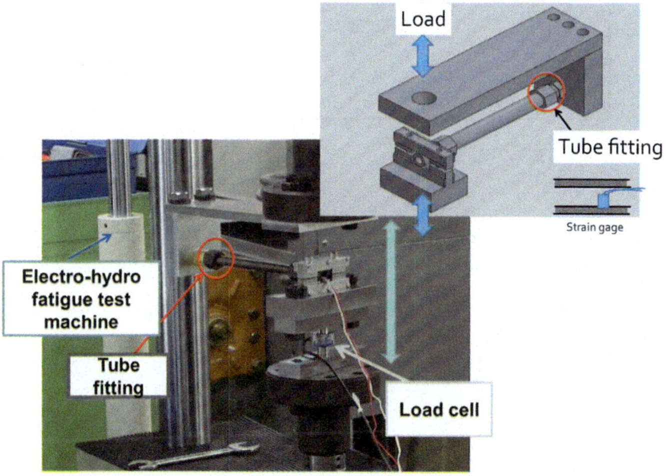

Fig. 7.8 Cyclic bending load jig

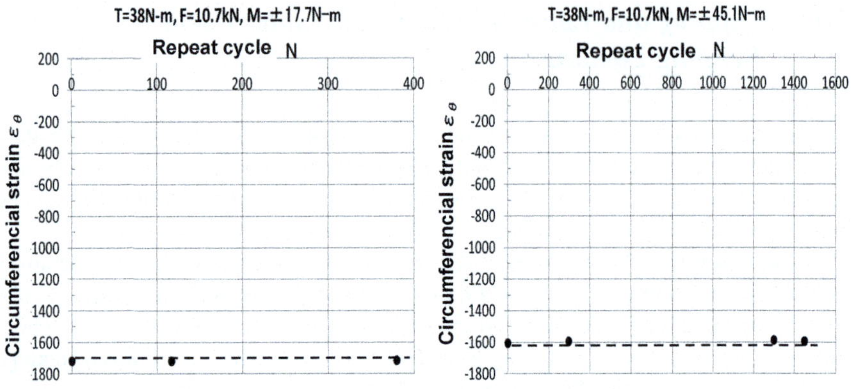

Fig. 7.9 Loosening behavior under cyclic bending loading (rated fastening conditions: torque 38 N-m, fastening force F_f 10.7 kN)

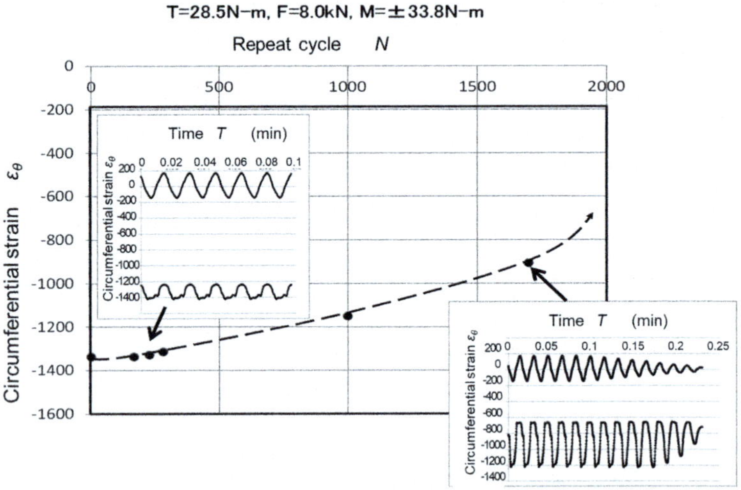

Fig. 7.10 Loosening behavior under cyclic bending loading (75% fastening condition; torque 33.8 N-m, tightening force F_f 8 kN)

$$M_c = \frac{2\,\mathrm{Pr}\,\mu}{\pi} \int_0^{\frac{\pi}{2}} \sqrt{\cos^2\theta + 3d\theta} \tag{7.3}$$

Here, P is the thread surface contact pressure, r is the effective thread radius, and μ is the friction coefficient.

$$P = \frac{F_f}{\cos\alpha} \tag{7.4}$$

Here $\boldsymbol{F_f}$ is the fastening force, and α is the thread half angle.

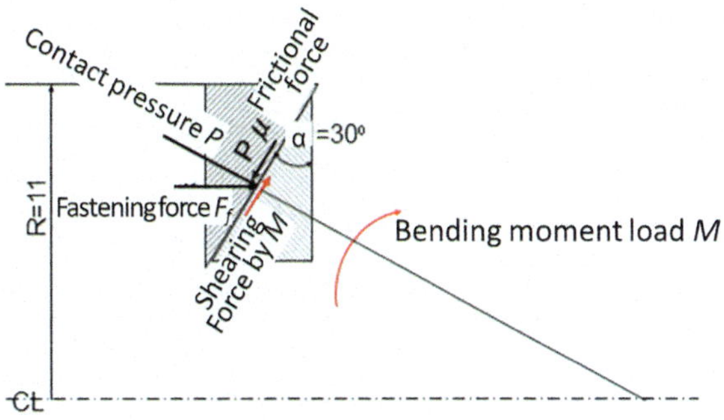

Fig. 7.11 Mechanics of nut/body fitting part under bending load

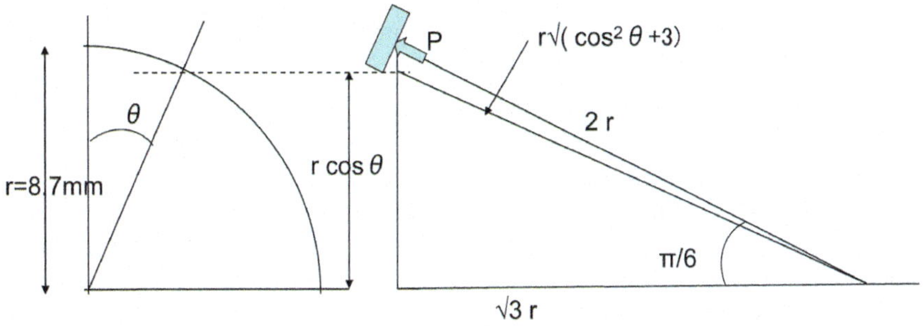

Fig. 7.12 Three-dimensional expansion model for integral around the entire circumference of the thread surface

Following above method, the critical slipping moment M_c at rated fastening can be estimated as 39.4 N-m.

The critical slipping moment in actual measurements was found to be 33.8 N-m at 75% fastening from Fig. 7.10, and then by converting this to rated fastening condition, we can estimate as follows,

$$33.8\ \text{N-m}/0.75 = 45.1\ \text{N-m}.$$

Although this is a little higher than the predicted result, relatively good agreement is seen.

7.4.3 Slipping and Loosening Behavior Under Cyclic Torsional Loadings

For the cyclic torsional loading test, we manufactured a torsion loading jig shown in Fig. 7.13 and observed loosening or damage on the test specimen by applying cyclic torsional loads.

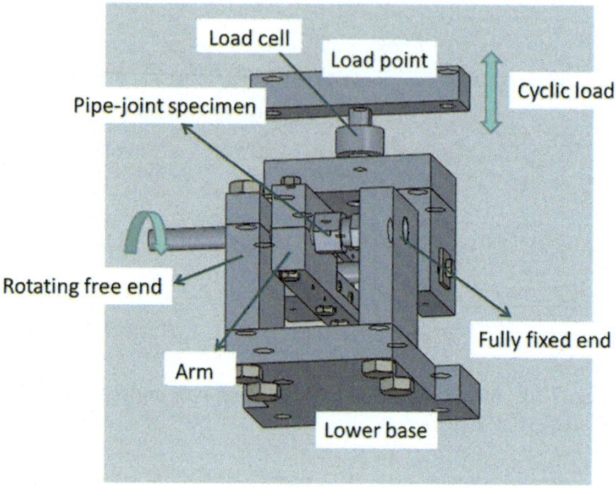

Fig. 7.13 Torsion loading jig for cyclic torsional loading test

The test results are shown in Fig. 7.14. In this case, in the process of increasing the cyclic torsional load amplitude, a significant increase in the slip amplitude (torsion angle) was observed rather than the loosening phenomenon, as shown in Fig. 7.14c. In addition, if this cyclic slipping state continued for a long time under constant displacement (torsion angle) amplitude, the load decreased and wear was observed at the tube/sleeve biting part, as shown in Fig. 7.15. Therefore, the slip occurrence limit under this cyclic torsional loading test can be considered to be 21 N-m.

Based on the experimental results, the analysis of slip under torsional load includes slipping at the nut joint as shown in Fig. 7.16 (points A and B) and slipping at the pipe/sleeve contact point as shown in Fig. 7.17 (point C) which were analyzed.

(a) Slippage at Nut Joint Point

In this case, the following prediction can be made using the contact pressure P of the contact surfaces of points A and B in Fig. 7.16 and the pitch diameter r.

$$\begin{aligned} T_{A;B} &= P\mu r = 9.88 \times 0.15 \times 8.7(\text{point A}) \\ &\quad + 14.2 \times 0.15 \times 7.2\ (\text{point B}) \\ &= 28.2\,\text{N-m} \end{aligned} \tag{7.5}$$

(b) Slipping at Pipe/Sleeve Biting Point

In this case, the surface pressure P_{C} of the biting point (C point) in Fig. 7.17 is calculated as follows.

$$\begin{aligned} P_{\mathrm{C}} &= F_f/(\cos\theta + \mu\sin\theta) \\ &= 10.5 \times (0.947/0.355) = 28.0\ \text{kN} \end{aligned} \tag{7.6}$$

$$T_{\mathrm{C}} = P_{\mathrm{C}}\mu r$$

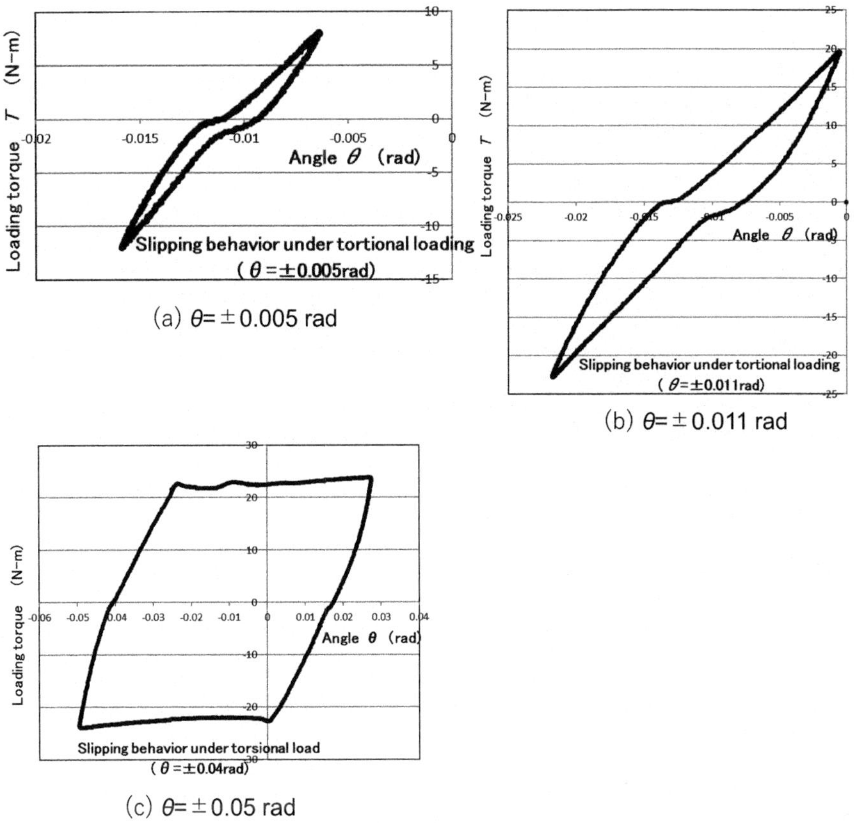

Fig. 7.14 Slipping behavior under cyclic torsional loadings (rating fastening condition: fastening torque T_f 38 N-m)

$$= 28 \times 0.15 \times 6 = 25.2 \text{ N-m} \tag{7.7}$$

Comparing the (a) and (b), the latter (the slipping torque at the pipe/sleeve biting point) is lower, so it can be predicted that slipping will occur here, this is also agree with the experimental results. Also, the analytical value of slipping torque at pipe/sleeve 25.2 N-m and those obtained from experiments 21 N-m were relatively consistent.

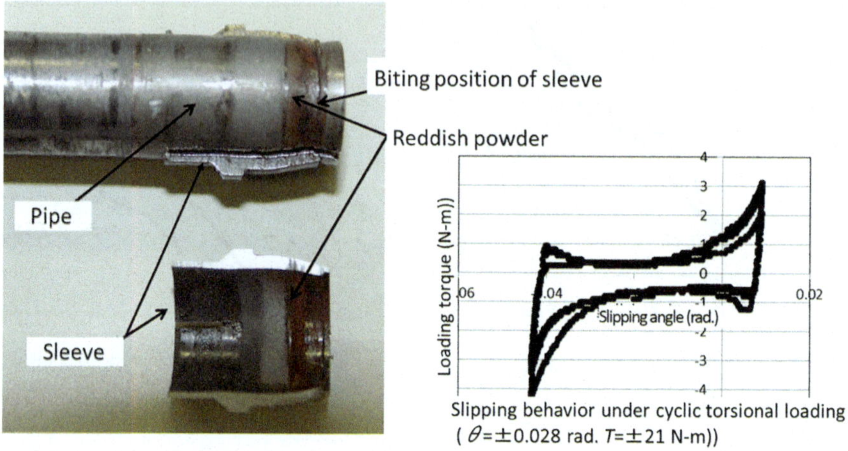

Fig. 7.15 Slipping and wearing conditions under cyclic torsional loading (rating fastening condition: fastening torque T_f 38 N-m)

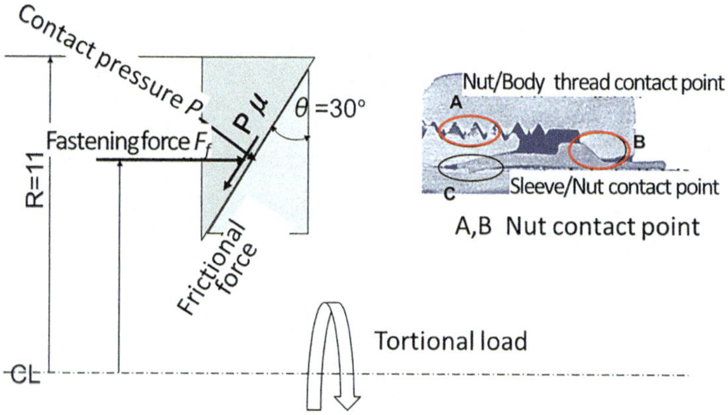

Fig. 7.16 Slipping analysis at nut contact points (A and B points)

7.4.4 Slipping and Loosening Behavior Under Bending and Torsional Combined Loadings

For the cyclic bending and torsion combined load test, a combined load test jig as shown in Fig. 7.18 was used. By aligning the load point with an arbitrary position on the load rod and load lever, it is possible to obtain arbitral ratio between the bending moment and torsion torque.

Figure 7.19 shows the experimental results for a given bending moment/torsion torque ratio. When the bending moment ratio is large, the nut joints (points A and B) slip. And when the torsional load ratio is large, slippage occurred at the tube/sleeve

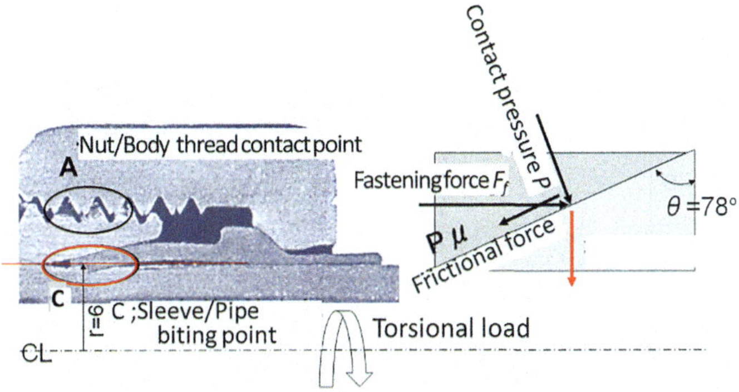

Fig. 7.17 Slipping analysis at sleeve/pipe biting point (C point)

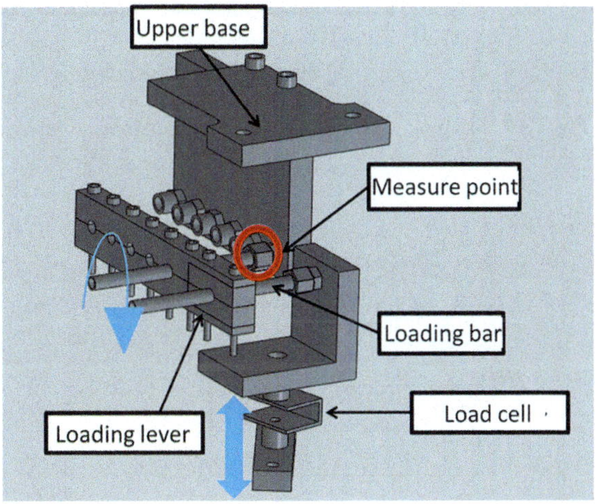

Fig. 7.18 Cyclic bending and torsion combined load test jig

biting point (point C). The experimental results are indicated by dotted line circle and solid line circle in the figure.

Slip analysis is performed separately for the nut joint (points A and B) and the pipe/sleeve biting point (point C).

(a) Slippage at Nut Joint Point (Points A and B)

The shearing force component on the thread contact surface under the bending load M indicated by the dotted arrow in Fig. 7.20 is calculated by referring to Eq. (7.3).

$$M / \int_{0}^{\frac{\pi}{2}} \sqrt{\cos^2\theta + 3d\theta} \tag{7.8}$$

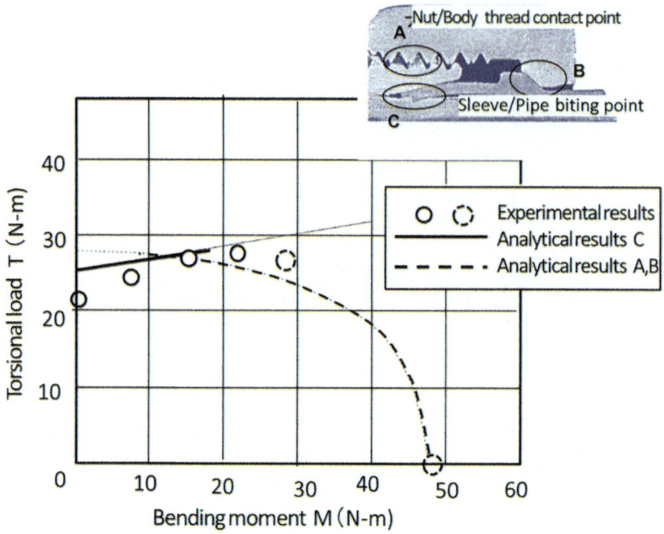

Fig. 7.19 Slipping criteria for bending and torsional combined loadings

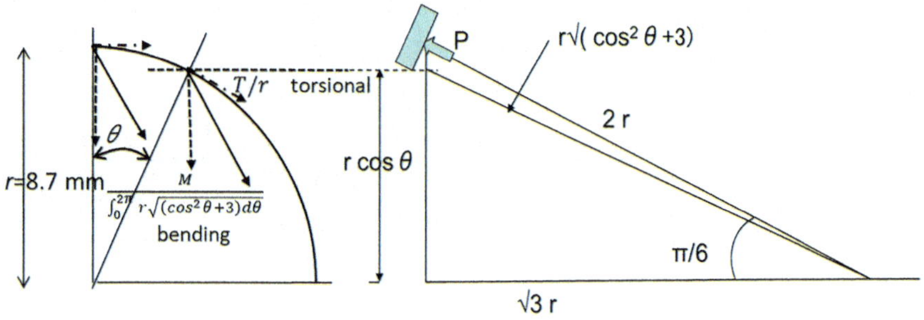

Fig. 7.20 Slipping analysis on the nut thread contact surface

And the shearing force component in the circumferential direction on the thread contact surface under the torsional load T indicated by the dash dot arrow is

$$T/r \tag{7.9}$$

The resultant combined shearing force shown by solid arrow is derived as follows,

$$\sqrt{\left(\frac{M}{\int_0^{2\pi} r\sqrt{\cos^2\theta + 3d\theta}}\right)^2 + \left(\frac{T}{r}\right)^2 + 2\left(\frac{M}{\int_0^{2\pi} r\sqrt{\cos^2\theta + 3d\theta}}\right)\left(\frac{T}{r}\right)\cos\left(\frac{\pi}{2} - \theta\right)} \tag{7.10}$$

Finally, the slipping criterion at the threaded joint under both loads is determined as follows,

$$\int_0^{2\pi} \sqrt{\left(\frac{M}{\int_0^{2\pi} r\sqrt{\cos^2\theta + 3d\theta}}\right)^2 + \left(\frac{T}{r}\right)^2 + 2\left(\frac{M}{\int_0^{2\pi} r\sqrt{\cos^2\theta + 3d\theta}}\right)\left(\frac{T}{r}\right)\cos\left(\frac{\pi}{2} - \theta\right)d\theta} = p\mu \tag{7.11}$$

This slipping criterion was shown in Fig. 7.19 by dotted line.

(b) Slipping at Pipe/Sleeve Biting Point

When only torsion torque T is applied to this bite joint, as shown in Fig. 7.21a, the surface contact pressure between the tube and sleeve remains unchanged as the surface contact pressure P at the time of fastening. But as shown in Fig. 7.21b, if the bending moment M is added, the fastening force F increases by ΔF due to the tension caused by tilting and slipping of sleeve, and it is predicted that the surface contact pressure between the tube and the sleeve will eventually increase by ΔP.

Strictly speaking, the analysis of this increase of fastening force ΔF is complicated as it is an indeterminate problem that also includes deformation analysis, but if we consider only the static mechanical balance as shown in Fig. 7.22, it can be roughly estimated as follows.

$$\Delta F = \frac{M}{\frac{7.2+7.9}{\cos\theta}}\cos\theta = \frac{M}{15.1}\cos^2\theta \tag{7.12}$$

The slip criteria of the tube/sleeve joint (point C) under a combined load of torsion torque and bending moment, including above effect, are shown as a solid line in Fig. 7.19.

Then I'll summarize the slip criteria or loosening criteria under combined loads of torsional torque and bending moment as follows.

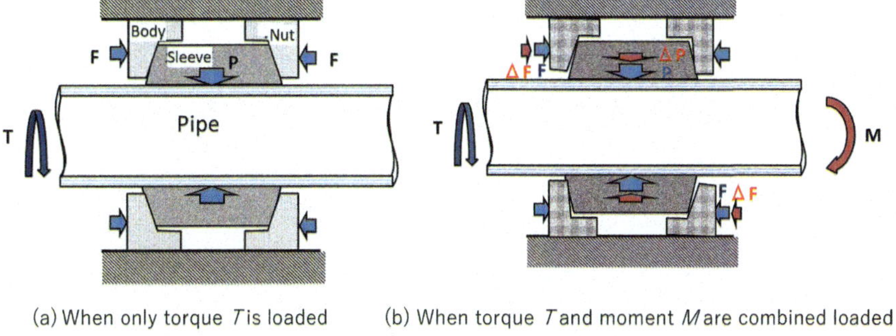

Fig. 7.21 Mechanism of contact pressure increase ΔP on pipe/sleeve contact surface. **a** When only torque T is loaded. **b** When torque T and moment M are combined loaded

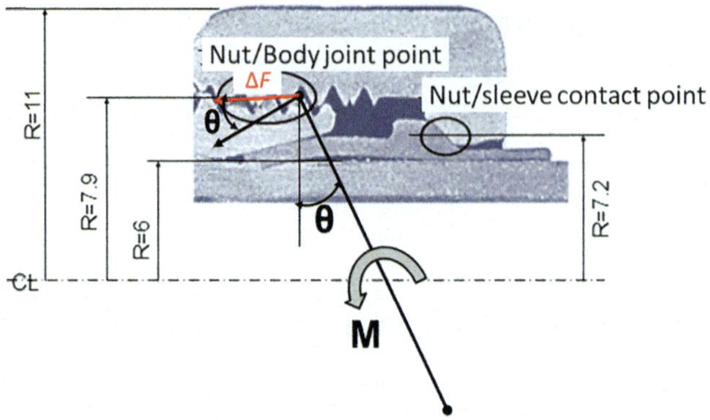

Fig. 7.22 Mechanical analysis of fastening force increase ΔF by application of bending moment M

By the mechanical analysis, it is predicted that slipping/loosening will occur based on the lower criterion of the following two slipping criteria: (a) at the threaded joint and (b) at the pipe/sleeve biting point. And the experimental results shown in Fig. 7.13 seem to match these predicted results fairly well.

The pipe joints discussed in this chapter are just a few examples of the many other types of pipe joints, but the simple mechanical analysis and experimental methods used here will help you develop and design your own new pipe joints in future. I think this will be a useful reference when making your own improvements and evaluations. Please make use of this.

References

1. http://www.ihara-sc.co.jp/. Ihara Science Corporation.
2. Bite Type Tube Fittings, Cat. No 0401, Ihara Science Corporation, (2008).

Chapter 8
Mechanisms and Countermeasures for Delayed Fracture

8.1 Mechanisms of Delayed Fracture

Delayed fracture is a problem that must be taken into account when using high-strength bolts with a tensile strength of 1.2 GPa or more. This delayed fracture occurs after a static tensile load or bending load is applied for a certain period of time at room temperature, without any apparent plastic deformation, a crack initiates from the bottom of the notch and gradually propagates, eventually resulting in an unstable fracture.

The mechanism of this delayed fracture [1–3] is generally known as hydrogen embrittlement, and as shown in Fig. 8.1 when the anodic dissolution of a metal occurs as a result of a corrosion reaction, in which hydrogen ions H^+ are deoxidized, and then this generated atomic hydrogen is adsorbed on the surface and then penetrates into the metal. The rate at which hydrogen diffuses in metals varies greatly depending on the crystal structure, and in bcc steel materials, hydrogen diffuses easily at room temperature. Hydrogen that diffuses through the metal accumulates at lattice defects such as dislocations and at interfaces between different phases such as grain boundaries and makes these position brittle and promoting the initiation and propagation of cracks.

An example of a delayed fracture curve is shown in Fig. 8.2. The vertical axis shows the amount of diffusible hydrogen, and the horizontal axis shows the delayed fracture time. As the amount of diffusible hydrogen decreases, the delayed fracture time increases, and it can be seen that fracture does not occur when the amount of diffusible hydrogen is below a certain level.

In addition, in this example, the critical diffusible hydrogen amount (the upper limit of the diffusible hydrogen amount at which delayed fracture does not occur) is greater than the hydrogen level in the actual environment, so if steel type A is used at a load stress of less than a, delayed fracture will not occur.

Figures 8.3, 8.4, 8.5 and 8.6 show the hydrogen embrittlement crack surface of high-strength, low-alloy steel 300 M (equivalent to AISI4340) under a tensile stress

© The Author(s), under exclusive license to Springer Nature Switzerland AG 2025
T. Hattori and T. Naruse, *Compendium for Design and Manufacturing of Bolted Joints*,
Solid Mechanics and Its Applications 280,
https://doi.org/10.1007/978-3-031-89628-6_8

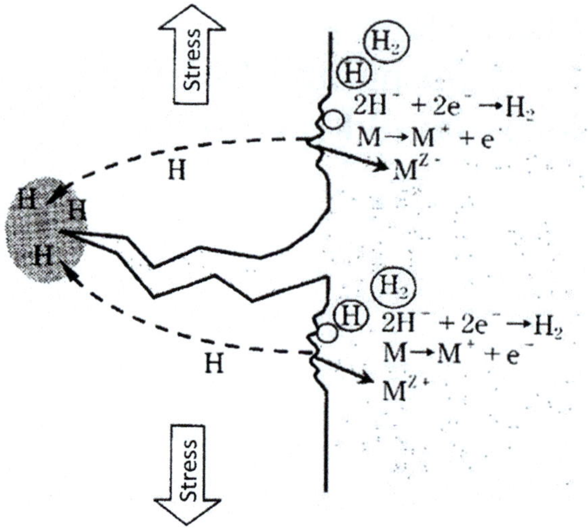

Fig. 8.1 Mechanism of delayed fracture (hydrogen embrittlement assist-type stress corrosion) [1]

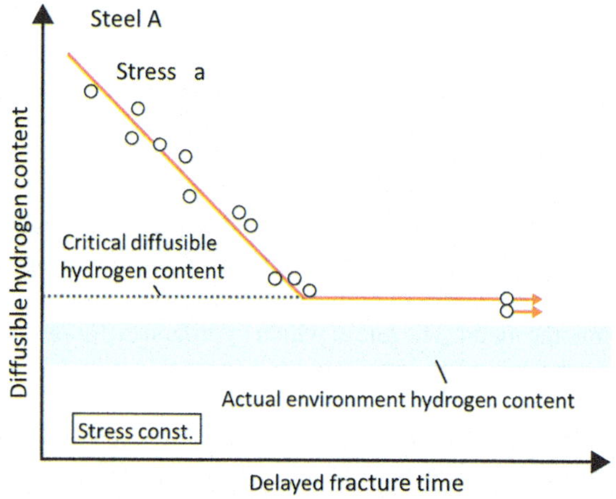

Fig. 8.2 Example of a delayed fracture curve

of 1269 MPa (75% of yield stress) and 57 days of rupture under exposure to a seaside environment [3, 4]. Figure 8.3 shows an optical photograph of the entire fracture surface of a notched specimen (stress concentration ratio 2.5). Figure 8.4 shows an SEM photograph of the hydrogen embrittlement crack at position 1 in Fig. 8.3. High-strength alloy steel is characterized by such intergranular crack fracture surfaces. Corrosion pits can be seen because this surface was exposed to a corrosive environment during the crack propagation process. Figure 8.5 shows the final fracture surface at position 2 in Fig. 8.3, and since it was exposed to a corrosive environment even after fracture, the dimples are also covered with such corrosion products. Figure 8.6 shows the notched machined surface at position 3 in Fig. 8.3, as this initial machined surface was exposed to a corrosive environment for a long time, the surface has dimple-like appearance, but these are the developed corrosion pits, and care must be taken not to mistake them for dimples on the fracture surface.

Fig. 8.3 Hydrogen embrittlement fracture surface of low-alloy steel 300 M [3, 4] (equivalent to AISI4340)

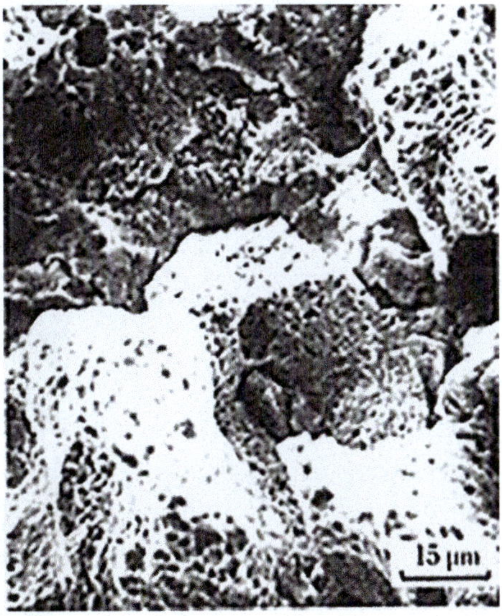

Fig. 8.4 SEM photograph of the hydrogen embrittlement crack at position 1 in Fig. 8.2 [3, 4]

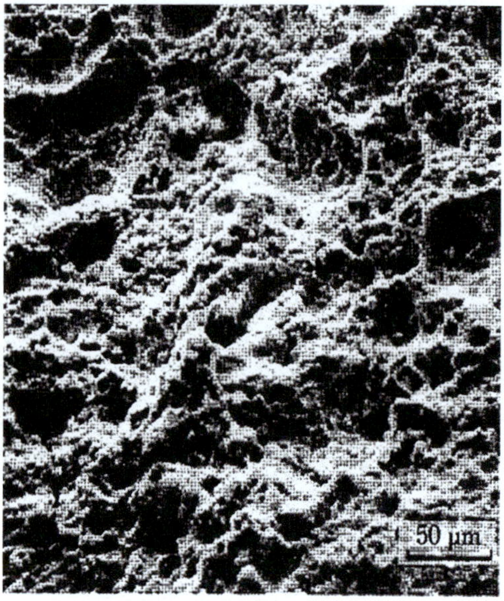

Fig. 8.5 SEM photograph of the hydrogen embrittlement crack at position 2 in Fig. 8.2 [3, 4]

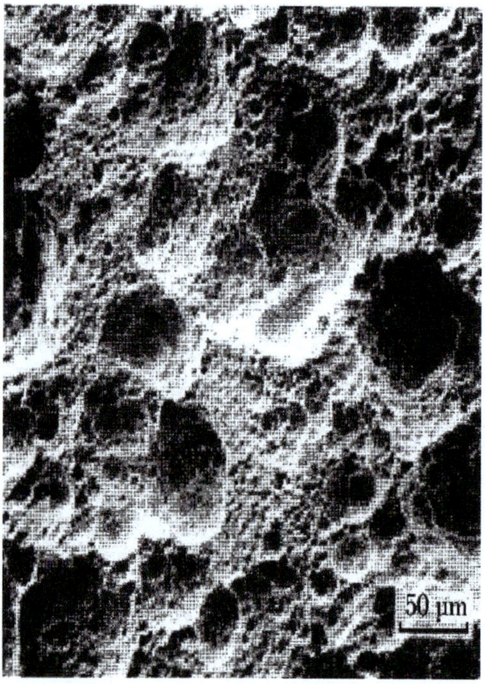

Fig. 8.6 SEM photograph of the notched machine surface at position 3 in Fig. 8.2 [3, 4]

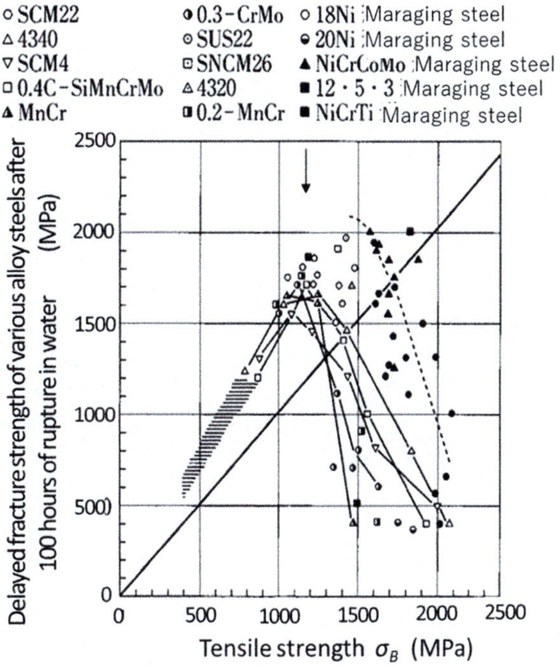

Fig. 8.7 Influence of tensile strength on delayed strength [4, 5]

8.2 Delayed Fracture Strength of Each Material

Figure 8.7 shows the delayed fracture strength of various alloy steels after 100 h of rupture in water [4, 5], and it can be seen that the delayed fracture strength of most alloys decreases rapidly when the tensile strength exceeds 130 MPa. Therefore, JIS B 1186 has come to state that bolts of *F* 11 T (tensile strength 110 MPa) or higher should not be used as much as possible. An example of an actual fracture surface of a high-tensile bolt is shown in Fig. 8.8, which is characterized by a typical intergranular fracture surface, which is the most important evidence of delayed destruction, in bolt accident analysis. Fractures often occur in stress concentrated areas and are characterized by the fact that delayed fracture of bolted joint occurs most frequently in areas with incomplete threads as shown in Fig. 8.9.

8.3 Delayed Fracture Tests [6]

In the delayed fracture test, the test specimens are prepared by fastening test bolt to the fastening body via a nut and washer as shown in Fig. 8.10. This fastening body is shaped like a ring and a strain gauge is attached as shown in this figure and is also used as a load cell for measuring the fastening force. The load cell and strain

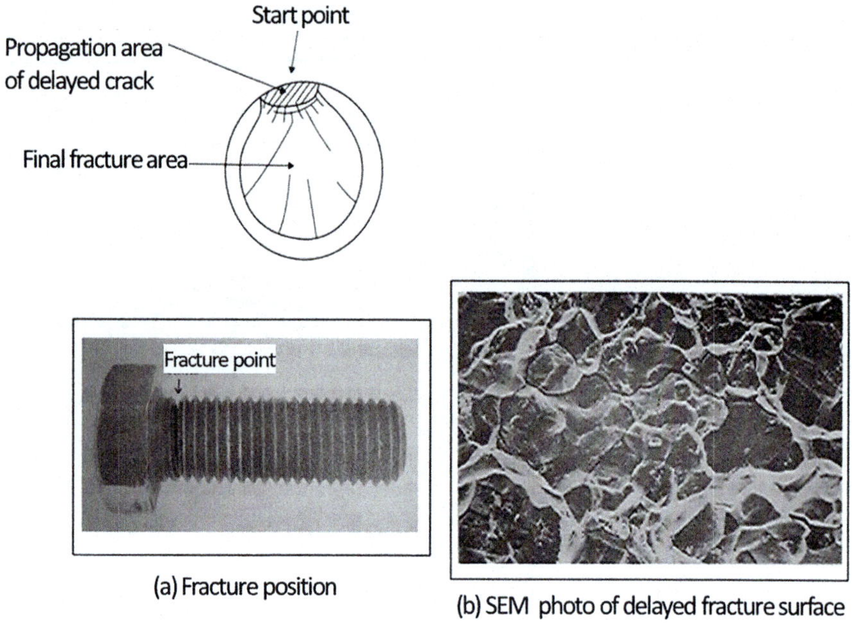

Fig. 8.8 Delayed fracture surfaces of high-strength bolt [4, 5]

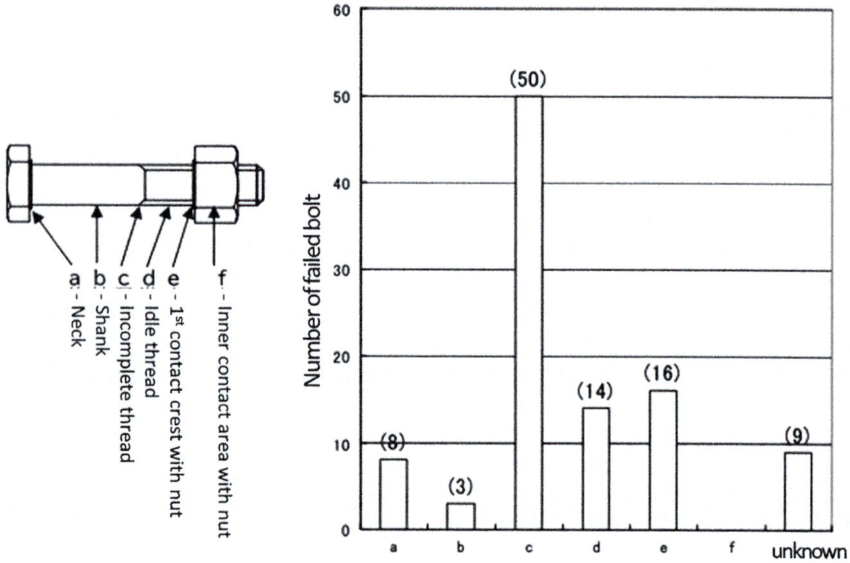

Fig. 8.9 Delayed fracture position in bolts

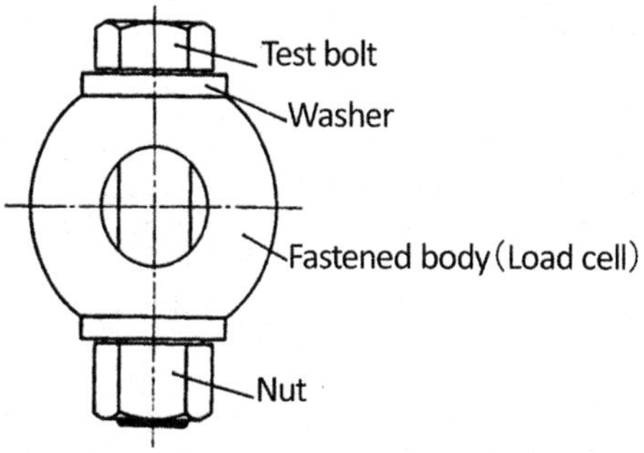

Fig. 8.10 Test specimen for delayed fracture strength test [6]

gauge are coated with a corrosion-resistant resin because they will be exposed to a corrosive environment for a long time.

In an exposure test under normal actual environments require a long time to break down, so tests are carried out in as harsh an environment as possible (for example, water, seawater, and a weak acid solution (pH 1–3)) so that equivalent evaluation can be done in a short period of time.

One of these is the immersion test shown in Fig. 8.11, where the material is immersed in a dip, and a filtration device is installed to prevent the liquid from changing or deteriorating due to long-term immersion.

As the above immersion method is far from a normal actual atmosphere, a Ferris wheel method is also used to repeat immersion and drying up and down in order to perform the test under more realistic conditions, as shown in Fig. 8.12. The feature of this test device is that if the cycle is repeated, for example, immersion in an water for 10 min and drying for 50 min, and if there are 50 rainy days in Japan in a year, then 50 immersion and drying cycles will be equivalent to one year's usage.

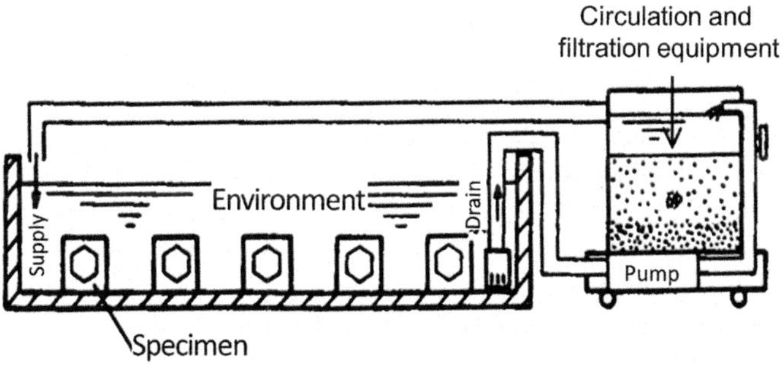

Fig. 8.11 Immersion test device [6]

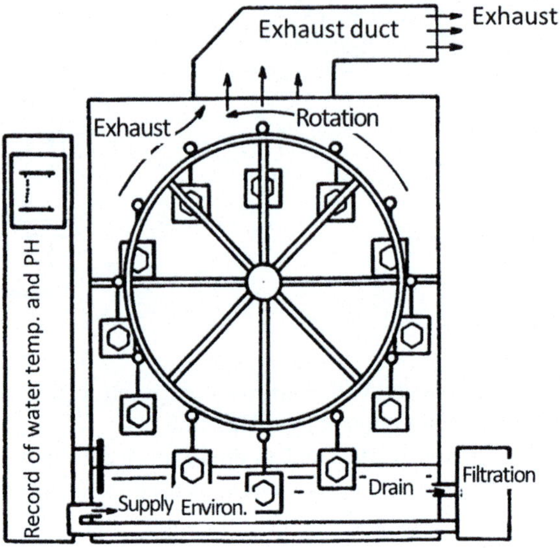

Fig. 8.12 Immersion and drying cycle test device (Ferris wheel) [6]

8.4 Countermeasures for Delayed Fracture

Since the cause of delayed fracture is thought to be caused by hydrogen that has intruded from the outside environment, the countermeasure is to prevent the intrusion of hydrogen from the outside or to remove the hydrogen that has intruded. In this case, the following methods are employed according to the time when hydrogen intrudes in the steel.

8.4.1 Surface Treatment Process (Plating Process)

Since hydrogen intrudes in the pickling process for electroplating or its pretreatment, dehydrogenation treatment (baking treatment) is performed for the purpose of removing the hydrogen that has intruded in the steel after plating. Even if such treatment is performed, there is a possibility that delayed fracture may occur depending on the plating method, so when electroplating is applied to a bolt, the bolt materials are limited to the strength classification less than 10.9.

8.4.2 Intrusion During Operation

If sufficient anti-rust treatment such as painting or oil application is applied, delayed destruction will be less likely to occur. Despite the fact that hexagon socket bolt belonging highest strength category of the JIS standard is commonly used, the occurrence of delayed fracture is extremely low because the environment is relatively good such as indoor, and sufficient rust prevention treatment is applied.

References

1. J. P. Hirth, Met. Trans. Vol.11A, No.6 (1980), p.861.
2. C. H. Lorrig, Behavior of Metals at Low Temperature, ASM (1953), p.71.
3. R. Dutton, K. Nuttal, M. P. Puls and L. A. Simpson, Mechanisms of hydrogen induced delayed cracking in hydride forming materials, Metallurgical transactions A, **8**, (1977) 1553–1562.
4. Battelle Columbus Laboratories, SEM/TEM Fractography Handbook, (1975), 350–352.
5. K. Nishiguchi, et al., New Edition Joining Handbook, Industrial Technology Service Center, (1994), 631–632.
6. Research Report for Standardization of Fastening Performance of High-Strength Bolts, The Japan Research Institute for Screw Threads and Fasteners (3rd Report), (1980).

Chapter 9
Mechanisms and Countermeasures for Fretting Fatigue of Joint Structures

9.1 Introduction

Fretting can occur when a pair of structural elements is in contact under a normal load while cyclic stress and relative displacement are forced along the contact surface. This condition can be seen in bolted or riveted joints [1, 2], in shrink-fitted shafts [3, 4], in the blade/dovetail joint of turbo machinery [5, 6], etc. During fretting process, the fatigue strength decreases to less than one-third of that without fretting [7, 8]. The strength is reduced because of concentrations of contact stresses, such as contact pressure and tangential stress at the contact edge, where fretting fatigue cracks initiate and propagate.

This concentration of stress can be calculated using the finite element method [9] or the boundary element method. Methods of estimation of the strength of fretting fatigue have been developed, which use the values of this stress concentration on a contact surface [3, 5]. However, the stress fields near the contact edges show singularity behavior, where the stresses at contact edges are infinite. Thus, maximum stresses cannot be used to evaluate fretting fatigue strength.

To date, authors presented the fretting crack initiation estimation method, using stress singularity parameters at contact edges [8–10], and fretting fatigue limit or life estimation methods using fracture mechanics [10, 12, 13]. But by using these fretting fatigue strength or life estimation methods, we could not predict the super-high-cycle fretting fatigue problems in the industrial field. For instance, a 660 MW turbogenerator rotor failed in England during the 1970s as a result of fretting fatigue cracking as shown in Fig. 9.1 [11]. In this case the number of loading cycles in just one year was about 1.6×10^9, and this problem was observed after many years of operation.

The problem is that all of the mechanical analyzes described above were based on the initial contact surface condition, but in reality, the occurrence and progression of local wear should be taken into account. So in this section we will presented the

© The Author(s), under exclusive license to Springer Nature Switzerland AG 2025

T. Hattori and T. Naruse, *Compendium for Design and Manufacturing of Bolted Joints*, Solid Mechanics and Its Applications 280,
https://doi.org/10.1007/978-3-031-89628-6_9

Fig. 9.1 Fretting fatigue failure example of turbogenerator rotor [11]

very-high-cycle fatigue life estimation methods considering fretting wear and change in the contact surface [14–17].

In general, fretting fatigue cases mainly involve damage under extremely high cycles, but in cases such as rotating machinery, it is also necessary to predict the strength and life in the low-cycle region due to repeated starting and stopping of that. So we develop a prediction method for low-cycle fretting fatigue strength and life using critical distance stress theory. We believe that this allows you to successfully explain the fretting strength life span of the entire range of actual machines, from low cycles to extremely high cycles.

9.2 Mechanical Consideration of Fretting Fatigue

In a mechanical model (shown in Fig. 9.2) in which two objects are pressed together with contact pressure P and tangential force Q acts simultaneously, if the relationship between the pressing force P and the tangential force Q is as follows, the two objects will slide completely. Surface damage behavior in these mechanical condition is treated as tribology.

$$Q = P\mu \tag{9.1}$$

In this equation μ is a friction coefficient of contact surface.

On the other hand, fretting damage refers to surface damage behavior in which two objects do not slip entirely, but only some parts of the object slip, and the rest kept sticking. The mechanical condition can be expressed as follows.

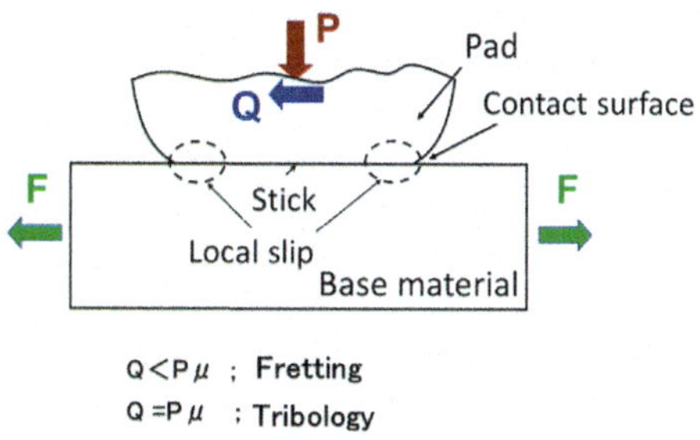

Fig. 9.2 Mechanical conditions for fretting damage

$$Q < P\mu \tag{9.2}$$

If we refer to a mechanical model such as that shown in Fig. 9.2, this local slip is generally a very small amount due to a mismatch between the elastic deformation of the pad and the base material. If this process continues repeatedly for a long period of time, microscopic wear will progress over a long period of time. This is called fretting wear. Furthermore, a large tangential force acts on the outermost periphery of the local slip area, causing fatigue cracks to occur. These cracks that occur in the base metal can propagate depending on mechanical conditions, causing fretting fatigue failure. Particularly, bolted bodies are subject to such mechanical conditions, and both the bolt and the fastened body are at risk, so enough attention is required.

9.3 Fretting Fatigues in Fastened Bodies

9.3.1 Fretting Fatigues When Subjected to a Transverse Load to the Bolt Axis

As shown in Fig. 9.3, when a cyclic external force F is applied in the direction perpendicular to the bolt axis, fretting cracks may occur at the contact edge or the boundary between the adhered area and the local slip area, depending on the surface pressure and loading conditions. The more detail mechanism of fretting fatigue crack initiation and propagation under such mechanical conditions is outlined in Fig. 9.4. The contact edge is a stress singularity state, and small cracks are easily generated due to significant high stress concentration. The criteria for micro-crack initiation can be evaluated with the stress singular parameters [8–10].

During this early period, fretting fatigue cracks tend to close and propagate very slowly especially in a low stress amplitude range due to the high contact pressure acting near this contact edge. But wear on the contact surface reduces the contact pressure near the contact edge, and cracks gradually open and start to propagate. Hence, the fretting fatigue life in the low stress amplitude range will be dominated by the propagation of these small cracks that initiate at the contact edge. To estimate the

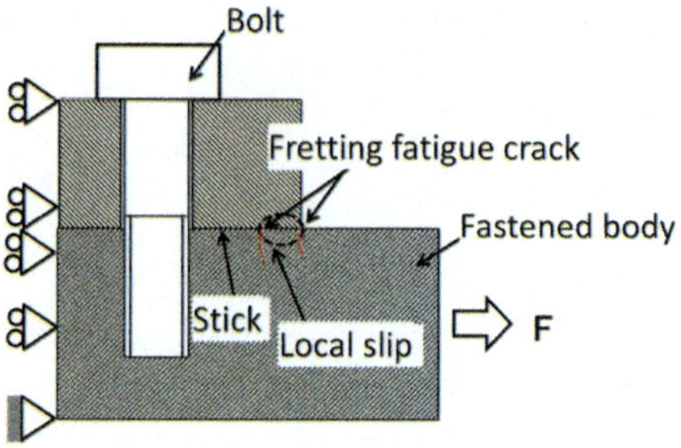

Fig. 9.3 Fretting damage of threaded fastener under load perpendicular to bolt axis

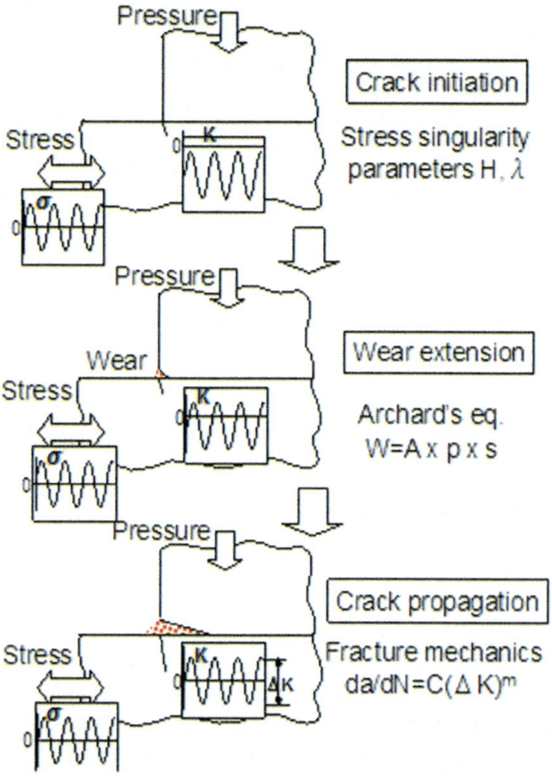

Fig. 9.4 Mechanism of fretting fatigue crack initiation and propagation

fretting fatigue strength or life in this low stress region, the precise estimation of the fretting wear progress is indispensable. So in this section, we present the estimation method of wear extension and then present the crack propagation estimation method considering wear extension.

Figure 9.5 shows the flow of fretting fatigue life analysis considering the extension of fretting wear. First the fretting wear amount is estimated using the contact pressure and relative slippage under each loading condition. Next, the shapes of the contact

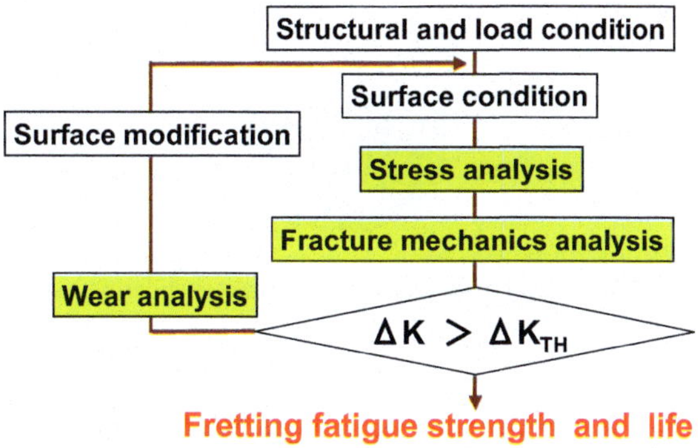

Fig. 9.5 Fretting fatigue process and flowchart for analysis of fretting fatigue life [14–17]

surfaces are modified following the fretting wear amount. Then fretting crack extension or arrest evaluation are performed using fracture mechanics; if the operating ΔK is higher than the threshold stress intensity factor range $\Delta K_{th,}$ we can estimate this load cycle as fretting life, and if the operating ΔK is lower than the threshold stress intensity factor range $\Delta K_{th,}$ fretting wear amount is estimated using new contact pressure, new relative slippage, and repeat these processes until operating ΔK reaches to the threshold stress intensity factor range ΔK_{th}.

(A) **Crack Initiation Evaluation Using Stress Singular Field Parameters**

Here we describe the prediction of fretting micro-crack initiation using stress singularity parameters and the optimization of the contact edge structure using these parameters.

In general, the stress distribution near the contact edge, as shown in Fig. 9.6, is represented by two stress singularity parameters H and λ. The order of stress singularity λ is determined by the elastic modulus of each material, the angle of the contact edge, and the friction coefficient [8–10]. Examples of stress analysis results with the simple fretting model of Fig. 9.7 are shown in Fig. 9.8. The intensity of stress singularity H is obtained by best-fitting the stress distribution near the contact edge to Eq. (9.3) below for contact edge angles of 90°, 80°, 60°, and 45°, respectively, which are shown in the figure.

$$\sigma = \frac{H}{r^{\lambda}} \tag{9.3}$$

By comparing these stress singularity parameters obtained in this way with the crack initiation limit criteria (Fig. 9.9; obtained from the fatigue limit σ_{w0} and fracture threshold stress intensity factor range (ΔK_{th}) for the material), the fretting crack initiation strength is obtained as shown in Fig. 9.10. This figure shows that the fretting fatigue crack initiation strength can be increased when the contact edge angle is less than $\pi/3$. Therefore, from the viewpoint of fretting fatigue strength, the shape of the contact edge on the right side of Fig. 9.11 is more appropriate than the shape on the left side of Fig. 9.11.

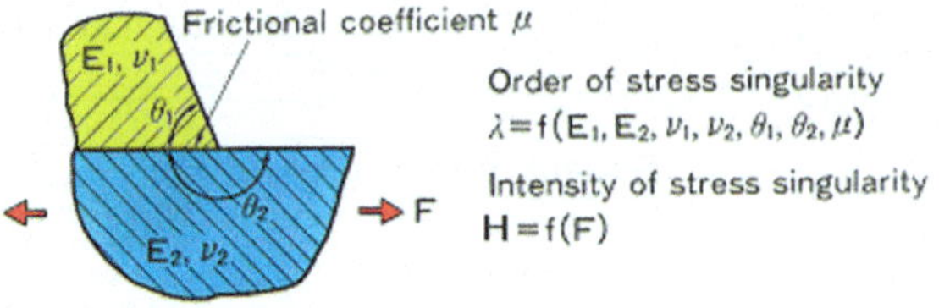

Fig. 9.6 Contact edge geometry and stress singularity parameters

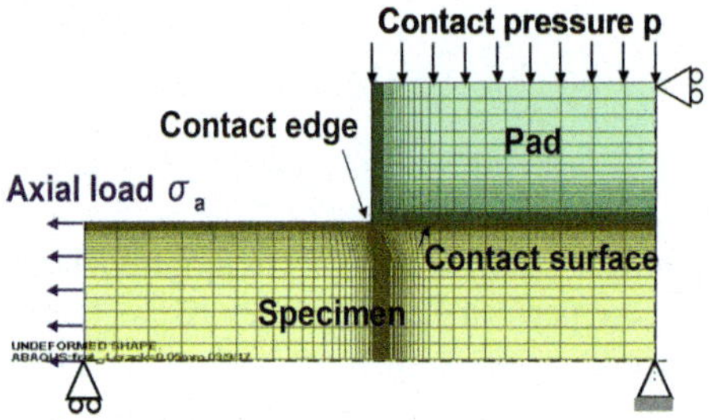

Fig. 9.7 FEM model for fretting fatigue crack initiation analysis

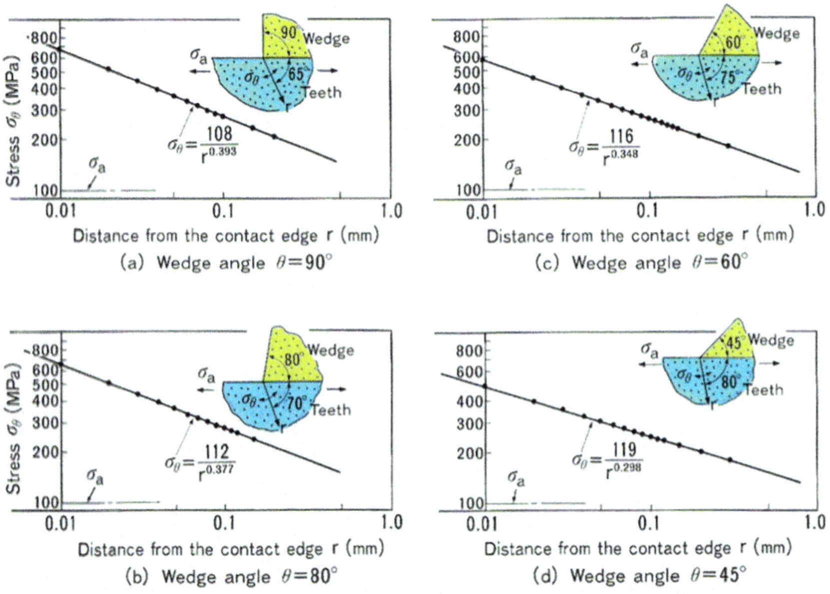

Fig. 9.8 Stress distributions near contact edges

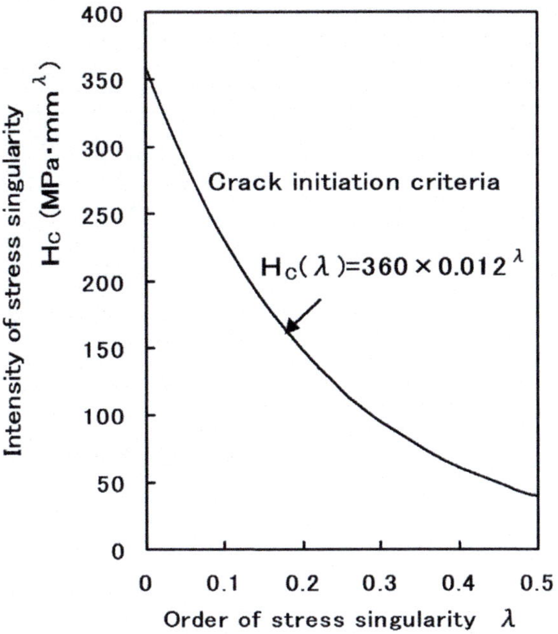

Fig. 9.9 Fretting fatigue crack initiation criteria (material Ni–Mo–V steel)

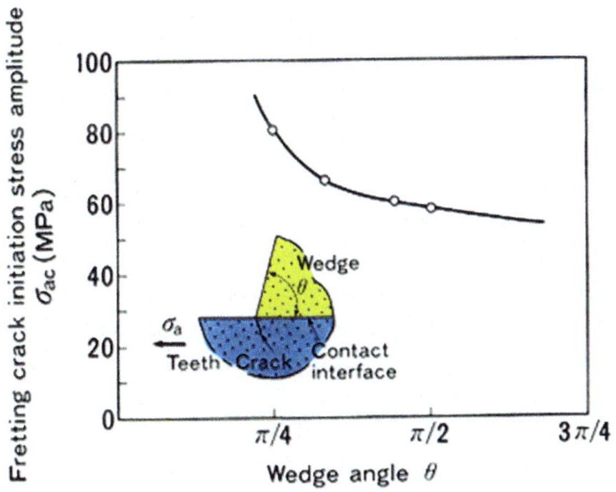

Fig. 9.10 Contact edge angle versus fretting fatigue crack initiation strength

(B) **Crack Propagation Estimation Considering Wear (High-Cycle Fatigue)** [14–17]

Next, we predict the propagation of micro-cracks at this contact edge as the wear progresses.

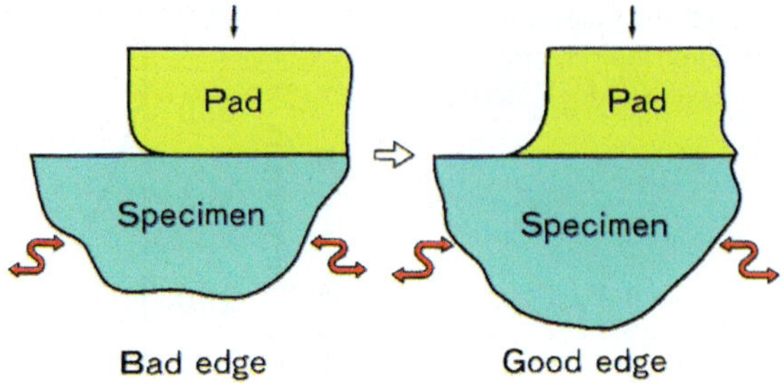

Fig. 9.11 Recommended shape of contact edge

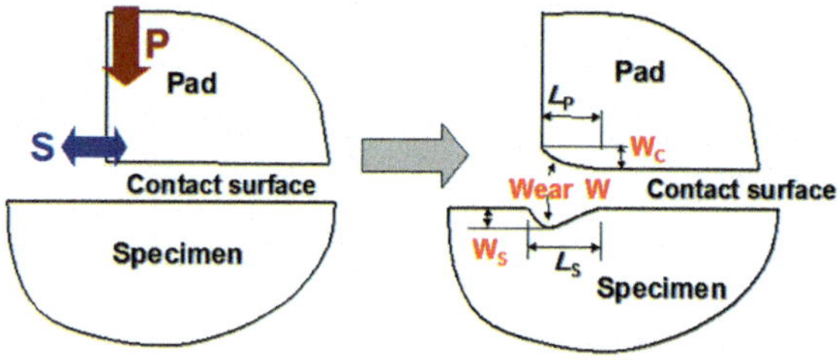

Fig. 9.12 Wear analysis using contact surface pressure and relative slip [14–17]

(B-1) **Wear Analysis**

According to the classical Archard's equation, contact surface wear is indicated as follows.

$$W = K \times P \times S \tag{9.4}$$

where W is the amount of wear, P is the contact pressure, S is the relative slip, and K is the wear coefficient. (See Fig. 9.12.)

FEM stress and deformation analyses are performed using the fretting model as shown in Fig. 9.7, and the wear progression W is predicted using the contact pressure P and relative sliding amount S obtained from these analyses and Eq. (9.5). From a comparison of the predicted amount of wear progression calculated in this way and the measured amount of fretting wear progression during the fretting fatigue test (an example is shown in Fig. 9.13), a wear coefficient K of 1.15×10^{-10} [mm^2/N] was obtained (material Ni–Cr–Mo steel).

Figure 9.14 shows the results of the deformation analysis of the specimen and pad under initial conditions (no wear). Using these results, the relative slip S of the specimen and pad at a half-load cycle (−50 to +50 MPa) is as shown in Fig. 9.15.

The wear amount W during half-load cycle (−50 to +50 MPa) (Fig. 9.16) can be obtained by the product of the contact pressure P, the relative slip amount S, and the wear coefficient K, as shown in Fig. 9.17.

The contact surface geometry is modified using this wear amount, W, and the stress and deformation analyses are repeated according to the flowchart in Fig. 9.5. Figures 9.18 and 9.19 show the calculated results of contact surface pressure P and

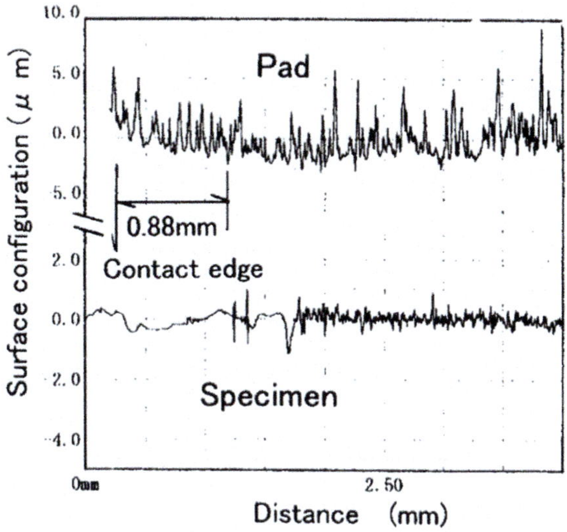

Fig. 9.13 Measured results of fretting wear (σ_a = 110 MPa, N = 2 × 10^7 (material Ni–Cr–Mo steel)) [14–17]

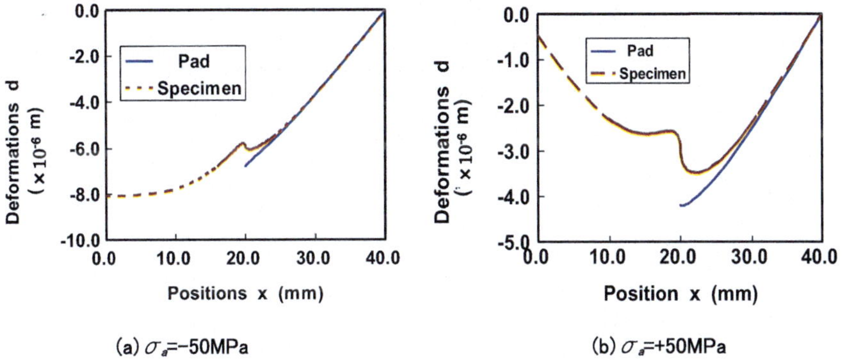

Fig. 9.14 Deformation of specimen and pad on contact surface for each load [14–17]

wear progress *W* during stress amplitude of 62 MPa and number of load cycles of 3.6 × 10^7 and 1.6 × 10^8.

(B-2) **Fracture Mechanics and Fatigue Life Analysis**

The progression of wear under various loading conditions was evaluated using the aforementioned method, and a fracture mechanics analysis was performed based on the stress analysis results for each wear condition. Figure 9.20 shows the analytical results of stress intensity factors for each axial stress σ_a at maximum wear of 0 (initial

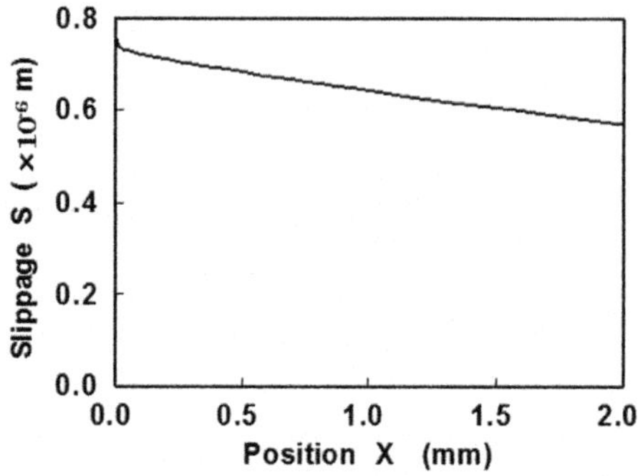

Fig. 9.15 Relative slip during half-load cycle on contact surface (no wear, −50 to +50 MPa) [14–17]

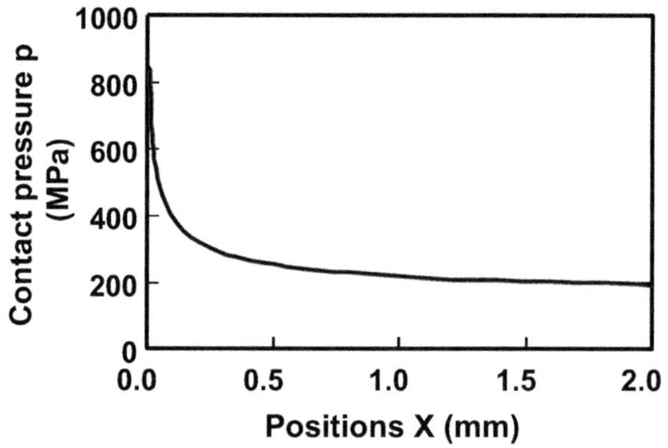

Fig. 9.16 Contact pressure near contact edge (no wear) [14–17]

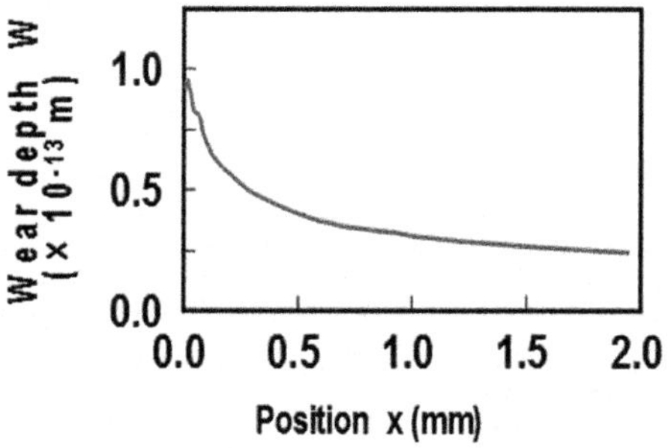

Fig. 9.17 Wear in half-load cycles (no wear, −50 to +50 MPa) [14–17]

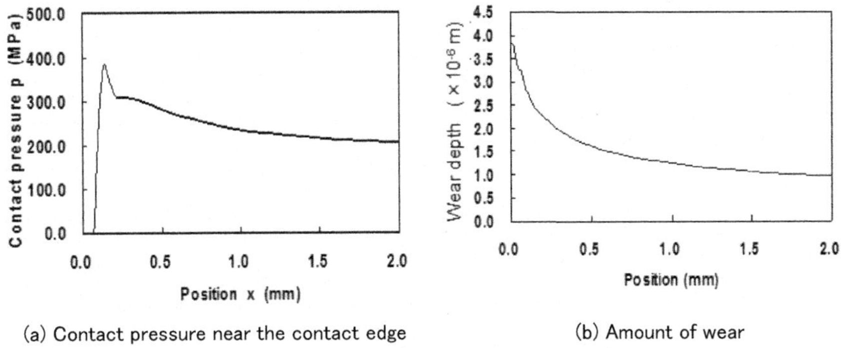

Fig. 9.18 Analysis of contact pressure and wear during fretting fatigue ($\sigma_a = 62$ MPa, $N = 3.6 \times 10^7$) [14–17]

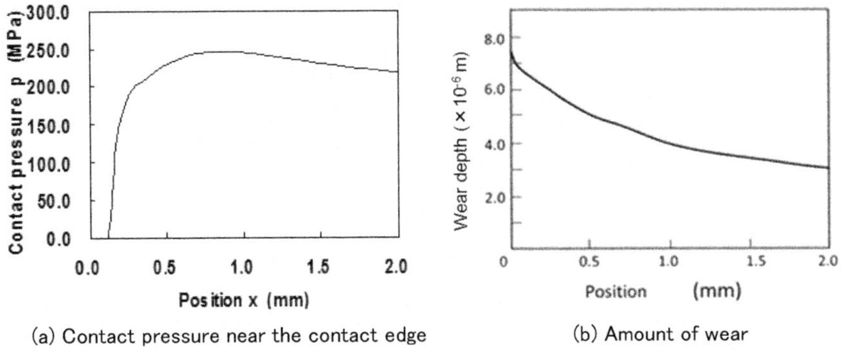

Fig. 9.19 Analysis of contact pressure wear during fretting fatigue ($\sigma_a = 62$ MPa, $N = 1.6 \times 10^8$) [14–17]

condition), 5 and 10 μm. In this case, the crack length was set to 0.25 mm, and this crack was inclined toward the contact side by 30° based on experimental results [10].

These results show that, in the initial state of zero wear, the crack does not open and has a negative stress intensity factor even under an axial tensile stress of 50 MPa, but as wear progresses, the crack tends to open and propagate more easily. Finally, the fretting fatigue life under each loading condition is determined by comparing the stress intensity factor range ΔK under operating load, which takes into account wear as analyzed in the previous section, with the threshold stress intensity factor range ΔK_{th} of the material. As an example, wear and fracture mechanics analyses were performed for axial stress amplitudes σ_a of 62, 110, and 140 MPa, and the fretting fatigue lives were predicted for each axial stress amplitude by comparing the calculated in-service stress intensity factor range ΔK with the threshold stress intensity factor range ΔK_{th}. The threshold stress intensity factor range ΔK_{th} was predicted considering crack length a and stress ratio R as described following equations [11–13].

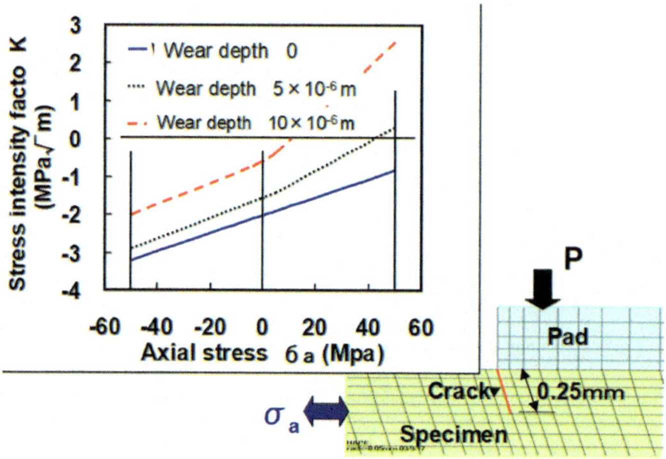

Fig. 9.20 Stress intensity factor analysis results for each wear amount and axial stress σ_a

$$\Delta K_{th(R,a)} = \Delta K_{th(R=0,a=\infty)}(1 - R)^{0.7}\sqrt{\frac{a}{a + a_0}} \quad (R \leq 0) \tag{9.5}$$

$$\Delta K_{th(R,a)} = \Delta K_{th(R=0,a=\infty)}(1 - R)^{\{0.5+\log(\frac{1-R}{10})\}}\sqrt{\frac{a}{a + a_0}} \quad (R > 0) \tag{9.6}$$

$$a_0 = \left(\frac{\Delta K_{th(a=\infty)}}{\Delta\sigma_{w0}}\right)^2 \frac{1}{\pi} \tag{9.7}$$

Figures 9.21, 9.22, and 9.23 show a comparison of the in-service stress intensity factor range ΔK and the threshold stress intensity factor range ΔK_{th} of the material for each of the operating loading conditions. Figure 9.21 shows the results with no wear as a reference. The result shows that the estimated result of fretting fatigue limit without wear is 140 MPa. Naturally, this value matched the prediction results made in the past without considering wear [10].

Figure 9.22 is for an axial stress amplitude σ_a of 110 MPa. Under this loading condition, the maximum depth of wear is 4 μm at a loading cycles of 2×10^7. Under this condition, small cracks generated near the contact edge begin to propagate. Therefore, the fatigue life at this stress amplitude $\sigma_a = 110$ MPa is predicted to be $N_f = 2 \times 10^7$.

Figure 9.23 shows the case of 62 MPa axial stress amplitude. Under this loading condition, the maximum wear depth is 10 μm at 1.6×10^8 cycles of loading, and the micro-cracks near the contact edge begin to propagate under this condition as well. Therefore, similarly, the fatigue life at this stress amplitude $\sigma_a = 62$ MPa is predicted to be $N_f = 1.6 \times 10^8$.

These results are summarized in Fig. 9.24, which shows a fretting fatigue life diagram (based on the number of cycles at the crack propagation limit) considering

Fig. 9.21 Comparison of in-service stress intensity factor range ΔK and threshold stress intensity factor range ΔK_{th} ($\sigma_a =$ 140 MPa, no wear) [14–17]

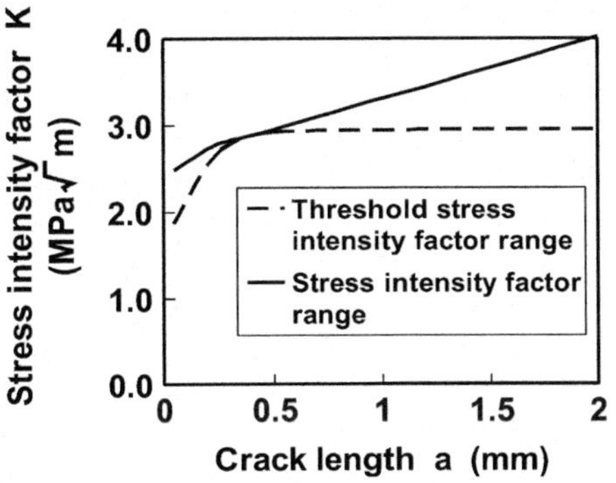

Fig. 9.22 Comparison of in-service stress intensity factor range ΔK and threshold stress intensity factor range ΔK_{th} ($\sigma_a =$ 110 MPa, $N_f = 2 \times 10^7$, wear 4 μm) [14–17]

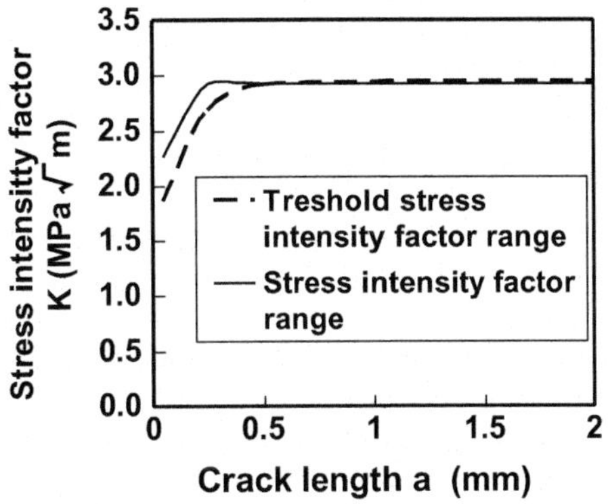

Fig. 9.23 Comparison of in-service stress intensity factor range ΔK and threshold stress intensity factor range ΔK_{th} ($\sigma_a =$ 62 MPa, $N_f = 1.6 \times 10^8$, wear 10 μm) [14–17]

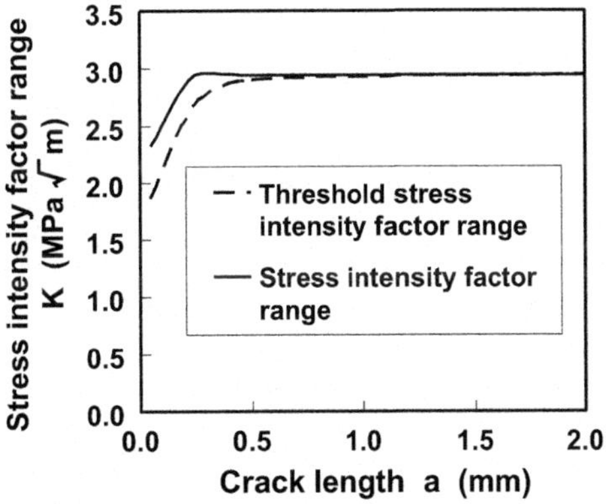

the fretting wear progress. The fatigue strength decreases in the high life region and final fretting fatigue limit $\sigma_{w0f} \approx 60$ MPa, which may well explain why there are many cases of fracture accidents during very-high-cycle regions such as 10^8 and 10^9 under actual fretting conditioning.

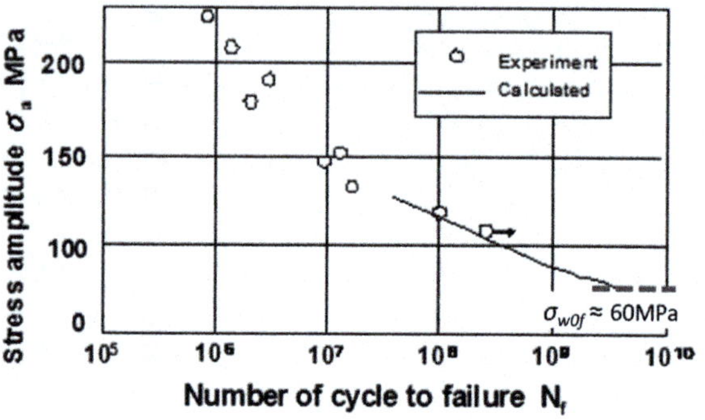

Fig. 9.24 Predicted fretting S–N curves considering wear (crack propagation limit condition) and comparison with experimental results [14–17]

9.3.2 Fretting Fatigues When Subjected to Bending Load

When a thread fastener is subjected to bending load as shown in Fig. 9.25, contact surface separation or slip occurs as shown in the figure depending on the surface pressure and loading conditions, and fretting fatigue cracks occur at the boundary between this separation or slip area and the stick area.

Figure 9.26 shows an example of fretting fatigue failure of a helical gear hub/bolt fastener. Because the helical gear tooth subjected to not only torque but also out-of-plane forces, the bolt fastener is subjected to repeated bending moments as shown in Fig. 9.25.

An example of FEM analysis of the surface pressure distribution of screw fastened bodies under bending moment loading is shown in Fig. 9.27. It shows the contact pressure distribution between the fastened bodies for each loading steps. In this case,

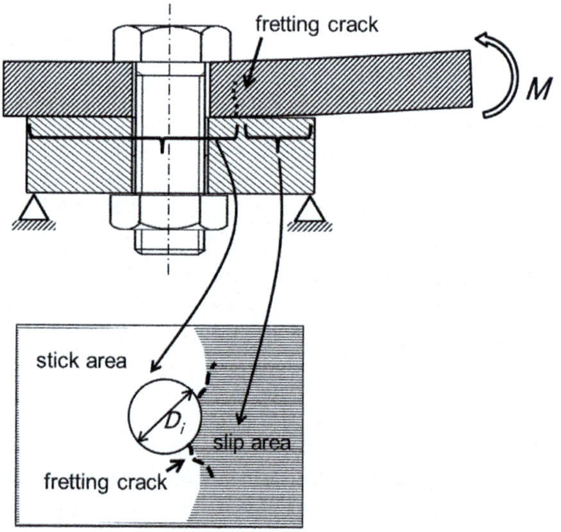

Fig. 9.25 Fretting damage of thread fastener under bending load

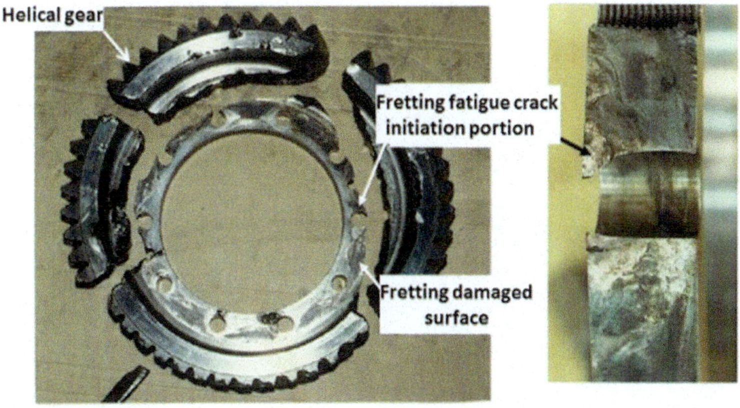

Fig. 9.26 Close-up view of a helical gear screw fastener with fretting fatigue failure and its fatigue crack initiation zone

fastened disk thickness is 6.0 mm and fastening bolt is M10. As the load moment increases, the contact pressure on the tensile side decreases, and in Fig. 9.27b, c, the contact pressure becomes zero and the contact surfaces separate.

The magnitude of the contact pressure on the compressive side near the separation is found to increase with increasing moment.

Figure 9.28 shows the results of a penetration test observation for a fretting fatigue crack in the helical gear screw fastening area. It can be clearly seen that the fretting fatigue cracks are generated at the boundary of the fixed area, which is subjected to high contact pressure.

This situation resulted in the breakage of the trailer hub and the Yurikamome hub, as shown in Sect. 2.7 of the case study.

Please refer to Chap. 10, Sect. 10.4, for a simplified strength evaluation method for such contact edges.

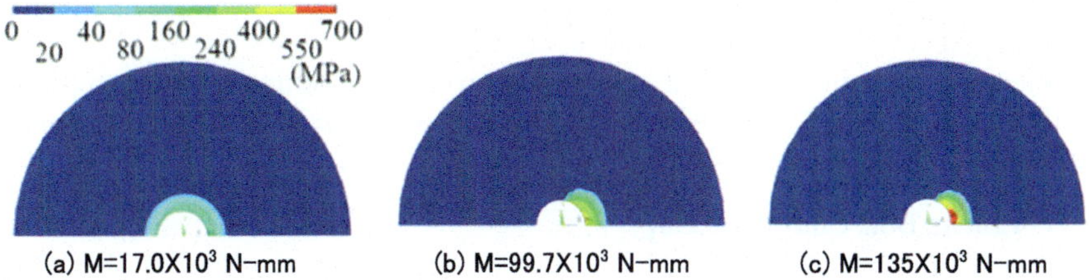

Fig. 9.27 Change in contact pressure distribution of disk/screw fastener under bending loading

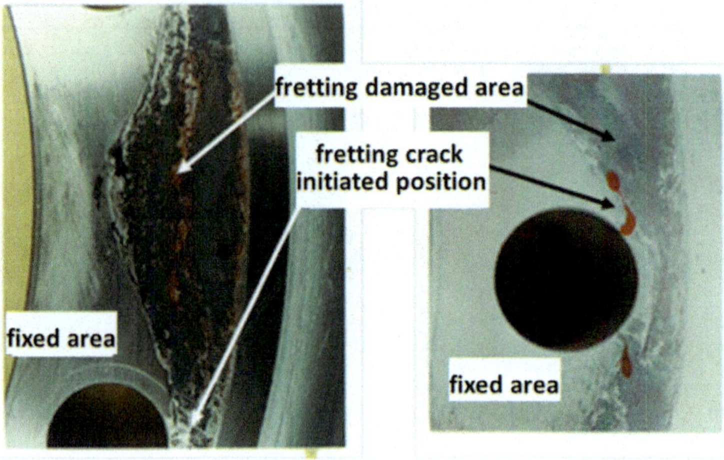

Fig. 9.28 Fretting damage area and fatigue crack initiation site of the helical gear screw fastener

9.4 Fretting Fatigue in Bolts

9.4.1 In the Case of Subjected to Eccentric External Load (Connecting Rod Bolt) [18]

The connecting-rod section of an automobile may be subjected to eccentric external forces as shown in Fig. 9.29, and the bolt and bolt hole may be partially interference fitted. In such cases, bending deformation of the bolt causes the high contact pressure at interference-fit end, which may cause fretting fatigue. In this section, we evaluate the fretting fatigue strength and life in such cases using the mechanical analysis under eccentric external load described in Chap. 3, Sect. 3.4, and the fretting fatigue strength and life analysis method described in previous Sect. 9.3 and compare them with the experimental results of Sakai et al. [18].

First, as shown in Fig. 9.29, the connecting-rod bolt fastening section is replaced with a simple prismatic structure subjected to eccentric external load for mechanical analysis. Each parameter is obtained as follows.

(A) **Linear Bolt Compliance δ_b**

$$\begin{aligned}\delta_b &= \frac{0.5d}{E_b\frac{\pi}{4}d^2} + \frac{l_g}{E_b\frac{\pi}{4}d_g^2} + \frac{l_s}{E_b\frac{\pi}{4}d_3^2} + \frac{0.5d}{E_b\frac{\pi}{4}d_3^2} + \frac{0.4d}{E_b\frac{\pi}{4}d^2}\\ &= \frac{9+26}{210\times 10^3\times\frac{\pi 10.3^2}{4}} + \frac{20+5}{210\times 10^3\times\frac{\pi 9.1^2}{4}}\\ &= 2.0\times 10^{-6} + 1.83\times 10^{-6} = 3.83\times 10^{-6}(\text{mm/N}) \end{aligned} \tag{9.8}$$

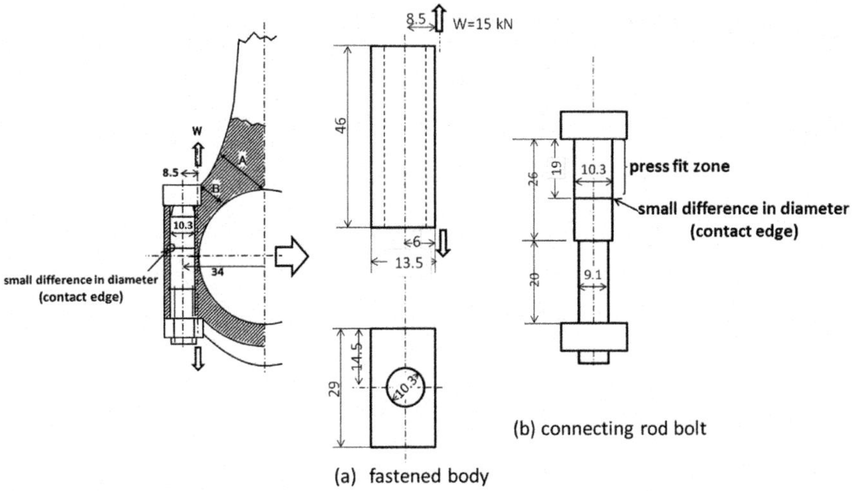

Fig. 9.29 Modeling of connecting-rod bolt fastening part [18]

Here,

Bolt large-diameter area length $l_g = 26$ mm.

Bolt thread area length $l_s = 20$ mm.

(B) **Linear Bolt Bending Compliance**

$$\begin{aligned}\beta_b &= \frac{0.5d}{E_b\frac{\pi}{64}d^4} + \frac{l_g}{E_b\frac{\pi}{64}d_g^4} + \frac{l_s}{E_b\frac{\pi}{64}d_3^4} + \frac{0.5d}{E_b\frac{\pi}{64}d_3^4} + \frac{0.4d}{E_b\frac{\pi}{64}d^4}\\ &= \frac{9+26}{210\times10^3\times\frac{\pi 10.3^4}{64}} + \frac{20+5}{210\times10^3\times\frac{\pi 9.1^4}{64}}\\ &= 3.02\times10^{-7} + 3.54\times10^{-7} = 6.58\times10^{-7}(1/\text{Nmm})\end{aligned} \tag{9.9}$$

(C) **Linear Fastened Body Compliance δ_c**

$$\begin{aligned}\delta_c &= \frac{l}{E_c A_c}\\ &= \frac{46}{210\times10^3\times\left(13.5\times29-\frac{\pi 10.3^2}{4}\right)}\\ &= 0.711\times10^{-6}\ (\text{mm/N})\end{aligned} \tag{9.10}$$

Here,

Fastened body width $b = 29$ mm.

Fastened body depth $a = 13.5$ mm.

(D) **Linear Fastened Body Bending Compliance β_c**

$$\beta = \frac{l_f}{E_c I_c} = \frac{12 \times 46}{210 \times 10^3 \times \left(13.5^3 \times 29\right)} = 3.68 \times 10^{-8}(1/\text{Nmm}) \quad (9.11)$$

And load factor Φ_e can be calculated as follows:

$$\Phi_e = \frac{F_b}{W} = \frac{\delta_{eW}}{\delta_b + \delta_{sF}} = \frac{\delta_c + \frac{s_B e_B l_f}{E_c I_T}}{\delta_b + \delta_c + \frac{s_B^2 l_f}{E_c I_T}} = 0.212 \quad (9.12)$$

Here,
Load position eccentricity $e_B = 8.5 + 0.75 = 9.25$ mm.
Bolt position eccentricity $s_B = 0.75$ mm.

(E) **One-Side Separation Limit W_s**

$$W_s = \frac{F_f}{\frac{(e_B - S_B)u}{\frac{I_T}{A_{Tc}} + S_B u} + (1 - \Phi_e)} = \frac{F_f}{3.14} \quad (9.13)$$

Here, u is the distance $a/2$ from the centroid to the end face of the object to be fastened, which is 6.75 mm. A_{Tc} is the contact area of the joint surface of the fastened body (excluding bolt holes), and I_T is the moment of inertia of the contact surface (including bolt holes).

(F) **Full Separation Limit**

$$F_f + F_b = \frac{\frac{a}{2} + e_B}{\frac{a}{2} + s_B} W = \frac{6.75 + 9.25}{6.75 + 0.75} W = 2.13W \quad (9.14)$$

Regarding the nonlinear behavior after one-sided separation, the new nonlinear parameters such as connected body's compliance δ'_c, the connected body's bending compliance β'_c, the internal force coefficient Φ'_e, and one-sided separation limit W'_s W'_s can be obtained by using the separation region length Δ in the contact surface as follows, u in each formula to $u - \Delta$, e_B to $e_B + \Delta/2$, and s_B to $s_B + \Delta/2$.

Therefore, the one-sided separation condition, i.e., the nonlinear W–F relationship is derived as follows.

$$W = \frac{F_f}{\frac{(e_B - S_B)(u - \Delta)}{\frac{I'_T}{A_{Tc}} + (S_B + \Delta/2)(u - \Delta)} + \left(1 - \Phi'_e\right)} \quad (9.15)$$

Figure 9.30 shows the W–F relationship obtained using these analytical formulas.

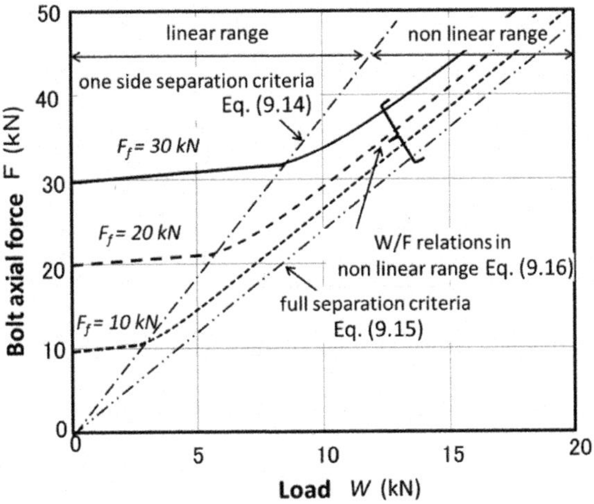

Fig. 9.30 Analytical results of W–F relationships

(G) **Fretting Fatigue Life Analysis of Connecting-Rod Bolt**

Furthermore, the fretting fatigue life at the contact edge of the press-fitting is predicted using the amplitude of the total combined stress $\sigma_{FM} = \sigma_F + \sigma_M$ on the outer surface of the cylindrical part of the bolt. Hear, σ_F is the axial stress and σ_M is the bending stress.

The axial stress variation σ_F of the bolt is obtained by the following equation.

$$\sigma_F = \frac{4F_b}{\pi {d_3}^2} = \frac{4_e W}{\pi {d_3}^2} \tag{9.16}$$

Here, diameter of bolt body $d_3 = 10.3$ mm

Also, the bending stress variation σ_M of the bolt is obtained by the following equation.

$$\sigma_M = \frac{32M_b}{\pi d_3^3} \tag{9.17}$$

Here, the bending moment M_b applied to the bolt is given by the following equation.

$$M_b = \frac{\left(e_B + \frac{\Delta}{2} - \left(s_B + \frac{\Delta}{2}\right)_e\right)W}{1 + \frac{\beta_b}{\beta_c}} \tag{9.18}$$

Figure 9.31 shows the calculated σ_{FM}–W relationship. The fretting fatigue life under cyclic fluctuating loads $W = 15$ and 10 kN per connecting-rod bolt joint under various initial fastening forces can be predicted from the fretting S–N curve in Fig. 9.32. The predicted results are shown in Fig. 9.33 by solid and dashed lines, and compared with experimental results shown by solid circle and dashed line circle.

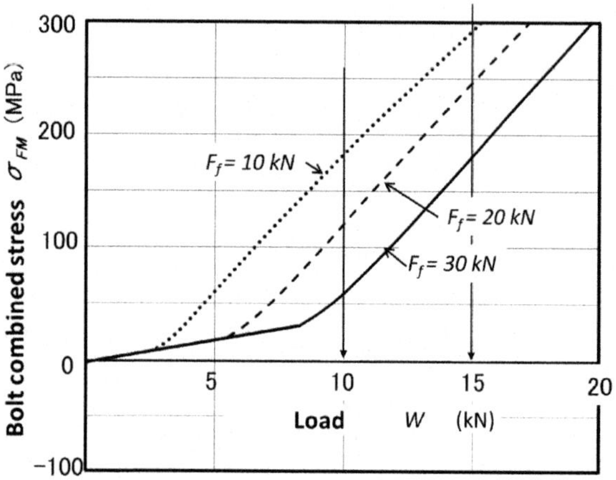

Fig. 9.31 Analytical results of total combined stress σ_{FM} of one connecting-rod bolt

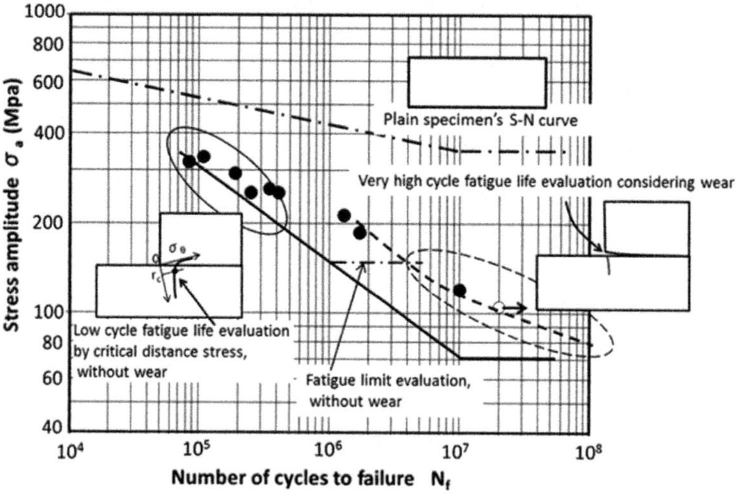

Fig. 9.32 Full-life range fretting S–N curve

A comparison of these prediction results with the experimental results of Sakai et al. [18] is also shown in Fig. 9.33.

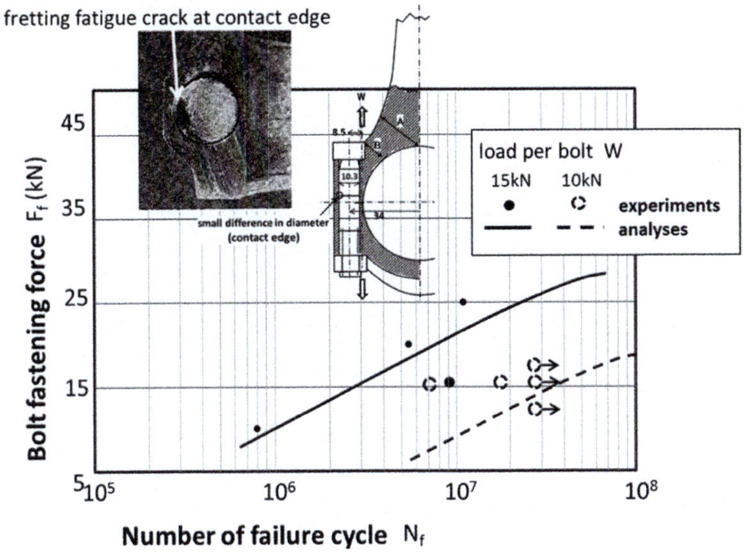

Fig. 9.33 Comparison of analytical and experimental results of fretting fatigue strength and life on connecting-rod bolt whole load amplitude $\Delta W = 20, 30$ kN (10, 15 kN per 1 bolt)

9.4.2 In the Case of a Large Structure Assembled from Parts with Incorrect Initial Dimensions (Large Wind Turbine Blade Fastening Bolts) [19]

(A) **Conditions of Occurrence**

When large-scale structures such as bridges and wind turbines are fastened with a large number of bolts, sometimes the threaded holes and bolt bodies come into strong contact due to inaccurate pitches of the both threaded holes in the fastened bodies, and fretting fatigue may occur in the fastening bolts due to the high local contact pressure. For example, in the large wind turbine system shown in Fig. 9.34, many large-diameter bolts are used in the blade/hub fastening area. As shown in Fig. 9.35, the misalignment (Δ) of the threaded holes between the fastened body A (blade side) and B (hub side) may cause fretting damages depending on the conditions, which may significantly reduce the fatigue strength. In particular, if the bolt is forcibly tapped without regard to the misalignment of each hole and a large contact pressure is generated at the contact point, pseudo-welding can easily occur at contact point and fretting damage such as small cracks can easily be generated.

(B) **Prevention Technology**

A typical example of such bolt shank fretting fatigue damage is described in Sect. 1.8, "Fretting fatigue failure of roller coaster fastening bolts", and countermeasures are also discussed in detail in that section. The groove moves the contact

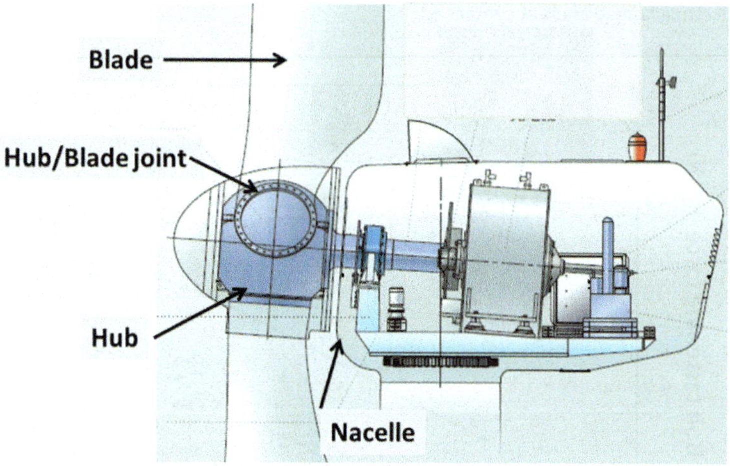

Fig. 9.34 Blade/hub fastening area in large wind turbine system [19]

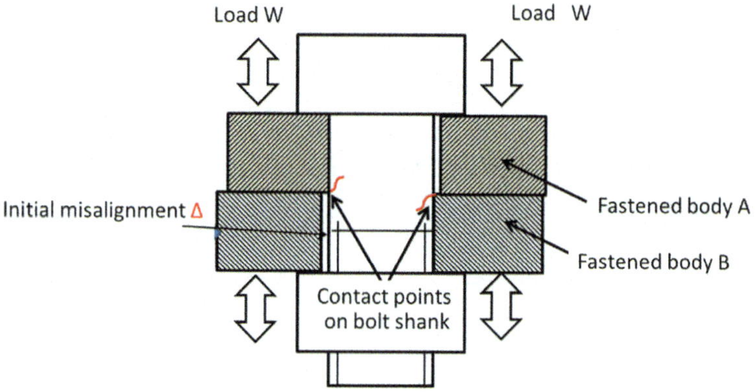

Fig. 9.35 Fretting fatigue damages due to contact between bolt shank and misaligned inner surfaces of threaded holes

edge to the corner of the groove, and the tangential force, which is the driving force for fretting fatigue crack initiation, is significantly reduced compared to the flat contact edge. However, there is a fear that the groove may reduce strength due to stress concentration at the bottom of the groove, and we confirmed the effectiveness of this measure through actual experiments and analysis (Fig. 9.36).

Figure 9.37 shows the test specimens for the fretting fatigue strength tests of bolts with such a stress-relief groove (b) and simple fretting model (a) [20]. The simple fretting model (without groove) was a model in which bridge pads were pressed from both sides of a round bar specimen by four bolts. The pressing load of the pad was set to 1.5 kN. In addition, as a fretting fatigue strength improvement model, as shown in the (b), a stress release groove structure was used in which an annular groove with

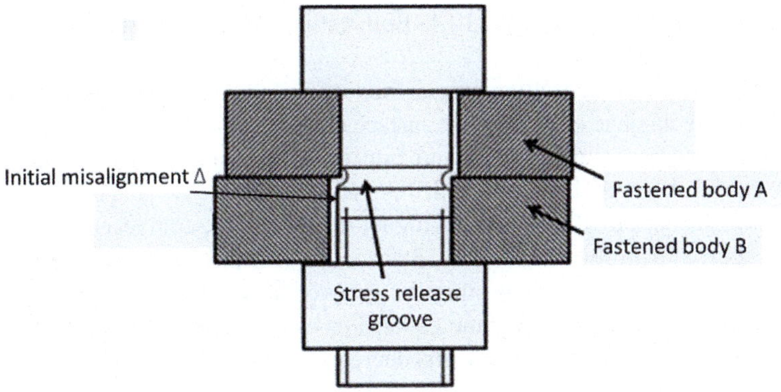

Fig. 9.36 Example of countermeasure by stress-relief groove (groove) [20]

radius *R* of 2.5 mm and a depth *d* of 1 mm near the contact edges on the specimen surface. Cyclic axial load was applied in this state. The material of specimens is SCM435.

The experimental results for both fretting tests are plotted by ▲ for grooved fretting specimens and ◆ for simple fretting specimens in Fig. 9.38. And it is known that the fretting fatigue limit can be improved by a factor of 1.5 with stress-relief groove comparing with plain fretting. In this S–N curves, the dashed line and dash-dot line are analytical results by critical distance stress theory, which will be explained

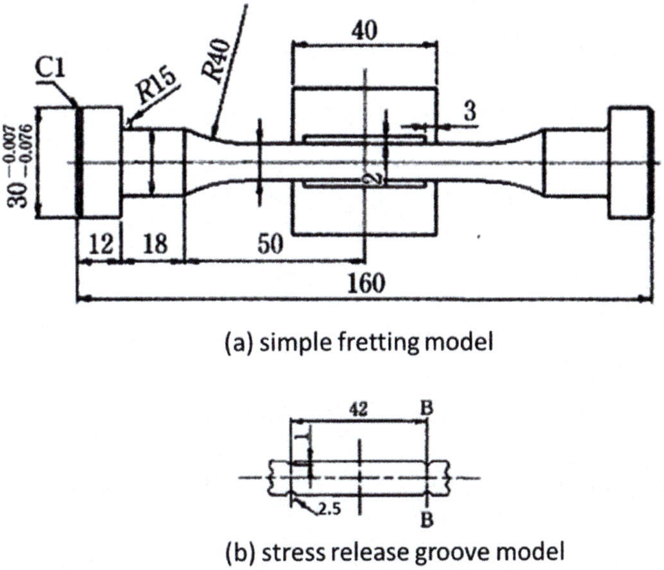

Fig. 9.37 Fretting fatigue strength of round bar material [21]

in detail later in Chap. 10, Sect. 10.4. Both estimated results coincided well with experimental results.

Figure 9.39 shows the fretting fatigue fracture surface of simple fretting specimen (a) and fretting damage of the contact surface on bridge pad (b). The fretting fatigue crack initiation from contact edge and fretting wear damage on the local contact surface can be observed. However, with this stress release groove shape ($R = 2.5$ mm), the stress at the bottom of the groove is higher than that at the contact edge and fatigue failure occurred at groove bottom. So, it is conceivable that the fretting fatigue limit can be further improved by adjusting the groove shape so that contact edge stress and groove bottom stresses become equal. For example, by increasing the groove bottom R from 2.5 to 3 mm or 4 mm and decrease groove bottom stress, the optimum groove shape can be obtained.

Then, we will show the optimizing example of stress release groove shape in the case of fretting fatigue under planar contact conditions with Ni–Cr–V steel as shown in Fig. 9.40. In this model a stress-relief grooves with the groove R is 5 mm, and the depth d is 1 mm and 2 mm are provided at the contact edge. And fatigue limit of each groove shape model is predicted as shown in Fig. 9.41. In this figure the predicted fretting fatigue limit at the contact edge is marked with ○ and the fatigue limit at the groove bottom is marked with ●, and failure will occur at the lower of these limits and the optimum groove depth can be estimated about $d = 1.5$ mm. The fatigue test results for the case of groove depth $d = 1$ mm are shown in Fig. 9.42 by mark Δ, which agrees well with predicted results by critical distance stress analysis [21] shown by dash-dot line. And by comparing with the experimental and analytical

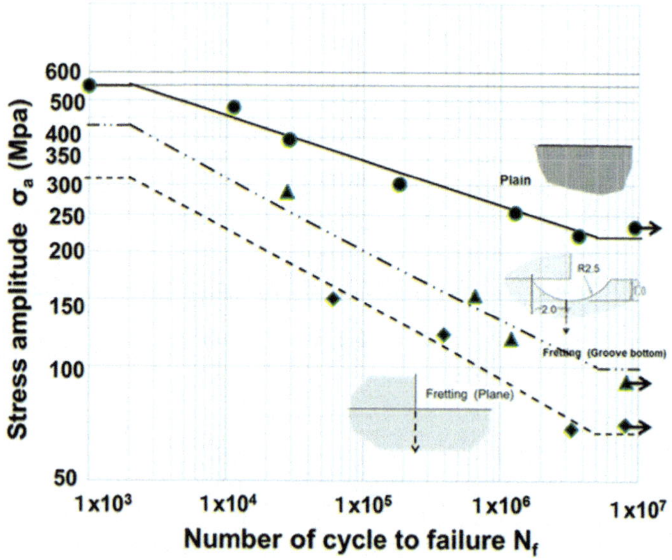

Fig. 9.38 Comparison of the estimated S–N curves of each fretting model with experimental results [21]

results for simple fretting model shown by ○ and dashed line, it can be seen that the fretting fatigue limit increases up to about 1.8 times. And as shown in Fig. 9.43, in the test with a groove depth of 1 mm, failure occurred at the contact edge, and in the test with a groove depth of 2 mm, failure occurred at the groove bottom, which also agrees well with the estimated results in Fig. 9.41.

9.5 Inspection and Maintenances for Fretting Fatigue [22]

In the case of fretting fatigue, as described above, the fretting crack propagation limit decreases with the wear progress at the contact edge, and therefore, the fatigue limit is related to the life time, so making fatigue limit design difficult. In other words, safe life design (design based on $N_f = 10^7$ fatigue limit and safety factor S_f) becomes difficult.

On the other hand, fretting fatigue cracks propagate very slowly as it follows the progress of wear, so this crack propagation characteristic is very suitable for damage tolerance design in which during machine operating time crack initiation and propagation are being monitored. Therefore, this system has been in operation on Japan's Shinkansen train axles for a long time.

And we will introduce it here [22]. As shown in Fig. 9.44, the axle of the Shinkansen has a 60 mm diameter hole in the center, which is used not only for axle weight reduction but also for ultrasonic testing for fretting fatigue cracks in the axle/wheel mating area. Although the axle is the most important part of a high-speed

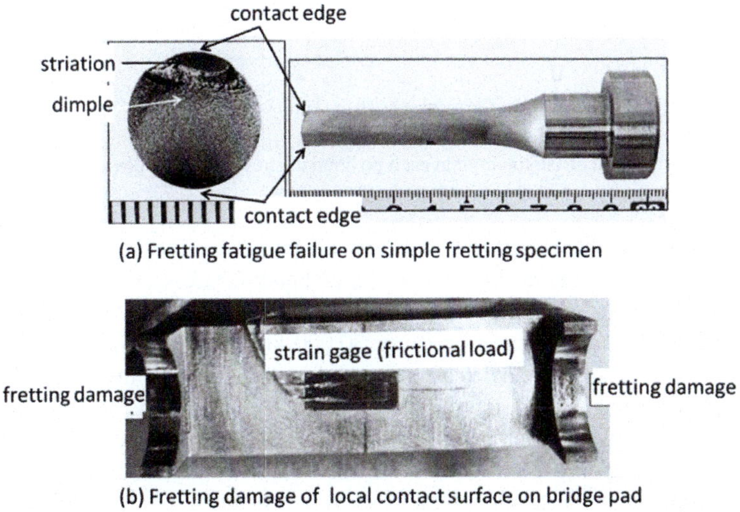

Fig. 9.39 Fretting fatigue fracture surface of the simple fretting specimen and fretting damage on the contact surface of the bridge pad [21]

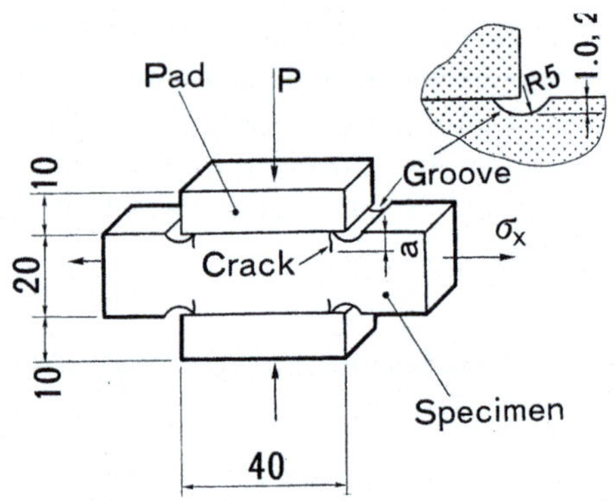

Fig. 9.40 Shape and dimensions of stress release grooved plane contact fretting model [21]

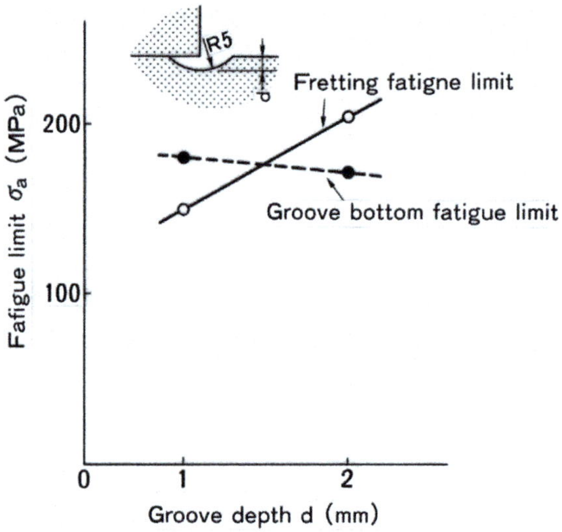

Fig. 9.41 Comparison of fatigue limit at each position in stress release groove models [21]

train that supports safe running, the axle/wheel mating area is used under very severe conditions with high surface pressure and high bending stress and is therefore prone to fretting fatigue cracks, as described above. Therefore, Shinkansen train axles are subjected to ultrasonic inspection every 30,000 km while assembled in the train. As shown in Fig. 9.44, the axles are subjected to oblique angle inspection from the inside of the boring. Furthermore, every 300,000 km, the axle is removed from the bogie at the factory and each part of the axle is inspected precisely. In particular, the axle/wheel mating area is exposed by shifting or removing the wheel, and magnetic particle inspection is performed. If the crack depth is greater than 0.15 mm, the wheel is repaired or replaced.

In the accident examples in Chap. 2, we introduced fretting damage in bolted joints such as trailer hubs, roller coaster axles, and oil field pipes, and we believe

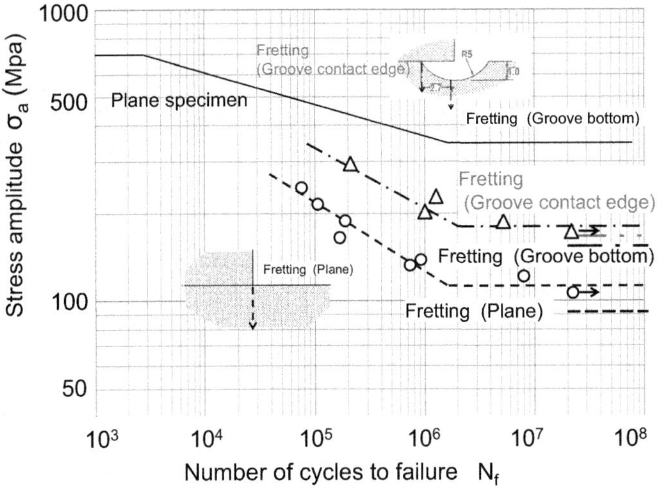

Fig. 9.42 Fretting fatigue test results and prediction results for stress release groove model under plane contact conditions [21]

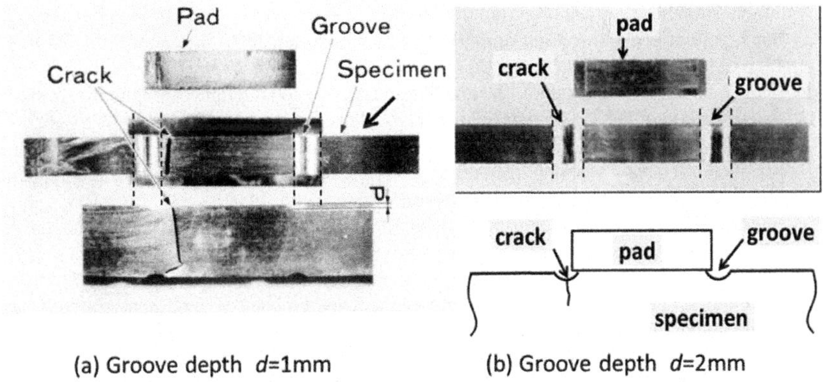

Fig. 9.43 Observation results of failure in fretting fatigue test of stress release groove model under plane contact conditions [21]

that even in these areas, periodic crack detection can be used to take advantage of the very slow growth rate of fretting fatigue cracks, as in the case of bullet train axles.

This technology to ensure consistent reliability throughout design and maintenance based on an accurate understanding of damage mechanisms and processes is likely to become important in future.

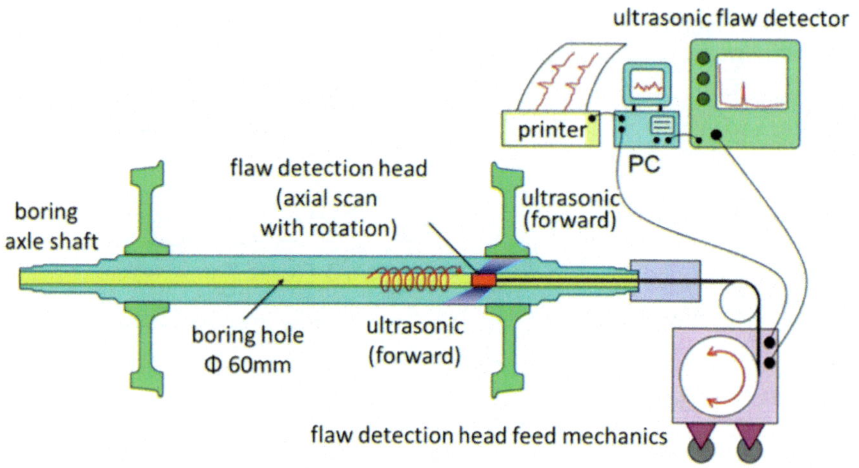

Fig. 9.44 Configuration of automatic boring axle flaw detector [22]

References

1. Gassner, E., The value of surface-protective media against fretting corrosion on the basis of fatigue strength tests, Laboratorium fur Betriebsfestigkeit TM19/67, 1967.
2. Buch, A., Fatigue and fretting of pin-lug joints with and without interference fit, Wear, 1977, 43, p. 9.
3. Hattori, T., Kawai, S., Okamoto, N. and Sonobe, T., Torsional fatigue strength of a shrink-fitted shaft, Bulletin of the JSME, 1981, 24, 197, p. 1893.
4. Cornelius, E. A. and Contag, D., Die Festigkeits-minderung von Wellen unter dem Einfluβ von Wellen-Naben- Verbindungen durch Lotung, Nut und Paβ feder, Kerbverzahnungen und Keilprofile bei wechselnder Drehung, Konstruktion, 1962, 14, 9, p. 337.
5. Hattori, T., Sakata, S. and Ohnishi, H., Slipping behavior and fretting fatigue in the disk/blade dovetail region, Proceedings, 1983 Tokyo Int. Gas Turbine Cong.
6. Johnson, R. L. and Bill, R. C., Fretting in aircraft turbine engines, NASA TM X-71606, 1974.
7. Okamoto N. and Nakazawa, M., Finite element incremental contact analysis with various frictional conditions, Int. 377.
8. Hattori, T., Sakata, H. and Watanabe, T., A stress singularity parameter approach for evaluating adhesive and fretting strength, ASME Book No. G00485 , MD-vol. 6, 1988, p. 43.
9. Hattori, T. and Nakamura, N., Fretting fatigue evaluation using stress singularity parameters at contact edges, Fretting Fatigue, ESIS Publication 18, 1994, p. 453.
10. Hattori, T., Nakamura, M. and Watanabe, T., Simulation of fretting fatigue life by using stress singularity parameters and fracture mechanics, Tribology International, 2003, 36, p. 87.
11. King, R. N .and Lindley, T. C., Fretting fatigue in a 3 1/2 Ni-Cr-Mo-V rotor steel, Proc. ICF5, 1980, p. 631.
12. Hattori, T. et al., Fretting fatigue analysis using fracture mechanics, JSME Int. J, Ser. 1, 1988, 31, p. 100.
13. Toshio Hattori, Masayuki Nakamura, Hiroshi Sakata, Takashi Watanabe, Fracture mechanics analysis of fretting fatigue, Transactions of the Japan Society of Mechanical Engineers (Part A), 53–492, (1987), pp. 1500–1507.
14. T. Hattori, M. Yamashita and N. Nishimura, Fretting fatigue strength and life estimation considering the fretting wear process, Proc. the 2nd JSME/ASME Int. Conf. on Materials and Processing 2005 pp.(ICS-01)-1-6 (2005).

15. T. Hattori, M. Yamashita and N. Nishimura, Fretting fatigue strength and life estimation in ultra high cycle region considering the fretting wear process, JSME Int. J., **48**-4, pp. 246–250 (2005).
16. T. Hattori and T. Watanabe, Fretting fatigue strength estimation considering the fretting wear process, Tribology International, **39**, pp. 1100–1105 (2006).
17. T. Hattori, N. Nishimura and M. Yamashita, Fretting fatigue strength and life estimation considering the fretting wear process, Proc. Of the 9th Int. Fatigue Congress 2006, pp. FT531 (2006).
18. Tomotsugu Sakai, Study of loosening characteristics of connecting rod caps, Transactions of the Japan Society of Mechanical Engineers (Part 3), Vol. 43, No. 368, (1977–4), pp. 1454–1561.
19. About the causes and countermeasures of Hibikinada wind power plant damage accident, Rep. No. 2-2, (2024).
20. T. Hattori, Fretting fatigue problems in structural design, Fretting Fatigue (Editors: R. B. Waterhouse and T. C. Lindley), pp. 437–451, Mechanical Engineering Publications (1994).
21. Muhammad Amiruddin Bin Ab Wahab, Niu Jie, Toshio Hattori and Minoru Yamashita, Fatigue Strength and Life Estimation Methods Using Critical Distance Stress Theory, Advanced Materials Research Vol. **694–697** pp. 853–863 (2013).
22. Hiromichi Ishizuka, 30 Years of Shinkansen Train Axles, Journal of the Japan Society of Mechanical Engineers (1992, 2), pp. 125–127.

Chapter 10
Mechanical Analyses and Strength Designs for Fastened Bodies

10.1 Introduction

In general, when designing the strength of bolted joints, attention tends to be paid to bolts rather than the fastened bodies, and this book also has basically discussed the strength design of bolts. However, recently, CFRP, aluminum alloy, titanium alloy, etc., have been used for fastened bodies to reduce weight, and in these cases, the characteristics and behavior of the fastened bodies may affect the load and strength of the bolt itself.

10.2 Bearing Surface Pressure and Embedment or Sink

⇒See Chap. 3 "Mechanical analysis and strength design under axial and bending loads" and Sect. 3.7 "Strength design".

Embedment is an attribute that must be recognized not as damage but as an unavoidable characteristic of fastening through the contact surface between each part. Basically, the empirical prediction formula for the amount of embedment $\boldsymbol{s}$ is as follows.

$$s = 3.29\left(\frac{l_f}{d}\right)^{0.34} \quad (\mu\text{m}) \quad \text{Chapter 3} \tag{3.113}$$

And initial embedment loosening amount $\boldsymbol{F_s}$ can be calculated using the embedment coefficient $Z = 1/(\delta + \delta_{bc})$ as follows.

$$F_s = \frac{1}{\delta_b + \delta_c} = Zs \quad \text{Chapter 3} \tag{3.112}$$

© The Author(s), under exclusive license to Springer Nature Switzerland AG 2025

T. Hattori and T. Naruse, *Compendium for Design and Manufacturing of Bolted Joints*, Solid Mechanics and Its Applications 280,
https://doi.org/10.1007/978-3-031-89628-6_10

Table 10.1 Critical distance for Ni–Mo–V steel [3]

$\Delta\sigma_{w0}$ ($R = 0$) [MPa]	ΔK_{th} ($R = 0$) [MPa m$^{1/2}$]	r_c [mm]	σ_B [MPa]	K_{IC} [MPa m$^{1/2}$]	r_c' [mm]
360	5.5	0.037	705	100	3.2

Unlike the above-mentioned embedment, the problem of sink is a macroscopic decrease in fastening force accompanied by compressive plastic deformation of bearing surface. In order to evaluate and take measures against seat sink, it is basically necessary to calculate the seat pressure p_w between the bolt head or nut and the fastened body shown below and then make sure that this value is lower than the critical surface pressure for each material shown in Table 10.1.

$$p_w = \frac{F_{f\,\max}}{A_w} = \frac{F_f + \Phi_n W_{a\,\max}}{A_w} < P_L \quad \text{Chapter 3} \tag{3.114}$$

In particular, when using CFRP, aluminum alloys, titanium alloys, or other materials with low yield strength for the fastening body for the purpose of reducing the weight of equipment, caution should be taken. As a countermeasure against unsatisfactory results, it is effective to increase the bearing surface area A_w by using a flanged nut or a thick large-diameter washer.

10.3 Mechanical Analyses Around Bolt Holes

In the past, stress analysis and strength evaluation around bolt holes and rivet holes were based only on simple material mechanics of the fastened plates and bolt, ignoring the initial fastening force and internal force coefficient. An example of these simple mechanical analysis case for rivet joint was the accident analysis of the JAL Jumbo jet pressure bulkhead, as shown in Fig. 1.11 in Chap. 1. However, there have been many accidents caused by damage in more complex loading configurations, and there is a need for more advanced mechanical analysis and strength evaluation. As shown in Figs. 1.49 and 1.50 in the accident investigation of the KLM aircraft panel fall accident in Chap. 1, it is now possible to elastomechanically analyze and evaluate damage of a fastened plate subjecting an out-of-plane load. So, in this section we I will explain summarizing these methods, including more advanced analytical techniques.

10.3.1 Transverse Loading Case

As shown in Fig. 10.1, for shear failure of bolts, strength evaluation is based on the following nominal shear stress calculated from the cross-sectional area of the bolt valley bottom.

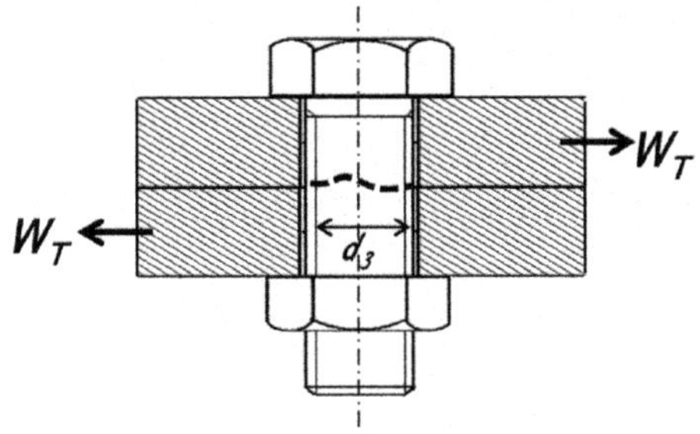

Fig. 10.1 Shearing failure of bolt

$$\tau_b = \frac{W_t}{\frac{\pi d_3^2}{4}} \tag{10.1}$$

The allowable stress τ_{ba} of τ_b is $1/\sqrt{3}$ of the vertical stress fatigue limit σ_{wk} of various bolted materials in Table 2.2 of Chap. 2 for fatigue limit design, and the following equation is used considering also the safety factor S_f.

$$\tau_{ba} \leq \frac{\sigma_{wt}}{\sqrt{3}S_f} \tag{10.2}$$

For shear failure of the plate to be fastened, as shown in Fig. 10.2, strength evaluation is based on the shear stress in the following equation, where e is the distance from the bolt hole to the edge of the plate and t is the plate thickness.

$$\tau_c = \frac{W_t}{2et} \tag{10.3}$$

The allowable stress τ_{ca} of τ_c for fatigue limit design can be estimated using following Eq. (10.4), using the shear stress fatigue limit τ_{w0} of the plate and the safety factor S_f.

$$\tau_{ca} \leq \frac{\tau_{w0}}{S_f} \tag{10.4}$$

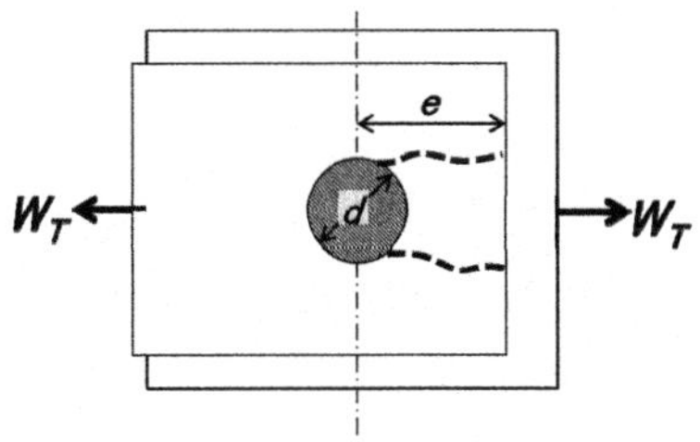

Fig. 10.2 Shearing failure of fastened plate

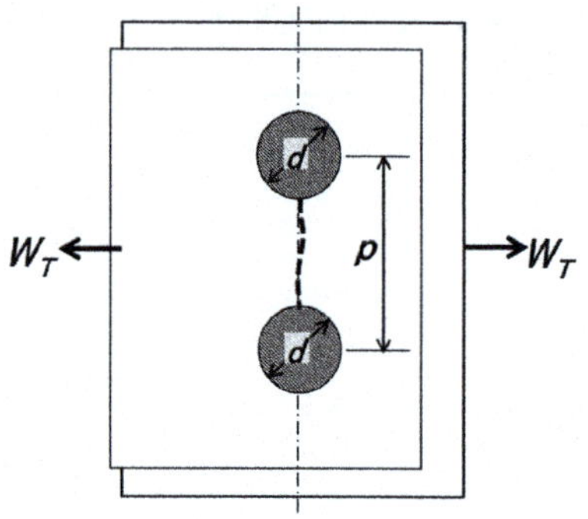

Fig. 10.3 Tension failure between bolt holes of the fastened plate

For the tension failure between bolt holes of the fastened plate as shown in Fig. 10.3, in which the distance between the bolt holes is p and the plate thickness is t, the strength evaluation can be performed using the normal stress in the following equation.

$$\sigma_c = \frac{W_t}{pt} \tag{10.5}$$

The allowable stress σ_{ca} for fatigue limit design ca be calculated using Eq. (10.6), considering the normal stress fatigue limit σ_{w0} of the plate and the safety factor S_f.

$$\sigma_{ca} \leq \frac{\sigma_{w0}}{S_f} \tag{10.6}$$

For crushing of the bolt hole or bolt shaft as shown in Fig. 10.4, contact pressure (nominal compressive stress) on the bolt shaft side or the side of the plate hole is calculated as follows.

$$\sigma_c' = \sigma_b' = \frac{W_t}{dt} \tag{10.7}$$

The design shall be made so that the above compressive stresses are lower than allowable stresses σ_{ca}' and σ_{ba}' which are calculated using yield stress σ_{cy} or σ_{by} of each material and the safety factor S_f as follows.

$$\sigma_c' \leq \sigma_{ca}' = \frac{\sigma_{cy}}{S_f}, \quad \sigma_b' \leq \sigma_{ba}' = \frac{\sigma_{by}}{S_f} \tag{10.8}$$

For the fracture between the bolt hole and the plate edge as shown in Fig. 10.5, the bending stress σ_c'' on the target area e is calculated by using Eq. (10.9), considering length between bolt holes p.

$$\sigma_c'' = \frac{M}{Z} = \frac{2W_t p}{(2e - d)^2 t} \tag{10.9}$$

Fig. 10.4 Crushing failure of bolt side or bolt hole sides of the fastened plate

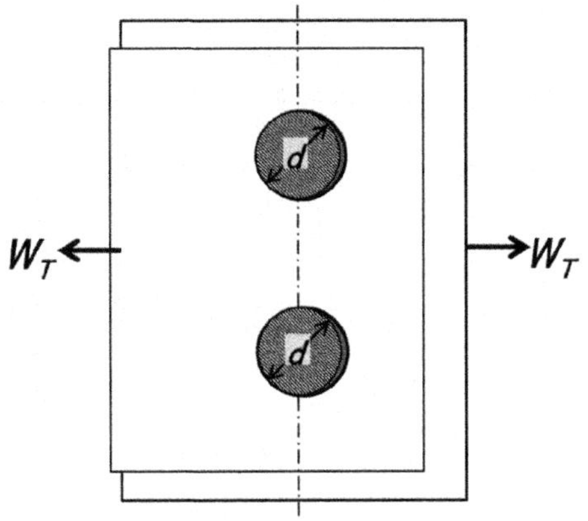

Fig. 10.5 Bending failure between the bolt hole and the plate edge

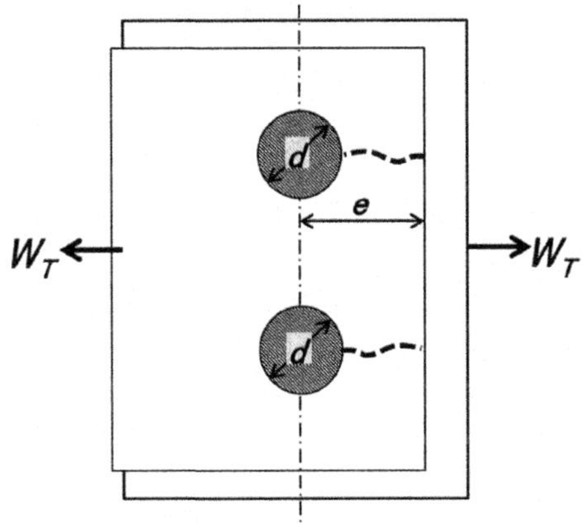

The allowable stress σ''_{ca} for fatigue limit design is obtained from the following equation, considering the normal stress fatigue limit σ_{w0} of the plate and the safety factor S_f.

$$\sigma''_c \le \sigma''_{ca} = \frac{\sigma_{w0}}{S_f} \tag{10.10}$$

10.3.2 Axial Loading Case

For shear failure of the fastened plate under axial loading W as shown in Fig. 10.6, the strength evaluation is based on the shear stress in the following equation.

Fig. 10.6 Shear failure of the fastened plate under axial loading

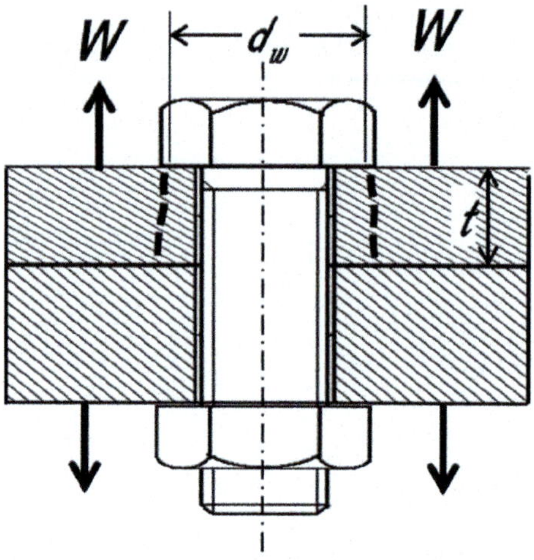

$$\tau_c' = \frac{W}{\pi d_w t} \tag{10.11}$$

In this equation d_w is the bolt or nut seating surface diameter and t is the plate thickness.

The allowable stress τ_{ca}' for fatigue limit design is derived using the shear stress fatigue limit τ_{w0} of the plate and the safety factor S_f as follows.

$$\tau_{ca}' \leq \frac{\tau_{w0}}{S_f} \tag{10.12}$$

However, even though the above case in Fig. 10.6 is similar in appearance to that in Fig. 10.7, when the pitch between bolts p increases, as in Fig. 10.7, the out-of-plane bending failure mode occurs at the bolt holes of the fastened plate, as described in Chap. 1, Sect. 1.6, investigation of the cause of the aircraft panel falling accident. However, since it is difficult to calculate this out-of-plane bending stress in terms of material mechanics, the fastening plate is equivalently replaced by a disk with an outer diameter D_c, inner diameter d, and simply supported circumference, as shown in Fig. 1.50 in Sect. 1.6 of Chap. 1, and is solved using theory of elasticity. The analytical result of circumferential bending stress σ_θ around the bolt hole is shown as follows (refer to Fig. 1.49 in Sect. 1.6).

$$\sigma_\theta = \frac{(M_\theta)_{r=d/2}}{\frac{t^2}{6}} = -\frac{6}{t^2}\frac{1+\nu}{4\pi}\left(-\frac{1-\nu}{1+\nu} + \frac{2p^2}{p^2-d^2}\ln\frac{p}{d}\right) \tag{10.13}$$

The allowable stress $\sigma_{\theta a}$ for fatigue limit design is derived using the normal stress fatigue limit σ_{w0} of the plate and the safety factor S_f as follows:

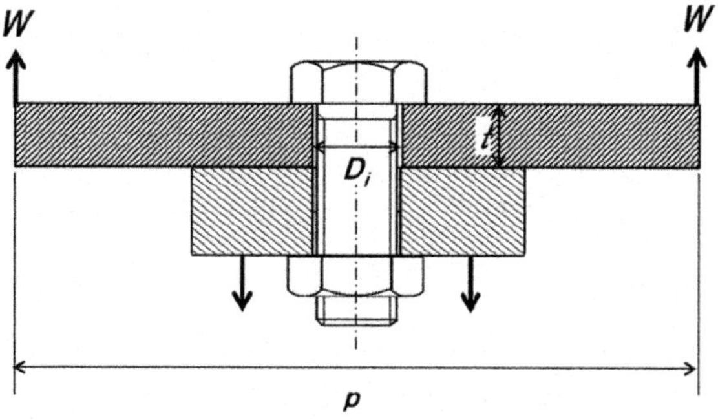

Fig. 10.7 Out-of-plane loading mode on fastened plate

$$\sigma_\theta \leq \sigma_{\theta a} = \frac{\sigma_{w0}}{S_f} \tag{10.14}$$

The solution of Eq. (10.13) is complicated from the viewpoint of a simple material mechanics solution and is not used very often anymore. But, it is important for a strength engineer to take the time to develop the ability to infer the flow of force by touching upon the basic mechanics such as material mechanics [1] and theory of elasticity [2] without easily running to finite element method (FEM) analysis. In other words, as shown in Fig. 10.8, a crack extending in the radial direction is observed on the back side of the fastened plate, and this is a characteristic of out-of-plane bending plate failure. A competent strength design engineer must be able to easily deduce this failure before starting FEM analysis.

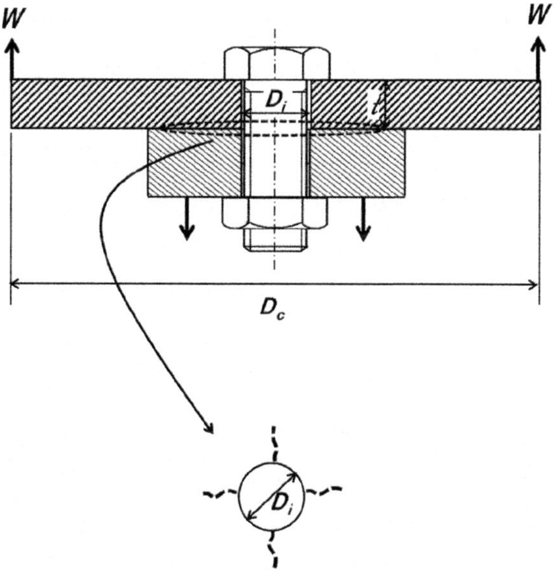

Fig. 10.8 Failure mode under out-of-plane loading on fastened plate

10.4 Fretting Fatigue at Contact Edges on Fastened Bodies

The complex mechanisms of fretting fatigue, such as micro-crack initiation, wear progression, micro crack propagation and final fracture after long life, were described in Chapter 9 Sect. 9.3. And here this mechanism is used here to elucidate the failure process at very long life (Nf>107) as it occurs in real fretting accidents.

10.4.1 Stress Distribution Near the Contact Edge and Strength Evaluation Method

Fretting fatigue failure in screw fasteners occurs at the boundary between the sticking and sliding areas on the contact surface between the fasteners, as shown in Chap. 1, Sect. 1.7: Failure of hub bolt fasteners, and Chap. 8, Sect. 8.3: Failure of helical gear flange bolt fasteners (Fig. 8.26). This is shown in Fig. 10.9. Figure 10.10 shows a schematic representation of the stress distribution in the direction of propagation near the crack initiation zone. The strength evaluation in the design site using this stress distribution is usually conveniently performed in following two steps.

(a) First, a rough material mechanics check is performed using macroscopic stresses to determine if there is a risk of fretting fatigue based on the following two points.

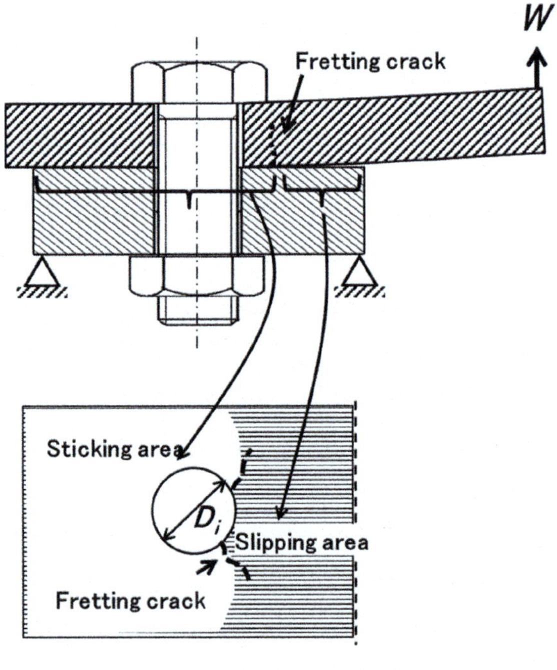

Fig. 10.9 Schematic diagram of fretting fatigue failure of fastened bodies

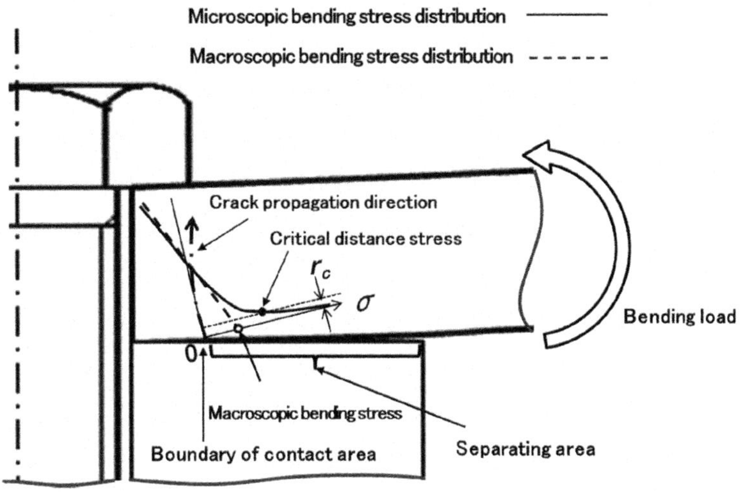

Fig. 10.10 Stress distribution on contact surface of fastened bodies and strength estimation by critical distance stress method

(a-1) Is the average surface pressure greater than 50 MPa?
(a-2) Is the amplitude of the cyclic bending stress greater than 50 MPa?

(b) Once the danger risk is confirmed, the next step is performed using microscopic stress distribution at the contact edge (contact boundary) using the finite element method, etc., and to evaluate the fretting fatigue limit and fatigue life using the critical distance stress method.

10.4.2 Evaluation by Average Surface Pressure and Average Bending Stress Amplitude

The average surface pressure p_m can be calculated using the cross-sectional area of the equivalent cylindrical body to be fastened and the initial tightening force F_f in Chap. 3, Sect. 3.3, Fig. 3.8.

$$p_m = \frac{F_f}{\frac{\pi\left(D_m^2 - D_i^2\right)}{4}} \tag{10.15}$$

The average bending stress σ_{mb} can be calculated using the acting bending moment M and the section modulus Z as follows.

$$\sigma_{mb} = \frac{M}{Z} \tag{10.16}$$

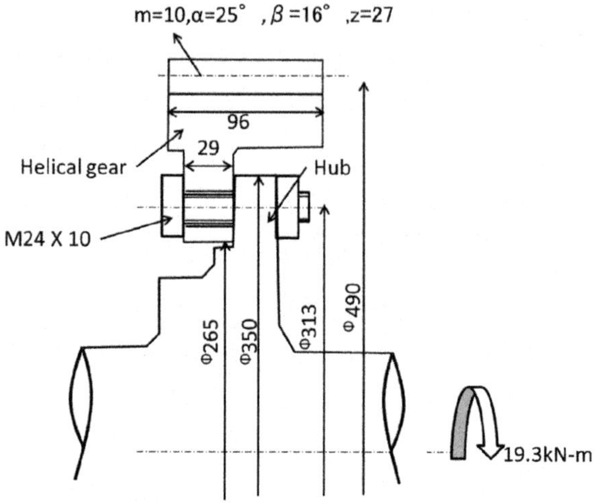

Fig. 10.11 Dimensional specifications of helical gears for evaluation

This method is applied to the evaluation of fretting fatigue strength of helical gear hub fasteners described above in Chap. 8, Sect. 8.3.2. Figure 10.11 shows the dimensional specifications of a helical gear used in a gear transmission as an example.

First,

(a-1) Is the average surface pressure greater than 50 MPa?

If the initial tightening force of all bolts is $F_f = 64.8$ kN, the average surface pressure at the contact surface between the gear web and the hub is estimated as $p_m = 76.7$ MPa, and this satisfies (a-1) criteria.

Next,

(a-2) Is the amplitude of the cyclic average bending stress greater than 50 MPa?

For the gear shown in Fig. 10.11, when the rated torque of 19.3 kN-m is applied to the gear in Fig. 10.11, the concentrated circumferential force $F_\theta = 78.8$ kN and concentrated axial force $F_a = 22.6$ kN act on the gear pitch diameter point, and an out-of-plane bending moment $M_a = 22.6 \text{ kN} \times 245 \text{ mm} = 5540 \text{ kN mm}$ acts on this gear. If we approximate this mechanical model to the model in the "Roark's formula collection; Warren C. Young, pp435" [2] with a fixed outer periphery and a bending load on the inner periphery, we will be able to analyze it as follows.

The following is an example of stress analysis of this circular plate model. Figure 10.12 shows the analytical graph following above Roark's formula. Here, the ratio of inner/outer diameter $b/a = 0.64$, resulting in a bending stress conversion factor $\beta = 0.93$,

$$\sigma_b = \frac{\beta M}{at^2} = \frac{0.93 \times 5540000}{245 \times 29^2} = 25 \text{ MPa}$$

Since this position is exactly where the bolt hole is located, we must consider a stress concentration factor of $\alpha = 3$.

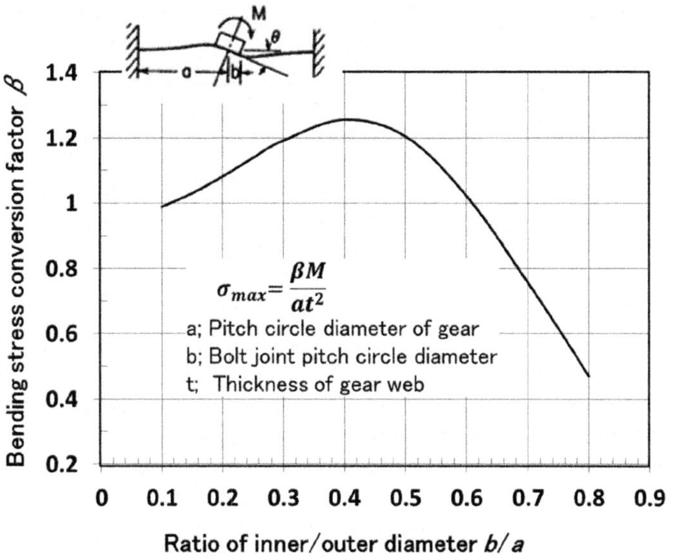

Fig. 10.12 Analytical graph for fixed outer periphery and a bending load on the inner periphery model [2]

So, the stress at this hole edge can be estimated as $\sigma_{\max} = 75$ MPa, which satisfies (a-2) criteria.

Here, we omit detailed analysis such as FEM and use the macroscopic stress value $\sigma_{\max} = 75$ MPa to predict the approximate fatigue life using the fretting fatigue S–N curve shown in Fig. 10.20 which will be described later and is estimated about $N_f = 2 \times 10^8$. If this transmission continues to operate at the rated 1475 rpm, the life is 94.2 days. This is a good explanation for the actual failures shown in Figs. 9.26 and 9.28 in Chap. 9 under the long life.

10.4.3 Fretting Fatigue Limit (Crack Initiation) and Fatigue Life Evaluation by Critical Distance Stress Theory

Until now, strength evaluation of members has been carried out using two methods:

(a) Material mechanical evaluation by maximum stress on concentration areas.
(b) Fracture mechanics evaluation for stress singular fields such as crack areas.

The details of the critical distance stress theory are given in Reference (10.3). However, with the recent spread of CAE technology such as the finite element method, strength evaluation using these two types of parameters separately has placed a great burden on strength design engineers and has become major labor obstacles. So it has become necessary to develop a method that avoids these drawbacks and allows evaluation of whole structures with a single parameter. Figure 10.13 illustrates a critical distance stress theory. In this figure the stress distribution on the fatigue limit of a smooth material is shown by dashed line and the stress distribution at the fatigue limit (crack propagation limit stress intensity factor range) of a cracked material is shown by solid line. We call the point where both stress distributions intersect, as "critical distance r_c and derived as Eq. (10.17)". In both of these members, there is no contradiction even if it is assumed that the condition in which the stress at the critical distance r_c reaches σ_{w0} is the fatigue limit.

$$r_c = \frac{1}{2\pi}\left(\frac{\Delta K_{th}}{\Delta\sigma_{w0}}\right)^2 \tag{10.17}$$

And we can easily understand that this assumption of critical distance stress theory can be extended for all general members.

Some may ask, "Isn't this too easy without any significant theoretical accumulation?" However, it is understood that fatigue failure of polycrystalline materials such as metals is caused by the accumulation of shear slip in single crystals on surface, and no matter how strictly the maximum stress at the extreme point is analyzed, it is meaningless, and the average stress in single crystals will determine the fatigue limit. Considering this, this critical distance r_c is equivalent to the grain size of steel materials, and also, when comparing low carbon steel and high tensile steel, both the specific position r_c and grain size are smaller for high tensile steel, which is consistent with the strength evaluation method that seems to make sense from a materials science perspective. So this method is widely used in recent product structural design.

Then I'll show the suitability of critical distance stress theory on the CAE tools such as finite element method. The finite element method has following weakness.

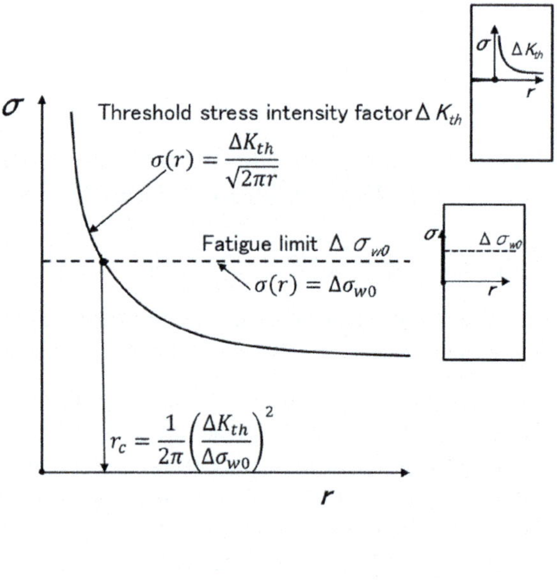

Fig. 10.13 Schematic diagram of the critical distance stress theory

(a) Poor accuracy of maximum stress at high stress concentration areas.
(b) The complexity of using different strength evaluation methods for various parts such as stress concentration members, crack members, contact members, etc. Stress concentration factor, fracture mechanics parameters, and stress singularity parameters have been used separately for each case.

This critical distance stress method can solve all of the above problems at once, and we hope that it will be applied not only to screw fastening structures but also to other applications with reference to Ref. [3] and others. The following are actual examples of analysis of fretting fatigue limit and fatigue life using this critical distance stress theory.

(A) **Fretting Fatigue Limit (Crack Initiation) Evaluation and Optimization of Shape of Contact Edge**

Table 10.1 shows the strength properties and critical distances for the material under consideration, Ni–Mo–V steel. An example of the fretting model used in the stress analysis is shown in Fig. 10.14. Figure 10.15 shows the results of stress analysis near the contact edge under conditions where the average surface pressure $p_m = 200$ MPa and axial force $\sigma_m = 100$ MPa were applied to this model. Using the stress value at a critical distance r_c in this figure, the fretting fatigue limit (crack initiation) can be derived. As a supplementary note, we show how the fretting fatigue crack initiation limit can be increased by changing the angle θ of the contact edge. Figure 10.16 shows the stress distribution near the contact edge for various contact edge angles θ. The estimated fretting fatigue crack initiation limit for each contact edge angles θ is shown in Fig. 10.17. By decreasing contact edge angles θ, we can increase the fretting fatigue crack initiation strength. We hope you will make use of this method for increasing fretting fatigue strength of your products.

(B) **Fretting Fatigue Life Evaluation**

The critical distance stress theory was originally proposed as a method for predicting fatigue limits as shown in previous section, but this method can also be extended to predict the static strength, low-cycle fatigue properties, and crack propagation rate of members of various shapes, by using σ_B of smooth materials, K_{IC} of cracked materials, the low-cycle fatigue properties of smooth materials, and crack propagation characteristics (da/dN) of a cracked material. Specifically, as shown before section,

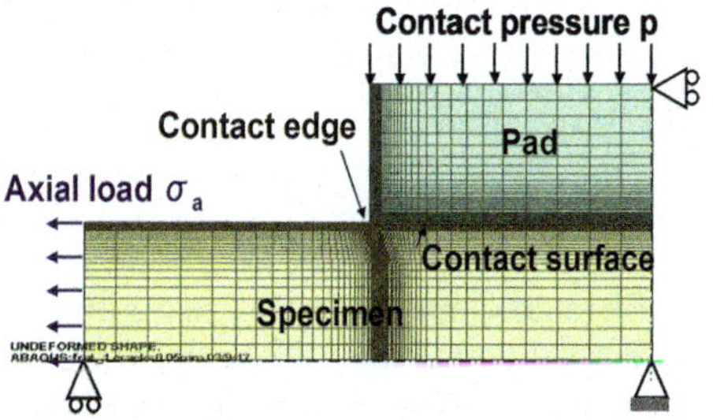

Fig. 10.14 FEM model for stress analysis under fretting condition [3]

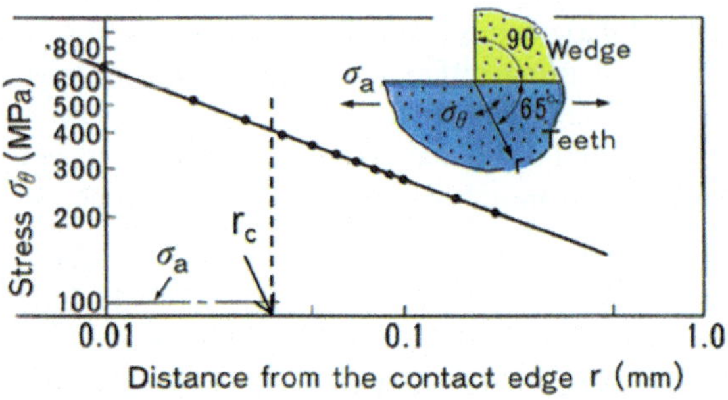

Fig. 10.15 Calculated results of stress distribution near at contact edge [3] (contact edge angle $\theta = \pi/2$, contact pressure $p_m = 200$ MPa, axial force $\sigma_m = 100$ MPa)

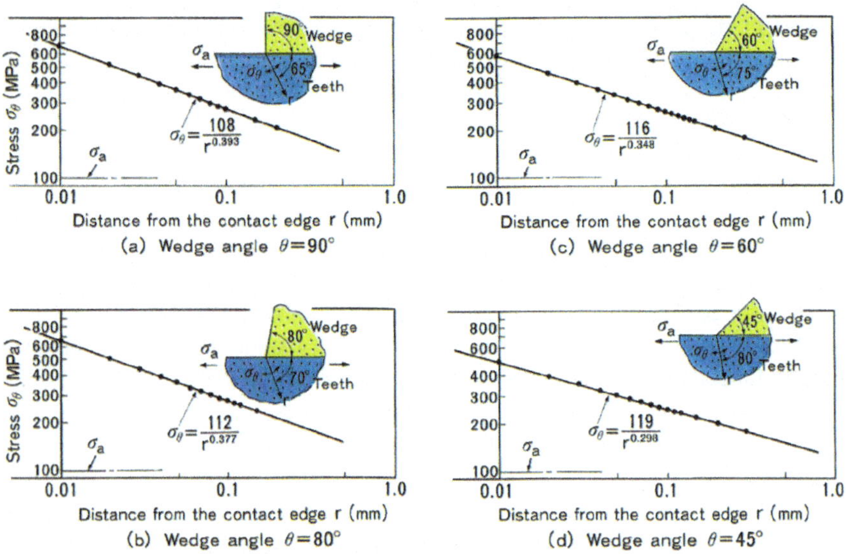

Fig. 10.16 Calculated results of stress distributions near at contact edges [3] (influence of contact edge angle θ)

the critical distance r_c corresponding to the fatigue limit can be derived using σ_{w0} and ΔK_{th} (see Eq. 10.17). And by the same way, the critical distance r_c' corresponding to static strength can be derived using σ_B and K_{IC} as follows.

$$r_c' = \frac{1}{2\pi}\left(\frac{K_{IC}}{\sigma_B}\right)^2 \tag{10.18}$$

By interpolating between these critical distances as shown on the right in Fig. 10.18 by dashed-dotted line, the critical distance at each fatigue life exists on this line. Therefore, if the tress distribution of the actual structural member is plotted as a

Fig. 10.17 Estimated results of fretting fatigue crack initiation strength σ_{ac} for each contact edge angles θ [3]

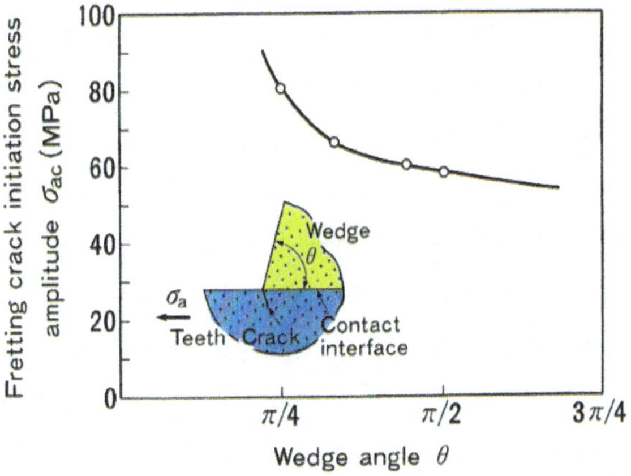

dashed line on this figure, then the reference stress σ_{ref} can be obtained from the intersection of this dashed-dotted line and the dashed line. By applying this σ_{ref} on the S–N curve of the smooth material in the upper left of Fig. 10.18, the failure life N_{fref} under this stress state can be predicted from this intersection point. Furthermore, if we apply the reference stress intensity factor range $\Delta K_{ref} = \sigma_{ref}(2\pi r_{cref})^{1/2}$ using this σ_{ref} and reference critical distance r_{cref} to the diagram of crack growth characteristics in the lower left of Fig. 10.18, we can estimate the reference crack growth rate $(da/dN)_{ref}$ of fatigue cracks in actual structural member can be estimated.

Figure 10.19 shows the application of this method to the fretting model described earlier (see Table 10.1 and Fig. 10.14). If the stress distribution near the contact edge for a nominal stress amplitude of $\boldsymbol{\sigma_a} = 200$ MPa is plotted as the dashed line in the right figure of Fig. 10.19 with reference to Fig. 10.15, the intersection point $\boldsymbol{\sigma_{ref}} = 500$ MPa can be obtained. And plotted this $\boldsymbol{\sigma_{ref}}$ on the S–N curve for smooth material in the left upper figure, the predicted fatigue life of the fretting model is obtained as $N_{fref} = 10^5$. Furthermore, if we apply the reference stress intensity factor range $\Delta K_{ref} = \sigma_{ref}(2\pi r_{cref})^{1/2}$ using this σ_{ref} and reference critical distance r_{cref} to the diagram of crack growth characteristics in the lower left of Fig. 10.18, we can estimate the reference crack growth rate $(da/dN)_{ref} = 3.0 \times 10^{-8}$ m/cycle of fretting fatigue crack initiated at contact edge of Fig. 10.14 can be estimated.

The low-cycle fatigue life of the fretting model predicted in this way is shown on the S–N curve in Fig. 10.20 by dashed line. It can be seen that the experimental points by ○ marks are coincided well with the predict result. The long-life regions indicated by the single-dotted and double-dotted lines are estimated results based on the prediction method that takes wear into account, as described in Chap. 9. It is considered that the whole region fretting fatigue S–N curve can be roughly predicted by connecting low-cycle fatigue region shown by dashed line and high-cycle fatigue region shown by single-dotted line. We hope that this method can be utilized for rough estimation of fretting fatigue strength and life when detailed analysis is difficult.

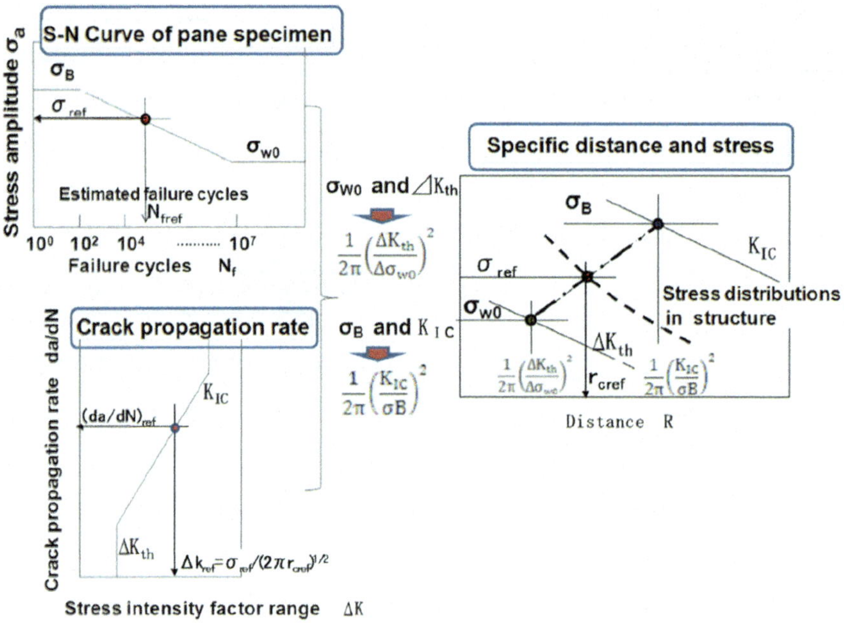

Fig. 10.18 Extension of critical distance stress theory on the low-cycle fatigue strength, life, and crack propagation rate estimations [3]

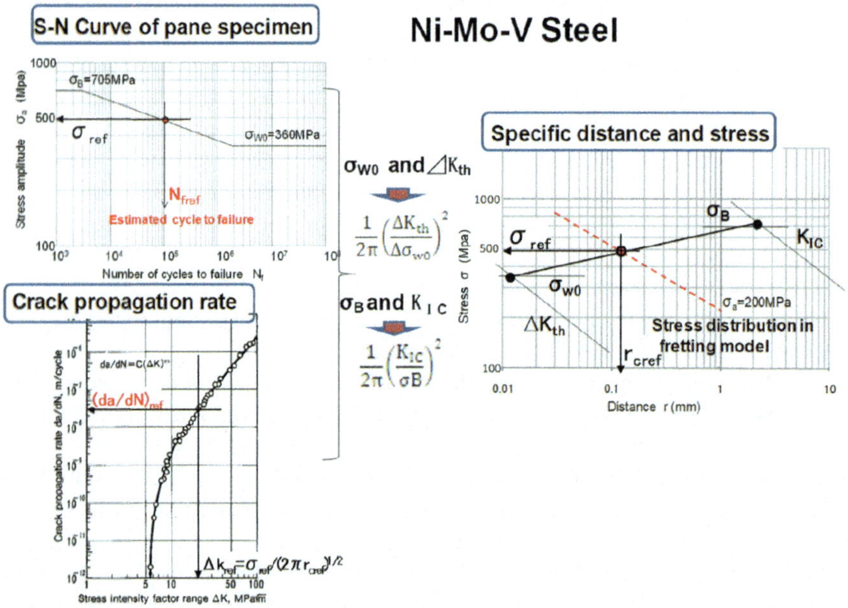

Fig. 10.19 Application of critical distance stress theory on the fretting fatigue strength, fretting fatigue life, and fretting crack propagation rate estimations [3]

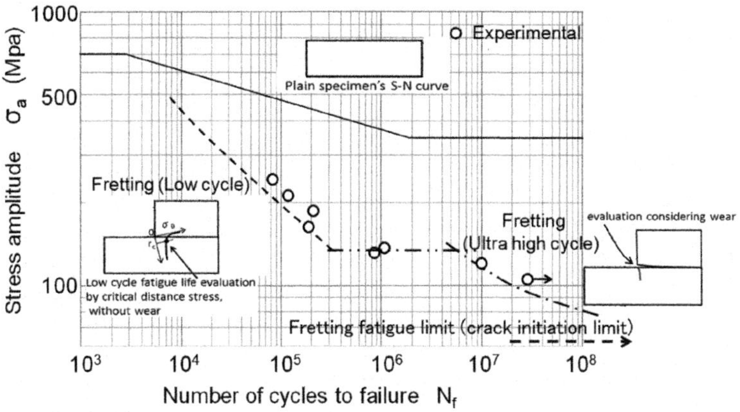

Fig. 10.20 Comparison of estimated results of fretting S–N curves with experimental results [3]

References

1. Seiichiro Seike, Fundamentals of Engineering: Mechanics of Materials, Newly revised edition, Kyoritsu Shuppan (2006-9) pp139.
2. ROARKS'S FORMULAS for STRESS and STRAIN, Table 11.2.
3. Toshio Hattori, Learning from Accident Cases: Fundamentals of Material Mechanics and Strength Design, Nikkan Kogyo Shimbun (2018–12) pp117.

Chapter 11
New Technologies Related to Fastened Joints

11.1 Plastic Region Fastening Method

11.1.1 Behavior of a Single Bolt

If we do not have the engineering concept that "the bolt will not bear all of the load applied to the bolted joint, as explained in Chap. 2, Sect. 2.4" we will mistakenly think that bolt will break by the loading on the bolted joint which was fastened with initial fastening force of the bolt higher than the yield stress. This mistake is caused by the misunderstanding that the bolt bears all of the external force applied to the bolted joint as shown in Fig. 11.1.

11.1.2 Behavior in a Bolted Joint

As explained in Chap. 2, Sect. 2.4, with the concept that the deformation of the bolt and the deformation of the fastened body must be same in the bolted joint, the deformation behavior of the bolted joint can be shown as Fig. 11.2. In this figure, when fastened at the high initial fastening force point ①, which exceeds the yield point, the bolt does not share the load at all for the first loading W and just continues plastic deformation and the bolt moves to point ② and the fastened body moves to point ②' and becomes stable. When the load W is removed from this state, the bolt deforms as an elastic body and moves to point ③, and the fastened body also moves to point ③. Therefore, this state becomes to the original point on the subsequent loading/unloading cycles.

So the deformation behavior of bolt after second loading cycles shows ③ ⇆ ②, and the fastened body shows ③ ⇆ ②' respectively. So, it can be seen that this plastic region bolting method utilizes 100% of the bolt material strength and can provide the

© The Author(s), under exclusive license to Springer Nature Switzerland AG 2025

T. Hattori and T. Naruse, *Compendium for Design and Manufacturing of Bolted Joints*,
Solid Mechanics and Its Applications 280,
https://doi.org/10.1007/978-3-031-89628-6_11

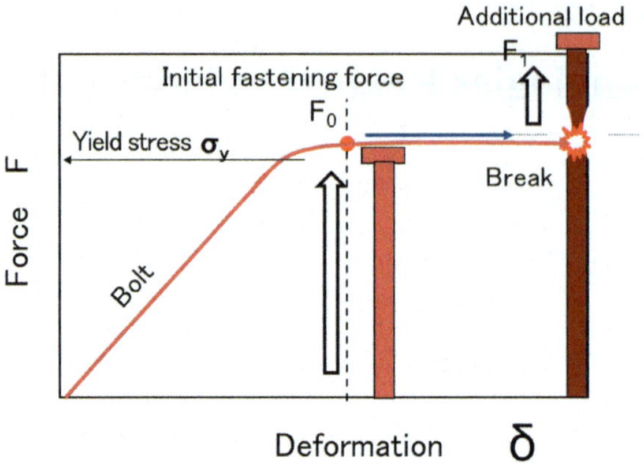

Fig. 11.1 Deformation behavior of bolt alone

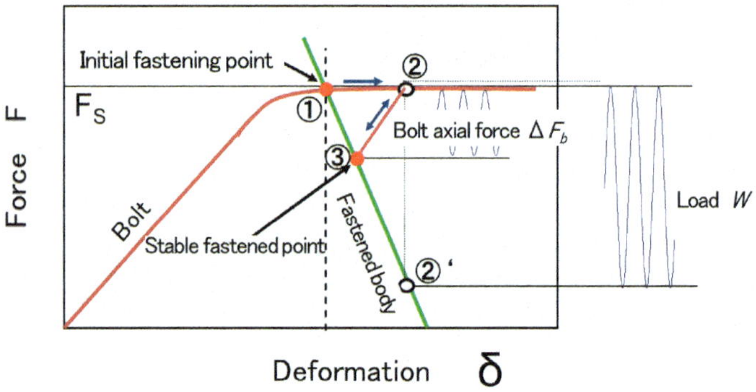

Fig. 11.2 Mechanical behavior of plastic region fastening method

highest fatigue strength. This method is currently used in various fields, especially in the automotive industry.

11.1.3 Application of Plastic Region Fastening Method to Eccentric External Loading Bolted Joints [1–3]

(A) **Initial fastening**

Figure 11.3 shows the model of the bolted joint subjecting eccentric external force. The bolt dimensions are M16XP1.5, under neck length 90 mm, bolt material is SCM435 steel (tensile strength >1000 MPa, strength class 10.9), tightening length l_f = 70 mm, and the initial tightening method is as described in Chap. 2, Sect. 2.3; the rotational angle method was used because it is more stable to control the elongation

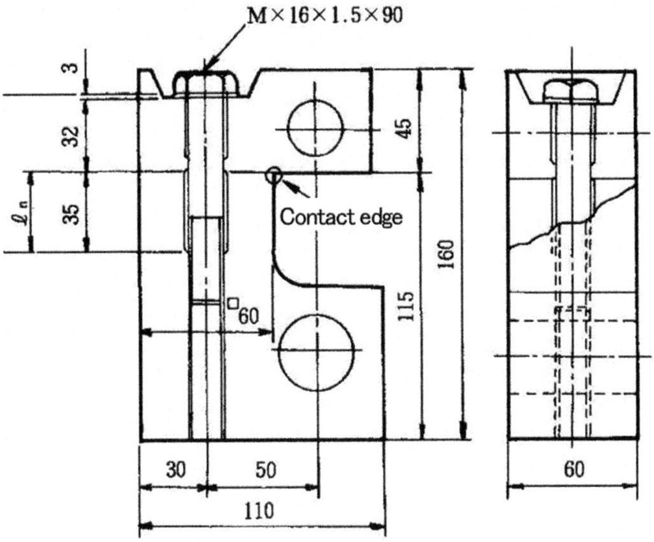

Fig. 11.3 Analytical and experimental model of the bolted joint subjecting eccentric external force [1–3]

by the tightening rotational angle than by the torque in the plastic range. Figure 11.4 shows the tightening conditions. The bolt fastening stress was measured by strain gages attached to the bolt shaft. A is the case when tightened using conventional elastic tightening conditions. B and C are the case where an excessive tightening rotation angle is applied to confirm the effect of this plastic region fastening method on fatigue strength and loosening behavior. The deformation behavior, fatigue strength, and loosening behavior of these three fastening levels A, B, and C under eccentric external force loading are shown below.

(B) **Analysis and experiment of deformation behavior**

Load test results of a bolted joint fastened under the aforementioned conditions are shown in Fig. 11.5 by dashed lines. The bolt shaft force F on the vertical axis is the result of measurement by a strain gage attached to the body of the bolt shaft. As shown in Sect. 3.4, there is a linear relationship between the bolt shaft force F and the eccentric external load W while the load W is small, but as the load increases, the bolt shaft force increases in a nonlinear manner. This point at which the relationship shifts to nonlinear seems to be the one-sided separation limit shown in Eq. (3.71) in Sect. 3.4.2. When further increasing the load W, it can be seen that this line approach to the full separation limit is shown in Eq. (3.81) of Sect. 3.4. From these results we can see that by plastic regeon fastening, the increase range of bolt force ΔF decrease for same sujecting load range ΔW. And we canunderstand the advantages of plastic regeon fastening.

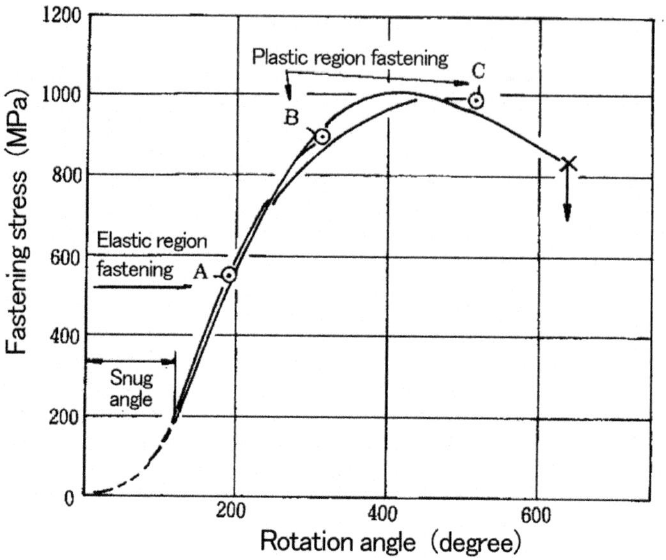

Fig. 11.4 Initial fastening by rotation angle method [1–3]

Fig. 11.5 Experimental and analytical results of bolted joints under eccentric loading

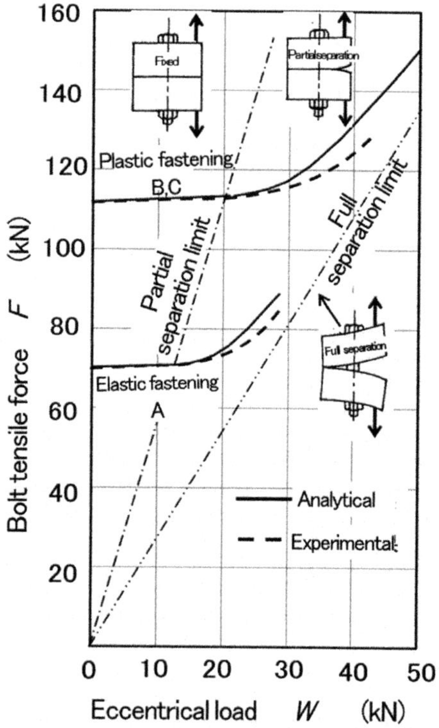

These behaviors are almost equal in the initial fastening conditions B and C. Also, as described below, there is no significant difference in fatigue strength and loosening, behavior, and there seems to be no particular problem even when fastening in a considerable deep plastic range as in C.

The calculated results of this deformation behavior by applying the mechanical analysis under eccentric fastening and eccentric loading, described in Sect. 3.4, are shown in Fig. 11.5 by solid lines. The calculated parameters for this model are shown below.

Contact surface dimensions; $a = b = 60$ mm
Fastening position; center fastening $s_B = 0$, $u = 25$ mm
External load position; eccentric external force $e_B = 50$ mm
Modification coefficient of external force action position; $n = 0.5$
Young's modulus of bolt and the fastened body; $E_b = E_c = 210$ GPa
Fastening length; $l_f = 70$ mm

And in Fig. 11.5 the one-sided separation limit obtained from Eq. (3.71) in Sect. 3.4 is shown as a single-dotted line, and the full separation limit obtained from Eq. (3.779) is shown as a double-dotted line. The F_b–W relationship in the linear area obtained from Eq. (3.62) and the $F_{b\Delta}$–W_Δ relationship in the nonlinear area obtained from Eqs. (3.74) and (3.75) are shown by solid lines. Both of these analytical results coincided well with the experimental results.

11.1.4 Fatigue Strength and Loosening Behavior of Plastic Region Bolted Joints

Figure 11.6 shows the total additional stress on the **a** side under eccentric loading (axial force and bending moment loading) at the bottom of the bolt thread valley (σ_{F+M}; Eq. (3.71) in the linear region Sect. 3.4) and ($\sigma_{F+M\Delta}$; Eq. (3.82) in the nonlinear region Sect. 3.4) by the solid lines. The additional stresses on the **b** side are also shown for reference by $\sigma_{F\text{-}M}$ in linear region and $\sigma_{F\text{-}M\Delta}$ in nonlinear region, both of which are in good agreement with the measured results shown by the dashed lines. The fatigue limits of this bolted joint subjected to eccentric external force are predicted as shown in Fig. 11.7, referring to the total additional stress on side **a** and the fatigue limit of the bolt (SCM435 steel, strength class 10.9) (see Table 3.3 in Chap. 3) σ_{wk} $= 60$ MPa. By this figure we can see that the experimental results in both cases are higher than estimated results. This may be due to the fact that the fatigue strength of the bolt (Table 3.3 in Chap. 3) will be the guaranteed value, and tested bolt material would be stronger than this guaranteed value.

Figure 11.8 shows the measurement results of bolt axial stress (tensile stress σ) and torsional stress (shear stress τ) during fatigue tests (load levels W are 42 kN and 33 kN, respectively) of eccentric loading, on the bolted joints fastened plastically ($F_f = 774$ MPa) and elastically ($F_f = 450$ MPa). There is no significant difference

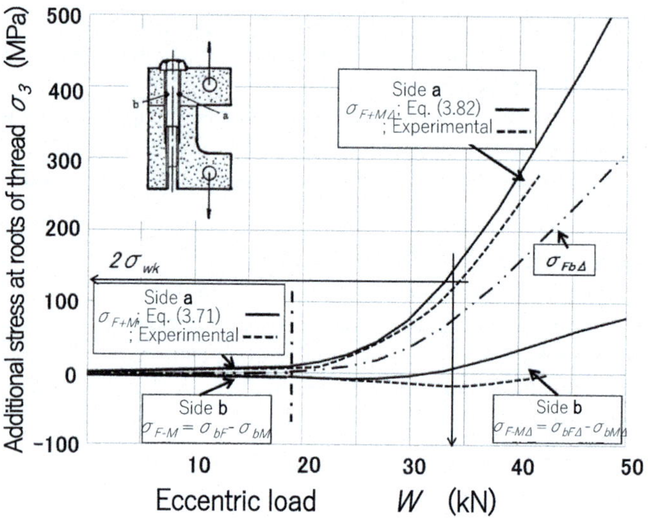

Fig. 11.6 Additional stress on the **a** side under eccentric loading (axial force and bending moment loading) at the bottom of the bolt thread valley

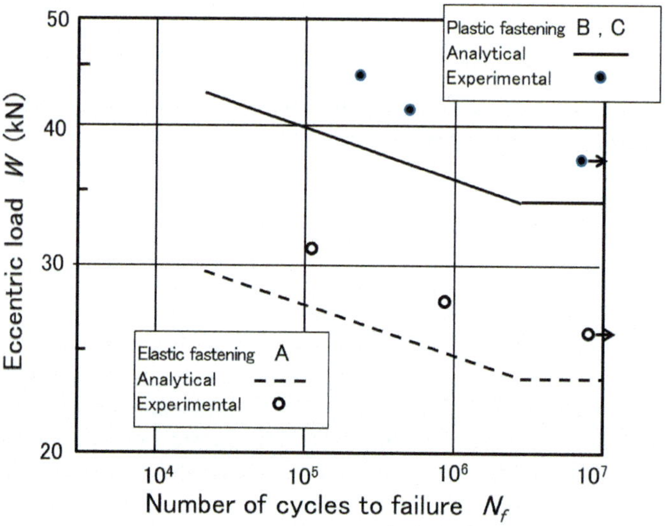

Fig. 11.7 S–N curves of bolted joints with plastic and elastic fastening under eccentric loading [1–3]

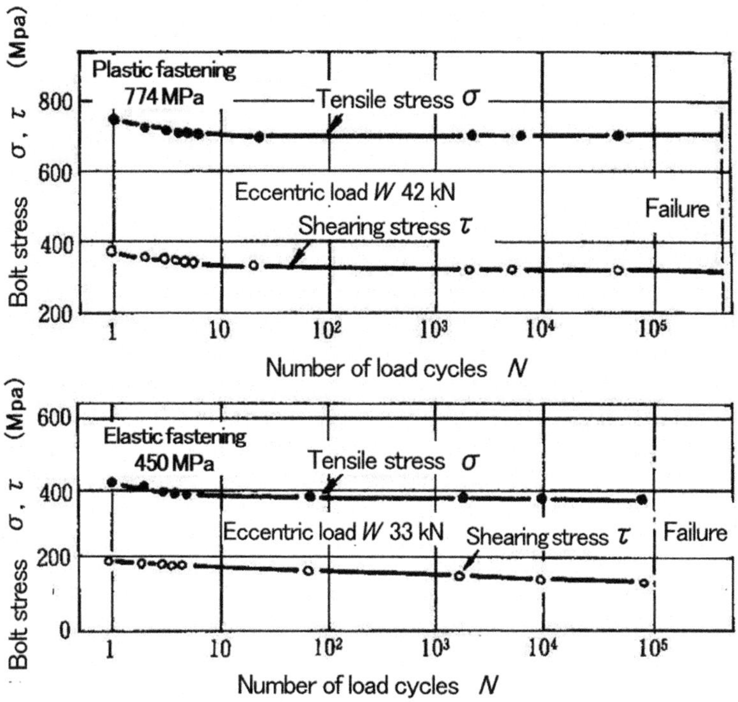

Fig. 11.8 Loosening behavior of bolted joints with plastic and elastic fastening under eccentric loading [1–3]

between elastic and plastic region fastening test results. This indicates that there is no danger of plastic deformation progressing or causing the bolt to loosen easily in plastic region fastening.

Both results exhibit a slight stress drop after the first 10 or so repetitions, but this is due to the micro wear (embedment) on the threaded surface and seat surface, and is not a problematic phenomenon (see Chap. 6).

11.2 Joints That Combine Adhesive and Thread Fastening

New lightweight and high-strength materials such as aluminum alloys and CFRP are beginning to be used in various fields of transportation equipment, including aircraft, high-speed trains, automobiles, and elevators, to achieve higher speeds and energy savings. However, welding and fastening with threads, which have been used with conventional steel materials, are difficult, so it has become essential to develop new fastening technologies to replace them. The most promising technology at present will be the adhesive joint, but although this joint certainly has sufficient rigidity, its

Table 11.1 Features of adhesive and screw/rivet joints

	Screw/rivet joint	Adhesive joint
Strong point	Stability of strength	High rigidity No hole drilling required
Cons	Low rigidity	Dispersion of strength

strength properties are sensitive to manufacturing processes such as surface treatment and working environment, and there are drawbacks such as large variations, poor overall reliability, and time-consuming processes [4]. In this section, we will experimentally and analytically understand the basic characteristics of each joining technology, and discuss how to combine them effectively, with the aim of "developing new joints that combine bolted fastening/adhesive joints" that skillfully combine the high rigidity of adhesives and the strength stability of bolted fastenings.

As mentioned in Chap. 2, Sect. 2.11, the discussion of environmental and aging degradation of adhesives is essential for evaluating the strength and durability of fasteners, and is also presented here.

11.2.1 Features and Hybridization of Adhesive and Bolted Joints

Since CFRP, aluminum alloys are difficult to weld like conventional steel materials, adhesives, bolts, etc., may be proposed as candidates. Conceivably, both joint structures have the characteristics shown in Table 11.1. In order to complement these advantages and disadvantages, we propose a hybrid joint of these two joints and will confirm mechanically and experimentally. Furthermore, this hybrid joint is applied not only for the above mechanical and strength properties, but also from the viewpoint of production cost, because this hybrid joint can drastically reduce the loss time of adhesive curing.

11.2.2 Strength Tests of Adhesive and Bolted Joints

As shown in Fig. 11.9, hybrid bolted joint/adhesive joints were made by sandwiching SS400 steel plate (②) between two CFRP laminates (①). Shear tensile loads were applied to this joint, and its deformation, failure status, and failure load were measured. The adhesive (③) was a methacrylate-based structural adhesive PLEXUS-MA425, and primer PC120 was used for pretreatment of the bonding surface. Bolts (⑤) and nuts (⑥) were made of SUS304, size M8.

The test conditions are shown in Fig. 11.10. Tensile load, deformation, and damage conditions were measured during the tests. The results of four representative tests are shown in Table 11.2. In all tests, the damage conditions were first delamination

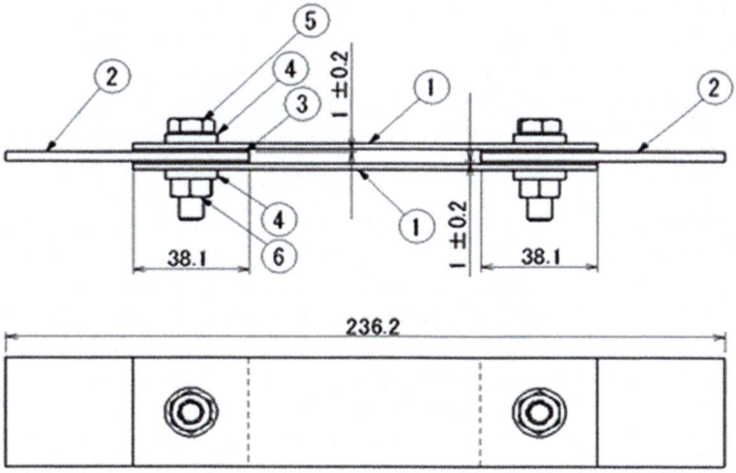

Fig. 11.9 Test specimen of bolted joint/adhesive hybrid joint

from the adhesive edge and finally bearing rupture of the CFRP plate hole, as shown in Figs. 11.11 and 11.12.

As shown in Table 11.2, the averages of both delamination strength and bearing strength are almost the same, but there is a large dispersion in delamination strength, indicating that the reliability of the product is highly problematic when applied as is.

11.2.3 Mechanical Studies and Application Considerations of Adhesive/Bolting Hybrid Joint

In this section we replace the experimental results in Fig. 11.11 to the simple mechanical model. First, the mean strength of the adhesive strength and bearing strength of CFRP are assumed to be identical and non-dimensionalized as 1.0. And the adhesive strength is assumed to vary from a minimum of 0.5 to maximum of 1.5. Then linear range stiffness of the bolted joint is assumed to be 1/3 of the stiffness of the adhesive joint (see Fig. 11.13). When hybridized bolted joints with the highest adhesive joining, the load/deformation characteristic will be shown as Fig. 11.14. But when hybridized bolted joints with the lowest adhesive joining, the load/deformation characteristic will be shown as Fig. 11.15. In this later case the strength in high stiffness region is very low and there is little effect of hybridization.

Therefore, in the case of hybridization with a single bolt in a single adhesive surface as described above is not applicable because the minimum strength of the joint will become too low as the variations of adhesive strength are very high.

To address this variation, here we propose a parallel arrangement of multiple hybrid joints.

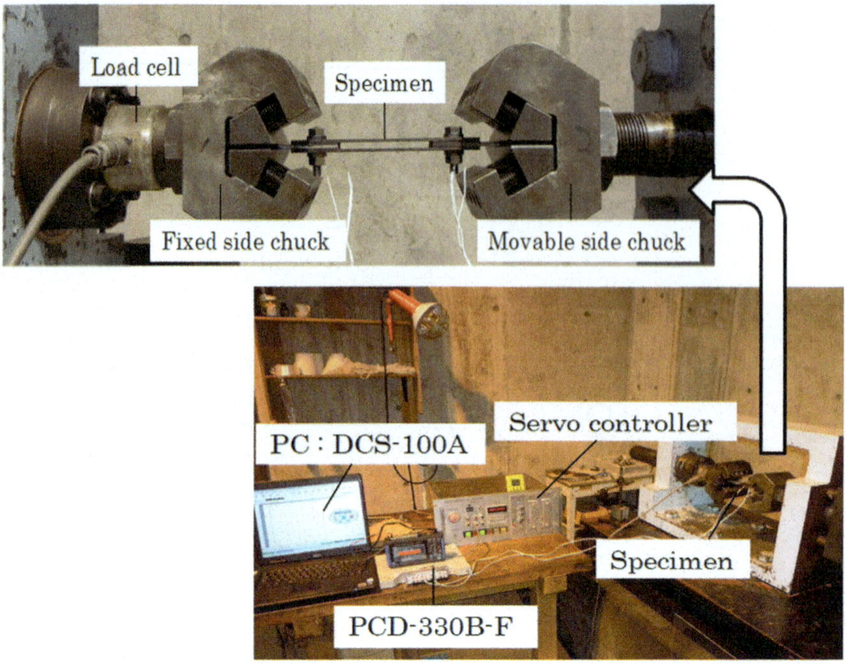

Fig. 11.10 Testing status of shear tensile tests for bolted/adhesive hybrid joints

Table 11.2 Test results of shear tensile tests of bolted/adhesive hybrid joints

	Test No. 1	Test No. 2	Test No. 3	Test No. 4
Adhesive strength (MPa)	7.18	5.81	13.7	11.8
Bearing strength (MPa)	9.42	9.48	9.31	9.32

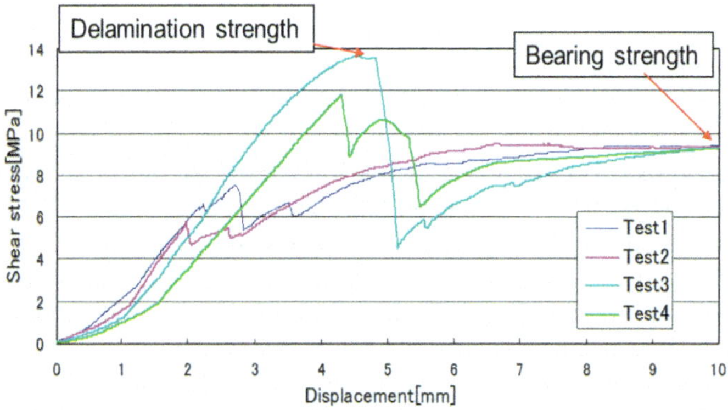

Fig. 11.11 Test results of shear tensile tests of bolted/adhesive hybrid joints

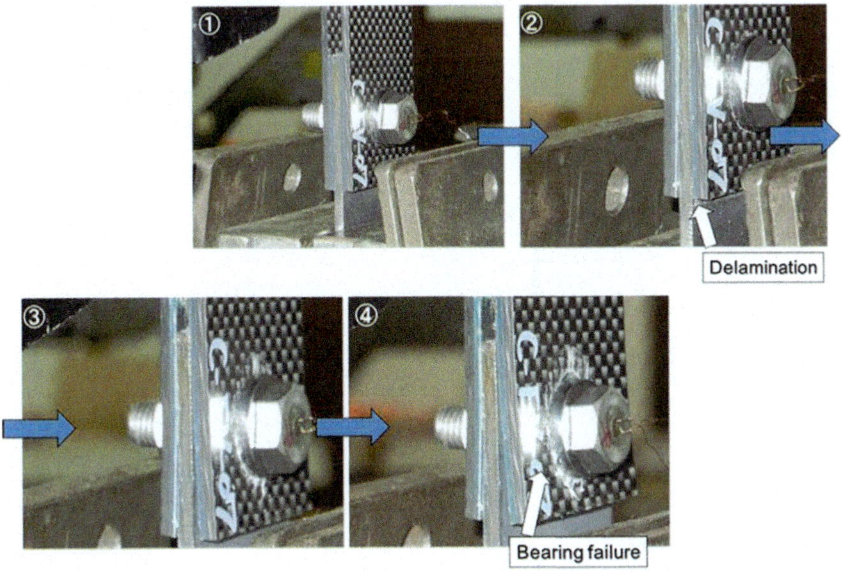

Fig. 11.12 Failure processes of shear tensile tests of bolted/adhesive hybrid joints

Fig. 11.13 Non-dimensional load/displacement curves for individual adhesive and bolted joints

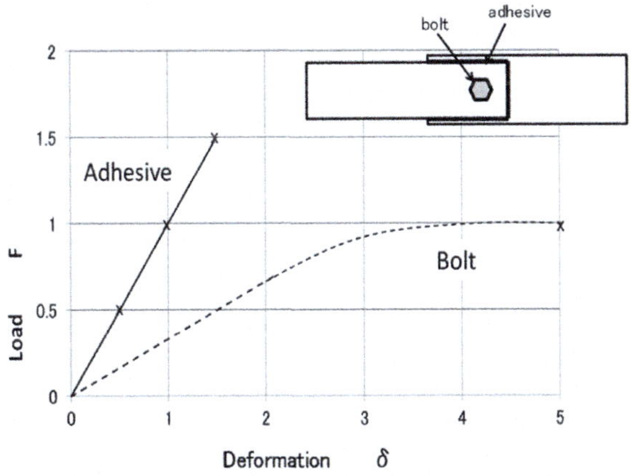

For reference, if we assume that within the adhesive surface, regions with dimensionless adhesive strengths of 0.5, 1.0, and 1.5 are evenly distributed, the total strength of adhesive strength can't be achieved as shown in Fig. 11.16. Moreover, as there is not much variation in bolt fastening itself, the multiparallel arranged by bolt can't increase strength or rigidity as shown in Fig. 11.17.

However, when multiple bolts are arranged in one hybrid joint as shown in Fig. 11.18, the range where rigidity and strength are secured will increase and the

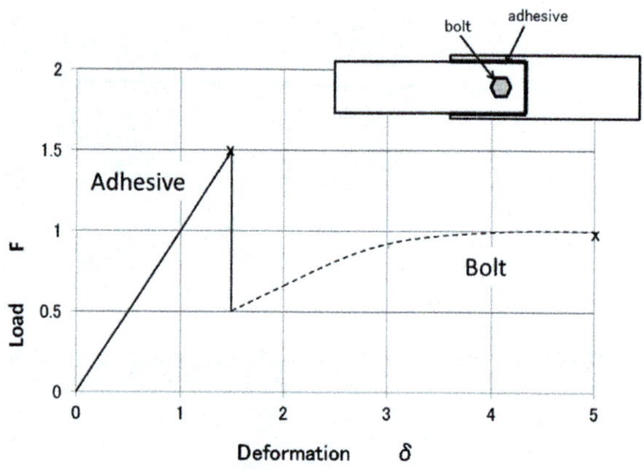

Fig. 11.14 Non-dimensional load/displacement curve for hybrid adhesive/bolt joint with high adhesive strength

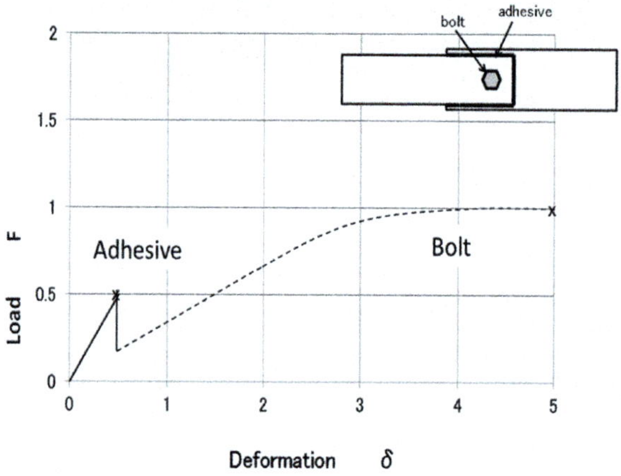

Fig. 11.15 Non-dimensional load/displacement curve for hybrid adhesive/bolt joint with low adhesive strength

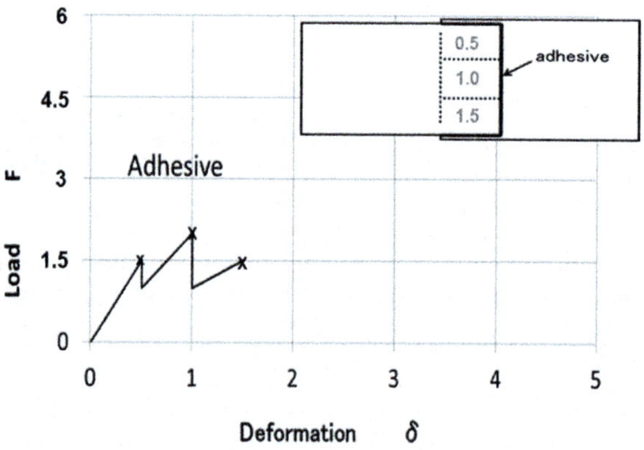

Fig. 11.16 Non-dimensional load/displacement curve for adhesive joint with considering the variation of adhesive strengths

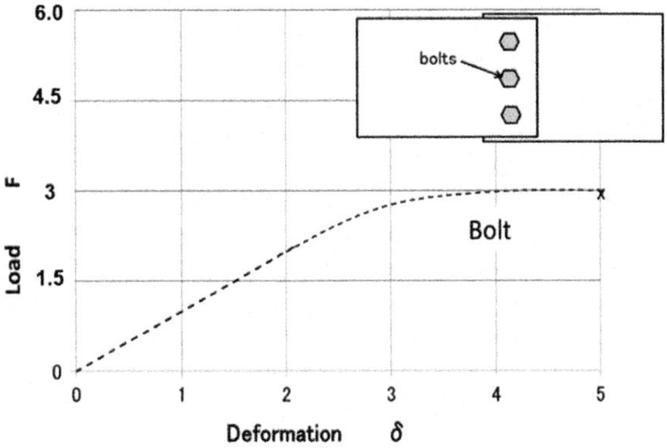

Fig. 11.17 Non-dimensional load/displacement curve for multi-arranged bolted joint

deformation capacity is also become large, and can show the sufficient effect of hybridization.

Therefore, for practical use of the proposed bolt/adhesive hybrid joint, we recommend a multiarray bolt fastening/adhesive hybrid joint as shown in Fig. 11.19, instead of hybridization with a single bolt as shown in the above Figs. 11.13, 11.14 and 11.15.

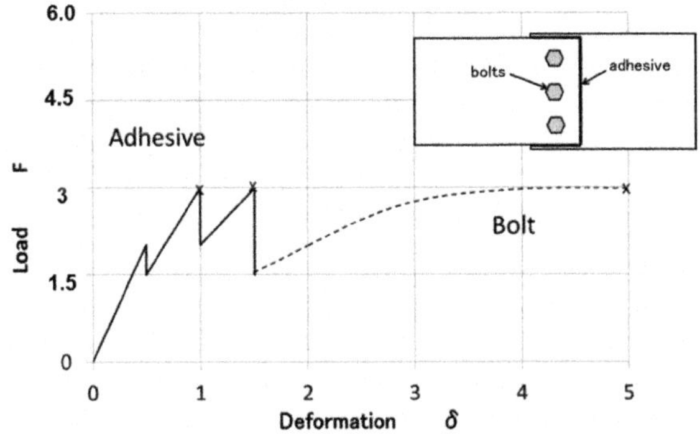

Fig. 11.18 Non-dimensional load/displacement curve for hybrid adhesive/bolt joint with multiple bolts

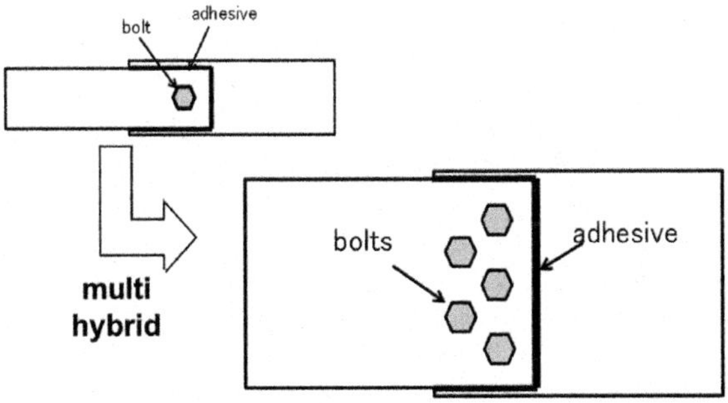

Fig. 11.19 Recommendation of multi-bolt/adhesive hybrid joint

11.2.4 Methods for Quantifying Aging Degradation of Adhesive Strength

In the previous section, we discussed the variation in strength as a practical problem with adhesive joint, then here we will discuss a further problem, aging and environmental deterioration of adhesive strength (refer Chap. 1, Sect. 1.11).

While degradation of bolted joints using ordinary metallic materials will be oxidative corrosion and delayed fracture (stress corrosion cracking) as described in Chap. 7, adhesive joints using mainly organic adhesives have the disadvantage of aging degradation due to temperature, humidity, and stress. Here, we introduce a method developed by Haraga et al. [5–7] for quantifying the safety factor that takes into account age-related degradation and variation in adhesive strength.

Figure 11.20 shows the concept and parameters of adhesive strength degradation over time. The adhesive strength shifts to the low strength side due to environmental degradation caused by the environment and stress during use. As mentioned above, there is a variation in adhesive strength, and this variation is expected to increase with age. The initial base value of adhesive strength is the adhesive strength at the highest temperature to which the adhesive area is exposed during product use. For instance, if the initial adhesive strength at room temperature is set as $F_{\mu R0}$ and the rate of degradation in adhesive strength at the highest service temperature is η_{T0}. The average initial strength at the highest temperature $F_{\mu R0} \times \eta_{T0}$ is set as the base strength. The degradation coefficients of adhesive strength for each of the various environments such as temperature, humidity, heat cycling, creep, and fatigue, are set as $\eta_1, \eta_2, \eta_3, \ldots \eta_n$, respectively. These individual degradation factors are determined by accelerated testing. For example, long-term temperature degradation is determined by an Arrhenius plot based on accelerated testing at high temperatures, humidity degradation is predicted by determining the concentration distribution of moisture in the adhesive by moisture diffusion [8, 9] or by an Arrhenius plot [10], and heat cycle degradation is predicted by an accelerated method over a wide temperature range. For creep degradation, a master curve based on the temperature–time conversion law [11] or the Larson-Miller master curve is used, and fatigue strength is obtained by extrapolating the S–N curve. The degradation factor under combined environmental and stress conditions is considered to be a multiplication of the degradation factors $\eta_1, \eta_2, \eta_3, \ldots \eta_n$ due to individual environmental and stress conditions, and η_y.

If the adhesive strength that satisfies the acceptable defective rate $[1 - R]$ at the end of the service life is called the effective adhesive strength F_y, and is expressed by the following equation.

$$F_y = F_{\mu R0} \times \eta_{T0} \times \eta_y \times D_y \tag{11.1}$$

where D_y is the variation rate over the service life and is the ratio of the effective adhesive strength to the average adhesive strength, as shown in Fig. 11.21. Variation in adhesive strength may increase due to degradation of adhesive properties. The variation rate D_y of adhesive strength after service life is calculated from Fig. 11.22

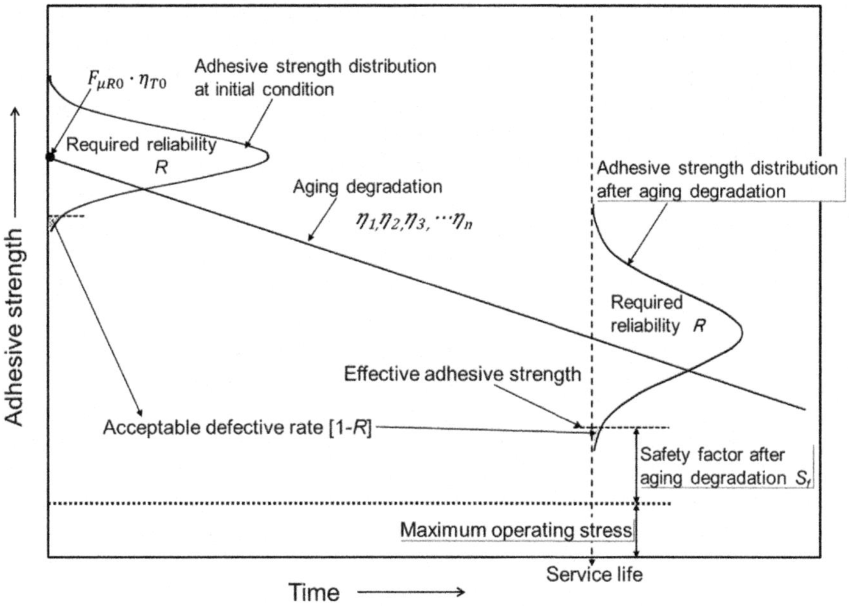

Fig. 11.20 Conceptual diagram and parameters of adhesive strength degradation over time [5–12]

as the coefficient of variation of initial adhesive strength, CV_0, which increases k times after service life, D_y for $k{\cdot}CV_0$. The value of k is assumed here to be $k = 1.5$ since there is not much data.

In Eq. (11.1), the degradation factors can be obtained by stress tests such as creep and fatigue tests at room temperature. If creep and fatigue tests at high temperatures are not too difficult, the test results can be directly used as the base strength for creep strength and fatigue strength F_{fT} after the service life and the effective adhesive strength F_y can be obtained by the following equation.

$$F_y = F_{fT} \times \eta_y \times D_y \tag{11.2}$$

If this F_{fT} is used, the initial room temperature adhesive strength $F_{\mu R0}$ and temperature degradation coefficient η_{T0} and the degradation coefficients for creep and fatigue strength η_i are not needed.

If the maximum stress applied to the adhesive joint during use is defined as the maximum generated stress $P_{\max}$. The safety factor S_y at the end of the service life is obtained from the following equation

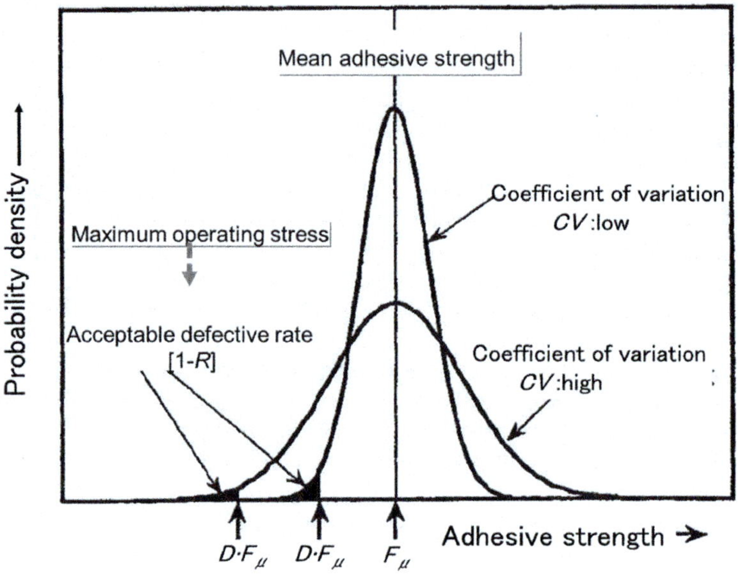

Fig. 11.21 Conceptual diagram of adhesive strength F_y, allowable failure rate $[1 - R]$ and variation rate D [5–12]

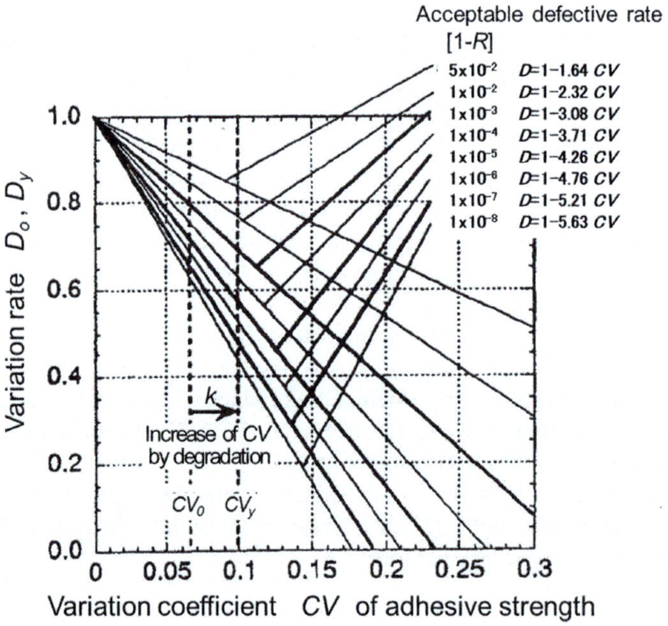

Fig. 11.22 Relations between coefficient of variation CV, allowable failure rate $[1 - R]$, and variation rate D [5–12]

$$S_y = \frac{F_y}{P_{\max}} \tag{11.3}$$

The basis of the strength design is to take this safety factor S_y to higher than 1.0.

11.2.5 Example of Evaluating Safety Factor After Aging on Adhesive Strength

The following is an example of evaluation of the safety factor on adhesive strength after 25 years of service life, satisfying the acceptable defective rate $[1 - R] = 1/100{,}000$ for adhesive joints subjected to fatigue and heat cycles at a maximum service temperature of 100 °C. The details of the requirements are shown in Table 11.3 and the evaluation results in Table 11.4. The results show that the safety factor $S_y = 2.51$ is sufficient, and it makes possible to reduce costs by such as simplifying the structure and improving workability.

For instances as shown in Table 11.5, the safety factor $S_y = 1.94$ is sufficient even with an acceptable defective rate $[1 - R] = 1 \times 10^{-8}$. Rationalization such as lowering the manufacturing standard for the area of adhesive application will be applicable. And this quantitative deterioration evaluation method can be used for an optimal design tool to achieve both reliability and cost.

Table 11.3 Required conditions on adhesive joints [5–12]

Required item		Conditions
Service life		25 years
Required reliability		0.99999
(Acceptable defective rate, $[1 - R]$)		(1×10^{-6})
Maximum generated stress		Shearing stress: 0.1 MPa
Loading condition		Number of cycles: 10^8
Temperature	During transportation	Winter: −40 °C–Summer: +40 °C Only once
	While stopping	Winter: −20 °C–Summer: +40 °C Winter: −20 °C–0 °C Summer: +20 °C–+40 °C
	While operating	Winter: −20 °C–Summer: +100 °C Winter: −20 °C–+60 °C Summer: +20 °C–+100 °C Thermal cycle: 1/day

Table 11.4 Calculation example of safety factor S_f for adhesive joints after 25 years of use on the basis of fatigue test results at 100 °C [5–12]

Fatigue strength at 100 °C; F_{f100}	0.58 MPa
Variation rate after 25 years; D_y	0.569
Temperature degradation coefficient (100 °C); η_{T0}	–
Degradation coefficient after service life; η_y ($\eta_y = \eta_2 \times \eta_3 \times \eta_4$)	0.76
Fatigue (25 °C); η_2	–
Maximum temperature change (each year); η_3	0.95
Maximum temperature change (each day); η_4	0.80
Thermal degradation; η_5	1.00
Effective adhesive strength ($F_y = F_{f100} \cdot D_y \cdot \eta_y$) F_y	0.251 MPa
Maximum generated stress $P_{\max}$	0.10 MPa
Safety factor after service life ($S_f = F_y/P_{\max}$) S_f	2.51

Table 11.5 Calculation example of change in safety factor S_f when changing acceptable defect rate [1 − R] area ratio and adhesive area ratio [5–12]

Acceptable defect rate [1 − R]		1×10^{-5}	1×10^{-7}	1×10^{-8}
Adhesive application area ratio	1	$S_f = 2.51$	$S_f = 2.11$	$S_f = 1.94$
	0.7	$S_f = 1.75$	$S_f = 1.47$	$S_f = 1.95$
	0.5	$S_f = 1.25$	$S_f = 1.05$	$S_f \leq 1.00$

11.2.6 Application Examples and Effectiveness of Riveted and Adhesive Hybrid Joints [4, 5]

The riveted/adhesive hybrid joints have already been used to fasten and join metal plate housings, for such as control panels, operation panels, and elevator housings, and have proven effective.

Figure 11.23 shows an example of a metal plate housing structure using a combined riveted and adhesive assembly. The outline, features, and effects of this assembly method are described in detail below.

(A) **Joining procedure**

Figure 11.24 shows the procedure for assembling rivet and adhesive hybrid joint. First, adhesive is applied to the one of two joint plates where the holes for rivets are drilled and the other parts are attached together. Next, before the adhesive has cured, the rivets are mounted on a riveter, inserted into the hole from one side of joint, and the stem is pulled to caulking and tighten rivet and plates, at the same time the stem is torn off. The rivet acts not only as a strength member but also as the fixing jig until the adhesive hardens and cured. So, by this hybrid method we can obtain the improvement of strength and reduction of man-hours.

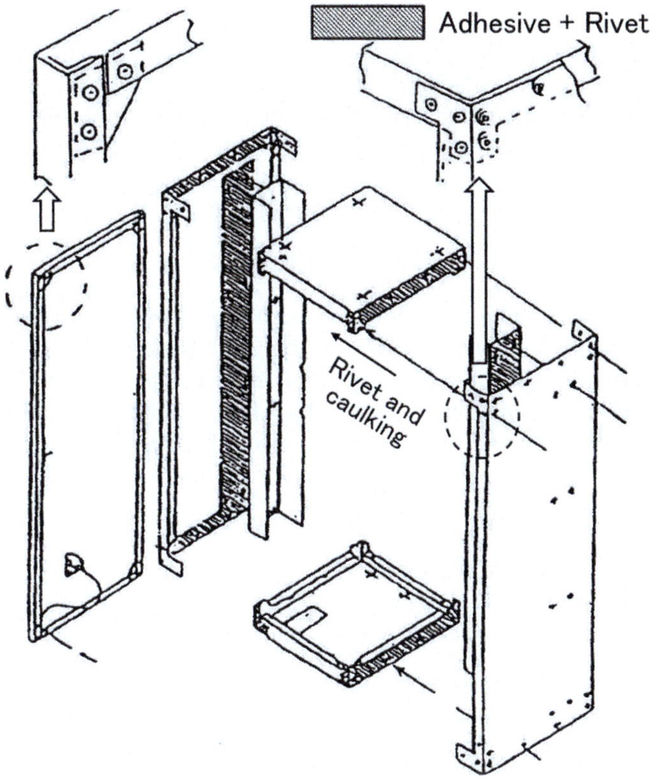

Fig. 11.23 Application example of adhesive/rivet hybrid assembly on elevator plate housing [5–12] (800 W × 800 D × 2300 H)

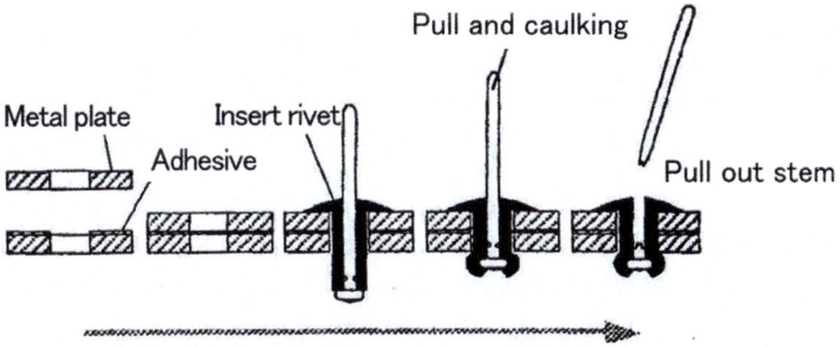

Fig. 11.24 Assembling procedure of adhesive/rivet hybrid joint [5–12]

(B) **Adhesives**

The adhesive used is Hard Lock M372 (manufactured by DENKI KAGAKU KOGYO, LTD.), a two-component organized acrylic adhesive. This adhesive has excellent oil-surface-resistant adhesion, so degreasing is basically unnecessary, and it cures in 15–20 min at room temperature. In terms of performance, it has high shear strength, peel strength, and impact strength, as well as excellent environmental resistance.

(C) **Role of adhesives and fasteners (rivets, bolts)**

The riveted/adhesive joint is the same as the bolt/adhesive joint described above. The joint strength, rigidity, and durability are basically dependent on the adhering process, and the rivets, screws, and other fasteners compensate for the disadvantages of adhesive bonding, such as variability. And it achieves both excellent overall workability and reliability. The respective roles of adhesives and fasteners (rivets and screws) are described below.

(C-1) Role of adhesives

(i) Ensure joint strength and rigidity
(ii) Ensuring sealing performance
(iii) Prevention of electrical corrosion in joining dissimilar materials.

(C-2) Role of fasteners (rivets, screws)

(i) Substitution of fixing jig until adhesive hardened
(ii) Easy and accurate positioning
(iii) Prevention of creep deformation
(iv) Maintains shape when adhesive is lost due to fire.

(D) **Performance and effectiveness**

Figure 11.25 compares the tensile shear strength of arc welded, spot welded, riveted, adhesive, and adhesive/riveted hybrid joints. The results show that the adhesive/rivet hybrid joint has strength equivalent to that of arc welding, and if the adhesive length is 50 mm, it becomes comparable to the rupture strength of the base steel plate. The strength of the adhesive-only and adhesive-riveted joints is equivalent, and there is no increase in strength due to the combination of the two. As mentioned in Sect. 11.2.3, it should be understood that the combined effect is not a mere addition of the two, but is expected as an effect of reducing the variation of the adhesive and the effect described in (C).

Table 11.6 compares the weight, man-hour, cost, construction period, and factory noise of welded and adhesive/riveted joints. The adhesive/riveted joints have sufficient strength and rigidity even in thin sheets due to stress dispersion at the joints, enabling thinner sheets and lighter weight. In addition, the use of adhesive-curing fixtures and curing are no longer necessary, resulting in significant benefits in terms of man-hour, cost, work period, and factory noise.

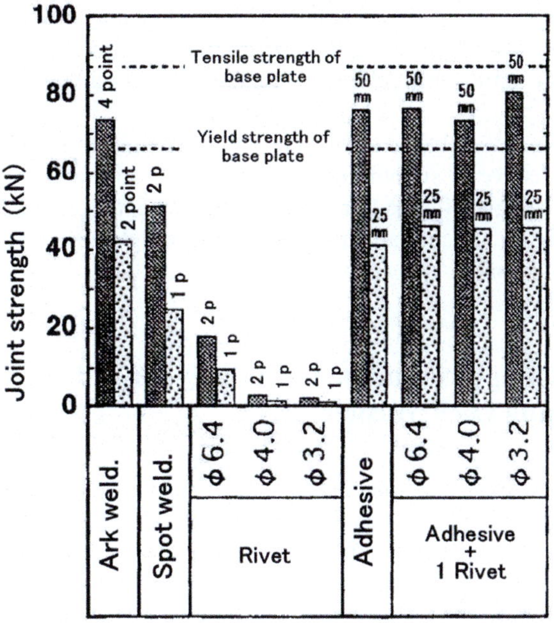

Fig. 11.25 Comparisons of fastening strengths by welding, riveting, adhesive and adhesive/rivet hybrid joints [5–12]

Table 11.6 Reduction effects of weight, man-hour, cost, work period, and factory noise by adhesive/rivet hybrid jointing [5–12]

	Welding	Adhesive/Rivet hybrid joint			
Surface treatment of steel plate	Steel plate		Plated steel plate		Painted steel plate
	Painted		Semi-painted		No painting
Plate thickness (mm)	3.2	2.3		1.6	
Housing weight (kgf) (Weight percentage)	146 (100%)	201 (82%)	201 (82%)	140 (57%)	140 (57%)
Man hour percentage	100%	69%	53%	58%	36%
Cost percentage	100%	80%	69%	70%	61%
Work period percentage	100%	82%	73%	73%	64%
Factory noise (hone)	80	80	80	80	80

(E) Conclusions

(1) The strength test results of the bolt/adhesive joint confirmed the superiority of the high rigidity of the adhesive part and the strength stability of the bolted part, but the strength of the adhesive part varied greatly, indicating that the hybridization of a single bolt/adhesive joint does not have a marked superiority.

(2) As a solution to the above problem, we found that a hybrid joint with multiple bolts, which is made by multiple-bolts array fastening and one adhesive surface, can provide sufficient rigidity and strength, and has a large final deformation capacity.
(3) In addition, not only strength, durability and rigidity improvements, it can also be effective in simplifying work processes and reducing weight, and is currently being applied to transportation equipment such as automobiles, rail-cars, elevators, and aircraft structures, where weight reduction is essential for SDGs' requirement.

11.3 Loosening Prevention Technology for Bolted Joint by Using Lock Washer [13]

In Chap. 6, we explained the anti-loosening technology in accordance with mechanics, and explained that double nuts, flange nuts, adhesives, etc., are effective, but washers are not always effective. However, there are washers that have been confirmed to be effective and are currently in use, and then here we will detail the mechanical analysis and experimental verification of lock washers that have proven in industry. In this section, one of the product examples, "HEICO-LOCK® Wedge Lock Washer", is taken up as an example and explained.

11.3.1 Overview of Lock Washers

The lock washers are composed of two pieces as shown in Fig. 11.26, and stack both cam surfaces against each other, which is then attached to the seating surfaces of bolts or nuts as follows:

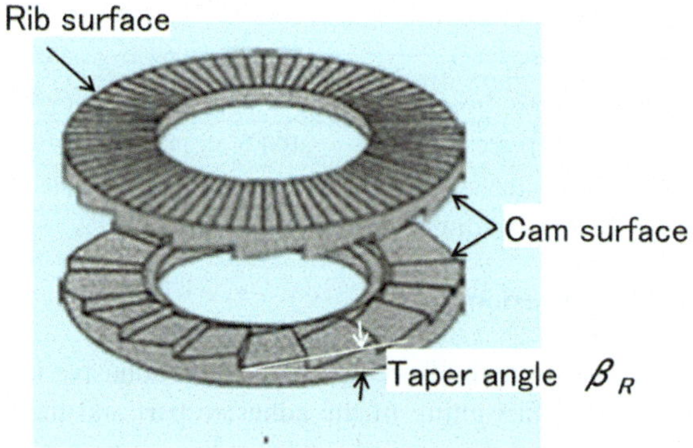

Fig. 11.26 Overview of lock washer "HEICO-LOCK® Wedge Lock Washer" [13]

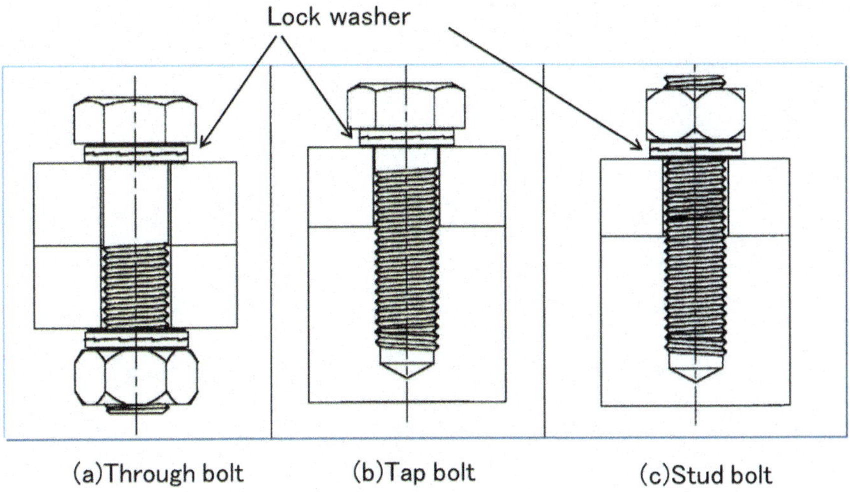

Fig. 11.27 Installation of lock washer [13]

(a) In the case of through bolts mount on both bearing surfaces of bolt head and nut.
(b) In the case of tap bolts mount on the bolt head bearing surface.
(c) In the case of nut tightening stud bolts mount on the nut bearing surface.

As shown in Fig. 11.27, the ribbed surface has a fine-jagged edge (radial teeth) that bite into the loosening side, resulting in a large coefficient of friction. The cam surface has a taper angle β_R which is larger than the bolt's lead angle β_s.

11.3.2 Experimental Verifications

(A) **Test bolts and washers**

For this test the M10 × 1.5 × 30, 10.9 bolts are used, and the fastening forces are measured by mounting bolt-gages (BTM-1C, by Tokyo Sokki Kenkyujyo Ltd.) in the central-holes $\Phi 1.0$ as shown in Fig. 11.28.

(B) **Test steel plate**

In this test, the two plates are used as shown in Fig. 11.29. Clamping thickness of both plates is 8.5 mm, so total clamping thickness is 17 mm. The bolt-head bearing surfaces are made to hexagonal shape as shown in Fig. 11.29 right. Molybdenum disulfide (MoS_2) was sprayed for lubrication on contact surface (see Fig. 11.30).

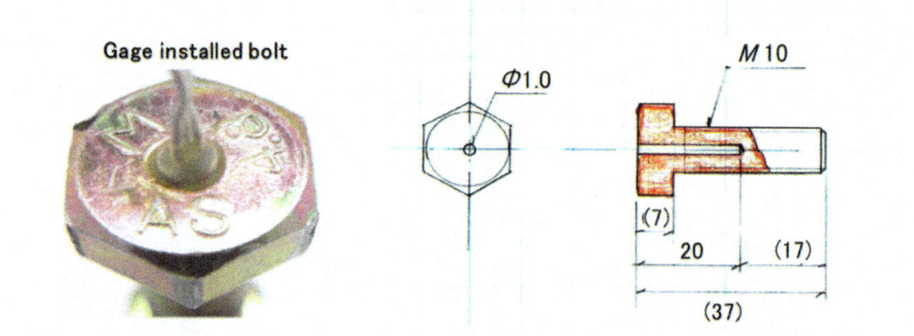

(a) Test bolt with gage installed (b) Test bolt with central hole

Fig. 11.28 Test bolt with central hole and bolt gage

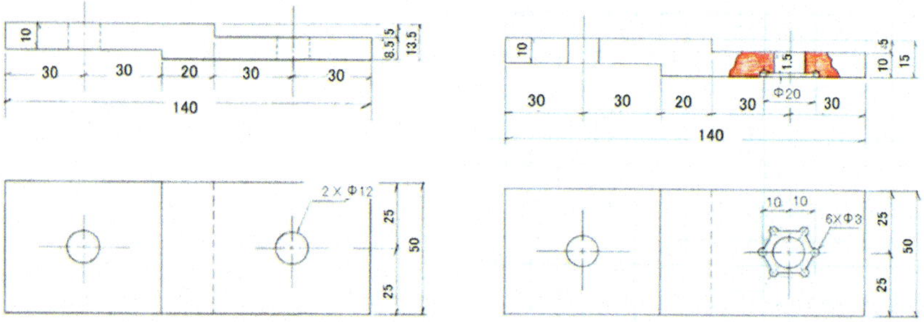

Fig. 11.29 Steel plates (SS400) for testing

(C) **Displacement sensor**

A clip gage (Clip Gage UB-5 by Tokyo Sokki) shown in Fig. 11.31 was used to monitor the amount of relative slip between the fastened plates.

(D) **Test apparatus and test conditions**

Figure 11.32 shows the installation of the fastened test specimen on the testing machine. The testing machine was an electro-hydraulic fatigue testing machine (Shimadzu SERVOPULSER, 50 kN capacity). The cyclic loading rate was set as 2.5 Hz due to heat generation limitations.

(E) **Test Results**

(E-1) Without lock washer

First, a loosening test was conducted using only standard bolts and nuts without lock washers. Figure 11.33 shows an example of the test results under an initial fastening force of 25 kN (standard fastening force for M10 bolts) and a cyclic relative

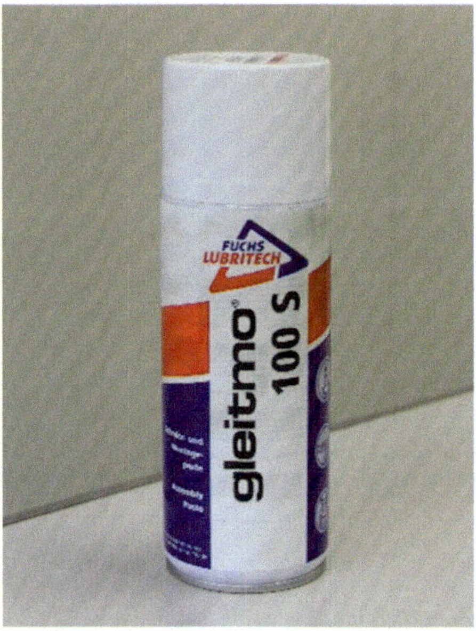

Fig. 11.30 Molybdenum disulfide (MoS_2)-based lubricants

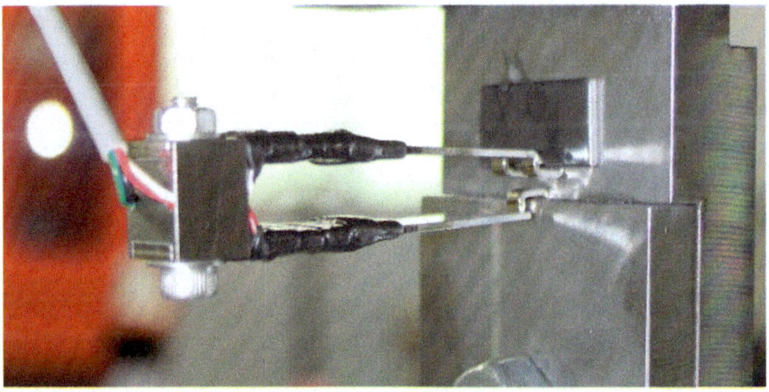

Fig. 11.31 Displacement sensor (Gap sensor UB-5)

slip of ±0.3 mm. In this case, the initial fastening force decreased rapidly and was completely lost after about 200 cycles.

(E-2) Using lock washer (HEICO-LOCK® Wedge Lock Washer)

Next, tests were conducted when lock washer (HEICO-LOCK® Wedge Lock Washer) was used in the nut seating area. Figure 11.34 shows an example of the test results under the same initial fastening force of 25 kN and cyclic relative slippage of ±0.3 mm as described above. In this case, the fastener loosened slightly in the initial stage of cyclic loading, but this loosening was completely stopped after about 100 cycles,

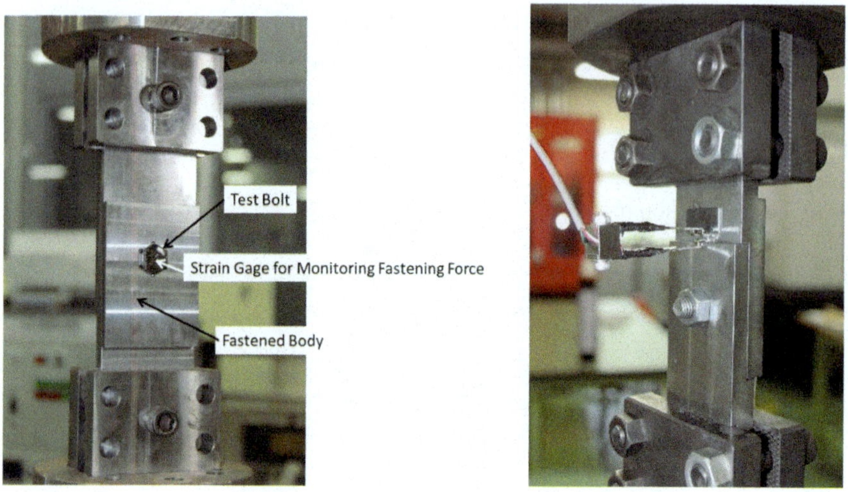

(a) Bolt head side with bolt gauge (b) Nut side with displacement sensor

Fig. 11.32 Installation of test specimen on testing machine

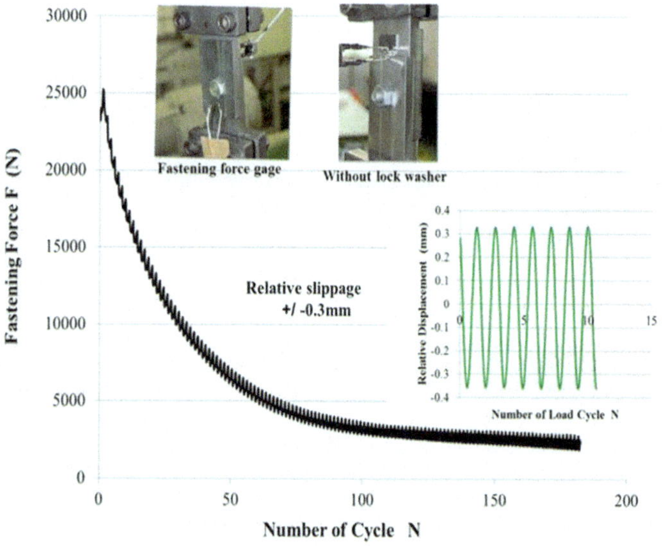

Loosening behavior without lock washer +/-0.3 mm

Fig. 11.33 Loosening behavior without lock washers (cyclic relative sliding ±0.3 mm)

and it was confirmed that more than 90% of the fastening force was still maintained even after 2,000 cycles.

(E-3) Comparison and Summary

The loosening behavior under the same conditions (M10 bolt, 25 kN initial fastening force, ±0.3 mm cyclic relative slip with and without the lock washer) can be seen from the comparison in Figs. 11.33 and 11.34. As can be seen from the comparison, without the lock washers the bolts completely loosen after 200 cycles, whereas with the lock washer, a significant difference was observed with securing more than 90% of the initial fastening force even after more than 2000 cycles of repeated loading. This shows that lock washer (HEICO-LOCK® Wedge Lock Washer) can make a significant contribution to improving the safety and durability of fasteners in machinery and structures in a wide range of industries.

11.3.3 Mechanical and Sliding Behaviors Analysis

Figure 11.35 shows the deformation and sliding behaviors of a structure in which a bolt, nut, locknut, and fastened body are subjected to an initial fastening force and a transverse load. The lower end of the bolt is fully constrained, and the end of the fastened body is free only in the x-direction (the direction of slip). As shown in Fig. 11.26, the lock washer has radial rib teeth with a high coefficient of friction on the ribbed surface, so that the two seating surfaces between the nut and lock washer and between the lock washer and the fastened body cannot slip. Therefore, the only sliding surfaces in this structure are the threaded surface between the bolt/nut and the lock washer cam surface. In Chap. 4, a mechanical analysis was performed under these conditions without the lock washer, and it was shown that under a transverse load, a torque T_{ss} (see Eqs. (4.23), (4.23′), and (4.23″)) was generated and the analysis shows that loosening occurs when this generated torque T_{ss} is greater than the nut loosening torque T_l (see Eq. (4.29) in Chap. 4). In this section, a similar mechanical and sliding analysis will be performed on the lock washer cam surface, and the loosening or fastening will be evaluated by integrating both the thread and cam surface analyses.

(A) **Sliding and mechanical analyses on lock washer cam surface**

Figure 11.35 shows the state of slippage perpendicular to the bolt axis on the lock washer cam face. This is the same as the mechanical analysis for the case of a square screw shown in Fig. 4.8 in Chap. 4, Sect. 4.4. Instead of the lead angle β for a square screw, the taper angle β_R of the lock washer cam face is applied. When slip occurs in the direction perpendicular to the bolt axis, the upper half of Fig. 11.36 show slide down on the cam face and the lower half show slide up on the cam face, and this mechanical asymmetry causes the nut and upper lock washer piece to generate a rotational torque T_{sR} in the direction shown in Fig. 11.35. Note, however, that

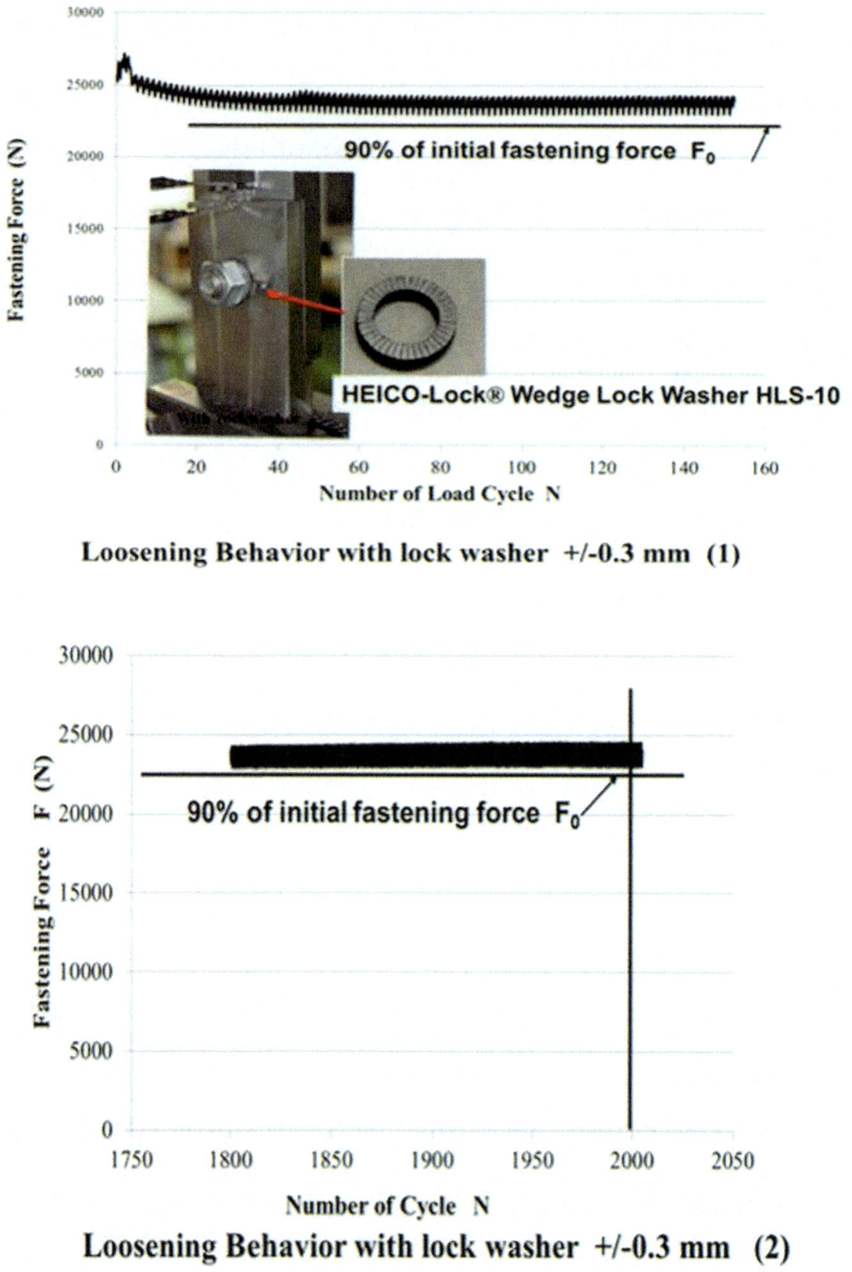

Fig. 11.34 Loosening behavior with lock washer (HEICO-LOCK® Wedge Lock Washer) (cyclic relative sliding ±0.3 mm)

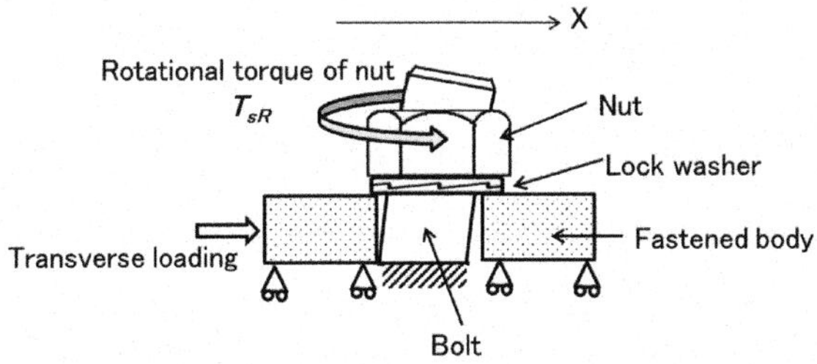

Fig. 11.35 Deformation and sliding behavior of bolt, nut, lock washer, and fastened body under transverse loading

this rotational torque T_{sR} is in the tightening direction for the lock washer, which increases the bolt tightening force. This T_{sR} is derived as follows with reference to Eq. (4.30) in Chap. 4.

$$T_{sR} = \frac{F_f \beta_R \left(D_o^2 + D_o D_i + D_i^2\right)}{6(D_o + D_i)} = \frac{F_f \beta_R D_R}{4} \quad (11.4)$$

where D_o: Lock washer outer diameter

D_i: Lock Washer inner diameter

D_R: Lock washer effective diameter $= \frac{2(D_o^3 - D_i^3)}{3(D_o^2 - D_i^2)}$.

The sliding limit torque in the direction of rotation corresponds to the expression for the tightening torque at the screw surface, and is derived as follows with reference to Eq. (4.31) in Chap. 4.

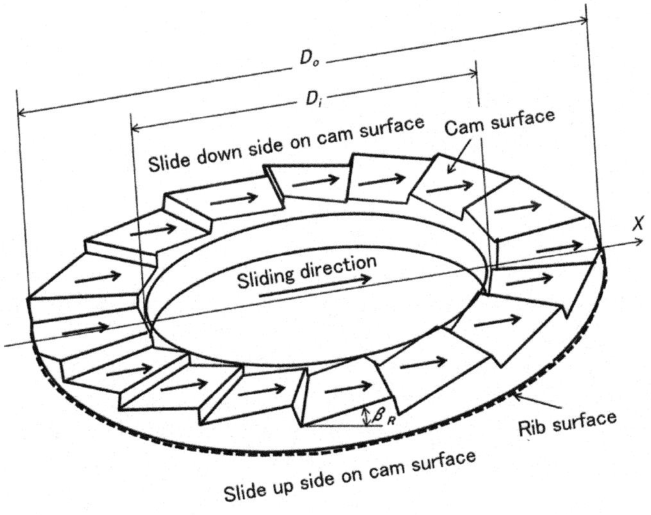

Fig. 11.36 Sliding behavior on cam face of lock washer

$$T_{fR} = \frac{F_f}{2}(D_w\beta_R + D_w\mu_R) = \frac{F_f D_w}{2}(\beta_R + \mu_R) \tag{11.5}$$

where μ_R is the friction coefficient of the cam surface.

(B) **Sliding and mechanical analysis that integrates the behavior of the cam surface and thread surface**

The generated torque T_{ss} in this rotational direction at the thread surface when sliding occurs in the transverse direction is derived from Eq. (4.41) in Chap. 4 as follows,

$$T_{ss} \cong \frac{F_f d_2 \tan\beta_s}{4\cos\alpha'} \tag{11.6}$$

and the slip generation limiting torque in this rotational direction on this thread surface corresponds to the expression for the loosening torque on the thread surface, which is derived as follows with reference to Eq. (4.32) in Chap. 4.

$$T_{ls} = \frac{F_f d_2}{2}\left(-\beta_s + \frac{\mu_s}{\cos\alpha}\right) \tag{11.7}$$

where β_s is the lead angle of the thread.

By synthesizing Eqs. (11.4)–(11.5), the rotational slip occurrence criteria of the nut under transverse sliding condition can be shown as follows.

$$T_{sR} + T_{ss} > T_{fR} + T_{ls}$$

$$\frac{F_f\beta_R D_R}{4} + \frac{F_f d_2 \tan\beta}{4\cos\alpha'} > \frac{F_f D_w}{2}(\beta_R + \mu_R) + \frac{F_f d_2}{2}\left(-\beta_s + \frac{\mu_s}{\cos\alpha}\right) \tag{11.8}$$

Assuming that the coefficient of friction between the thread and cam surfaces is equal, and rearranging Eq. (11.8), we obtain

$$\mu_s = \mu_R < \frac{d_2\beta_s\left(\frac{1}{\cos\alpha} + 2\right) - D_R\beta_R}{2\left(D_R + \frac{d_2}{\cos\alpha}\right)} \tag{11.9}$$

This was calculated with $d_2 = 9.026$ mm, $D_R = 16.3$ mm, and $\beta_s = 0.053$, corresponding to the $M10$, $P1.5$ screw in the previous section, and the taper angle of the cam face of the lock washer β_R, and is shown in Fig. 11.37.

In order for this lock washer to be tightened without loosening under transverse loading, the friction coefficient of the cam surface μ_R must be below Eq. (11.9) and cam taper angle must be $\beta_R > \beta_s = 0.053$ to tighten under the direction of rotation slip in Fig. 11.35. And as a conclusion, the hatched region in Fig. 11.37 is an effective range.

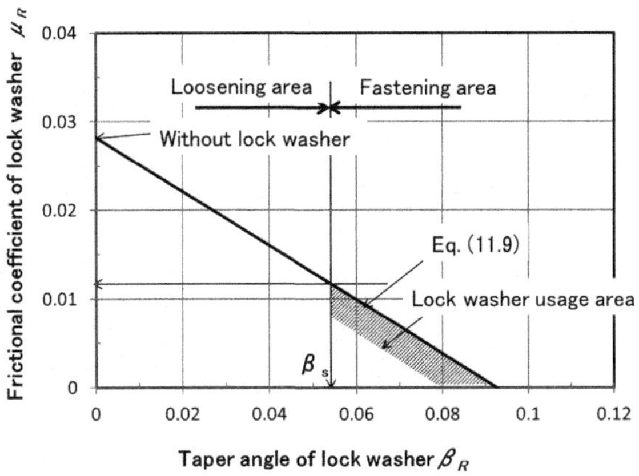

Fig. 11.37 Relation between the Taper angle of lock washer β_R and slip occurrence criteria on lock washer cam face

11.3.4 Comparative Discussion of Experimental and Analytical Results

The condition for slipping and tightening shown in Eq. (11.8) is that tightening is performed only with this torque and in this case the frictional coefficient of lock washer surface μ_R must lower than $\beta_R/2 = 0.027$ as shown in Fig. 11.38a, this will be impossible. But if slipping actually occurs in the direction perpendicular to the axis, the slipping will be superimposed and it will be possible to rotate in the tightening direction as shown in Fig. 11.38b.

However, at the same time, this torsional drive torque is very small, so it is not enough to increase the tightening force progressively, as shown in Fig. 11.34. However, compared to the loosening behavior without lock washers shown in Fig. 11.33 under the same loading conditions, this is an epoch-making improvement, which greatly verifies the validity of the mechanics and behavior analysis in the previous section and at the same time confirms the effectiveness of lock washers.

11.4 Fatigue Strength of Stud and Tap Bolts [14]

In general, in the strength designs of bolted joints, we tend to focus only on bolts rather than fastened bodies. However, recently, CFRP, aluminum alloys, titanium alloys, and other new materials are being used as fastened bodies to reduce weight, etc. The characteristics and behavior of the fastening element affect the stress distribution and strength of the bolt itself, or the strength of the fastened body itself. The number of cases of damage due to the strength of the fastened body itself has been increasing. Bolts and nuts are the most frequently used bolted joint parts, but there are many cases of large mechanical structures, such as engine blocks and pressure vessels, that are fastened by machining a female thread on the body side and tap bolt. It

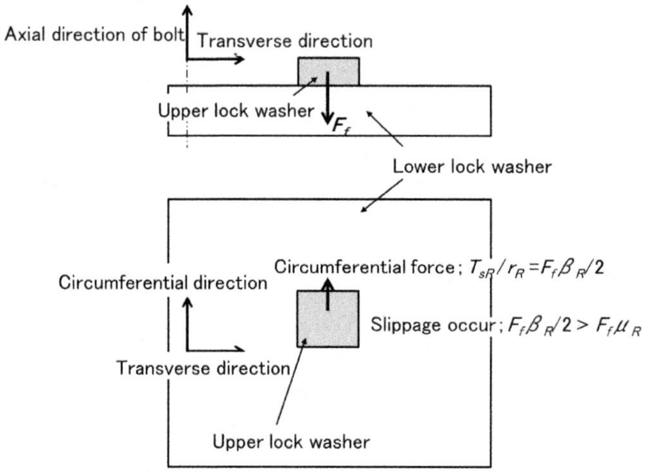

(a) Mechanical state with only torsional torque T_{sR}

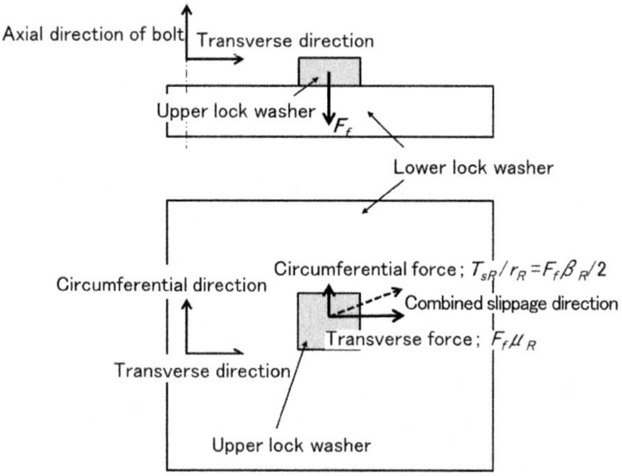

(b) Superposition behavior of torsional torque T_{sR} and transverse force $F_f\mu_R$

Fig. 11.38 Mechanical analysis and sliding behavior on opposing lock washer contact surface. **a** Mechanical state with only torsional torque T_{sR}. **b** Superposition behavior of torsional torque T_{sR} and transverse force $F_f\mu_R$

is well-known that the mechanical properties of such body-side threaded parts are considerably different from those of bolts and nuts due to their high rigidity around female thread. For example, fracture of a bolt/nut joint often occurs around the bottom of the first thread valley of the bolt that is closest to the nut seating surface. On the other hand, when an stud bolt or tap bolt is tightened by applying torque to the head of the bolt against a female thread machined in the body side, fatigue failure may occur from the valley bottom of the female thread at deep inside of the thread hole [14]. Similar accidents have been reported even in recent years. So, it is important to clarify the mechanical properties of fastening parts using tap bolt or stud bolt.

In this section, we will explain a research paper by Fukuoka et al. [14] that analyzed the strength around the female thread based on the finite element method stress analysis of stud or tap bolt joints. In the FEM analysis, a three-dimensional finite element model that faithfully reproduces the helical shape of the screw thread is used to clarify the phenomena caused by the "helix" of the screw thread. The main focus of the analysis is to elucidate the mechanism by which the stress amplitude increases at the bottom of the valley on the female thread at deep inside of the thread hole.

11.4.1 Mechanical Properties of Bolt/Nut Joints and Stud or Tap Bolt Joins

Figure 11.39 shows the general fatigue fracture pattern of a screw fastener. In the bolt/nut fastener shown in Fig. 11.39a, fatigue failure is most likely to occur around the bottom of the first thread valley bottom of the bolt, which is close to the nut seat surface. On the other hand, as shown in Fig. 11.39b, in the case of stud or tap bolt joints fatigue failure may also occur from the valley bottom of the female thread at deep inside of the thread hole [14]. This phenomenon is often become a problem in the connecting rod of diesel engines, as shown in Fig. 11.39c. In a photoelastic experiment a high stress amplitude was observed at the bottom of the valley on the female thread side at the bottom of the screw hole [15]. In view of the above, it is important to quantitatively clarify the stress concentration and distribution behaviors at the female threads from the viewpoint of the safety of stud or tap fastening parts. The following is a summary of the mechanical properties of stress concentration and stress amplitude at the bottom of the thread valley of a bolt/nut fastener obtained from previous studies for comparison with the stud or tap joint thread that is the main subject of this section [16, 17].

Figure 11.40 shows the calculated distribution of Mises stress along the bottom of the thread valley of a bolt and nut when subjected to axial force only, using a finite element model for an M16 coarse thread [16]. The vertical axis is the normalized Mises stress σ_{eq} divided by the bolt axial nominal stress σ_b, and the horizontal axis is the axial coordinate of the bolt where $z = 0$ represents the nut seating surface.

The screw pitch P is 2 mm and the number of threads in the fit z is 5. The maximum stress occurs at the bottom of the first thread valley of the bolt, 0.5 pitch away from the nut seating surface, and the stress at the bottom of the thread valley on the nut side is much lower than that on the bolt side. And the effect of the contact surface friction coefficient is small. Figure 11.41 shows the stress amplitude generated along the bottom of the bolt thread valley when a cyclic tensile load W is applied on the surface of the fastened body just outside the nut, 16 mm away from the bolt shaft

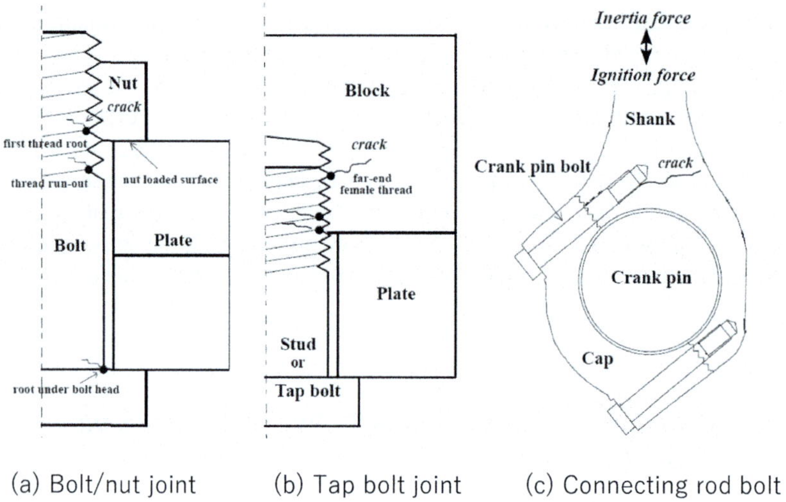

Fig. 11.39 Fatigue failure modes in each bolt joint structures [14]

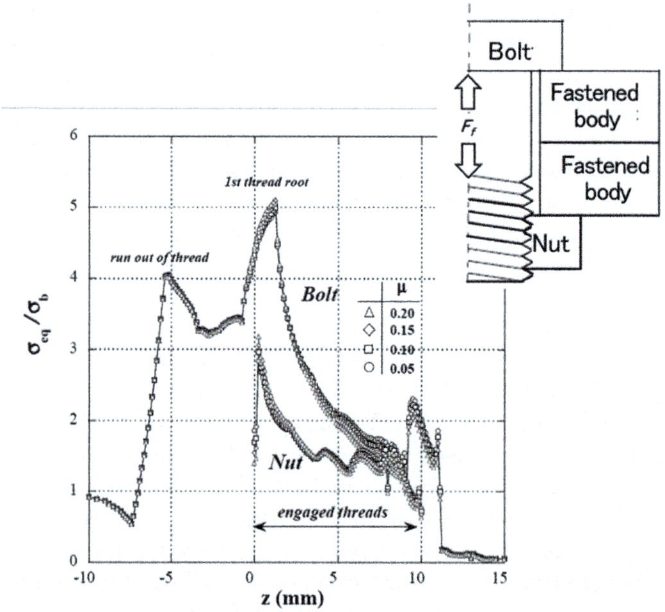

Fig. 11.40 Mises stress distributions along nut thread valley under initial fastening load (M 16, *P* 2.0) [14]

center, for the case of two hollow cylinders with thickness 24 mm and outer diameter 128 mm tightened by a single M16 bolt [17]. $W = 20.1$ kN in the figure is equivalent to the axial force when the bolt axial stress σ_b is 100 MPa. The stress amplitude is obtained by dividing the difference between the axial stress component σ_z at the time of external force and initial bolt tightening by 2. The stress amplitude is maximum at the bottom of the first thread valley of the bolt. γ in the figure represents the ratio of the radial contact lengths on the interface of the fastened body when the bolt is subjected to an external load W. When only the bolt-fastened force F_f is applied with no external load this ratio is $\gamma = 1$, and $\gamma = 0$ when a large external force is applied and the contact interface is completely separated. The analytical conditions and results for the stress distribution and stress amplitude along the thread valley bottom are described in detail in Refs. [15, 16]. Figure 11.41a clearly shows that the load factor Φ increases and the maximum stress amplitude rapidly increases as the external force W increases and the interface between the fastened bodies approaches to a state of complete separation. The stress amplitude at the bottom of the screw valley on the nut side shown in Figure 11.41b is much smaller than that on the bolt side, similar to the stress distribution at the bottom of the thread valley shown in Fig. 11.40. Comparing Figs. 11.40 and 11.41, it can be said that the both occurrence positions where high stress concentration and high stress amplitude occur in the bolt/nut joints are almost coincided at the bottom of the first valley of the bolt.

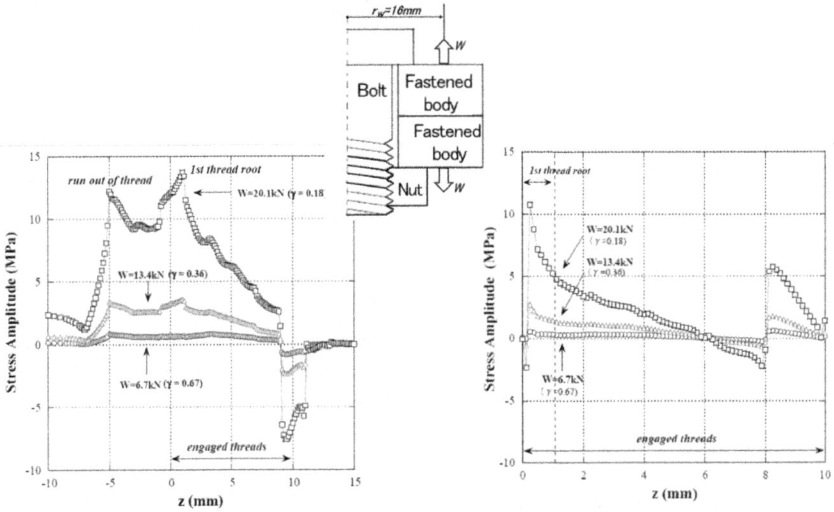

(a) Distributions of stress amplitudes along bolt thread valley

(b) Distributions of stress amplitudes along nut thread valley

Fig. 11.41 Stress amplitude distributions along bolt and nut thread valley under cyclic loading (M 16, P 2.0, $r_w = 16$ mm) [14]

11.4.2 Mechanical Analysis of Stud and Tap Bolt Fasteners

(A) **Analysis model**

In this section, in order to clarify the mechanical properties inherent to the body-side female threads caused by the "helical shape of the screw threads", analyses are conducted under all axisymmetric conditions except for the helical shape of the threads, such as the shape of the fastened body and the external force loading configuration. A large stress concentration occurs in a structure with sharp notches, such as a thread valley bottom, so the accuracy of the analytical model often makes a problem when performing numerical analysis. Therefore, in this section a helical model mesh pattern takes analytical accuracy and analytical efficiency into consideration as shown in Ref. [16].

Figure 11.42a schematically shows the dimensions of the analytical model. This bolted joint model is composed of a single bolt, fastened bodies with 48 mm thick hollow cylindrical plate and block body with female thread. The bolt is an M16 coarse thread bolt with a pitch of $P = 2$ mm, and is tightened with either a stud bolt or a tap bolt. The external load W is applied as a cyclic tensile load symmetrical to the center of the bolt axis. On the bottom face of the block body, axial displacement is constrained at all nodes and circumferential displacement is constrained at four nodes 90 degrees apart on the outer edge. Figure 11.42b and c shows the finite element model used in the analysis. To make the threads easier to see, the element mesh model is shown upside down compared to Fig. 11.42a. The number of threads with mating is 7.5. The total number of nodes and elements is 176,279 and 164,144, respectively. The method for creating the spiral mesh is as introduced in Ref. [16], so that each thread makes smooth contact.

(B) **Analysis conditions**

The initial fastening is performed either by directly applying tension using a hydraulic tensioner or by applying torque to the bolt head. In the former case, in the analytical model shown in Fig. 11.42b, the initial fastening bolt axial force is generated by applying an irregular amount of "bite" to the bolt head contact surface. In the latter case, the axial force is generated by applying a circumferential torsional moment to the outer edge of the bolt head. The initial fastening bolt axial stress σ_b is varied from 100 to 500 MPa. Here, fastening bolt axial stress $\sigma_b = 100$ MPa is equivalent to the tightening force of 20.1 kN. When calculating the stress amplitude, the initial fastening condition is just the case where the bolt is directly tensioned. The external load W is varied with $W/F_b = 1/3$, 2/3, 1, 4/3, and 2. For example, when the initial bolt axial stress σ_b is 100 MPa, the magnitudes of W are 6.7, 13.4, 20.1, 26.8, and 40.2 kN. The distance r_W from the center of the bolt shaft to the loading point of the external force W is set to 16 and 64 mm. Here, $r_W = 16$ mm is just outside of the outer circumference of the bolt head. All fasteners are assumed to be elastic, with Young's modulus of 200 GPa and Poisson's ratio of 0.3. The friction coefficients of the contact surfaces are varied from 0.05, 0.10, 0.15, and 0.2. Standard analysis conditions are as follows: bolt axial stress $\sigma_b = 100$ MPa, coefficient of friction $\mu =$

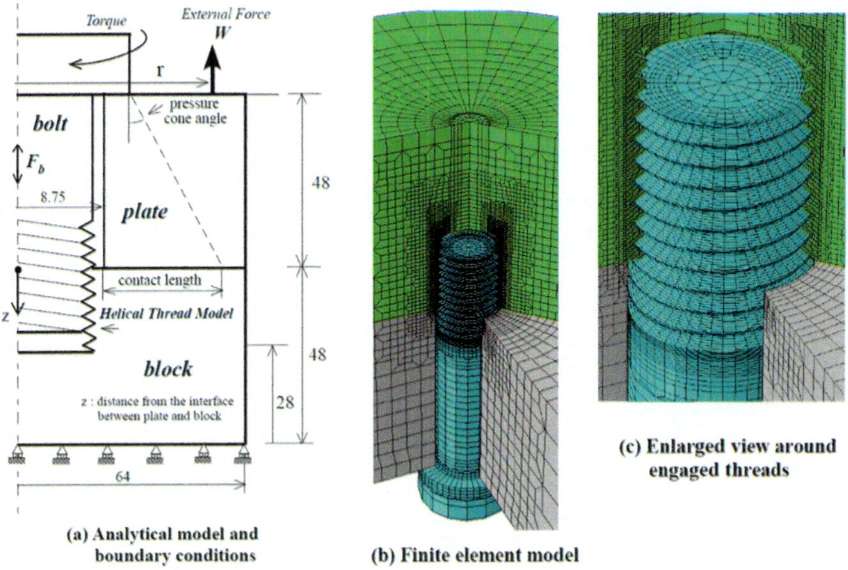

Fig. 11.42 FEM analytical model and boundary conditions for tap bolt joint [14, 16]

0:15, and $r_W = 16$ mm. Numerical analysis was performed using the general-purpose FEM analysis code ABAQUS Ver. 6.11.

11.4.3 Results of Mechanical Analysis of Stud and Tap Bolt Fasteners

(A) **Stress distribution and stress concentration at the bottom of the thread valley during initial fastening**

Figure 11.43a and b compares the stress distribution generated at the bottom of the male (bolt) and female (fastened body) thread valleys when the tap bolt was initially fastened by only axial stress of 100 MPa and by applying torque to achieve the same axial force. The coefficient of friction μ is set to 0.15. On the male thread side, the overall stress is higher when torque is applied due to the shear stress generated by the torsional moment. On the other hand, the opposite trend is observed on the female thread side.

(B) **Stress amplitude distributions along the bottom of the thread valley under cyclic loading**

Figure 11.44a and b shows the stress amplitudes generated along the valley bottom of the bolt thread (male) and block thread (female) under cyclic external force on

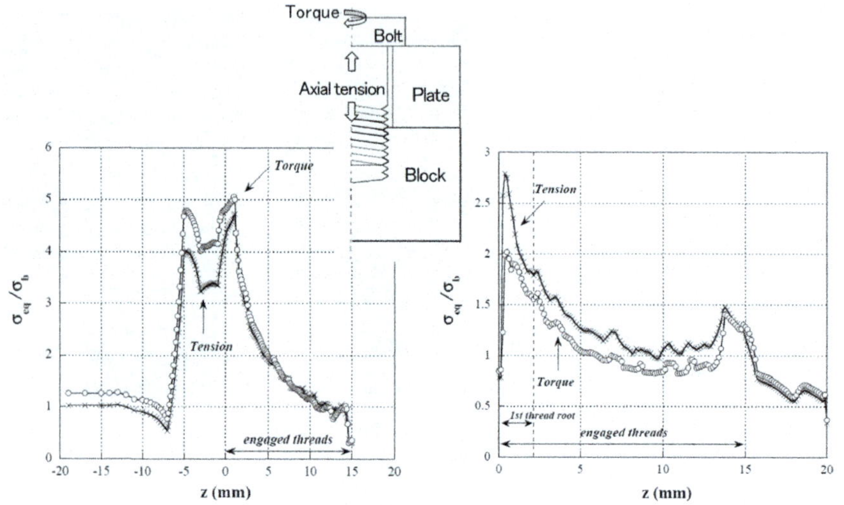

(a) Stress distributions along bolt thread valley under initial fastening

(b) Stress distributions along block thread valley under initial fastening

Fig. 11.43 Mises stress distributions along bolt thread valley and block thread valley of tap bolt fastened joint under initial fastening stress 100 MPa (M 16, P 2.0, $r_w = 16$ mm) [14]

the external force position $r_W = 16$ mm. Here, the interface between the block and the plate is in partial contact conditions, and γ in the figure is the ratio of the contact lengths defined in Sect. 11.4.1. As shown in Fig. 11.44a on the bolt thread side, when the external force W is small, the stress amplitude is maximum at the bottom of the first screw valley in no engagement region with block female, but as γ decreases with increasing W, the stress amplitude is maximum at about second thread on the engagement region with block female thread. When the value of γ approaches zero, the stress amplitude increases rapidly. Then as shown in Fig. 11.44b the distribution pattern of stress amplitude on the female thread side is even more characteristic. At the bottom of the thread valley at the far end, the maximum stress amplitude shows about 30 MPa, which is higher than that on the bolt thread side. This stress distribution behavior in female thread side is completely different from the stress distribution behavior in the bolt/nut fastener as shown in Fig. 11.41b.

Therefore, the results shown in Fig. 11.44b support the fatigue failure phenomenon of the actual machine as shown in Fig. 11.39b and c. These are the inherent mechanical properties on tap and stud bolt joints, and are considered to be important results revealed for the first time by applying the finite element method.

Figure 11.45 shows the stress amplitude along the bottom of the bolt thread valley when the loading position r_W is 64 mm, corresponding to Fig. 11.44 ($r_W = 16$ mm). The calculated results for the bolt threads show the same trend, except that the maximum stress amplitude occurs outside of the first thread becomes little decreased.

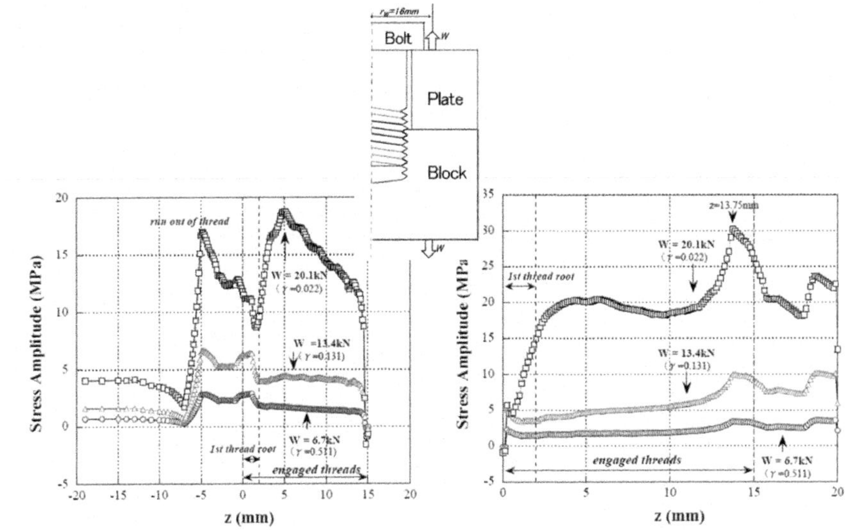

(a) Stress distributions along bolt thread valley under cyclic external force

(b) Stress distributions along block thread valley under cyclic external force

Fig. 11.44 Mises stress distributions along bolt thread valley and block thread valley of tap bolt fastened joint under cyclic loading (M 16, *P* 2.0, $r_w = 16$ mm) [14]. **a** Stress distributions along bolt thread valley under cyclic external force. **b** Stress distributions along block thread valley under cyclic external force

The results on the block thread side were omitted because the effect of r_W is even smaller than bolt side results.

11.4.4 Maximum Stress Amplitude Generation Mechanisms at the Bottom of the Female Thread Valley

In the analysis of the previous section, it was shown that the maximum stress amplitude occurs at the bottom of the female thread side valley at the edge of engaged region with bolt. In this section, we will explain the mechanism from a micro-perspective consideration and macro-perspective considerations.

(A) **Microscopic mechanisms**

As an example, consider the case shown in Fig. 11.44b where the magnitude of the external load *W* is 20.1 kN. The maximum stress amplitude slightly exceeds 30 MPa and occurs at $z = 13.75$ mm, close to the depth of the screw hole. Referring to Fig. 11.43b, the equivalent stress σ_{eq} during fastening shows a small peak at this location, and the axial stress σ_z has a similar distribution. This local stress

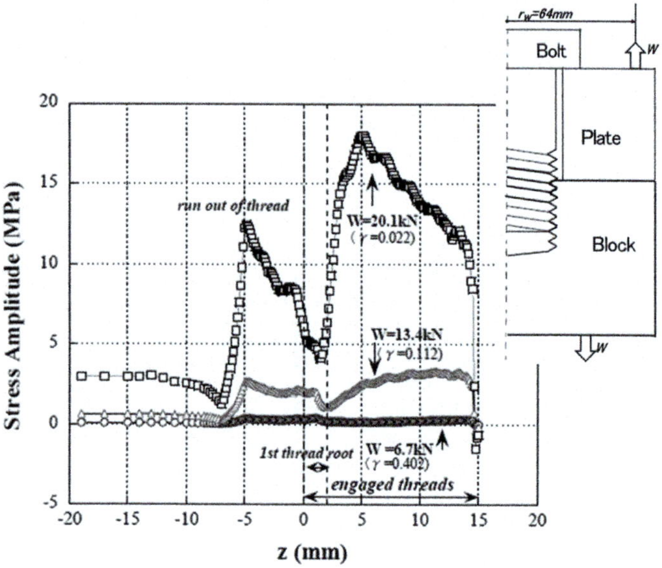

Fig. 11.45 Mises stress distributions along bolt thread valley of tap bolt fastened joint under cyclic loading (M 16, P 2.0, $r_w = 64$ mm) [14]

concentration will be discussed in terms of the microscopic deformation pattern in this area.

Figure 11.46a shows a schematic of the fitting area near the back of the screw hole. The aforementioned $z = 13.75$ mm corresponds to the position $0.625P$ from the tip end of the fit screw, which is the borderline position where the male thread and female thread cease to contact each other. As shown in this figure, since the surface pressure on the contact surface acts to push down each of the mail threads from the upside. Due to the continuity of similar deformation below than No.② concentration of deformation and stress of each thread will be alleviated. In contrast, at the bottom of the female thread valley ① concentration of deformation and stress become high as the contact pressure of the thread just above is 0 and so the opening of thread valley ① becomes larger. To verify this consideration we will show the deformation pattern on the female thread side around the back of the screw hole when the bolt axial stress $\sigma_b = 100$ MPa and the external force $W = 20.1$ kN, in Fig. 11.46b. The deformation is magnified by a factor of 1000. The figure shows that there is significant "mouth-opening deformation" around the bottom of the fit end female thread valley at $z = 13.75$ mm, compared to the mouth opening of the lower thread valley.

(B) **Macroscopic mechanisms**

In the previous section, the discussion was based only on the calculation results of the finite element method, but here we will provide a macroscopical and qualitative discussion based on the flow of force.

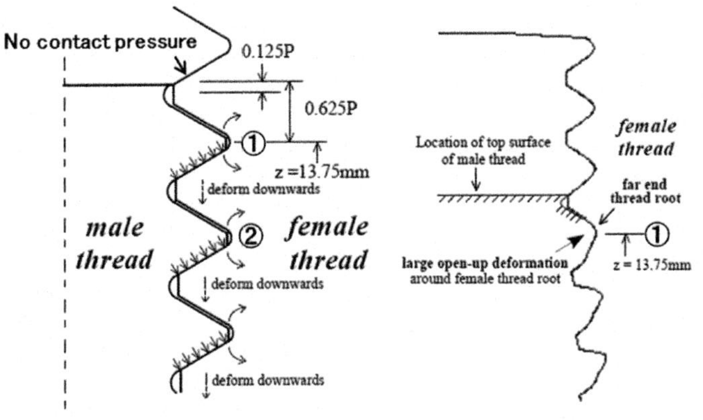

(a) open-up deformation around thread root (b) numerical result of female thread deformation

Fig. 11.46 Microscopic mechanism of deformation and stress concentration at fit end female thread valley in tap or stud bolt joint [14]

Figure 11.47 shows conceptual diagrams of the force flow and load point of each joint model, (a) is the bolt/nut-fastening structure and (b) is the tap or stud bolt fastener structure. In the bolt/nut-fastening structure of (a), the force flow passes through the first thread of the screw fit, so the nut valley bottom stress amplitude is also concentrated on the first thread side, as is generally explained. In the case of the bolt/nut-fastening structure shown in (b), if the external force loading point (restraining position) is on the top surface of the body as shown in the left figure of Fig. 11.47b, the valley bottom stress amplitude will also be concentrated on the first thread side as in the case of the bolt/nut-fastening structure shown in (a). In the case where the bottom surface of the body is restrained, as shown in the right figure of Fig. 11.47b, the force flow passes through the deepest thread of the screw fit area and is consistent with the above mentioned microscopic consideration.

(C) Summary

The effect of the body side restraint (loading point), load factor, and load sharing on the structure of such tap and stud bolt fastening bodies has not been discussed enough until now. So, it is essential for the designers themselves to construct a simple mechanical model that is close to the actual structure. For example, if the connecting rod-fastening bolt structure shown in Fig. 11.39c is the products subject to strength evaluation.

First of all, in order to be able to intuitively guess the parts at risk of failure, it is necessary to have the ability to visualize the flow of force as shown in Fig. 11.47b right; "Macroscopic Consideration".

Then, the quantitative mechanical analysis, such as FEM analysis, is performed to focus on this risk area.

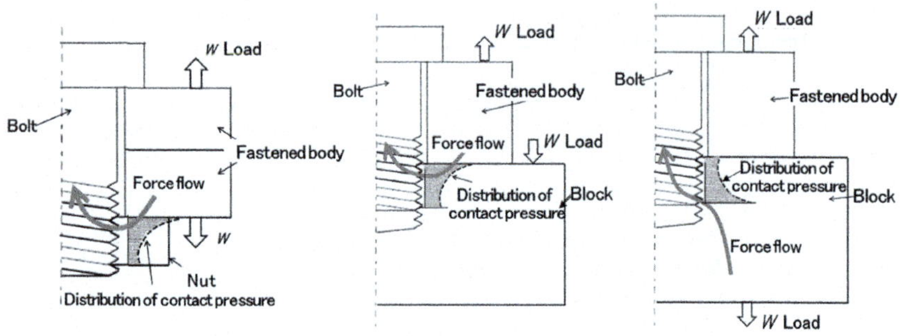

(a) Bolt/nut joint

/Loading at upside of block /Loading at depth of block
(b) Tap or stud bolt joint

Fig. 11.47 Influences of joint structure and loading point on the force flow in each joint structure. **a** Bolt/nut joint. **b** Tap or stud bolt joint

Finally, apply the detailed stress analysis results to critical distance diagram such as Figs. 10.3 and 10.18 to predict the fatigue limit, fatigue life, etc., with reference to the critical distance stress theory (detailed in Sect. 10.4.3); "Microscopic Consideration".

In the future, I hope that general-purpose simple parameters such as the load factor or load sharing parameter for bolt/nut joints will be created for such tap or stud bolt joint structures. This will eliminate the need for detailed stress analysis for each thread joint structure.

References

1. Toshio Hattori, Toshio Nonaka, Motoharu Taneda, Strength of fastened bolts in plastic zone, Pressure Technology, 23-1, pp. 7–13 (1985).
2. Nishiguchi, K. et al., New Edition of Joining Technology Handbook, Sangyo Gijutsu Service Center, pp. 636, (1994) (in Japanese).
3. Toshio Hattori, Learning from Accident Cases, Fundamentals of Material Mechanics and Strength Design, Nikkan Kogyo Shimbun, pp. 159, (2018-12)
4. Society of Automotive Engineers of Japan, JSAE Symposium No. 07-10, Joining Technology and Experimental Technology for Tomorrow's Automotive Strength and Reliability (2010).
5. Kosuke Haraga, Various Properties and Application Examples of Ribet Jointing, JSME Seminar No.08-113, Actual and Future Development of Fastening and Joining, pp. 17–22, (2008).
6. K. Haraga, Quantification of long-term reliability in adhesive bonding, Mate2003 (9th Symposium on "Microjoining and Assembly Technology in Electronics") pp. 139–144 (2003).
7. Kosuke Haraga, Proceedings of the 33rd Symposium on Reliability and Maintainability (Union of Japanese Scientists and Engineers), pp. 117–122 (2003,7).
8. K. Haraga, Distribution of water absorption and bond strength on the bonded area and its change with time, Journal of the Adhesion Society of Japan, 23(5)178 (1978).
9. K. Haruna, K. Haraga, Moisture Durability Prediction of Adhesively Bonded Joints Using Acrylic Adhesives, 8th Int. Conf. on Composites Engineering (2001).
10. Kosuke Haraga, Journal of the Adhesion Society of Japan, 15(12)568 (1979).

11. Tsutomu Hiranuma, Toyokazu Takeuchi, Hiroyuki Kayaki, Kunihisa Shimura, Report of the Special Committee on Automotive Structural Bonding Technology, Society of Automotive Engineers of Japan, pp. 56, (1992).
12. K. Haraga, N. Tani, S. Midorikawa, Proceedings of the 38th Annual Meeting of the Adhesion Society of Japan, pp. 7, (2000).
13. https://www.heico-group.com/ja/heico-lock/wedge-lock-washers/.
14. Toshimichi Fukuoka, Masataka Nomura, Yosuke Takeda, and Yuichiro Mori, "Finite Element Analysis of Tightening Characteristics and Stress Amplitude of Tap and Stud Bolts," Transactions of the Japan Society of Mechanical Engineers, A, Vol. 79, No. 799 (2013), pp. 313–326.
15. F. Böhm, "Progress in developing M.A.N. two-stroke and four-stroke Diesel-engines", ISME TOKYO'78 Technical Papers, (1978), pp. 157–169.
16. Toshimichi Fukuoka, Masataka Nomura, and Yuya Morimoto, "Accurate Modeling and Finite Element Analysis of Thread Spiral Geometry," Proc. of the Japan Society of Mechanical Engineers, Vol. 72, No. 723 (2006), pp. 1639–1645.
17. Toshimichi Fukuoka, Masataka Nomura, Takashi Fuchigami, "Fatigue Strength Evaluation of Bolted Fasteners Using Threaded Spiral Model", Proc. of the Japan Society of Mechanical Engineers, Vol. 75, No. 759 (2009), pp. 1570–1576.

Chapter 12
Outline of Standards Related to Bolted Joints

12.1 Overview of International Standards (ISO) and Japanese Standards (JIS)

12.1.1 History of Standards

Screw fasteners are a versatile and important component of industrial products and machinery, and as such, they have been standardized globally starting from the early stages of the industrial history [1]. The historical changes in international standards are shown below. Traditionally, there have been three main standards for screws: the Whitworth screw, which was established in the UK, the Unified screw, which was created in the United States, and the metric screw, which originated in Europe. The ISO metric screw, established by the International Organization for Standardization (ISO), has currently become the dominant standard. The establishment, revision, and abolition of ISO standards related to screws are primarily carried out by two Technical Committees of TC1 (Screws) and TC2 (Fasteners).

In 1841, Whitworth in the UK developed the Whitworth screw.

In 1864, Sellers in the US developed the Sellers screw.

In 1868, the Sellers screw was adopted as the US screw (American standard).

In 1882, the Whitworth screw was adopted as BSW (British standard).

In 1894, France established the SF screw (metric screw).

In 1898, the SF screw developed into the SI screw through discussions among France, Germany, and Switzerland.

In 1918, the Society of Automotive Engineers (SAE) announced the SAE screw.

In 1926, the International Federation of the National Standardizing Associations (ISA) was established.

In 1933, the American Standards Association integrated the US screw, SAE screw, and others (referred to as the American screw).

© The Author(s), under exclusive license to Springer Nature Switzerland AG 2025

T. Hattori and T. Naruse, *Compendium for Design and Manufacturing of Bolted Joints*, Solid Mechanics and Its Applications 280,
https://doi.org/10.1007/978-3-031-89628-6_12

In 1940, the ISA metric screw was established from the SI screw (the first international standard).

In 1947, ISO was established, inheriting ISA.

In 1948, the Unified screw was established in the US, UK, and Canada, and replaced the American screw.

In 1962, ISO established the ISO metric screw and ISO inch screw (based on the Unified screw).

In Japan, the first metric screw, designated as JES standard, was established in 1924. Since then, it has been continuously revised while aligning with the current JIS standard and ISO. The following shows the transition of standardization in Japan [2]. For further details on the subsequent transitions, please refer to the Ref. [4].

In 1924, the first metric screw, known as JES standard No. 13, was established (metric coarse thread). Afterward, metric fine thread screws, the Whitworth screws No. 1, and Whitworth fine thread screws were standardized.

In 1939, the temporary Japanese Industrial Standard (JIS) was established.

In 1945, the Industrial Standardization Council was established, and the Japanese Standard (New JIS) was established.

In 1949, the Japanese Industrial Standard (JIS) was established, and the JES standard and temporary JES standard were transitioned to the JIS standard.

In 1952, the standards for metric coarse thread screws and inch coarse thread screws were established.

In 1965, ISO screws were introduced.

In 1968, ISO standard metric screws were adopted, and Whitworth screws were discontinued.

Currently, the standardization committee established by the Japan Research Institute for Screw Threads and Fasteners (JFRI) is primarily responsible for the establishment, revision, and abolition of JIS standards for screws [3, 4]. The standardization committee of JFRI also participates in and contributes to ISO/TC1 and ISO/TC2 [5, 6]. In recent years, with the globalization of economic activities, it has become necessary for JIS standards to align with international standards. For screw-related standards, the degree of conformity to international standards is indicated in three categories based on ISO/IEC Guide 21:1999 (Adoption of International Standards as Regional or National Standards): IDT (identical), MOD (modified), and NEQ (not equivalent). The harmonization of JIS standards with international standards is progressing accordingly. Tables 12.1a, 12.1b, 12.1c, 12.1d, 12.1e, 12.1f, 12.1g, 12.1h, 12.1i, 12.1j, and 12.1k shows the status of JIS standards as listed in the 2019 JIS Handbook [7, 8]. The number of modified international standards (MOD) has gradually decreased, while the number of identical standards (IDT) has increased. This indicates a trend toward revising JIS standards to be identical to ISO standards.

Table 12.1a Standards for screws of JIS and ISO

Category	JIS number: established or revised year	Name	ISO number: revised year
Terms, designation and drawing	B 0002-1:1998	Technical drawings—Screw threads and threaded parts—Part 1: General conventions	ISO 6410-1:1993
	B 0002-2:1998	Technical drawings—Screw threads and threaded parts—Part 2: Screw thread inserts	ISO 6410-2:1993
	B 0002-3:1998	Technical drawings—Screw threads and threaded parts—Part 3: Simplified representation	ISO 6410-3:1993
	B 0101:2013	Screw threads and fasteners—Vocabulary	**ISO 1891, 5408:2009**
	B 0123:1999	Designation system for screw threads	
	B 0143:2013	Fasteners—Bolts, screws, studs and nuts—Symbols and descriptions of dimensions	ISO 225:2010
	B 0147:2004	Blind rivets—Terminology and definitions	ISO 14588:2000
	B 1010:2003	Designation system for fasteners	ISO 8991:1986
Basics (general use)	B 0201:2017	Miniature screw threads	ISO 1501:2009
	B 0202:1999	Parallel pipe threads	**ISO 228-1:1994**
	B 0203:1999	Taper pipe threads	**ISO 7-1:1994**
	B 0205-1:2001	ISO general purpose metric screw threads—Part 1:Basic profile	ISO 68-1:1998
	B 0205-2:2001	ISO general purpose metric screw threads—Part 2:General plan	ISO 261:1998

(continued)

(continued)

Category	JIS number: established or revised year	Name	ISO number: revised year
	B 0205-3:2001	ISO general purpose metric screw threads—Part 3:Selected sizes for screws, bolts and nuts	ISO 262:1998
	B 0205-4:2001	ISO general purpose metric screw threads—Part 4:Basic dimensions	ISO 724:1993
	B 0206:1973	Unified coarse screw threads	**ISO 263:1973**
	B 0208:1973	Unified fine screw threads	**ISO 263:1973**

ISO standards in bold indicate that ISO standards are modified in JIS standards to localize in Japan. ISO standards in regular text show that JIS standards are identical to ISO standards

Table 12.1b Standards for screws of JIS and ISO

Category	JIS number: established or revised year	Name	ISO number: revised year
Basics (general use)	B 0209-1:2001	ISO general purpose metric screw threads—Tolerances—Part 1: Principles and basic data	ISO 965-1:1998
	B 0209-2:2001	ISO general purpose metric screw threads—Tolerances—Part 2: Limits of sizes for general purpose external and internal screw threads—Medium quality	ISO 965-2:1998
	B 0209-3:2001	ISO general purpose metric screw threads—Tolerances—Part 3: Deviations for constructional screw threads	ISO 965-3:1998
	B 0209-4:2001	ISO general purpose metric screw threads—Tolerances—Part 4: Limits of sizes for hot-dip galvanized external screw threads to mate with internal screw threads tapped with tolerance position H or G after galvanizing	ISO 965-4:1998
	B 0209-5:2001	ISO general purpose metric screw threads—Tolerances—Part 5: Limits of sizes for internal screw threads to mate with hot-dip galvanized external screw threads with maximum size of tolerance position h before galvanizing	ISO 965-5: 1998

(continued)

(continued)

Category	JIS number: established or revised year	Name	ISO number: revised year
	B 0210:1973	Limits of sizes and tolerances for unified coarse screw threads	
	B 0212:1973	Limits of sizes and tolerances for unified fine screw threads	
	B 0216-1:2013	ISO metric trapezoidal screw threads—Part 1: Basic profiles and maximum material profiles	ISO 2901:1993
	B 0216-2: 2013	ISO metric trapezoidal screw threads—Part 2: General plan	ISO 2902:1977

ISO standards in bold indicate that ISO standards are modified in JIS standards to localize in Japan. ISO standards in light style show that JIS standards are identical to ISO standards

Table 12.1c Standards for screws of JIS and ISO

Category		JIS number: established or revised year	Name	ISO number: revised year
Basics (general use)		B 0216-3:2013	ISO metric trapezoidal screw threads—Part 3: Basic dimensions	ISO 2904:1977
		B 0217-1:2013	ISO metric trapezoidal screw threads—Tolerances—Part 1: Principles and basic data	**ISO 2903:1993**
		B 0217-2:2013	ISO metric trapezoidal screw threads—Tolerances—Part 2: Limits of sizes for external and internal screw thread	
Limit gauges for threads		B 0251:2008	Limit gauges for metric screw threads	**ISO 1502:1996**
		B 0253:1985	Gauges for taper pipe threads	**ISO 72:1982**
		B 0254:2011	Limit gauges for parallel pipe threads	**ISO 228-2:1987**
		B 0255:1998, 2008	Limit gauges for unified screw threads	
		B 0261:2004	Parallel screw threads gauges—Measuring method	
		B 0262:1989	Gauges for taper pipe threads—Measuring method	

(continued)

(continued)

Category		JIS number: established or revised year	Name	ISO number: revised year
		B 0271:2018	Wires for measuring screw threads	**ISO 16239:2013**
		B 3102:2001	Gauge blanks for screw threads limit gauges	
Components	Dimensions	B 1001:1985	Diameter of clearance holes and counterbores for bolts and screws	**ISO 273:1979**
		B 1002:1985	Dimensions of widths across flats	**ISO 272:1982, 2343:1972, 4762:1977**
		B 1003:2014	Fasteners—Ends of parts with external ISO metric thread	ISO 4753:2011
		B 1004:2009	Hole sizes before threading	
		B 1005:2003	General purpose bolts and screws—Metric series—Radii under the head	ISO 885:2000

ISO standards in bold indicate that ISO standards are modified in JIS standards to localize in Japan. ISO standards in regular text show that JIS standards are identical to ISO standards

Table 12.1d Standards for screws of JIS and ISO

Category		JIS number: established or revised year	Name	ISO number: revised year
Components	Dimensions	B 1006:2009	Fasteners—Thread run-outs for external thread in accordance with ISO general purpose metric screw threads	**ISO 3508:1976**
		B 1007:2015	Tapping screws thread	ISO 1478:1999
		B 1008:2009	Fasteners—Bolts and studs—Split pin holes and wire holes	**ISO 7378:1983**
		B 1009:2014	Fasteners—Bolts, screws and studs—Nominal lengths and thread lengths	ISO 888:2012
		B 1012:1985	Cross recesses for screws	**ISO 4757:1983**
		B 1013:1994	Countersunk head screws—Head configuration and gauging	ISO 7721:1983

(continued)

(continued)

Category		JIS number: established or revised year	Name	ISO number: revised year
		B 1014:1994	Countersunk flat head screws—Part 2: Penetration depth of cross recesses	ISO 7721-2:1990
		B 1015:2018	Hexalobular internal driving feature for bolts and screws	ISO 10664:2014
		B 1016:2006	Gauging of hexagon sockets	ISO 23429:2004
		B 1017:2008	Countersinks for countersunk head screws	ISO 15065:2005
		B 1018:2009	Fasteners—Thread undercuts for external metric ISO threads	ISO 4755: 1983
		B 1021:2003	Tolerances for fasteners—Part 1: Bolts, screws, studs and nuts—Product grades A, B and C	ISO 4759-1:2000
		B 1022:2018	Tolerances for fasteners—Washers for bolts, screws and nuts—Product grades A, C and F	ISO 4759-3:2016
		B 1041:1993	Fasteners—Surface discontinuities—Part 1: Bolts, screws and studs for general requirements	ISO 6157-1:1988
		B 1042:1998	Fasteners—Surface discontinuities—Part 2:Nuts	ISO 6157-2:1995

ISO standards in bold indicate that ISO standards are modified in JIS standards to localize in Japan. ISO standards in regular text show that JIS standards are identical to ISO standards

Table 12.1e Standards for screws of JIS and ISO

Category		JIS number: established or revised year	Name	ISO number: revised year
Components	Dimensions	B 1043:1993	Fasteners—Surface discontinuities—Part 3: Bolts, screws and studs for special requirements	ISO 6157-3:1988
	Surface treatment	B 1044:2001	Fasteners—Electroplated coatings	ISO 4042:1999
		B 1045:2001	Fasteners—Preloading test for the detection of hydrogen embrittlement—Parallel bearing surface method	ISO 15330:1999

(continued)

(continued)

Category		JIS number: established or revised year	Name	ISO number: revised year
		B 1046:2005	Fasteners—Non-electrolytically applied zinc flake coatings	ISO 10683:2000
		B 1047:2006	Passivation of corrosion-resistant stainless-steel fasteners	ISO 16048:2003
		B 1048:2007	Fasteners—Hot dip galvanized coatings	**ISO 10684:2004**
	Mechanical properties	B 1051:2014	Mechanical properties of fasteners made of carbon steel and alloy steel—Bolts, screws and studs with specified property classes—Coarse thread and fine pitch thread	ISO 898-1:2013
		B 1052-2:2014	Mechanical properties of fasteners made of carbon steel and alloy steel—Part 2: Nuts with specified property classes—Coarse thread and fine pitch thread	ISO 898-2:2012
		B 1053:2014	Mechanical properties of fasteners made of carbon steel and alloy steel—Set screws and similar threaded fasteners with specified hardness classes—Coarse thread and fine pitch thread	ISO 898-5:2012
		B 1054-1:2013	Mechanical properties of corrosion-resistant stainless steel fasteners—Part 1:Bolts, screws and studs	ISO 3506-1:2009
		B 1054-2:2013	Mechanical properties of corrosion-resistant stainless steel fasteners—Part 2:Nuts	ISO 3506-2:2009

ISO standards in bold indicate that ISO standards are modified in JIS standards to localize in Japan. ISO standards in regular text show that JIS standards are identical to ISO standards.

Table 12.1f Standards for screws of JIS and ISO

Category		JIS number: established or revised year	Name	ISO number: revised year
Components	Mechanical properties	B 1054-3:2013	Mechanical properties of corrosion-resistant stainless steel fasteners—Part 3: Set screws and similar fasteners not under tensile stress	ISO 3506-3:2009
		B 1054-4:2013	Mechanical properties of corrosion-resistant stainless steel fasteners—Part 4: Tapping screws	ISO 3506-4:2009
		B 1055:2015	Heat-treated steel tapping screws—Mechanical properties	ISO 2702:2011
		B 1056:2011	Fasteners—Prevailing torque steel nuts—Functional properties	**ISO 2320:2008**
		B 1057:2001	Mechanical properties of non-ferrous metal fasteners	**ISO 8839:1986**
		B 1058:1995	Mechanical properties of fasteners Part 7: Torsional test and minimum torques for bolts and screws with nominal diameters 1–10 mm	ISO 898-7:1992
		B 1059:2001	Drilling screws with tapping screw thread—Mechanical and functional properties	ISO 10666:1999
		B 1060:2003	Mechanical and performance requirements of case hardened and tempered metric thread rolling screws	ISO 7085:1999
	Examination and inspection	B 1071:2010	Fasteners—Methods of verification for size and geometry	

(continued)

(continued)

Category		JIS number: established or revised year	Name	ISO number: revised year
		B 1081:1997	Threaded fasteners—Axial load fatigue testing—Test methods and evaluation of results	ISO 3800:1993
		B 1082:2009	Stress area and bearing area for threaded fasteners	
		B 1083:2008	General rules for tightening of threaded fasteners	
		B 1084:2007	Fasteners—Torque/ clamp force testing	ISO 16047:2005
		B 1085:1995	Cone proof load test on nuts	ISO 10485:1991
		B 1086:1998	Widening test on nuts	ISO 10484:1997

ISO standards in bold indicate that ISO standards are modified in JIS standards to localize in Japan. ISO standards in regular text show that JIS standards are identical to ISO standards

Table 12.1g Standards for screws of JIS and ISO

Category		JIS number: established or revised year	Name	ISO number: revised year
Components	Examination and inspection	B 1087:2004	Blind rivets—Mechanical testing	ISO 14589:2000
		B 1091:2003	Fasteners—Acceptance inspection	ISO 3269:2000
		B 1092:2006	Fasteners—Quality assurance system	ISO 16426:2002
		B 1099:2012	Fasteners—General requirements for bolts, screws, studs and nuts	ISO 8992:2005
Screw threads components (including washers) for general use		B 1101:2017	Slotted head screws	**ISO 1207, 1580, 2009, 2010:2011**
		B 1107:2018	Hexalobular socket head screws	**ISO 14580:2011, 14581, 14582:2013, 14583, 14584:2011**
		B 1111:2017	Cross-recessed head screws	**ISO 7045–7048:2011**

(continued)

(continued)

Category		JIS number: established or revised year	Name	ISO number: revised year
		B 1116:2009	Slotted head screws for precision instruments	
		B 1119:1995	Machine screws and nuts for spectacle frames	**ISO 11381:1994**
Screw threads components (including washers) for general use		B 1130:2012	Screw and washer assemblies made of steel with plain washers—Washer hardness classes 200 HV and 300 HV	**ISO 10644:2009**
		B 1188:2017	Cross-recessed head screws with captive washer	
		B 1117:2010	Slotted set screws	**ISO 4766, 7434–7436:1983**
		B 1118:2010	Square head set screws	
		B 1177:2007	Hexagon socket set screws	**ISO 4026–4029:2003**
		B 1115:2015	Slotted head tapping screws	**ISO 1481–1483:2011**
		B 1122:2015	Cross-recessed head tapping screws	**ISO 7049–7051:2011**

ISO standards in bold indicate that ISO standards are modified in JIS standards to localize in Japan. ISO standards in regular text show that JIS standards are identical to ISO standards

Table 12.1h Standards for screws of JIS and ISO

Category	JIS number: established or revised year	Name	ISO number: revised year
Screw threads components (including washers) for general use	B 1123:2015	Hexagon head tapping screws	**ISO 1479:2011**
	B 1124:2015	Drilling screws with tapping screw thread	**ISO 15480–15483:1999**
	B 1125:2015	Self-drilling tapping screws	
	B 1126:2015	Hexagon washer head tapping screws	**ISO 7053:2011**
	B 1127:2015	Hexagon flange head tapping screws	**ISO 10509:2012**

(continued)

(continued)

Category	JIS number: established or revised year	Name	ISO number: revised year
	B 1128:2004	Hexalobular socket head tapping screws	**ISO 14585–14587:2001**
	B 1129:2004	Tapping screw washer assemblies with plain washers	ISO 10510:1999
	B 1112:1995	Cross-recessed head wood screws	
	B 1135:1995	Slotted head wood screws	
Bolts	B 1136:2014, 2015	Hexalobular socket head cap screws	**ISO 14579:2011**
	B 1166:2009, 2015	T-slot bolts	
	B 1168:1994	Eyebolts	
	B 1171:2005, 2015	Cup head square neck bolts	**ISO 8678:1988**
	B 1173:2010, 2015	Studs	
	B 1174:2017	Hexagon socket button head screws	**ISO 7380-1, -2:2011**
	B 1175:1988	Hexagon socket head shoulder screws	**ISO 7379:1983**
	B 1176:2014, 2015	Hexagon socket head cap screws	**ISO 4762:2004, 12474:2010**
	B 1178:2009, 2015	Foundation bolts	
	B 1179:2009, 2015	Flat head bolts	
	B 1180:2014	Hexagon head bolts and hexagon head screws	**ISO 4014:2011, 4015–4018:1979, 8676, 8765:2011**

ISO standards in bold indicate that ISO standards are modified in JIS standards to localize in Japan. ISO standards in regular text show that JIS standards are identical to ISO standards

Table 12.1i Standards for screws of JIS and ISO

Category	JIS number: established or revised year	Name	ISO number: revised year
Bolts	B 1182:2009, 2015	Square head bolts	
	B 1184:2010	Wing bolts	
	B 1187:2017	Hexagon head screws with captive washer	

(continued)

(continued)

Category	JIS number: established or revised year	Name	ISO number: revised year
	B 1189:2014, 2015	Hexagon bolts with flange	**ISO 15071:2011, 15072:2012**
	B 1194:2006	Hexagon socket countersunk head screws	**ISO 10642:2004**
	B 1195:2009, 2015	Weld bolts	
Nuts	B 1163:2009	Square nuts	
	B 1167:2011	T-slot nuts	
	B 1169:1994	Eyenuts	
	B 1170:2011	Hexagon slotted and castle nuts	
	B 1181:2014	Hexagon nuts and hexagon thin nuts	**ISO 4032–4036, 8673–8675:2012**
	B 1183:2010	Domed cap nuts	
	B 1185:2010	Wing nuts	
	B 1190:2014	Hexagon nuts with flange	**ISO 4161, 10663:2012**
	B 1196:2010	Weld nuts	
	B 1199–1:2001, 2008	Prevailing torque type nuts—Part 1: Hexagon nuts with non-metallic insert	**ISO 7040:1997, 7041:2002, 10511, 10512:1997**
	B 1199-2:2001	Prevailing torque type nuts—Part 2: All-metal hexagon nuts	**ISO 7042, 7719, 7720, 10513:1997**
	B 1199–3:2001	Prevailing torque type nuts—Part 3: Hexagon nuts with flange with non-metallic insert	**ISO 7043, 12125:1997**

ISO standards in bold indicate that ISO standards are modified in JIS standards to localize in Japan. ISO standards in regular text show that JIS standards are identical to ISO standards

Table 12.1j Standards for screws of JIS and ISO

Category	JIS number: established or revised year	Name	ISO number: revised year
Nuts	B 1199-4:2001	Prevailing torque type nuts—Part 4: All-metal hexagon nuts with flange	**ISO 7044, 12126:1997**
	B 1200:2018	Fasteners—Hexagon weld nuts with flange	**ISO 21670:2014**
	B 1216:2006	Push-on spring nuts	
Washers	B 1250:2008	Plain washers for metric bolts, screws and nuts for general purposes—General plan	ISO 887:2000

(continued)

(continued)

Category	JIS number: established or revised year	Name	ISO number: revised year
	B 1251:2018	Spring lock washers	
	B 1256:2008	Plain washers	**ISO 7089–7094:2000**
	B 1257:2004	Plain washers for tapping screw and washer assemblies—Normal and large series—Product grade A	ISO 10669:1999
	B 1258:2012	Plain washers for screw and washer assemblies—Small, normal and large series—Product grade A	**ISO 10673:2009**
Pins and rivets	B 1351:1987	Split pins	**ISO 1234:1976**
	B 1352:1988, 2006	Taper pins	**ISO 2339:1986**
	B 1353:1990	Taper pins with split	
	B 1354:2012	Parallel pins	**ISO 2338:1997**
	B 1355:2012	Dowel pins (Parallel pins, hardened)	**ISO 8734:1997**
	B 1358:1990	Taper pins with thread	**ISO 8736, 8737:1986**
	B 1359:2012	Parallel pins with internal thread	**ISO 8733, 8735:1997**
	B 1360:2006	Snap pins	
	B 2808:2013	Spring pins	**ISO 8748:2007, 8749:1986, 8750, 8751:2007, 8752, 13337:2009**

ISO standards in bold indicate that ISO standards are modified in JIS standards to localize in Japan. ISO standards in regular text show that JIS standards are identical to ISO standards

Table 12.1k Standards for screws of JIS and ISO

Category	JIS number: established or revised year	Name	ISO number: revised year
Pins and rivets	B 1213:1995	Cold-headed rivets	
	B 1214:1995	Hot headed rivets	
	B 1215:1976	Semi-tubular rivets	
Screw threads components for special use	A 5540:2008	Turnbuckle for building	
	A 5541:2008	Body of turnbuckle for building	
	B 1186:2013	Sets of high strength hexagon bolt, hexagon nut and plain washers for friction grip joints	

(continued)

(continued)

Category	JIS number: established or revised year	Name	ISO number: revised year
	B 1198:2011	Headed studs	
	B 1220:2015	Set of anchor bolt for structures	
	B 1217:2018	Bolting for use with pipe flanges	
	B 4633:1998	Screwdrivers for cross-recessed head screws	**ISO 8764-2:1992, ISO/DIS 8764-1:1996**
	B 4648:2008	Hexagon socket screw keys	**ISO 2936:2001**
	B 4652:2008	Hand torque tools—Requirements and test methods	**ISO 6789:2003**

ISO standards in bold indicate that ISO standards are modified in JIS standards to localize in Japan. ISO standards in regular text show that JIS standards are identical to ISO standards

12.1.2 Key Standards Related to Screws

(A) **Metric threads**

The actual cross-sectional shape of a screw is determined by the basic profile of one pitch of the screw thread. The basic profile of the metric screw thread is specified in JIS B 0205-1, which is identical to ISO 68, and the basic dimensions of metric screw threads are specified in JIS B 0205-2, which is identical to ISO 724. Figure 12.1 shows the basic profile of a metric screw thread, where the thread angle is 60° for a triangular thread, and the outer diameter of the external thread and the root diameter of the inner thread are referred to as the major diameter d and minor diameter D, respectively. The effective diameter d_2, root diameter d_1, effective diameter D_2, and inner diameter D_1 are calculated based on the pitch P, with $H = \sqrt{3}/2 \times P$.

The nominal diameter and pitch of metric screws are specified in JIS B 0205-2, which is identical to ISO 261, ranging from 1 to 300 mm. Table 12.2 shows the nominal diameter and pitch for the range of 5–30 mm specified in standards. When a user selects a screw, the nominal diameter in the first column is selected first, and if necessary, the second or third columns are selected in order. For each nominal diameter, both coarse and fine pitches are specified. Fine pitch screws are a general term for screws with a finer pitch than coarse pitch screws, and some nominal diameters may have multiple fine pitch options. Coarse pitch screws are normally used, and fine pitch screws are selected on the basis of the application.

The tolerance of screw threads is defined by the ISO 965, to which JIS B 0209 is identical. Figure 12.2 shows the tolerance notation for M10 screws and nuts. The

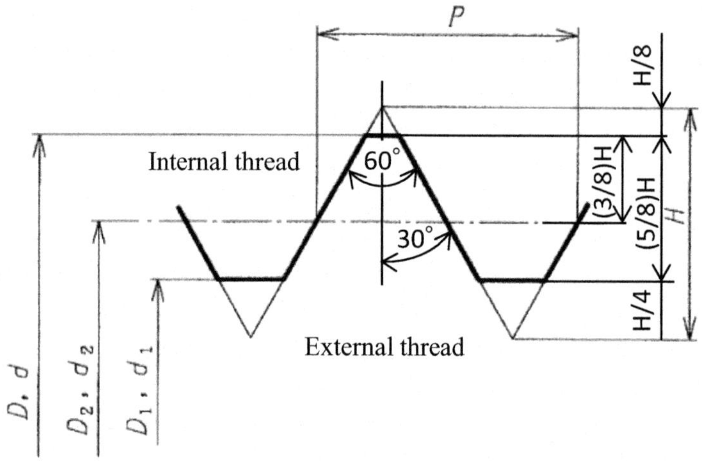

D is the basic major diameter internal of internal thread (nominal diameter)

d is the basic major diameter internal of external thread (nominal diameter)

D_2 is the basic pitch diameter of internal thread

d_2 is the basic pitch diameter of external thread

D_1 is the basic minor diameter of internal thread

d_1 is the basic minor diameter of external thread

H is the height of fundamental triangle

P is the pitch

Fig. 12.1 Basic profile of metric coarse screw threads (Reproduction from Fig. 1 of JIS B 0205-4:2001 (identical to ISO 724:1998))

notations 6g and 6H represent the tolerance classes, where the numbers indicate the tolerance grade representing the range of tolerance relative to the reference dimension. The letters with e, f, g, and h for screws and G and H for nuts represent the position of the tolerance zone relative to the reference dimension and indicate that tolerance deviation from the reference dimension increases in alphabetical order. Figures 12.3 and 12.4 schematically illustrate the dimensional tolerances for screws and nuts, respectively. Tables 12.3, 12.4, and 12.5 provide the tolerances for each nominal diameter and pitch. It can be observed that even for the same 6g tolerance class, larger nominal diameters or pitches result in larger absolute deviations from the tolerance range and reference dimension.

By combining the tolerance positions in Table 12.3 with the tolerance grades in Tables 12.4 and 12.5, it is possible to create multiple tolerance class ranges. However, ISO 965-1, to which JIS B 0209-1 is identical, specifies that there are 18 types for

Table 12.2 Combination of diameter and pitch of metric coarse and fine screw threads (Excerpt from Table 2 of JIS B 0205-2:2001 (identical to ISO 261:1998))

Nominal diameter *D*, *d*			Pitch *P*	
Column 1	Column 2	Column 3	Coarse	Fine
1st Selection	2nd Selection	3rd Selection		
5			0.8	0.5
		5.5		0.5
6			1	0.75
	7		1	0.75
8			1.25	1, 0.75
		9	1.25	1, 0.75
10			1.5	1.25, 1, 0.75
		11	1.5	1, 0.75
12			1.75	1.5, 1.25, 1
	14		2	1.5, 1.25[a], 1
		15		1.5, 1
16			2	1.5, 1
		17		1.5, 1
	18		2.5	2, 1.5, 1
20			2.5	2, 1.5, 1
	22		2.5	2, 1.5, 1
24			3	2, 1.5, 1
		25		2, 1.5, 1
		26		1.5
	27		3	2, 1.5, 1
		28		2, 1.5, 1
30			3.5	(3), 2, 1.5, 1

[a] It may be used exclusively as the sparking plug for internal combustion engines

screws (as shown in Table 12.6) and nine types for nuts (as shown in Table 12.7). It is recommended to select the appropriate fit classification from the specified tolerance class ranges for the intended application. The fit classifications are defined as follows: "precision" for precision screws, "medium" for general use, and "coarse" for cases where some clearance is necessary to avoid manufacturing difficulties.

(B) **Unified screw threads**

Unified screw threads are triangular threads with a thread angle of 60°. The dimensional definition is in the inch system, but other than that, the shape is the same as the metric threads shown in Fig. 12.1. Unified inch coarse threads are specified by JIS B 0206, which is modified from ISO 263. For example, when a unified thread is called 3/8-16 UNC, 3/8, which is 3/8 of an inch (9.525 mm), represents the major diameter

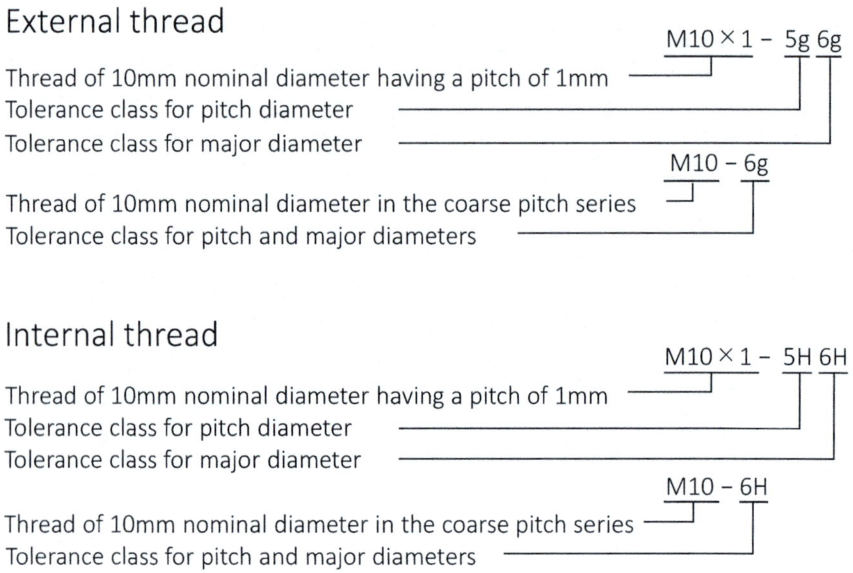

Fig. 12.2 Designation of screw thread and tolerance class. (Reproduction from JIS B 0209-1:2001 (identical to ISO 965-1:1998))

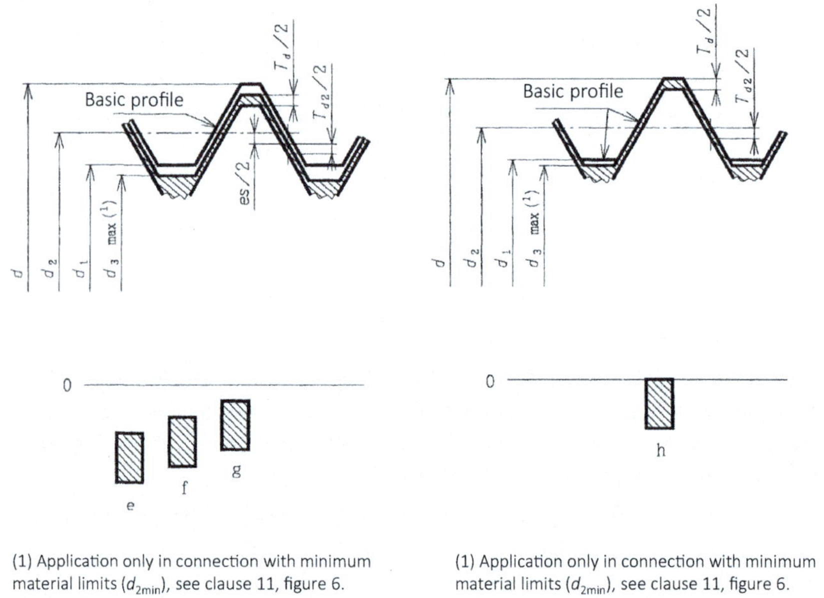

(a) Tolerance position of e, f, g (b) Tolerance position of h

Fig. 12.3 External threads with tolerance positions. (Reproduction from Figs. 4 and 5 of JIS B 0209-1:2001 (identical to ISO 965-1:1998))

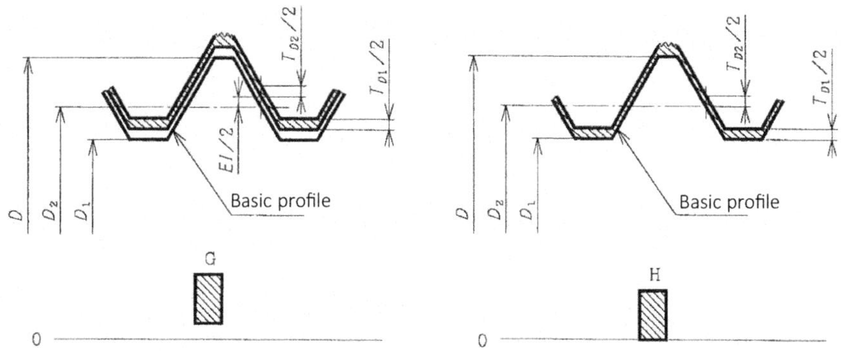

(a) Tolerance position of G (b) Tolerance position of H

Fig. 12.4 Internal threads with tolerance positions. (Reproduction from Figs. 2 and 3 of JIS B 0209-1:2001 (identical to ISO 965-1:1998))

Table 12.3 Fundamental deviations for internal threads and external threads. (Excerpt from Table 1 of JIS B 0209-1:2001 (identical to ISO 965-1:1998))

Pitch *P*	Internal thread D_2, D_1		External thread d, d_2			
	G EI	H EI	e es	f es	g es	h es
mm	μm	μm	μm	μm	μm	μm
0.5	+20	0	−50	−36	−20	0
0.75	+22	0	−56	−38	−22	0
1	+26	0	−60	−40	−26	0
1.25	+28	0	−63	−42	−28	0
1.5	+32	0	−67	−45	−32	0
1.75	+34	0	−71	−48	−34	0
2	+38	0	−71	−52	−38	0
2.5	+42	0	−80	−58	−42	0
3	+48	0	−85	−63	−48	0

of external thread or the internal thread, and 16 indicates a pitch of 16 threads per inch. Thread sizes smaller than 1/4 of an inch are designated by numbers such as No. 10. Unified inch fine threads are specified by JIS B 0208, which is modified from ISO 263, and are denoted as UNF to distinguish them from coarse threads.

(C) **Trapezoidal screw threads**

The trapezoidal screw is used for transmission and movement because it has less friction compared to the triangular screw. The angle of the screw differs between metric trapezoidal screws and inch trapezoidal screws, at 30° and 29° respectively,

Table 12.4 Tolerance mean diameter of external thread. (Excerpt from Table 6 of JIS B 0209-1:2001 (identical to ISO 965-1:1998))

Large diameter *d* mm		Pitch *P* mm	Tolerance grade						
Over	Up to and including		3	4	5	6	7	8	9
			Tolerance [μm]						
5.6	11.2	0.75	50	63	80	100	125	–	–
		1	56	71	90	112	140	180	224
		1.25	60	75	95	118	150	190	236
		1.5	67	85	106	132	170	212	265
11.2	22.4	1	60	75	95	118	150	190	236
		1.25	67	85	106	132	170	212	265
		1.5	71	90	112	140	180	224	280
		1.75	75	95	118	150	190	236	300
		2	80	100	125	160	200	250	315
		2.5	85	106	132	170	212	265	335
22.4	45	1	63	80	100	125	160	200	250
		1.5	75	95	118	150	190	236	300
		2	85	106	132	170	212	265	335
		3	100	125	160	200	250	315	400
		3.5	106	132	170	212	265	335	425
		4	112	140	180	224	280	355	450
		4.5	118	150	190	236	300	375	475

as specified in JIS B 0221 and JIS B 0222. However, both standards were abolished after the establishment of JIS B 0216, which is identical to ISO 2901, to align with ISO metric trapezoidal screws. The alignment work with ISO has continued until recent years, and JIS B 0216-1 to 0216-3 and JIS B 0217-1 to 0217-2 were revised in 2013.

Figure 12.5 shows the standard profile of a trapezoidal screw. The screw is denoted, for example, as Tr20 × 4, where 20 represents the outer diameter of the screw and 4 represents its pitch.

(D) **Pipe threads**

There are two types of pipe threads used in pipe connections and so on. One is the parallel pipe thread, as shown in Fig. 12.6 and specified in JIS B 0202, which is modified ISO 228-1, and the other is the taper pipe thread, as shown in Fig. 12.7 and specified in JIS B 0203, which is modified ISO 7-1. Both types are defined in inch dimensions. The former does not provide a tight seal at the threaded portion and is often used in conjunction with a gasket on the sealing surface, such as for attaching pressure gauges. The latter provides a tight seal at the threaded portion,

Table 12.5 Tolerance mean diameter of internal threads. (Excerpt from Table 5 of JIS B 0209-1:2001 (identical to ISO 965-1:1998))

Large diameter *D* mm		Pitch *P* mm	Tolerance grade				
Over	Up to and including		4	5	6	7	8
			Tolerance [μm]				
5.6	11.2	0.75	85	106	132	170	–
		1	95	118	150	190	236
		1.25	100	125	160	200	250
		1.5	112	140	180	224	280
11.2	22.4	1	100	125	160	200	250
		1.25	112	140	180	224	280
		1.5	118	150	190	236	300
		1.75	125	160	200	250	315
		2	132	170	212	265	335
		2.5	140	180	224	280	355
22.4	45	1	106	132	170	212	–
		1.5	125	160	200	250	315
		2	140	180	224	280	355
		3	170	212	265	335	425
		3.5	180	224	280	355	450
		4	190	236	300	375	475
		4.5	200	250	315	400	500

Table 12.6 Recommended tolerance classes for external threads. (Reproduction from Table 9 of JIS B 0209-1:2001 (identical to ISO 965-1:1998))

Quality tolerance	Tolerance position e			Tolerance position f			Tolerance position g			Tolerance position h		
	S	N	L	S	N	L	S	N	L	S	N	L
Gentle	–	–	–	–	–	–	–	(4g)	(5g4g)	(3h4h)	**4h**	(5h4h)
Medium	–	**6e**	(7e6e)	–	**6f**	–	(5g6g)	**6g**	(7g6g)	(5h6h)	6h	(7h6h)
Rough	–	(8e)	(9e8e)	–	–	–	–	8g	(9g8g)	–	–	–

Table 12.7 Recommended tolerance classes for internal threads. (Reproduction from Table 8 of JIS B 0209-1:2001 (identical to ISO 965-1:1998))

Quality tolerance	Tolerance position G			Tolerance position H		
	S	N	L	S	N	L
Gentle	–	–	–	4H	5H	6H
Medium	(5G)	**6G**	(7G)	**5H**	**6H**	**7H**
Rough	–	(7G)	(8G)	–	7H	8H

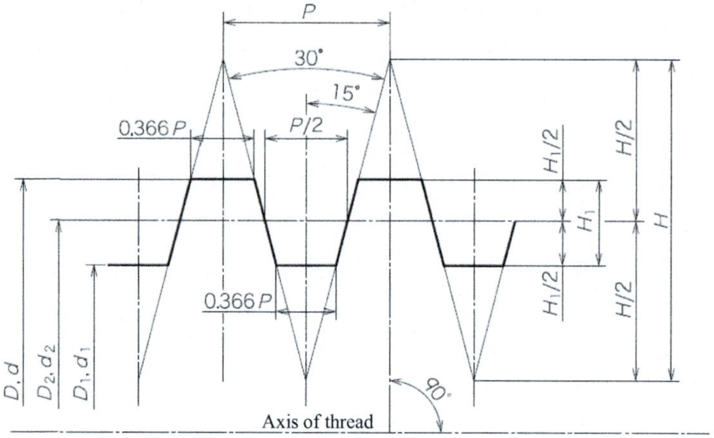

Fig. 12.5 Basic profile of trapezoidal screw. (Reproduction from Fig. 1 of JIS B 0216-1:2003 (identical to ISO 2902-1:1977))

and in practice, sealing tape or sealant is applied to the threaded surface to prevent leakage.

(E) **Tapping screws threads**

A tapping screws thread is the external thread of a screw that is tightened into a pre-drilled hole without cutting the thread, after which the internal thread of the hole is formed by the external thread. Figure 12.8 shows the thread end of a tapping screw. The thread end is gradually shaped from a parallel thread to a sharp needle in order to cut internal thread on the pre-drilled hole. The external thread on the parallel section is triangular threads with a 60° thread angle, similar to metric threads. Tapping screws

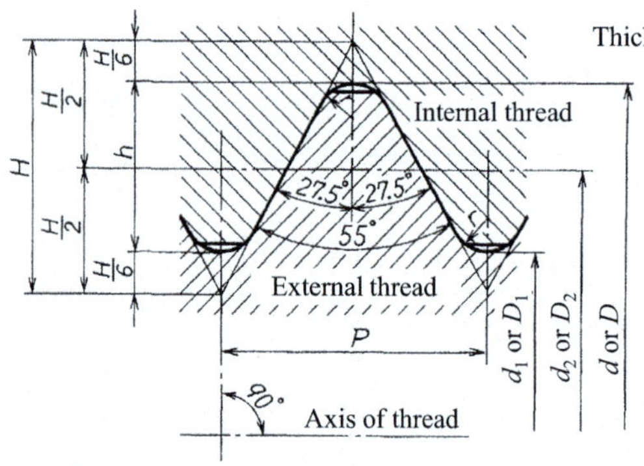

Thick continuous line shows basic profile.

$$P = \frac{25.4}{n}$$

$$H = 0.960\,491\ P$$

$$h = 0.640\,327\ P$$

$$r = 0.137\,329\ P$$

$$d_2 = d - h \qquad D_2 = d_2$$

$$d_1 = d - 2h \qquad D_1 = d_1$$

Fig. 12.6 Basic profile of parallel pipe thread. (Reproduction from attached Table 1 of JIS B 0202:1999 (modified from ISO 228-1:1994))

Basic profile applied for taper external and taper internal threads

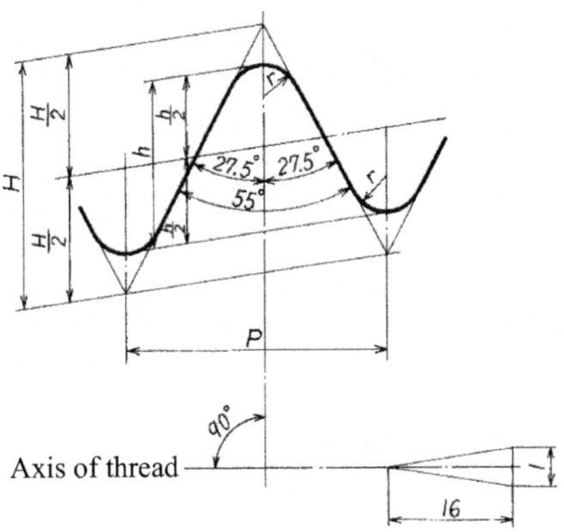

Thick continuous line shows basic profile.

$$P=\frac{25.4}{n}$$

$$H=0.960\,237\ P$$

$$h=0.640\,327\ P$$

$$r=0.137\,278\ P$$

Basic profile applied for parallel internal threads

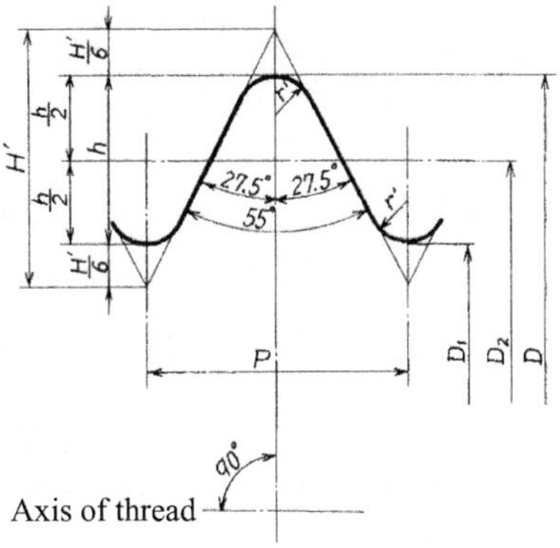

Thick continuous line shows basic profile.

$$P=\frac{25.4}{n}$$

$$H'=0.960\,491\ P$$

$$h=0.640\,327\ P$$

$$r'=0.137\,329\ P$$

Fit between taper external thread and taper internal or parallel internal thread

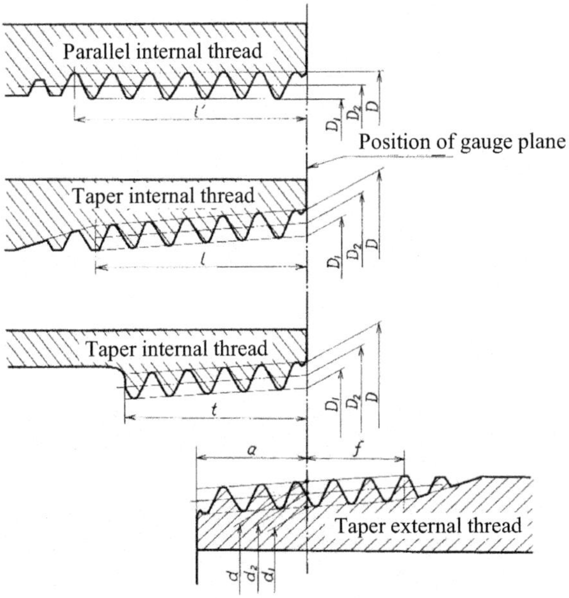

Fig. 12.7 Basic profile of tapper pipe thread. (Reproduction from attached Table 1 of JIS B 0203:1999 (modified from ISO 7-1:1994))

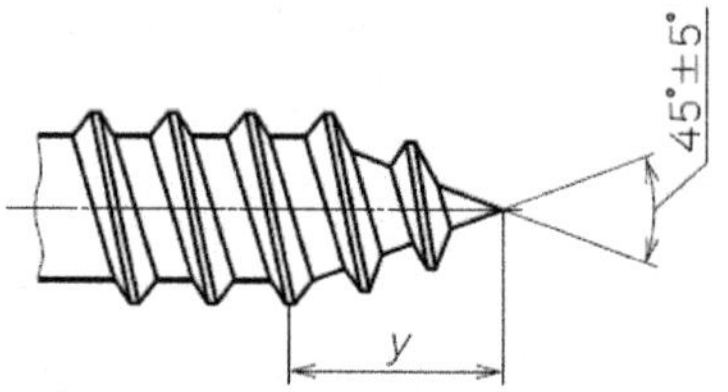

Type C a) (**Cone end**)

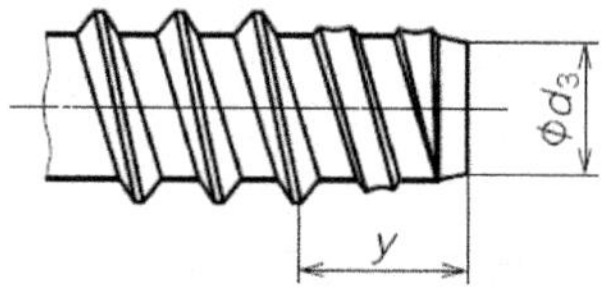

Type F (**Flat end**)

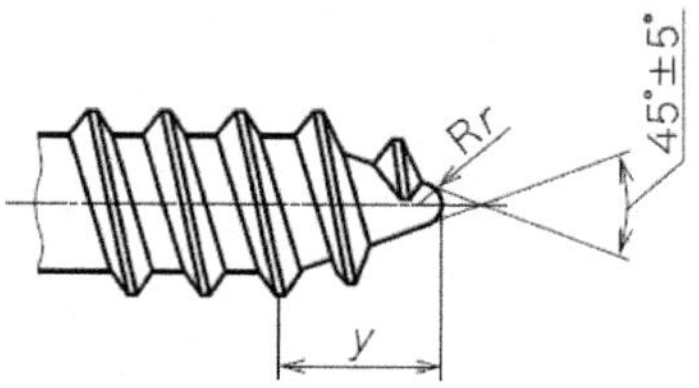

Type R (**Rounded end**)

a) No extrusion of excess metal beyond the apex of the type C cone end resulting from thread rolling shall be permissible. A slight rounding or truncation of the point is desirable.

Fig. 12.8 Thread end of tapping screw. (Reproduction from Fig. 2 of JIS B 1007:2015 (identical to ISO 1478:1999))

are represented, for example, as ST1.9, where ST is the designation for tapping screws and 1.9 represents the nominal diameter.

(F) Screw heads and nuts (hexagon, cross recess, with flange, etc.)

The most common type of screw head is the hexagon bolt shown in Fig. 12.9. Nominal diameter hexagon head bolts and continuous thread hexagon head bolts are specified in three product grades (A, B, and C) according to JIS B 1021. The shape of hexagon bolts is also specified in three product grades (A, B, and C) according to JIS B 1180, which is modified from ISO 4014 through 4018, 8676, and 8765. Table 12.8 shows the dimensional values for product grades A and B. JIS B 1180 also provides dimensional values for parts grade C, which is specified in ISO 4018 and fine thread screws, which are specified in ISO 8676 and 8765.

Five styles of hexagon nuts, which are paired with hex bolts, are specified in JIS B 1181: style 1 (ISO 4032), style 2 (ISO 4033), hexagon nut-C (product grade in ISO 4034), hexagon thin nut chamfered on both sides (ISO 4035), and hexagon thin nut unchamfered (ISO 4036). The structure of the hexagon regular nut (style 1) is shown in Fig. 12.10, and Table 12.9 shows its dimensional values. Hexagon high nuts (style 2) are higher than style 1. Figures 12.11 and 12.12 show hexagon thin nuts chamfered (style 0) and unchamfered (style 0).

There are two other types of screw head: the hexagon bolts with flange, which is shown in Fig. 12.13 (JIS B 1186, which is modified from ISO 15071 and 15072), and

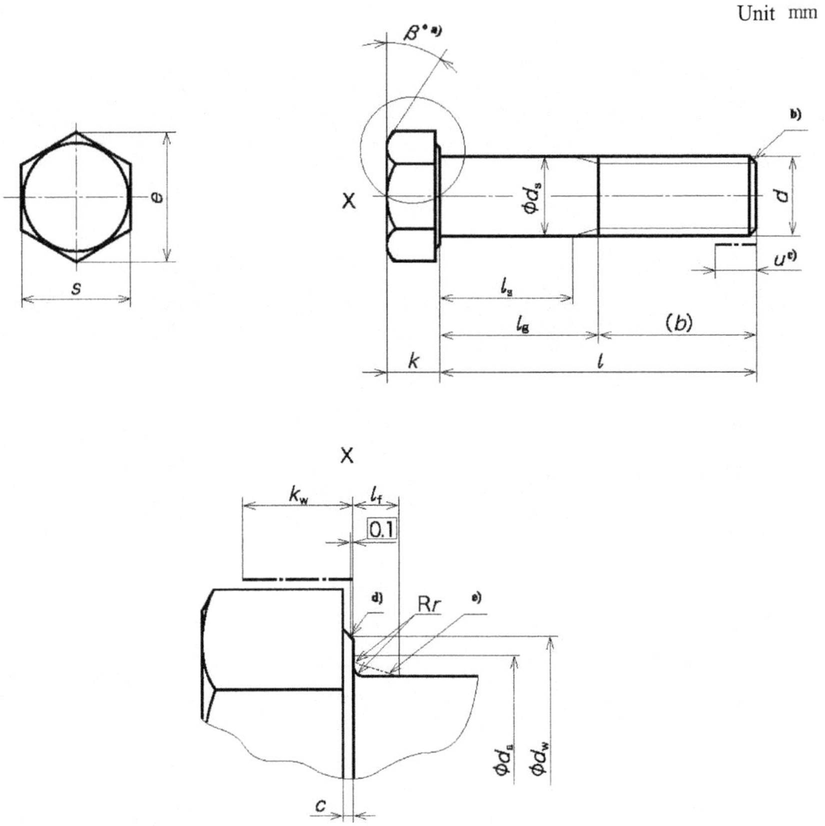

a) β=15° to 30°
b) Point shall be chamfered or for thread ≤ M4 may be as-rolled (sheared end) in accordance with JIS 1003 (ISO 4753).
c) Incomplete thread $u \leq 2P$.
d) Reference datum for d_w.
e) Maximum underhead fillet.

Fig. 12.9 Shapes of nominal diameter hexagon head bolts. (Reproduction from Fig. 1 of JIS B 1180:2014 (modified from ISO 4014:2011))

the hexagon socket head cap screw, which is shown in Fig. 12.14 (JIS B 1176, which is modified from ISO 4762 and 12474). JIS B 1101 specifies small screws tightened with a flathead screwdriver, such as slotted countershunk flat head screws (shown in Fig. 12.15), slotted cheese head screws, slotted pan head screws, and slotted raised countershunk head screws. Small screws with a cross recess are specified in JIS B 1111, including countershunk flat head screws (shown in Fig. 12.16), cheese head screws, pan head screws, and raised countershunk head screws. The cross recess is specified as either the Phillips H-type or the Pozidriv Z-type. Only the Phillips

Table 12.8 Dimensions of nominal diameter hexagon head bolts. (Excerpt from Table 3 of JIS B 1180:2014 (modified from ISO 4014:2011))

Designation of screw thread *d*				M8	M10	M12	M16	M20	M24
P				1.25	1.5	1.75	2	2.5	3
b (reference)			b)	22	26	30	38	46	54
			c)	28	32	36	44	52	60
			d)	41	45	49	57	65	73
c			max	0.60	0.60	0.60	0.8	0.8	0.8
			min	0.15	0.15	0.15	0.2	0.2	0.2
d_a			max	9.2	11.2	13.7	17.7	22.4	26.4
d_s	nom = max			8.00	10.00	12.00	16.00	20.00	24.00
	Product grade	A	min	7.78	9.78	11.73	15.73	19.67	23.67
		B		7.64	9.64	11.57	15.57	19.48	23.48
d_w	Product grade	A	min	11.63	14.63	16.63	22.49	28.19	33.61
		B		11.47	14.47	16.47	22	27.7	33.25
E	Product grade	A	min	14.38	17.77	20.03	26.75	33.53	39.98
		B		14.20	17.59	19.85	26.17	32.95	39.55
l_f			max	2	2	3	3	4	4
K	nom			5.3	6.4	7.5	10	12.5	15
	Product grade	A	max	5.45	6.58	7.68	10.18	12.715	15.215
			min	5.15	6.22	7.32	9.82	12.285	14.785
	Product grade	B	max	5.54	6.69	7.79	10.29	12.85	15.35
			min	5.06	6.11	7.21	9.71	12.15	14.65
k_w	Product grade	A	min	3.61	4.35	5.12	6.87	8.6	10.35
		B		3.54	4.28	5.05	6.8	8.51	10.26
R			min	0.4	0.4	0.6	0.6	0.8	0.8
s	nom = max			13.00	16.00	18.00	24.00	30.00	36.00
	Product grade	A	min	12.73	15.73	17.73	23.67	29.67	35.38
		B		12.57	15.57	17.57	23.16	29.16	35.00

H-type was available in the past, but since it has a tendency to slip out when high torque is applied. The Pozidriv Z-type was developed to improve this issue.

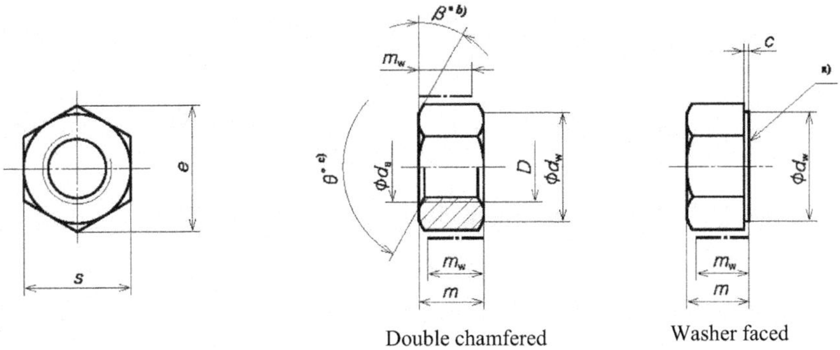

a) The shape of nuts, unless otherwise designated, shall be "double chamfered", and the "washer faced" shall be as designated by the purchaser.
b) β=15° to 30°
c) θ=90° to 120°

Fig. 12.10 Hexagon regular nuts (style 1). (Reproduction from Fig. 1 of JIS B 1181:2014 (modified from ISO 4032:2012))

Table 12.9 Dimensions of nominal diameter hexagon head bolts. (Excerpt from Table 3 of JIS B 1181:2014 (modified from ISO 4032:2012))

Designation of screw thread *D*		M8	M10	M12	M16	M20	M24
P		1.25	1.5	1.75	2	2.5	3
c	max	0.60	0.60	0.60	0.80	0.80	0.80
	min	0.15	0.15	0.15	0.20	0.20	0.20
d_a	max	8.75	10.80	13.00	17.30	21.60	25.90
	min	8.00	10.00	12.00	16.00	20.00	24.00
d_w	max	11.60	14.60	16.60	22.50	27.70	33.30
e	min	14.38	17.77	20.03	26.75	32.95	39.55
m	max	6.80	8.40	10.80	14.80	18.00	21.50
	min	6.44	8.04	10.37	14.10	16.90	20.20
m_w	min	5.20	6.40	8.30	11.30	13.50	16.20
s	nom = max	13.00	16.00	18.00	24.00	30.00	36.00
	min	12.73	15.73	17.73	23.67	29.16	35.00

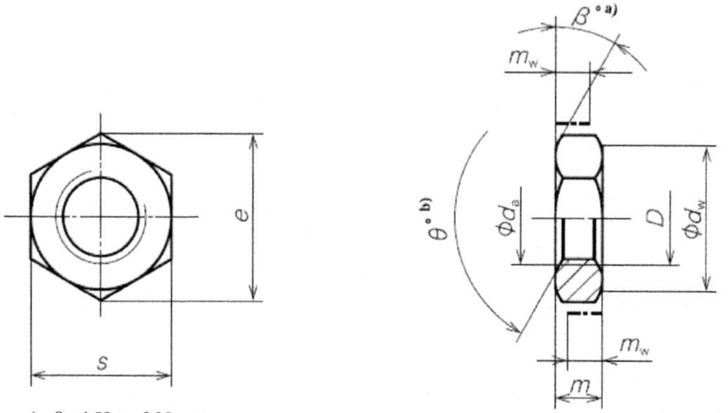

a) β=15° to 30°

b) θ=110° to 120°

Fig. 12.11 Hexagon thin nuts chamfered (style 0). (Reproduction from Fig. 6 of JIS B 1181:2014 (modified from ISO 4035:2012))

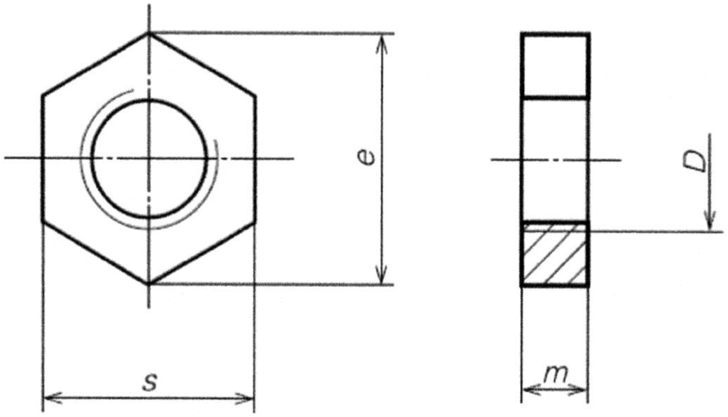

Fig. 12.12 Hexagon thin nuts unchamfered (style 0). (Reproduction from Fig. 8 of JIS B 1181:2014 (modified from ISO 4036:2012))

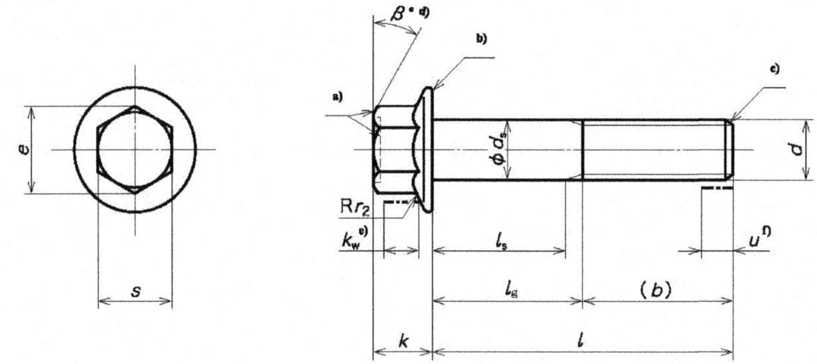

a) The top of the head shall be either full form or indented at the manufacturer's discretion and shall be either chamfered or rounded. The minimum diameter of the chamfer circle or start of rounding shall be the maximum width across flats minus 15 %. If the top of the head is indented, the periphery may be rounded.
b) Contour of the edge at the discretion of the manufacturer
c) Chamfered end (as specified in JIS B 1003 (ISO 4753))
d) β=15° to 30°
e) k_w is the wrenching height (see note of corresponding table)
f) Incomplete thread $u \leq 2P$.

Fig. 12.13 Hexagon bolts with flange—Full shank—Standard type. (Reproduction from Fig. 1 of JIS B 1189:2014 (modified from ISO 15071:2011))

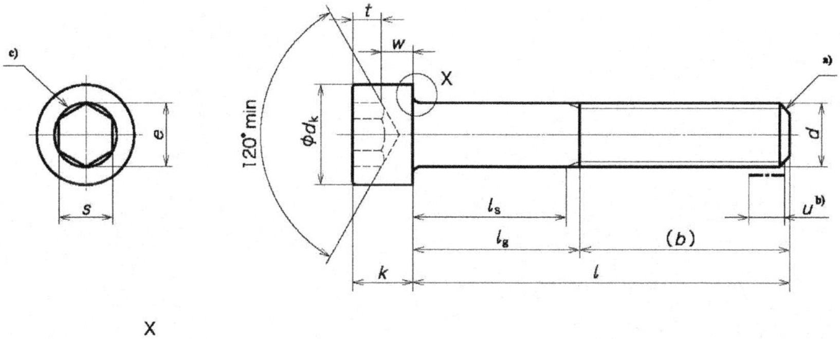

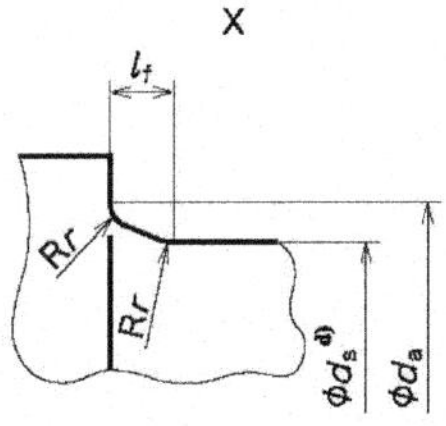

Maximum underhead filled $l_{f,\max} = 1.7\, r_{\max}$

$$r_{\max} = \frac{d_{a,\max} - d_{s,\max}}{2}$$

$r_{\min}$, see corresponding table

Fig. 12.14 Hexagon socket head cap screws. (Reproduction from Fig. 1a of JIS B 1176:2014 (modified from ISO 4762:2004))

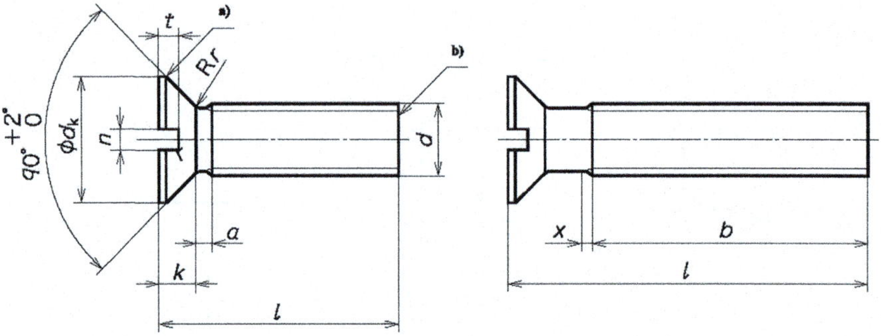

a) edge (rounded or flat)
b) as-rolled end (RL) in JIS B 1003 (ISO 4753)

Fig. 12.15 Slotted countershunk flat head screws—Product grade A. (Reproduction from Fig. 3 of JIS B 1101:2017 (modified from ISO 2009:2011))

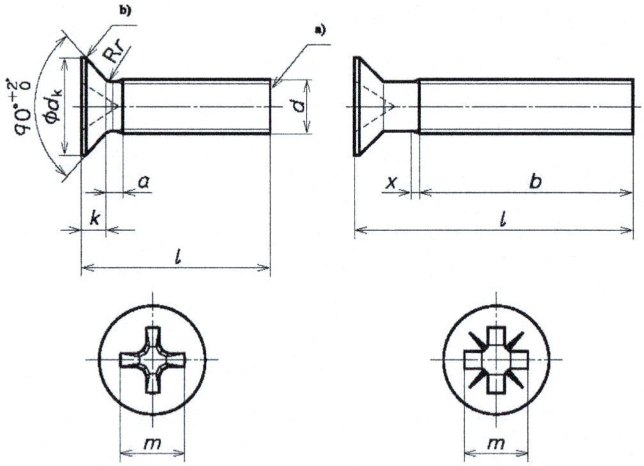

a) Cross recess – type H b) Cross recess – type Z

The shank diameter is approximately equal to the pitch diameter or equal to the permissible major thread diameter.

a) edge (rounded or flat)
b) as-rolled end (RL) in JIS B 1003 (ISO 4753)

Fig. 12.16 Countersunk flat head screws (common head style)—Product grade A. (Reproduction from Fig. 2 of JIS B 1111:2017 (modified from ISO 7046-1, 7046-2:2011))

One type of hexagon nuts is the hexagon nut with flange, as shown in Fig. 12.17 (JIS B 1190, which is modified from ISO 4161 and 10663). Another type is the hexagon-domed cap nuts as shown in Fig. 12.18. ISO does not standardize the hexagon-domed cap nuts.

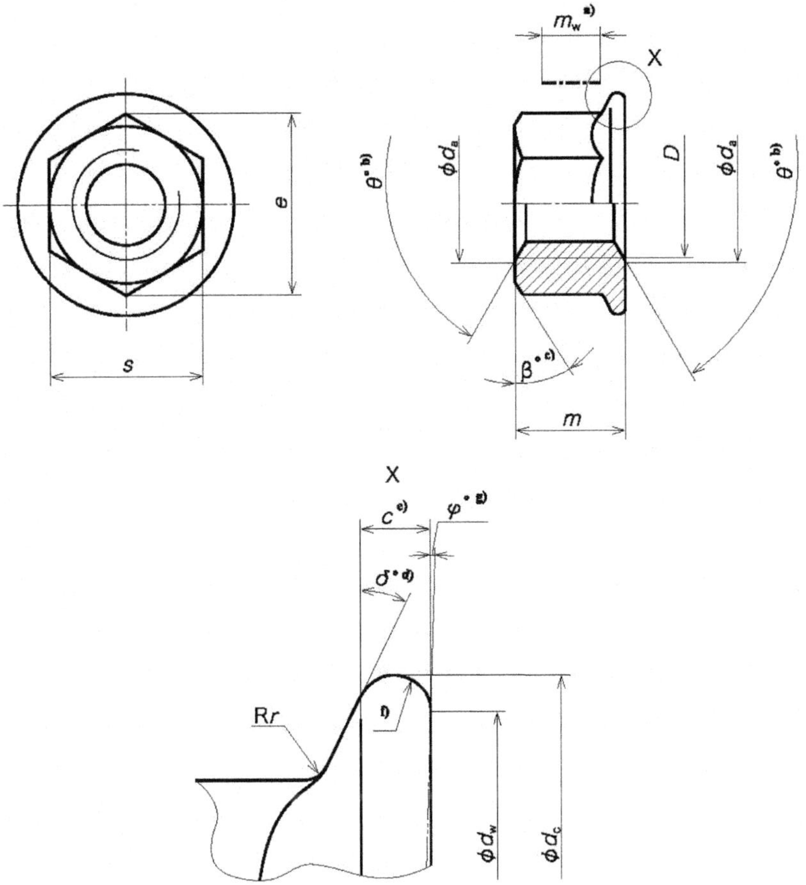

a) m_x is the wrenching height; see note to corresponding table.
b) θ=90° to 120° .
c) β=15° to 30° .
d) δ=15° to 15° .
e) c is measured at $d_{w,\ min}$.
f) Contour of the edge is at the discretion of the manufacturer.

Fig. 12.17 Hexagon nuts with flange. (Reproduction from Fig. 1 of JIS B 1190:2014 (modified from ISO 7046-1, 7046-2:2011))

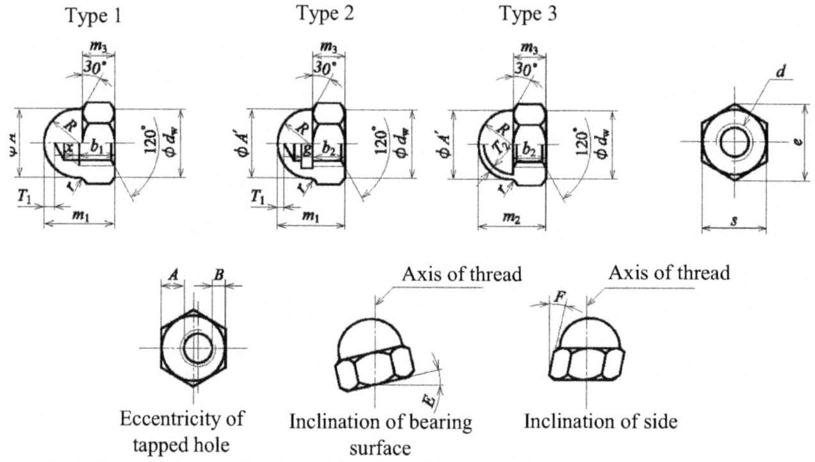

x is the length of incomplete thread that should approximately be 1.5P.
The diameter of thread chamfer thread is slightly larger than the major thread diameter. However, it is permissible to omit the chamfer when purchaser specifies.

Fig. 12.18 Hexagon-domed cap nuts. (Reproduction from Table 2 of JIS B 1183:2010)

(G) Materials and mechanical properties of fasteners

The material specifications for screw parts made of steel, stainless steel, and non-ferrous metals are defined in standards. For steel fasteners, mechanical and physical properties are defined in JIS B 1051, which is identical to ISO 898-1, and organized into property classes represented by integer values and one decimal place, as shown in Table 12.10. The integer value represents the nominal value of 1/100 of the tensile strength in the unit of MPa, and the decimal represents the ratio of yield strength or proof strength to tensile strength. The mechanical properties for steel nuts are defined in JIS B 1052, which is identical to ISO 898-2. These properties are specified by strength grades with integer values of 4, 5, 6, 8, 9, 10, and 12. It is important to note that a nut with higher strength than the screw should be selected.

Materials and mechanical properties of stainless steel fasteners are defined in JIS B 1054-1, -2, -3, -4, which are identical to ISO 3506-1, -2, -3, -4 respectively, as shown in Table 12.11. The comparison between grade of stainless steel grades and equivalent materials in JIS is shown in Table 12.12. The equivalent materials of JIS are only referred to in JIS G 4315 as cold-rolled stainless steel.

Table 12.10 Mechanical and physical properties of bolts, screws, and studs made of carbon steel and alloy steel. (Excerpt from Table 3 of JIS B 1051:2014 (identical to ISO 898-1:2013))

No.	Mechanical or physical property		Property class									
			4.6	4.8	5.6	5.8	6.8	8.8		9.8	10.9	12.9/12.9
								d ≤ 16mm	d > 16mm	d ≤ 16mm		
1	Tensile strength, R_m, MPa	nom	400		500		600	800		900	1000	1200
		min	400	420	500	520	600	800	830	900	1040	1220
2	Lower yield strength, R_{eL}, MPa	nom	240	–	300	–	–	–	–	–	–	–
		min	240	–	300	–	–	–	–	–	–	–
3	Stress at 0.2% nonproportional elongation, $R_{p0.2}$, MPa	nom	–	–	–	–	–	640	640	720	900	1080
		min	–	–	–	–	–	640	660	720	940	1100
5	Stress under proof load, S_p, MPa	nom	225	310	280	380	440	580	600	650	830	970
6	Percentage elongation after fracture for machined test pieces, A, %	min	22	–	20	–	–	12	12	10	9	8
7	Percentage reduction of area after fracture for machined test pieces, Z, %	min	–					52		48	48	44
8	Vickers hardness, HV F ≧ 98N	min	120	130	155	160	190	250	255	290	320	385
		max	220				250	320	335	360	380	435

Table 12.11 Mechanical and physical properties of corrosion-resistant stainless steel fasteners. (Excerpt from Table 3 of JIS B 1054-1 and Table 3 of JIS B 1054-2:2013 (identical to ISO 3506-1, -2:2009))

Group			Austenitic grade			Ferritic grade		Martensitic grade					
Property grade	Grade		A1, A2, A3, A4, A5			F1		C1			C3	C4	
	Property class		50	70	80	45	60	50	70	110	80	50	70
Bolts, screws and studs	Tensile strength, R_m, MPa	min	500	700	800	450	600	500	700	1100	800	500	700
	Stress at 0.2% permanent strain, $R_{p0.2}$, MPa	min	210	450	600	250	410	250	410	820	640	250	410
	Elongation after fracture, mm	min	0.6*d*	0.4*d*	0.3*d*	0.2*d*		0.2*d*					
Nuts	Stress under proof load, S_p, MPa	$m \geq 0.8D$	500	700	800	450	600	500	700	1100	800	500	700

Table 12.12 Equivalent materials in JIS for stainless steel grade of fasteners. (Excerpt from Table JA.1 of JIS B 1054-1:2013)

Group	Austenitic		Ferritic	Martensitic
Grade	A2	A4	F1	C1
Equivalent materials in JIS	SUS 304 SUS 304L SUS 304J3 (SUS 305) SUS 305J1 SUS XM7	SUS 316 SUS 316L	SUS 430 SUS 434	(SUS403) (SUS410)

(H) **Blind rivets**

Rivets are not a type of screw, but they are introduced here as one of the fastening elements. The blind rivet is described as a "mechanical fastener with the capability to fasten together the component parts of an assembly even though the access for its installation and setting may be limited to one side only" in JIS B 0147, which is identical to ISO 14588. Figure 12.19 shows some example structures of blind rivets. In the tightening procedure, the mandrel is pulled out from mandrel end side, the blind rivet deforms from the rivet end, and the tightening force is applied to clamped members. In blind rivets, in which the mandrel or its head is pulled and broken, the breaking load is equivalent to the tightening force. Thus, it is an advantage of blind rivets that the variation of tightening force is small. However, blind rivets cannot be disassembled and retightened, as screw fasteners can.

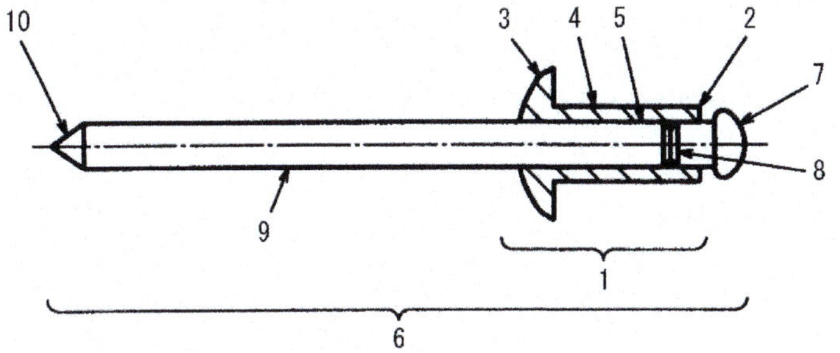

1 Blind rivet body	2 Blind rivet end
3 Blind rivet head	4 Blind rivet shank
5 Blind rivet core	6 Mandrel
7 Mandrel head	8 Break area
9 Mandrel shank	10 Mandrel end

(a) Blind rivet elements

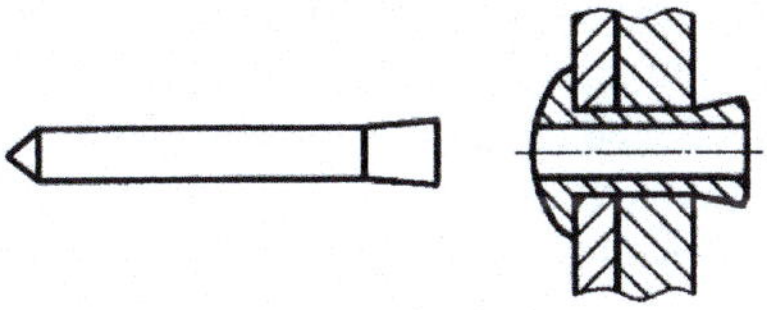

(b) Pull through mandrel

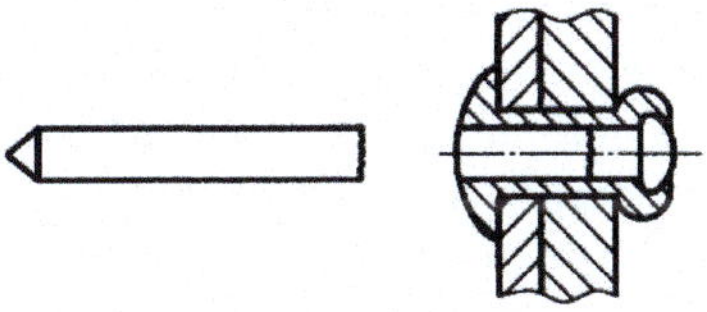

(c) Break pull mandrel

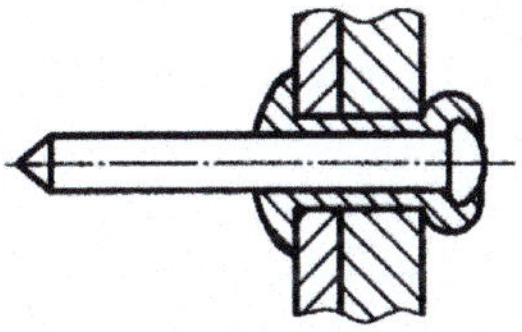

(d) Non-break pull mandrel

(e) Drive pin

Fig. 12.19 Blind rivets. (Reproduction from Figs. 1, 4, 5, 7 and 10 of JIS B 0147:2004 (identical to ISO 14588:2000)

12.2 History of VDI 2230

VDI 2230 Systematische Berechnung hochbeanspruchter Schraubenverbindungen/ Systematic Calculation of Highly Stressed Bolted Joints is a guidebook on the design of bolted joints issued by the Association of German Engineers (Verein Deutscher Ingenieure, VDI). It has been recognized worldwide as a systematic compilation of strength design for bolted joints, even after 40 years since its first publication. In Japan, the Japan Research Institute for Screw Threads and Fasteners (JFRI) has obtained permission to translate and publish the 1974, 1977, 1986, 2003, and 2014 edition into Japanese [10].

In the 1986 edition of VDI 2230, a preface was added that Part 2 for the calculation of multiple bolted joints was being planned, and the subtitle "Joints with one cylindrical bolt" was added to Part 1. For about three decades, only Part 1 has been published. In the 2014 edition, Part 2 titled "Multibolted joints" was first published as a separate booklet.

The following explanation of the architecture of the guidebook is based on VDI 2230 2014 [10]. In Part 1, titled "Joints with one cylindrical bolt", the evaluation method for assessing items of bolted joints (such as separation of clamped members, static strength, and fatigue strength evaluation of bolts) and the strength evaluation of bearing surfaces is summarized. These evaluations are based on the loads and geometric parameters of a single bolted joint, taking the initial tightening force, internal and external force ratio (load factor), and the axial and transverse loads and bending moments applied to the bolted joint into consideration.

Multiple bolted joints share the external load, and the proportional load in multiple bolted joints is needed to evaluate the strength of a bolt. In Part 2, titled "Multibolted joint", the method for evaluating the proportional load in multiple bolted joints is explained. First, the location of the bolted joint with the maximum load and the magnitude of that maximum load is estimated by using the method for determining the proportional load. Next, the evaluation procedure of Part 1 is applied to the bolted joint subjected to the maximum load. Three methods for determining the proportional load are described in VDI 2230 Part2. In the first method, the proportional load is calculated from the equilibrium equations of forces and moments, assuming the clamped structure as a rigid body. In the second method, the proportional load on the bolted joint is evaluated by taking the elastic deformation of the clamped members into consideration using solid mechanics formulae for beams, plates, shells, and so on. The third method uses the finite element method (FEM) to calculate the load on the bolted joint.

In Japan, the Research Committee of JFRI was launched in 2011, and the committee is acting to establish design methods for screw thread fasteners. As a result of the activities of the committee, the first edition of "Design Method for Threaded Fasteners" was published in 2016. Since then, the committee has continuously revised the book, and the second edition was distributed in 2022 [11].

Below, we will provide an overview of VDI 2230 Parts 1 and 2, dividing into seven sections ((A) through (G)). The overview of VDI2230 Part 1 covers (A) bolt stiffness, (B) stiffness of clamped parts, (C) internal–external force ratio (load factor) and the load induction factor, and (D) evaluation of eccentric loads. The overview of VDI2230 Part 2 covers methods for determining the distribution of loads, including (E) the method with the rigid body model, (F) the method with the model of solid mechanics, and (G) the finite element method (FEM).

(A) **Bolt stiffness**

As mentioned in Sect. 3.3, in order to estimate the load state of a bolted joint, it is necessary to understand the joint diagram shown in Fig. 12.20. The joint diagram is drawn with the stiffnesses (compliances) of the bolt and the clamped parts. From the 2003 edition to the latest 2014 edition of VDI 2230, the compliance of the bolt is shown for both through-bolted joints, in which the bolt is tightened with a nut, and tapped-thread joints, in which the bolt is tightened into a tapped-thread component. The compliance δ_b of the through bolt shown in Fig. 12.21, is expressed by the following equation.

$$\delta_b = \frac{0.5d}{E_b \frac{\pi}{4} d^2} + \frac{l_g}{E_b \frac{\pi}{4} d_g^2} + \frac{l_s}{E_b \frac{\pi}{4} d_3^2} + \frac{0.5d}{E_b \frac{\pi}{4} d_3^2} + \frac{0.4d}{E_b \frac{\pi}{4} d^2} \quad (12.1)$$

Here, d is the nominal diameter of the bolt, d_3 is the minor diameter, E_b is the modulus of the longitudinal elasticity of the bolt, l_g is the length of the bolt cylindrical section with the diameter of d_g, and l_s is the thread length in the fastening range. The first item on the right side is the compliance of the bolt head. The second item is the compliance of the bolt cylindrical section. The third item denotes the compliance of

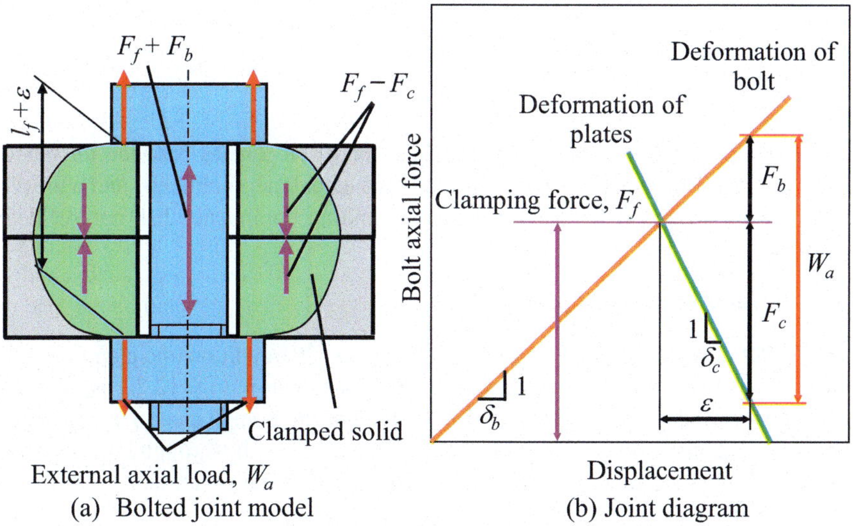

Fig. 12.20 Stiffnesses of the bolt and clamped part, and load sharing

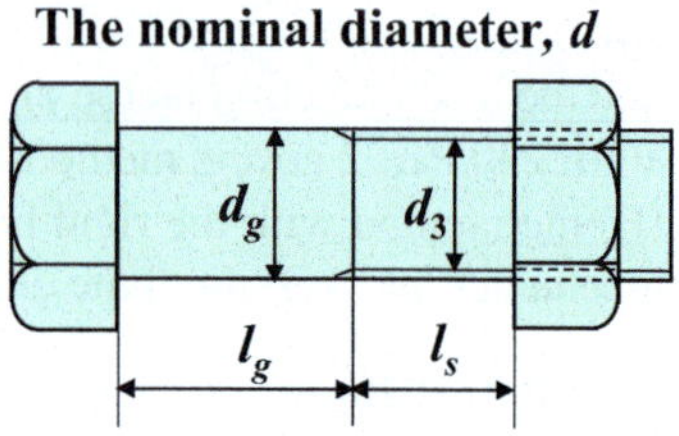

Fig. 12.21 Shape of the bolts

the unengaged loaded part of the thread. The force item refers to the compliance of the engaged bolt thread. The fifth item is the compliance of the deformation of the nut. In the first item on the right side of Eq. (12.1), the numerator of $0.5d$ is for the case of hexagon heads. For the case of hexagon socket heads, after 2003 edition $0.4d$ is used instead of $0.5d$. In the 1986 edition, $0.4d$ is used for both cases of hexagon heads and hexagon socket heads.

In the 1974 edition, the numerator for the bolt head and the nut is unified to $0.4d$, and the compliance of the engaged bolt thread is omitted, as the following equation.

$$\delta_b = \frac{0.4d}{E_b \frac{\pi}{4} d^2} + \frac{l_g}{E_b \frac{\pi}{4} d_g^2} + \frac{l_s}{E_b \frac{\pi}{4} d_3^2} + \frac{0.4d}{E_b \frac{\pi}{4} d^2} \tag{12.2}$$

When eccentric loads or bending loads are applied to a bolted joint, it is necessary to take into consideration the bending stiffness of each part. The bending compliance of the bolt, β_b has been mentioned since the 2003 edition. In the case of a through bolt, as shown in Fig. 12.21, the bending compliance β_b is calculated as the sum of the bending compliance of each part, β_i, using the following equation.

$$\beta_b = \sum_i \beta_i = \frac{0.5d}{E_b \frac{\pi}{64} d^4} + \frac{l_g}{E_b \frac{\pi}{64} d_g^4} + \frac{l_s}{E_b \frac{\pi}{64} d_3^4} + \frac{0.5d}{E_b \frac{\pi}{64} d_3^4} + \frac{0.4d}{E_b \frac{\pi}{64} d^4} \tag{12.3}$$

However, VDI 2230 states that there may be discrepancies between the calculated results of this bending compliance and the actual measurements, and a universally accepted method for determining the bending compliance has not yet been established.

(B) **Stiffness of clamped parts**

In the VDI 2230 from the 2003 edition to the latest 2014 edition, the clamped solid (compressive body) under compressive stress, as shown in Fig. 12.22, is modeled with an equivalent conical deformation body, and the compliance of the clamped body, δ_c, is then calculated using the conical angle φ. For the through bolt shown in Fig. 12.22, the conical angle φ is obtained by the following equation.

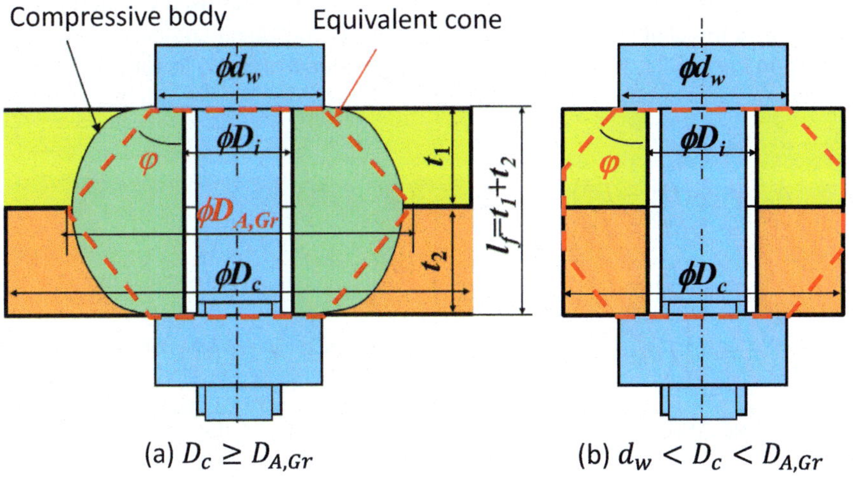

Fig. 12.22 Equivalent cone of the clamped parts (VDI 2230 (2003) (2014))

$$\tan\varphi = 0.362 + 0.032\ln\left(\frac{l_f}{2d_w}\right) + 0.153\ln\left(\frac{D_c}{d_w}\right) \tag{12.4}$$

The maximum outer diameter of the cone, $D_{A,Gr}$, can be calculated as follows.

$$D_{A,Gr} = d_w + wl_f \tan\varphi \tag{12.5}$$

Here, w is called the joint coefficient, and $w = 1$ is used for through bolts. The compliance of the clamped parts, δ_c, is calculated by the following equation.

$$\begin{aligned} \delta_c &= \frac{2\ln\left[\frac{(d_w+D_i)(d_w+wl_f\tan\varphi-D_i)}{(d_w-D_i)(d_w+wl_f\tan\varphi+D_i)}\right]}{wE_c\pi D_i\tan\varphi} \quad (D_c \geq D_{A,Gr}) \\ &= \frac{\frac{2}{wD_i\tan\varphi}\ln\left[\frac{(d_w+D_i)(D_c-D_i)}{(d_w-D_i)(D_c+D_i)}\right]}{E_c\pi} \\ &\quad + \frac{\frac{4}{D_c^2-D_i^2}\left[l_f-\frac{(D_c-d_w)}{w\tan\varphi}\right]}{E_c\pi} \quad (d_w < D_c < D_{A,Gr}) \end{aligned} \tag{12.6}$$

Here, E_c represents the modulus of the longitudinal elasticity of the clamped parts. However, in the equation of VDI 2230, the difference in thickness between the two clamped plates cannot be considered, so the following equation [12] has been proposed as a modification to Eq. (12.4).

$$\tan\varphi = 0.323 + 0.032\ln\left(\frac{l_f}{2d_w}\right) + 0.153\ln\left(\frac{D_c}{d_w}\right) + 0.0717\ln\left(\frac{t_1}{t_2}\right) \quad (t_1 \leq t_2) \tag{12.7}$$

In the 1974 edition of VDI 2230, a simpler equation was used to determine the stiffness of the clamped body by representing the equivalent compressive body, as shown in Fig. 12.23, with a cylinder of cross-sectional area A_{eq} as follows.

$$\begin{aligned}\delta_c &= \frac{l_f}{E_c A_{eq}}\\ A_{eq} &= \tfrac{\pi}{4}\left(D_c^2 - D_i^2\right) && (D_c < d_w)\\ &= \tfrac{\pi}{4}\left(d_w^2 - D_i^2\right)\\ &\quad + \tfrac{\pi}{8}\left(\tfrac{D_c}{d_w} - 1\right)\left(\tfrac{d_w l_f}{5} + \tfrac{l_f^2}{100}\right) && (d_w < D_c < 3d_w,\ l_f \le 8d)\\ &= \tfrac{\pi}{4}\left[\left(d_w + \tfrac{l_f}{10}\right)^2 - D_i^2\right] && (D_c \ge 3d_w,\ l_f \le 8d)\end{aligned} \tag{12.8}$$

The 1986 edition of VDI 2230 improved Eq. (12.8) as follows.

$$\begin{aligned}\delta_c &= \frac{l_f}{E_c A_{eq}}\\ A_{eq} &= \tfrac{\pi}{4}\left(D_c^2 - D_i^2\right) && (D_c < d_w)\\ &= \tfrac{\pi}{4}\left(d_w^2 - D_i^2\right)\\ &\quad + \tfrac{\pi}{8} d_w (D_c - d_w)\left[(x+1)^2 - 1\right] && (D_c \ge d_w)\\ x &= \sqrt[3]{\frac{l_f d_w}{D_c^2}}\\ &\text{When } D_c \ge d_w + l_f,\ D_c = d_w + l_f\end{aligned} \tag{12.9}$$

The bending stiffness is required when eccentric loads or bending loads are applied to a bolted joint. The cross-sectional quadratic moment of the clamped parts, which represents the stiffness of the clamped part, is described in the 2003 edition or later of VDI2230, as follows.

$$\begin{aligned}I_{Bers} &= \frac{3\pi}{64}\frac{(D_c - d_w) d_w^3 D_c^3}{D_c^3 - d_w^3}\\ &\text{where the cross-sectional quadratic moment}\\ &\text{of the bolt hole } \tfrac{\pi}{64} D_i^4 \text{ is included}\end{aligned} \tag{12.10}$$

If the center of the clamped body and the center of the bolt in the axial direction are eccentric, they are modified as follows, taking the eccentricity s_{sym} into account.

$$I_{Bers}^{Ve} = I_{Bers} + s_{sym}^2 \frac{\pi}{4} D_A^2 \tag{12.11}$$

The bending compliance of the clamped body is defined by using Eq. (12.10) or (12.11) as follows.

$$\beta_c = \frac{l_f}{E_c I_{Bers}} \tag{12.12}$$

(C) **Internal–external force ratio (load factor) and the load induction factor**

When an axial external force W is applied to the bolted joint, the increase of the bolt axial load, F_b, and the reduction of compressive force in the clamped body, F_c, take place, as shown in Fig. 12.20. The ratio of the increase of the bolt axial load, F_b, to the external force W is called the load factor. When the axis of the external force W coincides with the bolt center axis and the external force is applied on the bearing surfaces, the load factor Φ can be obtained by using the compliances of the bolt and the clamped body, δ_b and δ_c, as shown in the following equation.

$$\Phi = \frac{\delta_c}{\delta_b + \delta_c} \tag{12.13}$$

When the axis of the external force W is eccentric from the center axis of the bolt, or when the external force is applied inside the clamped body from the bearing surface, the load factor Φ in Eq. (12.13) is converted by using the load introduction factor $n < 1$, and the converted load factor is denoted as Φ_n.

$$\Phi_n = n\Phi \tag{12.14}$$

The load introduction factor highly depends on the geometric shape of the bolted joint, and thus, it is difficult to formulate for arbitrary shapes. From the 2003 edition to the latest edition of VDI 2230, the load introduction factor n is categorized into SV1 to SV6 based on the position of the load, as shown in Fig. 12.24 and Table 12.13. In addition, there are two other methods for calculating the load introduction factor mentioned in the editions from 2003 onward.

In the VDI 2230 editions from 1974 and 1986, the calculation method for the load introduction factor n was provided under more limited conditions. As mentioned above, since the load introduction factor varies with the geometric shape of the bolted joint, a universally applicable calculation method has not yet been established.

(D) **Evaluation method under eccentric load**

The evaluation method under an eccentric load W is described in detail in Sect. 3.4. Here, we discuss the evolution of the evaluation method under eccentric load in VDI 2230.

The mechanical model under an eccentric load, as mentioned in Sect. 3.4, is shown in Fig. 12.25. The compliance of the clamped body under an eccentric load, δ_{eW}, is expressed by the following equation.

$$\delta_{eW} = \delta_c + \frac{s_B e_B l_f}{E_c I_B} \tag{12.15}$$

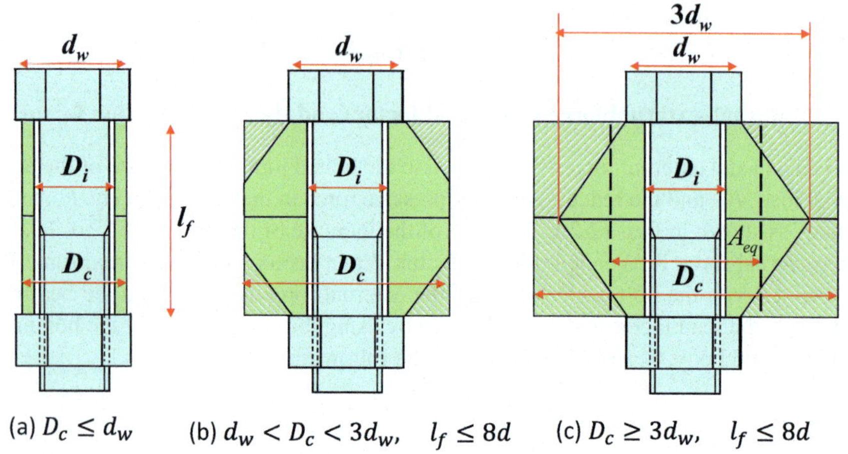

Fig. 12.23 Equivalent cylinder of the clamped parts (VDI 2230 (1974))

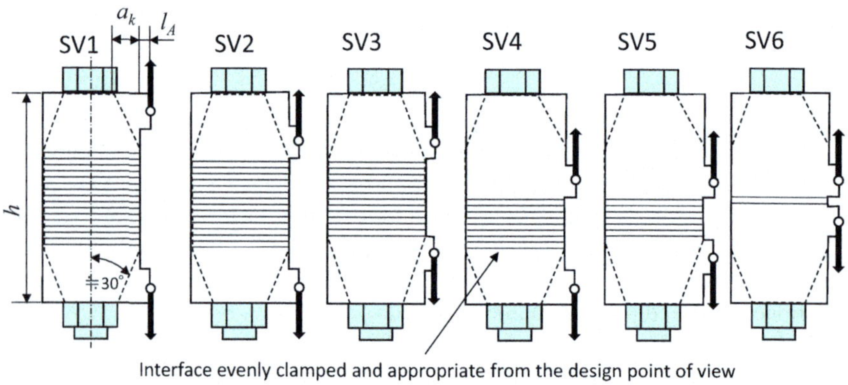

Fig. 12.24 Categorization of load introduction factor. (Created by the author based on Fig. 22 of VDI 2230 Part 1 (2014))

This equation has not been changed from the 1974 edition to the 2014 edition of VDI 2230.

In versions from 2003 onward, an evaluation for the case, in which an external moment M_B is applied in addition to an eccentric load W, has been appended. That is, the additional axial force of the bolt, F_{bab}, which causes the separation of the clamped surface under the eccentric load W and the external moment M_B, is given by the following equation.

$$F_{bab} = W\left(\frac{I_B + e_B u A_D}{I_B + s_B u A_D} - \Phi_{en}\right) + M_B\left(\frac{u A_D}{I_B + s_B u A_D} - \Phi_m\right) \tag{12.16}$$

Table 12.13 Load introduction factor for joint types SV1 to SV6. (Created by the author based on Table 2 of VDI 2230 Part 1 (2014))

l_A/h	0.00				0.10				0.20				≥0.30			
a_k/h	0.00	0.10	0.30	≥0.5	0.00	0.10	0.30	≥0.5	0.00	0.10	0.30	≥0.5	0.00	0.10	0.30	≥0.5
SV1	0.70	0.55	0.30	0.13	0.52	0.41	0.22	0.10	0.34	0.28	0.16	0.07	0.16	0.14	0.12	0.04
SV2	0.57	0.46	0.30	0.13	0.44	0.36	0.21	0.10	0.30	0.25	0.16	0.07	0.16	0.14	0.12	0.04
SV3	0.44	0.37	0.26	0.12	0.35	0.30	0.20	0.09	0.26	0.23	0.15	0.07	0.16	0.14	0.12	0.04
SV4	0.42	0.34	0.25	0.12	0.33	0.27	0.16	0.08	0.23	0.19	0.12	0.06	0.14	0.13	0.10	0.03
SV5	0.30	0.25	0.22	0.10	0.24	0.21	0.15	0.07	0.19	0.17	0.12	0.06	0.14	0.13	0.10	0.03
SV6	0.15	0.14	0.14	0.07	0.13	0.12	0.10	0.06	0.11	0.11	0.09	0.06	0.10	0.10	0.08	0.03

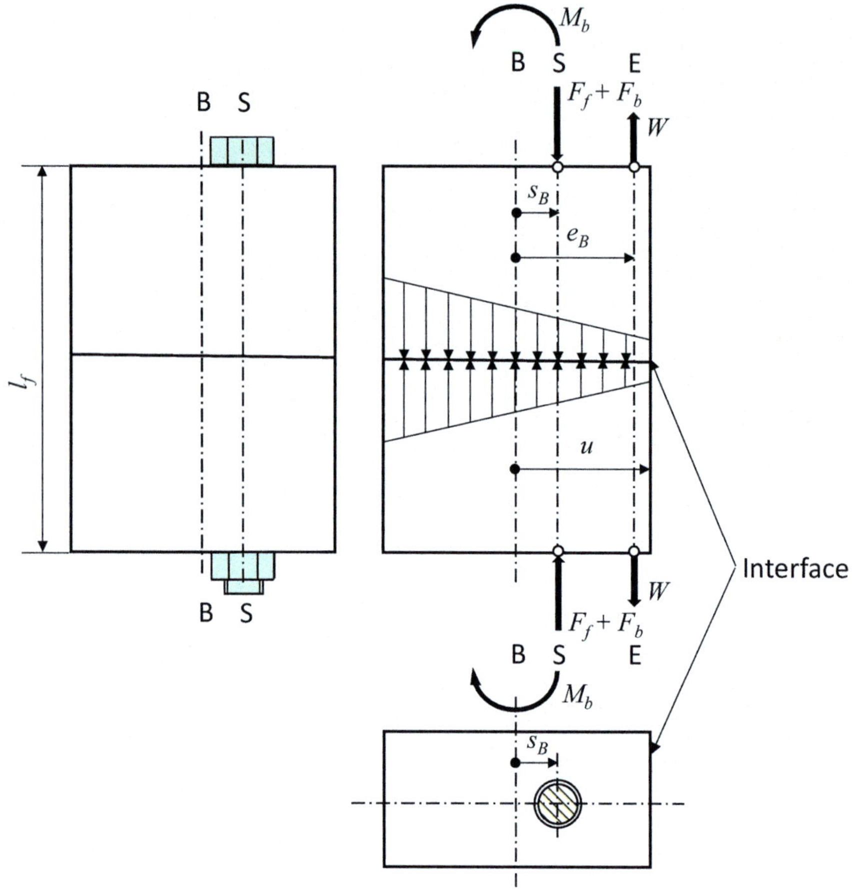

Fig. 12.25 Eccentric clamping and eccentric loading of a bolted joint. (Modified by the author based on Fig. 25 of VDI 2230 Part 1 (2014))

Here, A_D represents the sealing area between the clamped parts, Φ_{en} represents the load factor for eccentric load, and Φ_m represents the load factor for the external moment. In the 1974 and 1986 versions, the external moment M_B was not taken into consideration, so it was identical to $M_B = 0$ in Eq. (12.16).

Furthermore, from the 2003 version, a method for approximating the nonlinear behavior between the eccentric load and the additional bolt axial force after opening the sealing area between the clamped parts is provided as shown in Fig. 12.26. The prediction method for this nonlinear behavior was described in the main text in the 2003 version, but in the 2014 version, it is moved to the appendix.

(E) Prediction of proportional load in multiple bolted joints with rigid model

VDI 2230 published only Part 1 titled "Joints with one cylindrical bolt" until the 2003 edition, but added Part 2 titled "Multibolted joints" from the 2014 edition. In

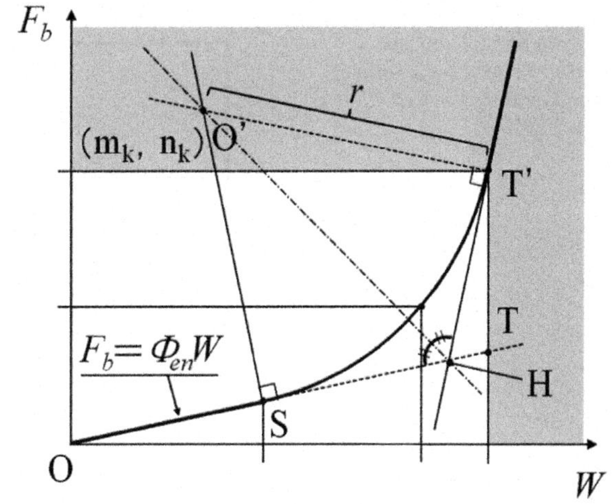

Fig. 12.26 Nonlinear behavior under eccentric loading after opening joints. (Created by the author based on Fig. D1 of VDI 2230 Part 1 (2014))

the evaluation of multiple bolt connections, the load applied to multiple bolted joints is shared with each bolt, and the strength evaluation of the bolt with the maximum proportional load is performed using the method for the single bolted joint of Part 1.

The simplest method to determine the proportional load in multiple bolted joints is to model the clamped bodies modeled as rigid bodies. In the rigid body model, the proportional load in multiple bolted joints can be calculated only by the equilibrium of the load.

The centroid of the bolt array, the total area of the bolt array, and the second moment of area are determined from the arrangement of bolts, which share the external load, in the rigid body model. For the layout of bolted joints shown in Fig. 12.27, the coordinates of the centroid of the bolt array and the second moment of area around the x-axis and z-axis are calculated from the diameter d_i and cross-sectional area A_i of each bolt as follows.

$$x_s^* = \frac{\sum_{i=1}^{n_s} A_i x_i^*}{\sum_{i=1}^{n_s} A_i} \tag{12.17}$$

$$z_s^* = \frac{\sum_{i=1}^{n_s} A_i z_i^*}{\sum_{i=1}^{n_s} A_i} \tag{12.18}$$

$$I_{xx} = \sum_{i=1}^{n_s} \left(\frac{\pi d_i^4}{64} + z_i^2 \frac{\pi d_i^2}{4} \right) \tag{12.19}$$

$$I_{zz} = \sum_{i=1}^{n_s} \left(\frac{\pi d_i^4}{64} + x_i^2 \frac{\pi d_i^2}{4} \right) \tag{12.20}$$

Next, the load distribution of the bolts is calculated. Here, we will explain the case in which an eccentric tensile force F_B is applied, as shown in Fig. 12.28, based on

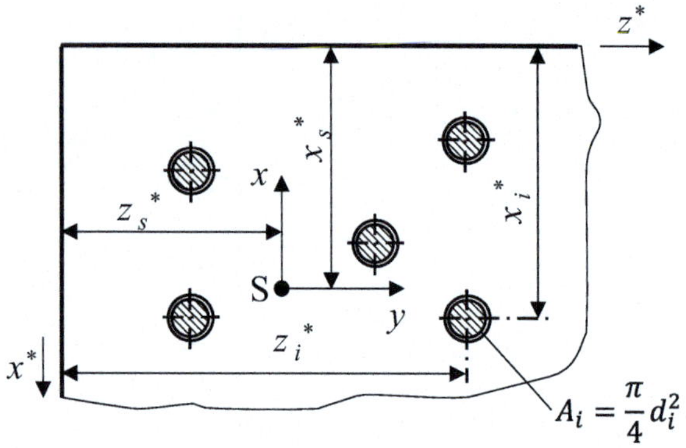

Fig. 12.27 Centroid of bolt array in multiple bolted joints. (Created by the author based on Fig. 3 of VDI 2230 Part 2 (2014))

the 2014 edition of VDI 2230 Part 2. Each bolt is subjected to the axial load F_B/n_s, which is obtained by dividing the tensile force F_B by the number of bolts n_s, and is also subjected to the moments of $M_x = F_B\,|z_S|$ and $M_z = F_B\,|x_S|$, which are generated due to the eccentricity of the external force F_B from the centroid of the bolt array, S. Therefore, each bolt is subjected to the proportional load of $F_{A(M)i}$ as follows.

$$F_{A(M)i} = \frac{M_z A x_i}{I_{zz}} + \frac{M_x A z_i}{I_{xx}} \tag{12.21}$$

Here, all bolts are calculated with the same cross-sectional area A. Therefore, the maximum axial load of the bolt, in which the load is distributed among the multiple bolted joints, can be obtained by the following equation.

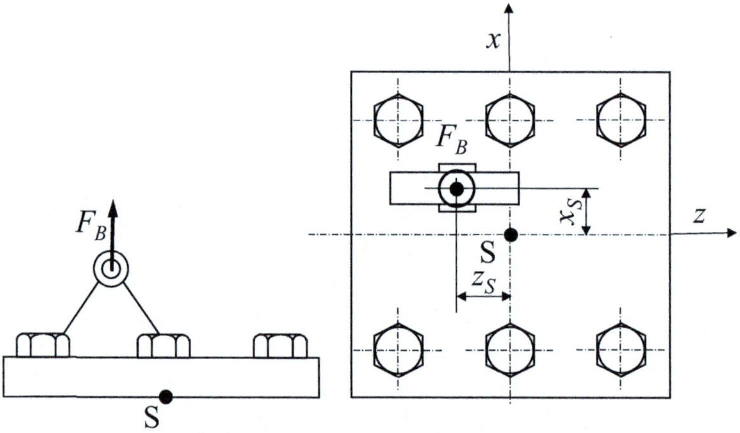

Fig. 12.28 Application of an axial eccentric tensile load of a multiple bolted joints. (Modified by the author based on Figs. 9 and 10 of VDI 2230 Part 2 (2014))

$$F_{A\max} = \frac{F_B}{n_s} + \max\left(F_{A(M)i}\right) \tag{12.22}$$

Once the maximum proportional load is obtained, the strength evaluation of the bolt with the maximum proportional load is carried out with the method of Part 1 titled "Joints with one cylindrical bolt".

In the 2014 edition of VDI 2230 Part 2, there are also cases in which transverse external forces and moment are applied as shown in Fig. 12.29, and a combination of axial and transverse forces as an eccentric force from a diagonal direction is applied as shown in Fig. 12.30, in addition to the example of eccentric tensile force mentioned above. Furthermore, cases are illustrated in Figs. 12.31 and 12.32, in which axial external force and moment are applied to circular flanges such as piping, and circular flanges are subjected to torque assumed by axial couplings, etc.

(F) Prediction of proportional load in multiple bolted joints with equivalent elastic model

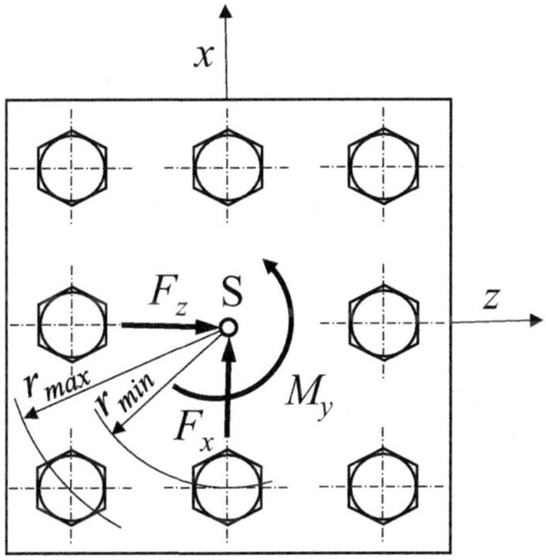

Fig. 12.29 Transverse forces and moment around the perpendicular axis in multiple-bolted joints. (Modified by the author based on Figs. 7 and 8 of VDI 2230 Part 2 (2014))

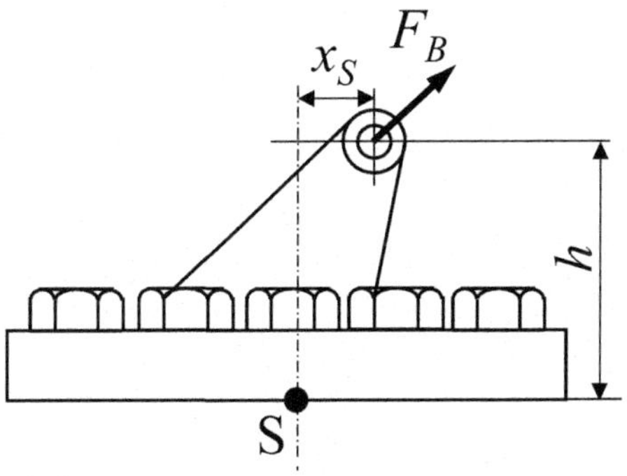

Fig. 12.30 Application of an eccentric force from a diagonal direction in multiple-bolted joints. (Modified by the author based on Fig. 14 of VDI 2230 Part 2 (2014))

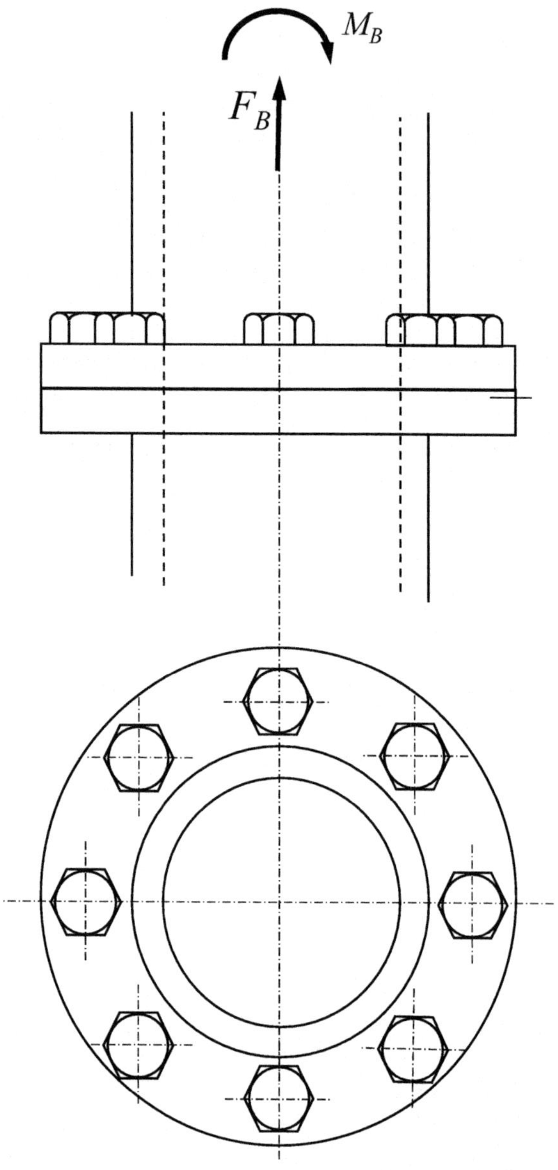

Fig. 12.31 Flanged connection under a moment and a longitudinal force. (Created by the author based on Fig. 18 of VDI 2230 Part 2 (2014))

In a structure fastened with multiple-bolted joints, the rigid body model (as mentioned above) tends to cause errors in the calculation of load distribution when the deformation of the structure is relatively larger than the rigid body model. To obtain a more accurate value of the proportional load, the structure can be modeled as an elastic body (such as a beam, plate, or shell), which is considered as a type of equivalent model in solid mechanics. The proportional load can be obtained by solving the

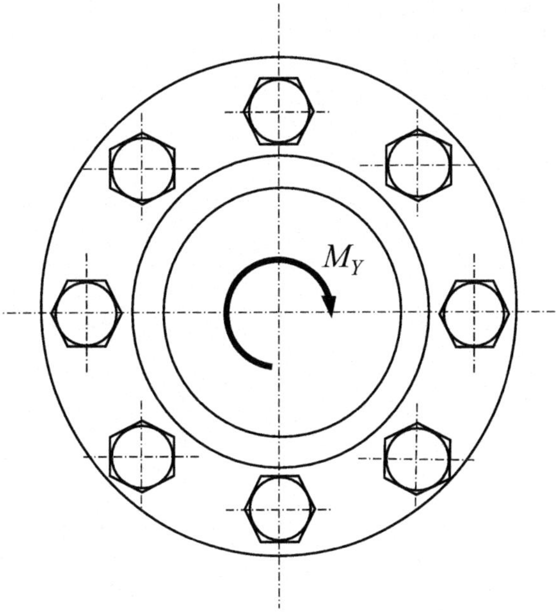

Fig. 12.32 Flanged connection under a torque. (Modified by the author based on Fig. 17 of VDI 2230 Part 2 (2014))

deformation and load equilibrium. It is described as elastomechanics in the 2014 edition of VDI 2230.

Here, similar to the model shown in Fig. 3.3, we explain an example model in Fig. 12.33, in which a bolt array and an elastic beam are subjected to bending. Other examples of models, including the one shown in Fig. 12.34, are also provided. While there seems to be a limited amount of detailed explanations, it appears that the user requires both knowledge and experience to effectively apply these models to actual structures.

(G) Evaluation of multiple-bolted joints by using finite element analysis

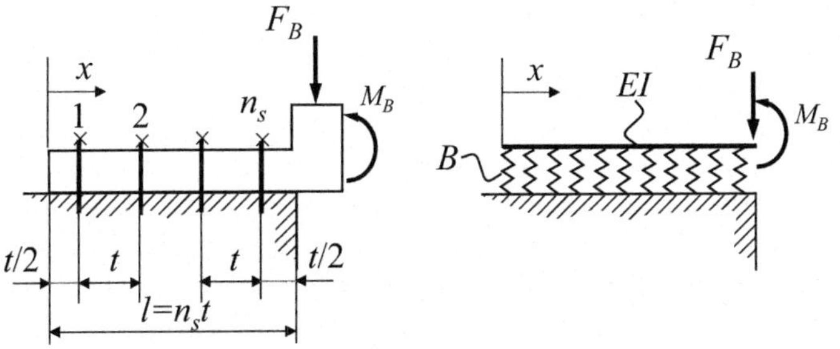

Fig. 12.33 Bolt array and corresponding model of an elastically bedded bending beam. (Created by the author based on Fig. 30 of VDI 2230 Part 2 (2014))

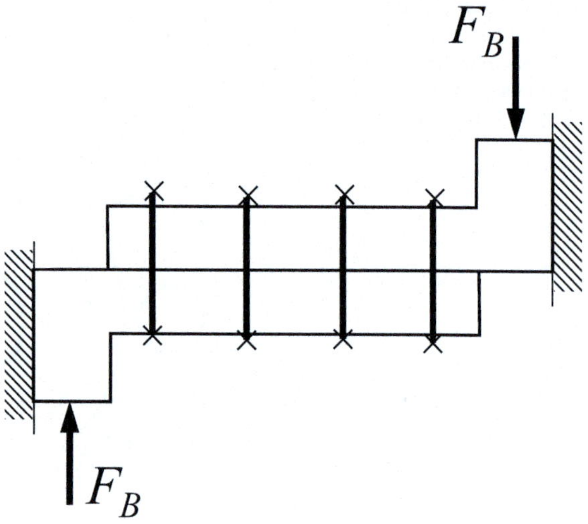

Fig. 12.34 Point-symmetric through-bolt joint under an external force. (Created by the author based on Fig. 34 of VDI 2230 Part 2 (2014))

The above solid mechanics model in multiple-bolted joints cannot be applied to arbitrary shapes, so the finite element analysis (FEA) is typically used for complex shapes. In addition, to evaluate phenomena such as separation of contact surfaces, it is necessary to use a nonlinear FEA that takes contact condition into account. In recent years, with the development of computers, 3D CAD and FE solvers, it has become possible for people other than FEA specialists to perform FEA on their own personal computers. However, FEA heavily relies on the modeling skills and ability of the user for evaluating the obtained result. Thus, it is important to use FEA carefully, as it can lead to incorrect judgments.

In the 2014 edition of VDI 2230, modeling guidelines to use FEA for bolted joints are provided. Depending on the level of detail, models are organized into classes I to IV, as shown in Tables 12.14 and 12.15.

Model class I is a finite element model, in which clamped parts are modeled in a single continuum without separately modeling the bolts and the joint surfaces. The sealing surfaces are treated as a single entity within the dimensions of G or G' specified in VDI 2230 Part 1. This model is utilized to determine internal forces such as forces and moments at the sealing surfaces.

Model class II is a model that uses bolt elements or spring elements, etc., in bolted joints. The bolt head and nut bearing surfaces of clamped parts are constrained by coupling connections with the bolt elements. This model takes or does not take contact condition of the sealing surfaces between the clamped bodies into consideration. With this method, it is possible to directly calculate the nominal stress of the bolt section. Additionally, this model takes or does not take initial tightening force into consideration.

Model class III does not model the screw but models the bolt itself with solid elements. The stiffness of the screw section and bolt head can be adjusted to match the stiffness of the actual bolt by modifying the shape and material model. In this

Table 12.14 Overview of FEA model classes. (Created by the author based on Table 1 of VDI 2230 Part 2 (2014)) [9]

	Model class I	Model classes II and III	Model class IV
	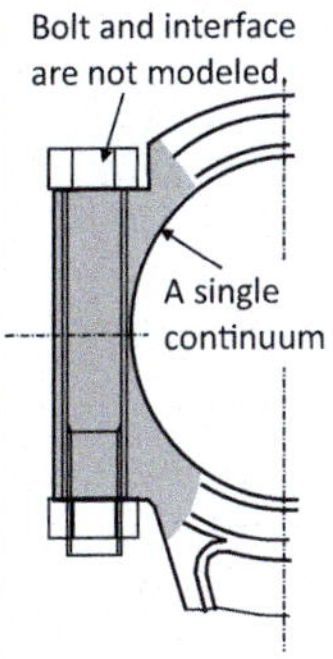	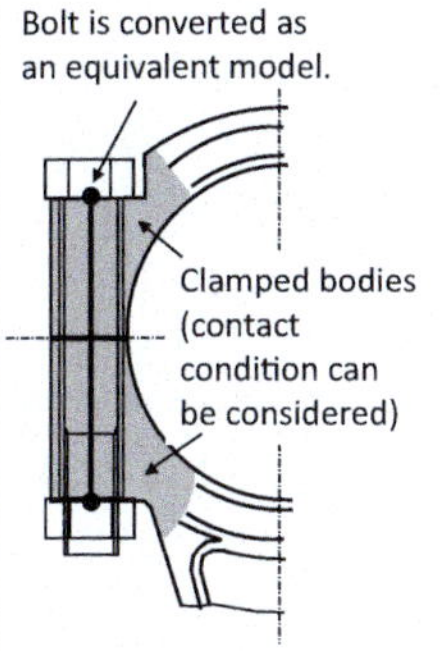 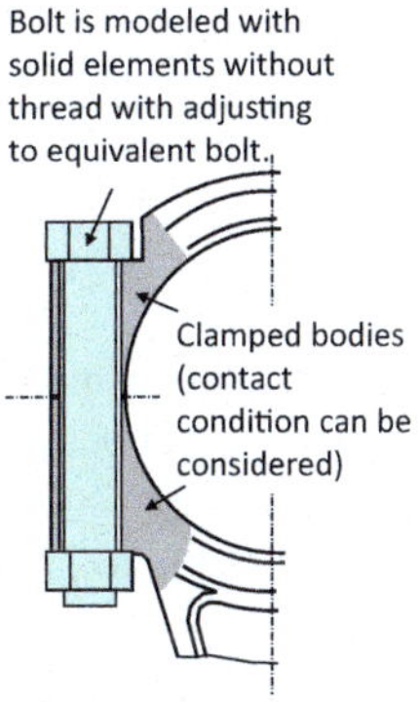	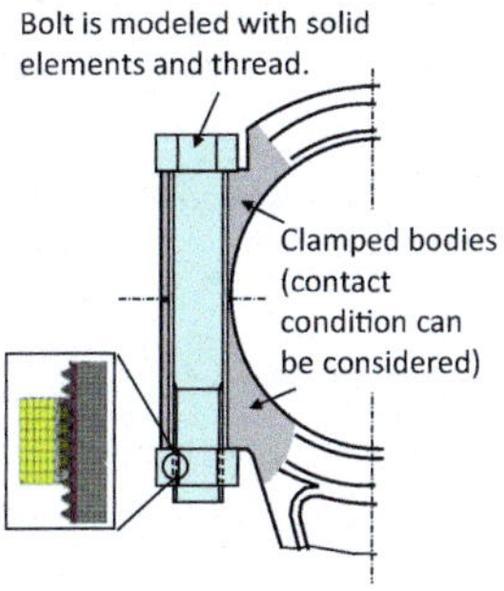
Characteristic	Bolt and interface are not taken into consideration (preload is not included in representation)	Bolt is covered as an equivalent model (beam or equivalent volume) (preload and interface contact can be taken into consideration)	Bolt is fully modeled, including thread, preload, and interface contact
Objective of calculation	Internal forces in the interface as input variables for calculation following VDI 2230 Part 1 (cf. therein Example B4) or deformations of the bearing areas	Internal forces (nominal sizes) of the bolt	Internal forces (nominal sizes) of the bolt; local stress in the bolt and in the clamping area

Table 12.15 Classification of FE models with regard to modeling and the parameters required. (Created by the author based on Table 2 of VDI 2230 Part 2 (2014)) [9]

Model class	I	II and III	IV
Modeling the bolted joints (BJ)			
Effort	Low	Medium	High
Idealization of the bolt	Not modeled	Simplified	Modeled in detail
Contact conditions in the interface	Not modeled	Modeled	Modeled
Preload	Without	With	With
Required parameters from VDI 2230 Part 1 (relevant section number is indicated)			
Compliance of the bolt δ_S	5.1.1	5.1.1/included in the model	Included in the model
Compliance of the plates δ_P (δ_P^*, δ_P^{**})	5.1.2[a]	Included in the model	Included in the model
Load application factor n	5.2.2[a]	Included in the model	Included in the model
Tightening factor α_A	5.4.3	5.4.3	5.4.3
Amount of embedment f_Z	5.4.2.1	5.4.2.1	5.4.2.1

[a] With restriction

model, the contact condition of the bolt head and nut bearing surfaces and that of the sealing surfaces of the clamped bodies can be taken into consideration. Therefore, it is possible to directly evaluate the nominal stress in the bolt section. Additionally, similar to Model class II, it is also possible to provide initial tightening force to the model.

Model class IV is a model with screw threads that can evaluate contact condition of the thread. Similar to Model class III, it can also take the initial tightening force and the contact state of the bolt and nut bearing surfaces and the sealing surfaces of the clamped bodies into account. Additionally, it can evaluate phenomena such as rotational loosening, as it takes the contact condition of the thread into account. However, due to its detailed mesh, it requires a significant number of elements. While it is possible to model a single bolted joint, applying Model class IV to the structure of multiple bolted joints is not practical because there will be so many elements.

In VDI 2230, there is only a brief overview of these modeling methods and evaluation techniques, and there is little specific information. How to model and evaluate is left to the user. As mentioned in (F) and (G), Part 2 added in the 2014 edition of VDI 2230 has limited specific information, and it seems to heavily depend on the user's knowledge and experience to use it appropriately for actual design. We hope that it will be improved in the future while being used.

References

1. Akira Yamamoto, Theory and Calculation of Screw Fastening, Yokendo (2004), in Japanese.

2. Isam Yoshimoto, Key Point of Designing Screw Fastening Design, Japanese Standards Association (2018), in Japanese.
3. Nobutoshi Ohashi, State of JIS concerning screw threads and fasteners in 1999 to 2008 [1/3], Journal of the Japan Research Institute for Screw Threads and Fasteners, vol. 40(2009), pp. 127, in Japanese.
4. Nobutoshi Ohashi, State of JIS concerning screw threads and fasteners in 2009 to 2018 [1/4], Journal of the Japan Research Institute for Screw Threads and Fasteners, vol. 40(2009), pp. 127, in Japanese.
5. Masaya Hagiwara, The activities of the international standardization for 10 years on ISO/TC2 relation in 1999 to 2008, Journal of the Japan Research Institute for Screw Threads and Fasteners, vol. 40(2009), pp. 99, in Japanese.
6. Masaya Hagiwara, The activities of the international standardization for 10 years on ISO/TC2 relation in 2009 to 2018, Journal of the Japan Research Institute for Screw Threads and Fasteners, vol. 50(2019), pp. 138, in Japanese.
7. Japanese Standards Association, JIS Handbook Fasteners & Screw Threads I, Japan Standards Association (2019), in Japanese.
8. Japanese Standards Association, JIS Handbook Fasteners & Screw Threads II, Japan Standards Association (2019), in Japanese.
9. Editorial committee of the Japan Research Institute for Screw Threads and Fasteners, VDI 2230 Systematic calculation of high duty bolted joints, The Japan Research Institute for Screw Threads and Fasteners (2018), in Japanese.
10. Toshiyuki Sawa, Activity report of research committee, Journal of the Japan Research Institute for Screw Threads and Fasteners, vol. 49(2018), pp. 112, in Japanese.
11. Research committee of the Japan Research Institute for Screw Threads and Fasteners, Analysis and Design for Bolted Joints (Second Edition), the Japan Research Institute for Screw Threads and Fasteners (2022), in Japanese.
12. Tomohiro Naruse and Yoji Shibutani, Higher Accurate Estimation of Axial and Bending Stiffnesses of Plates Clamped by Bolts, Journal of Solid Mechanics and Materials Engineering, 6–5 (2012), 397–406. https://doi.org/10.1299/jmmp.6.397

Chapter 13
Origin and History of Screws

13.1 Origin of Screws in the World

The concept of a screw can be thought of as the "slope mechanics" of a helical helix wrapped around a cylindrical rod. The origin of such a concept can be traced back to the construction of the ancient Egyptian pyramids in the twenty-seventh century B.C. [1]. Although there is no precise description of the construction process at that time, it is believed that stones weighing an average of 2.5 tons (maximum 7 tons) were manually piled up to a height of 140 m by "rollers and sleds" and "human power" while making a slope with soil, as shown in Fig. 13.1. The stones are thought to have been lifted by "rollers and sleds" and "human power" while building a slope with soil as shown in Fig. 13.1. This primitive mechanical concept of slope mechanics is thought to have led to the "screw" mechanism, which generates a large power with a small force.

It is said that the screw form (cylindrical helix) was used in the pumping pump of Archimedes around 250 B.C. (Fig. 13.2) or in the presses for grapes, olives, etc., at that time (Fig. 13.3).

It is thought that the screws of wooden screws, such as those used in such presses, were spiraled on the surface of the cylindrical material with string, etc., and the threads were processed by hand along them. It is believed that a threading machine was used for the wooden internal thread as shown in Fig. 13.4. The lead thread part is used as a guide jig, and when the lead angle guide pin (5 pins) combined with the lead angle of the screw valley is inserted and the thread part is rotated, the iron tool attached to the bottom of the guide can be used to cut the internal thread into the workpiece under the guide (pilot hole drilled in advance). It seems that the cutting amount was a mechanism to increase the amount by hammering an iron tool from below with a wedge.

© The Author(s), under exclusive license to Springer Nature Switzerland AG 2025

T. Hattori and T. Naruse, *Compendium for Design and Manufacturing of Bolted Joints*,
Solid Mechanics and Its Applications 280,
https://doi.org/10.1007/978-3-031-89628-6_13

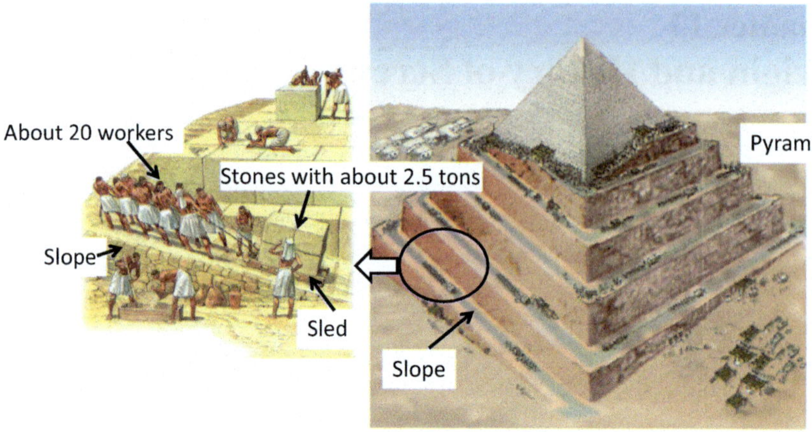

Fig. 13.1 Ancient pyramid construction, probably using a slope

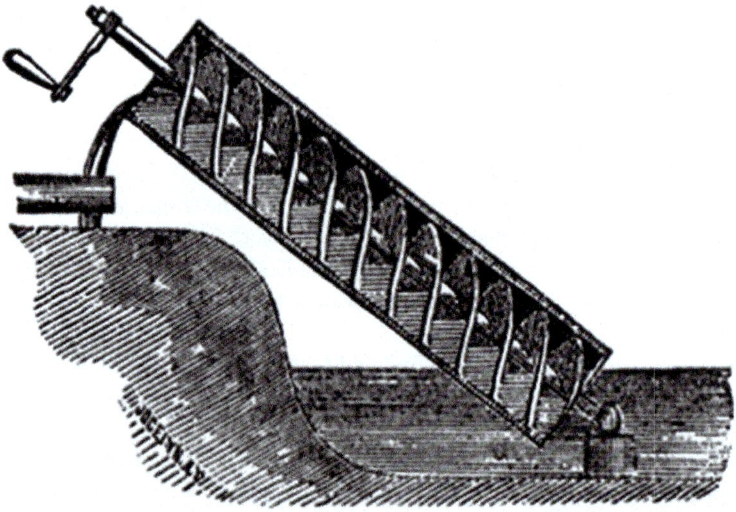

Fig. 13.2 Archimedes pumping screw pump [2]

Wooden screw threads like those used in these presses are thought to have been handmade by drawing a spiral on the surface of a cylindrical piece of wood using string, etc., and then machining along the spiral by hand. The threaded part that serves as the parent thread is used as a guide jig, and when the threaded part is rotated along a guide into which five lead angle guide pins are inserted to match the lead angle of the thread valley, an iron bite attached to the bottom of the guide can cut a female thread into the work piece (drilled in advance) below the guide. The amount of cut is calculated by wedging the steel bite. The amount of cutting was apparently increased by tapping the iron bite from below with a wedge.

Fig. 13.3 Grapes and olives pressing machine [3]

German Gutenberg invented "the type printing technology" called three greatest inventions of the renaissance period along with a compass, gunpowder in 1450, and "structure of the screw-type press" is used here (Fig. 13.5).

Furthermore, the birth of fastening screws that tighten two or more things is said to be from around 1500 just when Leonardo da Vinci studying a screw. Leonardo da Vinci leaves the sketching ideas of geometric shape of the screw and a thread cutting machine.

The practical use invention of the threading lathe was accomplished by Jack Besson at the mid-sixteenth century. However, Henry Mozley developed the lathe (Fig. 13.6) with the feed stand, and the lathe "threading lathes" which could process screw PHS constantly here spread [5]. It is said to be around 1800.

These "Threading lathes" which could process screw pitch constantly spread industrial field. It is said to be around 1800.

In this way, unification, the standard of the screw, the compatibility of bolt/nut keep being epoch-making and they right contribute to mass production of the Industrial Revolution in the eighteenth century very much and reach it in the present age.

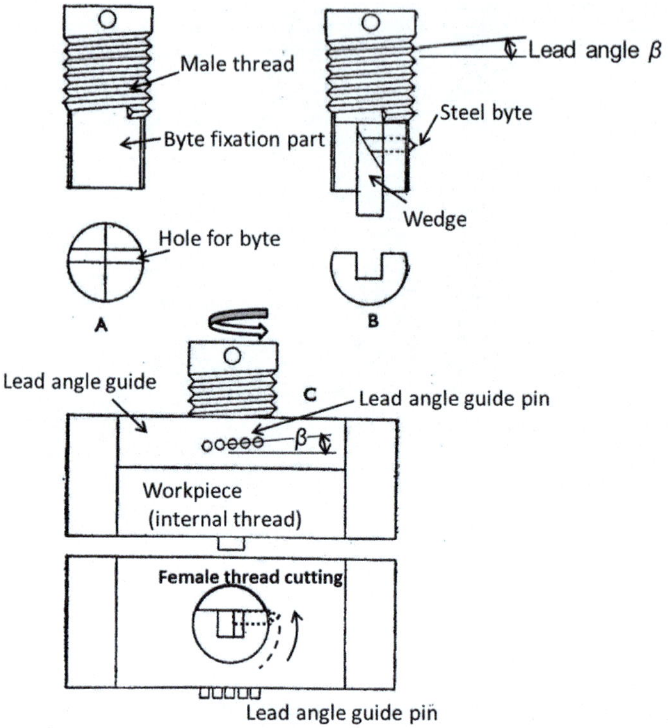

Fig. 13.4 Restoration of Heron's threading machine [3]. A: Side and bottom of the male thread. There is a hole for a rotating handle at the upper end and a hole for plugging a byte at the lower end. B: A square groove is provided in the lower part, and an iron byte is inserted. C: The wooden male screws shown in A and B are rotated along the holes of the outer diameter of the male screw and the five lead angle guide pins. Then, the female thread with the lead angle of the screw is cut into the workpiece (which becomes the internal thread) along a pilot hole. C below: This is a view of a thread cutting machine from below, and if worker pushes the byte with a wedge while rotating the internal thread, the internal thread will be cut

13.2 Transmission and Import of Screws to Japan

It is said that the origin of the screw in Japan will be matchlock gun which was brought from Portuguese ship drifted in Tanegashima Island around the mid-1500s. A screw was used for the breechblock which put the gunpowder of the matchlock in and out (Fig. 13.7).

Feudal lord Tokitaka Tanegashima of Tanegashima purchased this matchlock gun and order Kinbei Yasaka, who was a swordsmith, to make the same thing.

But in that time there was not the technique to make a spiral. In Japan at the time, the only tools that used slope mechanics were wedges (Fig. 13.8). Therefore, the biggest challenge for Mr. Kinbei Yasaka was recreating the screws used for the tail plug. And one year later, he succeeded in the production of the screw.

Fig. 13.5 Structure of the screw-style press [4] at the time of the invention

Fig. 13.6 Lathe with the feed stand which Henry Mozley invented

13.3 Commencement of Production of Screws in Japan

The manufacture of screws in Japan began about 300 years after the production of screws introduced by matchlock gun. In 1860, Mr. Oguri Uenosuke, who was an envoy to the United States, brought back screws from the United States (Fig. 13.9) and, with the help of his French employer Verny, built a steel mill and a shipyard in

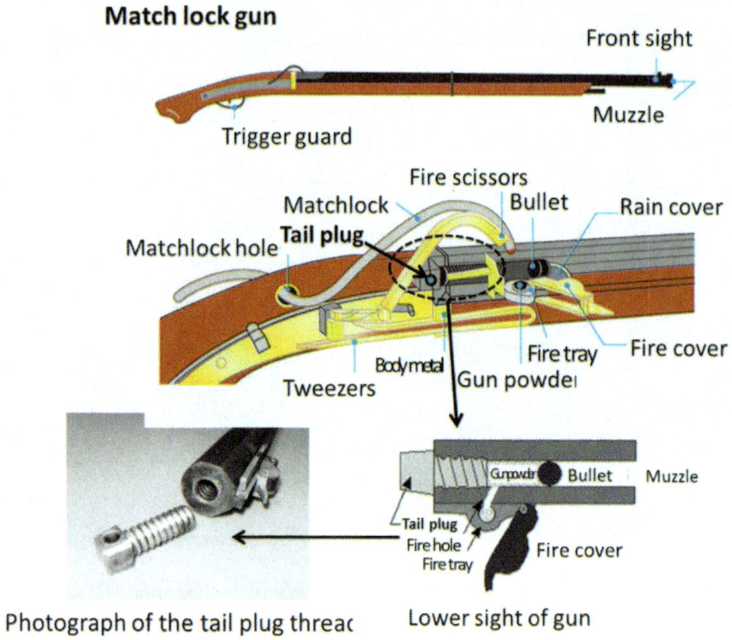

Fig. 13.7 Matchlock gun introduced to Tanegashima and tail plug screw [6]

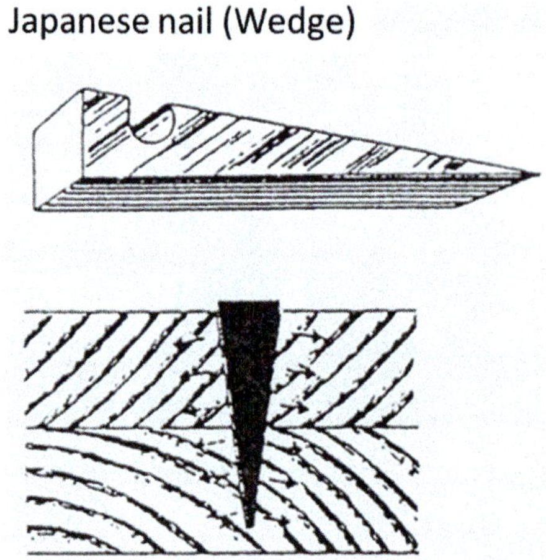

Fig. 13.8 Japanese nail (wedge) and it's uses

Yokosuka, and played a major role in the development of industry in Japan. Screws are sometimes called Vis in Japanese, which means screw in French, and it is said that it is a remnant of the fact that screws were imported with the cooperation of France at that time.

Fig. 13.9 Screws brought back by Oguri Uenosuke, the U.S. envoy (Tozenji Temple, Takasaki City, Gunma Prefecture) [7]

13.4 The Road to JIS and ISO Standards

Japan's Industrial Standards (JIS) were established in Japan when the Industrial Standardization Law was promulgated on June 1, 1946. After the war, due to the influence of the United States, JIS had three standards: meter, wit, and unify. Currently, there are two standards in JIS: metric and unified. The shape of the screw head, the diameter of the thread, the angle of the thread, and the dimension of the distance between the threads (pitch) are all specified by the standard, and it has contributed to the development of the industry, Japan the unification of the meshing of male with female threads. Currently, the JIS standard for screws is established, revised, and abolished mainly by the standardization committee established by the Japan Screw Research Association [8, 9]. He also participated in and contributed to ISO/TC1 and ISO/TC2 from the Standardization Committee [10, 11]. In recent years, with the globalization of economic activities, JIS standards are also required to be harmonized with international standards. And standards related to threads are also being promoted based on ISO/IEC Guide 21;1999 (Adoption of international standards for regional or national standards) to determine the 3 degrees of compliance with international standards in IDT: identical, MOD: modified, and NEQ: not equivalent.

References

1. Hitoshi Takeuchi, On the Origins of Civilization, Kyokusha, (1988).
2. Archimedes' screw, Public domain, from Chamber's Encyclopedia (Philadelphia: J. B. Lippincott Company, (1875) Added to illustrate article en: Archimedes.
3. A. P. Usher, A history of mechanical inventions, McGraw Hill (1929).
4. Grolier Eric de, Histoire du livre, Published by Editions P.U.F, (1954).
5. The Science Museum, London.
6. New edition joining technology overview, Editorial committee, Industrial technology service center, (1994), pp. 615.

7. Nittoseiko HP, https://www.nittoseiko.co.jp/nitto-blog/2017/12/05/285.
8. Nobutoshi Ohashi, Status of JIS on screws in 1999-2008 [1/3], Journal of the Japan Screw Research Association, vol. 40, 127 (2009).
9. Nobutoshi Ohashi: Status of JIS on screws in 2009~2018 [1/4], Journal of the Japan Screw Research Association, vol. 49, 323 (2018).
10. Masaya Hagiwara: 10 Years of International Standardization of ISO/TC2, -1999-2008-, Journal of the Japan Screw Research Association, vol. 40, 99 (2009).
11. Masaya Hagiwara: 10 Years of International Standardization Related to ISO/TC2 -2009-2018-, Journal of the Japan Screw Research Association, vol. 50, 138 (2019).

The manufacturer's authorised representative in the EU is Springer Nature Customer Service Centre GmbH, Europaplatz 3, 69115 Heidelberg, Germany. If you have any concerns regarding our products, please contact ProductSafety@springernature.com

Printed and bound by CPI Group (UK) Ltd, Croydon, CR0 4YY
30/09/2026
02225571-0018